analysis, algebra, and computers in mathematical research

PURE AND APPLIED MATHEMATICS

A Program of Monographs, Textbooks, and Lecture Notes

LECTURE NOTES IN PURE AND APPLIED MATHEMATICS

1. *N. Jacobson,* Exceptional Lie Algebras
2. *L.-Å. Lindahl and F. Poulsen,* Thin Sets in Harmonic Analysis
3. *I. Satake,* Classification Theory of Semi-Simple Algebraic Groups
4. *F. Hirzebruch, W. D. Newmann, and S. S. Koh, Differentiable Manifolds and Quadratic Forms*
5. *I. Chavel,* Riemannian Symmetric Spaces of Rank One
6. *R. B. Burckel,* Characterization of C(X) Among Its Subalgebras
7. *B. R. McDonald, A. R. Magid, and K. C. Smith,* Ring Theory: Proceedings of the Oklahoma Conference
8. *Y.-T. Siu,* Techniques of Extension on Analytic Objects
9. *S. R. Caradus, W. E. Pfaffenberger, and B. Yood,* Calkin Algebras and Algebras of Operators on Banach Spaces
10. *E. O. Roxin, P.-T. Liu, and R. L. Sternberg,* Differential Games and Control Theory
11. *M. Orzech and C. Small,* The Brauer Group of Commutative Rings
12. *S. Thomier,* Topology and Its Applications
13. *J. M. Lopez and K. A. Ross,* Sidon Sets
14. *W. W. Comfort and S. Negrepontis,* Continuous Pseudometrics
15. *K. McKennon and J. M. Robertson,* Locally Convex Spaces
16. *M. Carmeli and S. Malin,* Representations of the Rotation and Lorentz Groups: An Introduction
17. *G. B. Seligman,* Rational Methods in Lie Algebras
18. *D. G. de Figueiredo,* Functional Analysis: Proceedings of the Brazilian Mathematical Society Symposium
19. *L. Cesari, R. Kannan, and J. D. Schuur,* Nonlinear Functional Analysis and Differential Equations: Proceedings of the Michigan State University Conference
20. *J. J. Schäffer,* Geometry of Spheres in Normed Spaces
21. *K. Yano and M. Kon,* Anti-Invariant Submanifolds
22. *W. V. Vasconcelos,* The Rings of Dimension Two
23. *R. E. Chandler,* Hausdorff Compactifications
24. *S. P. Franklin and B. V. S. Thomas,* Topology: Proceedings of the Memphis State University Conference
25. *S. K. Jain,* Ring Theory: Proceedings of the Ohio University Conference
26. *B. R. McDonald and R. A. Morris,* Ring Theory II: Proceedings of the Second Oklahoma Conference
27. *R. B. Mura and A. Rhemtulla,* Orderable Groups
28. *J. R. Graef,* Stability of Dynamical Systems: Theory and Applications
29. *H.-C. Wang,* Homogeneous Branch Algebras
30. *E. O. Roxin, P.-T. Liu, and R. L. Sternberg,* Differential Games and Control Theory II
31. *R. D. Porter,* Introduction to Fibre Bundles
32. *M. Altman,* Contractors and Contractor Directions Theory and Applications
33. *J. S. Golan,* Decomposition and Dimension in Module Categories
34. *G. Fairweather,* Finite Element Galerkin Methods for Differential Equations
35. *J. D. Sally,* Numbers of Generators of Ideals in Local Rings
36. *S. S. Miller,* Complex Analysis: Proceedings of the S.U.N.Y. Brockport Conference
37. *R. Gordon,* Representation Theory of Algebras: Proceedings of the Philadelphia Conference
38. *M. Goto and F. D. Grosshans,* Semisimple Lie Algebras
39. *A. I. Arruda, N. C. A. da Costa, and R. Chuaqui,* Mathematical Logic: Proceedings of the First Brazilian Conference
40. *F. Van Oystaeyen,* Ring Theory: Proceedings of the 1977 Antwerp Conference
41. *F. Van Oystaeyen and A. Verschoren,* Reflectors and Localization: Application to Sheaf Theory
42. *M. Satyanarayana,* Positively Ordered Semigroups
43. *D. L Russell,* Mathematics of Finite-Dimensional Control Systems
44. *P.-T. Liu and E. Roxin,* Differential Games and Control Theory III: Proceedings of the Third Kingston Conference, Part A
45. *A. Geramita and J. Seberry,* Orthogonal Designs: Quadratic Forms and Hadamard Matrices
46. *J. Cigler, V. Losert, and P. Michor,* Banach Modules and Functors on Categories of Banach Spaces

47. *P.-T. Liu and J. G. Sutinen,* Control Theory in Mathematical Economics: Proceedings of the Third Kingston Conference, Part B
48. *C. Byrnes,* Partial Differential Equations and Geometry
49. *G. Klambauer,* Problems and Propositions in Analysis
50. *J. Knopfmacher,* Analytic Arithmetic of Algebraic Function Fields
51. *F. Van Oystaeyen,* Ring Theory: Proceedings of the 1978 Antwerp Conference
52. *B. Kadem,* Binary Time Series
53. *J. Barros-Neto and R. A. Artino,* Hypoelliptic Boundary-Value Problems
54. *R. L. Sternberg, A. J. Kalinowski, and J. S. Papadakis,* Nonlinear Partial Differential Equations in Engineering and Applied Science
55. *B. R. McDonald,* Ring Theory and Algebra III: Proceedngs of the Third Oklahoma Conference
56. *J. S. Golan,* Structure Sheaves Over a Noncommutative Ring
57. *T. V. Narayana, J. G. Williams, and R. M. Mathsen,* Combinatorics, Representation Theory and Statistical Methods in Groups: YOUNG DAY Proceedings
58. *T. A. Burton,* Modeling and Differential Equations in Biology
59. *K. H. Kim and F. W. Roush,* Introduction to Mathematical Consensus Theory
60. *J. Banas and K. Goebel,* Measures of Noncompactness in Banach Spaces
61. *O. A. Nielson,* Direct Integral Theory
62. *J. E. Smith, G. O. Kenny, and R. N. Ball,* Ordered Groups: Proceedings of the Boise State Conference
63. *J. Cronin,* Mathematics of Cell Electrophysiology
64. *J. W. Brewer,* Power Series Over Commutative Rings
65. *P. K. Kamthan and M. Gupta,* Sequence Spaces and Series
66. *T. G. McLaughlin,* Regressive Sets and the Theory of Isols
67. *T. L. Herdman, S. M. Rankin III, and H. W. Stech,* Integral and Functional Differential Equations
68. *R. Draper,* Commutative Algebra: Analytic Methods
69. *W. G. McKay and J. Patera,* Tables of Dimensions, Indices, and Branching Rules for Representations of Simple Lie Algebras
70. *R. L. Devaney and Z. H. Nitecki,* Classical Mechanics and Dynamical Systems
71. *J. Van Geel,* Places and Valuations in Noncommutative Ring Theory
72. *C. Faith,* Injective Modules and Injective Quotient Rings
73. *A. Fiacco,* Mathematical Programming with Data Perturbations I
74. *P. Schultz, C. Praeger, and R. Sullivan,* Algebraic Structures and Applications: Proceedings of the First Western Australian Conference on Algebra
75. *L Bican, T. Kepka, and P. Nemec,* Rings, Modules, and Preradicals
76. *D. C. Kay and M. Breen,* Convexity and Related Combinatorial Geometry: Proceedings of the Second University of Oklahoma Conference
77. *P. Fletcher and W. F. Lindgren,* Quasi-Uniform Spaces
78. *C.-C. Yang,* Factorization Theory of Meromorphic Functions
79. *O. Taussky,* Ternary Quadratic Forms and Norms
80. *S. P. Singh and J. H. Burry,* Nonlinear Analysis and Applications
81. *K. B. Hannsgen, T. L. Herdman, H. W. Stech, and R. L. Wheeler,* Volterra and Functional Differential Equations
82. *N. L. Johnson, M. J. Kallaher, and C. T. Long,* Finite Geometries: Proceedings of a Conference in Honor of T. G. Ostrom
83. *G. I. Zapata,* Functional Analysis, Holomorphy, and Approximation Theory
84. *S. Greco and G. Valla,* Commutative Algebra: Proceedings of the Trento Conference
85. *A. V. Fiacco,* Mathematical Programming with Data Perturbations II
86. *J.-B. Hiriart-Urruty, W. Oettli, and J. Stoer,* Optimization: Theory and Algorithms
87. *A. Figa Talamanca and M. A. Picardello,* Harmonic Analysis on Free Groups
88. *M. Harada,* Factor Categories with Applications to Direct Decomposition of Modules
89. *V. I. Istrăţescu,* Strict Convexity and Complex Strict Convexity
90. *V. Lakshmikantham,* Trends in Theory and Practice of Nonlinear Differential Equations
91. *H. L. Manocha and J. B. Srivastava,* Algebra and Its Applications
92. *D. V. Chudnovsky and G. V. Chudnovsky,* Classical and Quantum Models and Arithmetic Problems
93. *J. W. Longley,* Least Squares Computations Using Orthogonalization Methods
94. *L. P. de Alcantara,* Mathematical Logic and Formal Systems
95. *C. E. Aull,* Rings of Continuous Functions
96. *R. Chuaqui,* Analysis, Geometry, and Probability
97. *L. Fuchs and L. Salce,* Modules Over Valuation Domains

98. *P. Fischer and W. R. Smith,* Chaos, Fractals, and Dynamics
99. *W. B. Powell and C. Tsinakis,* Ordered Algebraic Structures
100. *G. M. Rassias and T. M. Rassias,* Differential Geometry, Calculus of Variations, and Their Applications
101. *R.-E. Hoffmann and K. H. Hofmann,* Continuous Lattices and Their Applications
102. *J. H. Lightbourne III and S. M. Rankin III,* Physical Mathematics and Nonlinear Partial Differential Equations
103. *C. A. Baker and L, M. Batten,* Finite Geometrics
104. *J. W. Brewer, J. W. Bunce, and F. S. Van Vleck,* Linear Systems Over Commutative Rings
105. *C. McCrory and T. Shifrin,* Geometry and Topology: Manifolds, Varieties, and Knots
106. *D. W. Kueker, E. G. K. Lopez-Escobar, and C. H. Smith,* Mathematical Logic and Theoretical Computer Science
107. *B.-L. Lin and S. Simons,* Nonlinear and Convex Analysis: Proceedings in Honor of Ky Fan
108. *S. J. Lee,* Operator Methods for Optimal Control Problems
109. *V. Lakshmikantham,* Nonlinear Analysis and Applications
110. *S. F. McCormick,* Multigrid Methods: Theory, Applications, and Supercomputing
111. *M. C. Tangora,* Computers in Algebra
112. *D. V. Chudnovsky and G. V. Chudnovsky,* Search Theory: Some Recent Developments
113. *D. V. Chudnovsky and R. D. Jenks,* Computer Algebra
114. *M. C. Tangora,* Computers in Geometry and Topology
115. *P. Nelson, V. Faber, T. A. Manteuffel, D. L. Seth, and A. B. White, Jr.,* Transport Theory, Invariant Imbedding, and Integral Equations: Proceedings in Honor of G. M. Wing's 65th Birthday
116. *P. Clément, S. Invernizzi, E. Mitidieri, and I. I. Vrabie,* Semigroup Theory and Applications
117. *J. Vinuesa,* Orthogonal Polynomials and Their Applications: Proceedings of the International Congress
118. *C. M. Dafermos, G. Ladas, and G. Papanicolaou,* Differential Equations: Proceedings of the EQUADIFF Conference
119. *E. O. Roxin,* Modern Optimal Control: A Conference in Honor of Solomon Lefschetz and Joseph P. Lasalle
120. *J. C. Díaz,* Mathematics for Large Scale Computing
121. *P. S. Milojević,* Nonlinear Functional Analysis
122. *C. Sadosky,* Analysis and Partial Differential Equations: A Collection of Papers Dedicated to Mischa Cotlar
123. *R. M. Shortt,* General Topology and Applications: Proceedings of the 1988 Northeast Conference
124. *R. Wong,* Asymptotic and Computational Analysis: Conference in Honor of Frank W. J. Olver's 65th Birthday
125. *D. V. Chudnovsky and R. D. Jenks,* Computers in Mathematics
126. *W. D. Wallis, H. Shen, W. Wei, and L. Zhu,* Combinatorial Designs and Applications
127. *S. Elaydi,* Differential Equations: Stability and Control
128. *G. Chen, E. B. Lee, W. Littman, and L. Markus,* Distributed Parameter Control Systems: New Trends and Applications
129. *W. N. Everitt,* Inequalities: Fifty Years On from Hardy, Littlewood and Pólya
130. *H. G. Kaper and M. Garbey,* Asymptotic Analysis and the Numerical Solution of Partial Differential Equations
131. *O. Arino, D. E. Axelrod, and M. Kimmel,* Mathematical Population Dynamics: Proceedings of the Second International Conference
132. *S. Coen,* Geometry and Complex Variables
133. *J. A. Goldstein, F. Kappel, and W. Schappacher,* Differential Equations with Applications in Biology, Physics, and Engineering
134. *S. J. Andima, R. Kopperman, P. R. Misra, J. Z. Reichman, and A. R. Todd,* General Topology and Applications
135. *P Clément, E. Mitidieri, B. de Pagter,* Semigroup Theory and Evolution Equations: The Second International Conference
136. *K. Jarosz,* Function Spaces
137. *J. M. Bayod, N. De Grande-De Kimpe, and J. Martínez-Maurica,* p-adic Functional Analysis
138. *G. A. Anastassiou,* Approximation Theory: Proceedings of the Sixth Southeastern Approximation Theorists Annual Conference
139. *R. S. Rees,* Graphs, Matrices, and Designs: Festschrift in Honor of Norman J. Pullman
140. *G. Abrams, J. Haefner, and K. M. Rangaswamy,* Methods in Module Theory

141. *G. L. Mullen and P. J.-S. Shiue*, Finite Fields, Coding Theory, and Advances in Communications and Computing
142. *M. C. Joshi and A. V. Balakrishnan*, Mathematical Theory of Control: Proceedings of the International Conference
143. *G. Komatsu and Y. Sakane*, Complex Geometry: Proceedings of the Osaka International Conference
144. *I. J. Bakelman*, Geometric Analysis and Nonlinear Partial Differential Equations
145. *T. Mabuchi and S. Mukai*, Einstein Metrics and Yang–Mills Connections: Proceedings of the 27th Taniguchi International Symposium
146. *L. Fuchs and R. Göbel*, Abelian Groups: Proceedings of the 1991 Curaçao Conference
147. *A. D. Pollington and W. Moran*, Number Theory with an Emphasis on the Markoff Spectrum
148. *G. Dore, A. Favini, E. Obrecht, and A. Venni*, Differential Equations in Banach Spaces
149. *T. West*, Continuum Theory and Dynamical Systems
150. *K. D. Bierstedt, A. Pietsch, W. Ruess, and D. Vogt*, Functional Analysis
151. *K. G. Fischer, P. Loustaunau, J. Shapiro, E. L. Green, and D. Farkas*, Computational Algebra
152. *K. D. Elworthy, W. N. Everitt, and E. B. Lee*, Differential Equations, Dynamical Systems, and Control Science
153. *P.-J. Cahen, D. L. Costa, M. Fontana, and S.-E. Kabbaj*, Commutative Ring Theory
154. *S. C. Cooper and W. J. Thron*, Continued Fractions and Orthogonal Functions: Theory and Applications
155. *P. Clément and G. Lumer*, Evolution Equations, Control Theory, and Biomathematics
156. *M. Gyllenberg and L. Persson*, Analysis, Algebra, and Computers in Mathematical Research: Proceedings of the Twenty-First Nordic Congress of Mathematicians
157. *W. O. Bray, P. S. Milojević, and Č. V. Stanojević*, Fourrier Analysis: Analytic and Geometric Aspects
158. *J. Bergen and S. Montgomery*, Advances in Hopf Algebras

Additional Volumes in Preparation

analysis, algebra, and computers in mathematical research

proceedings of the twenty-first Nordic congress of mathematicians

edited by

Mats Gyllenberg
University of Turku
Turku, Finland

Lars Erik Persson
Luleå University of Technology
Luleå, Sweden

Marcel Dekker, Inc. **New York • Basel • Hong Kong**

Library of Congress Cataloging–in–Publication Data

Nordic Congress of Mathematicians (21st : 1992 : Luleå University of Technology)
Analysis, algebra, and computers in mathematical research: Proceedings of the Twenty-first Nordic Congress of Mathematicians / edited by Mats Gyllenberg, Lars Erik Persson.
p. cm. -- (Lecture notes in pure and applied mathematics ; v. 156)
Includes bibliographical references (p. -)
ISBN 0—8247—9217—3 (acid-free)
1. Mathematical analysis--Congresses. 2. Algebra--Congresses. 3. Stochastic processes--Congresses. 4. Mathematics--Research--Data processing--Congresses. I. Gyllenberg, Mats. II. Persson, Lars Erik. III. Title. IV. Series.
QA299.6.N67 1992
510--dc20 94-2464
CIP

The publisher offers discounts on this book when ordered in bulk quantities. For more information, write to Special Sales/Professional Marketing at the address below.

This book is printed on acid-free paper.

MARCEL DEKKER, INC.
270 Madison Avenue, New York, New York 10016

Current printing (last digit):
10 9 8 7 6 5 4 3 2 1

PRINTED IN THE UNITED STATES OF AMERICA

Preface

The Twenty-first Nordic Congress of Mathematicians was held at Luleå University of Technology. One day was devoted to a special program on computers in mathematical research.

The organizers of the congress were Mats Gyllenberg and Lars Erik Persson from Luleå University of Technology, Tord Ganelius and Jan-Erik Roos from Stockholm University, and Peter Sjögren from Chalmers University of Technology. The local organizing committee consisted of the following members of Luleå University of Technology: Andrejs Dunkels, Ove Lindblom, Torsten Lindström, Lars Erik Persson, Margareta Oldgren, Ann-Catrin Wallin, Lena Wasserman, and Sven Öberg.

The congress was sponsored by Apple Computer AB, Canon Svenska AB, Computer Solutions Europe AB, DoubleClick AB, Hewlett–Packard, HP (Kista), IBM Sweden, Sun Microsystems AB, TK-House, the Kempe Foundation, Trygg-Hansa, the Swedish Natural Science Research Council (NFR), the Swedish Research Council for Engineering Sciences (TRF), the Lettersted Society, and the Lundberg Foundation.

In the main program of the congress there were 13 invited lectures given in joint sessions. The other lectures were given in three parallel sessions. There were more than 100 participants. The special program on computers in mathematical research was organized by Jan-Erik Roos from Stockholm University and Björn von Sydow from Chalmers University of Technology. In particular, Torsten Ekedahl from Stockholm University gave a crash course in Mathematica™ programming and Michael B. Monagan from ETM Centrum, Zürich, a crash course in MAPLE programming for mathematicians. This volume consists of 26 papers selected from the lectures of the participants.

The organizers of this congress are indebted to many persons for the success of the congress and the publication of this volume. First of all we express our sincere gratitude to the speakers for their well prepared lectures and to the session chairmen for maintaining the schedule. We also want to thank the local organizing committee for their attention to logistics. Particular thanks go to Lena Wasserman and Ann-Catrin Wallin for their careful handling of all correspondence and financial matters. Finally, we thank our sponsors for financial support and God for the marvelous weather during the congress and the excursion to the Polar Circle and Jokkfall waterfall.

Mats Gyllenberg
Lars Erik Persson

Contents

Contributors

E. SPARRE ANDERSEN Københavns University, København, Denmark

MATS E. ANDERSSON Uppsala University, Uppsala, Sweden

ANDERS BARRLUND University of Umeå, Umeå, Sweden

BJÖRN BIRNIR University of California, Santa Barbara, California, and University of Iceland, Reykjavik, Iceland

GÖRAN BJÖRCK Stockholm University, Stockholm, Sweden

TIMO ERKAMA University of Joensuu, Joensuu, Finland

JUHANI FISKAALI University of Oulu, Linnanmaa, Oulu, Finland

RALF FRÖBERG Stockholm University, Stockholm, Sweden

ERLAND GADDE University of Umeå, Umeå, Sweden

CHRISTER GLADER Åbo Akademy, Åbo, Finland

CHRISTIAN GOTTLIEB University of Stockholm, Stockholm, Sweden

MATS GYLLENBERG Luleå University of Technology, Luleå, Sweden (Current affiliation: University of Turku, Turku, Finland)

HEIKKI HAAHTI University of Oulu, Linnanmaa, Oulu, Finland

BERND HERZOG University of Stockholm, Stockholm, Sweden

GÖRAN HÖGNÄS Åbo Akademy, Åbo, Finland

IAN KIMING University of Essen, Essen, Germany

TIMO KOSKI Luleå University of Technology, Luleå, Sweden

LARRY LAMBE Kent State University, Kent, Ohio (current affiliation: Rutgers University, Piscataway, New Jersey)

MOGENS ESROM LARSEN Københavns University, København Ø, Denmark

PETER LINDQVIST Norwegian Institute of Technology, Trondheim, Norway

TORSTEN LINDSTRÖM Luleå University of Technology, Luleå, Sweden

ANDERS MELIN Lund Institute of Technology, Lund, Sweden

RAGNI PIENE University of Oslo, Blindern, Oslo, Norway

IGOR RIVIN Institute for Advanced Study, Princeton, New Jersey

MIKAEL STENLUND University of Umeå, Umeå, Sweden (Current affiliation: Mãlardalen University, Vãsterås, Sweden)

JAN-OLOV STRÖMBERG University of Tromsø, Tromsø, Norway

THOMAS STRÖMBERG Luleå University of Technology, Luleå, Sweden

TORBJÖRN TAMBOUR Stockholm University, Stockholm, Sweden

HERMANN THORISSON University of Iceland, Reykjavik, Iceland

BERNT ØKSENDAL University of Oslo, Blindern, Oslo, Norway

PATRIC R. J. ÖSTERGÅRD Helsinki University of Technology, Espoo, Finland

Combinatorial Summation Identities

E. SPARRE ANDERSEN AND MOGENS ESROM LARSEN

Matematisk Institut
Københavns Universitet
Universitetsparken 5
DK–2100 København Ø Danmark

Abstract. In this paper we present the most frequently appearing combinatorial sums, i.e. the binomial sum, the Chu–Vandermonde convolution, the Kummer–, Gauß– and Bailey–sums, and for one more complicated example, the Pfaff–Saalschütz formula. As these formulas usually appear in disguise, we give a method of recognizing such formulas.

You never know when you will encounter a binomial coefficient sum
Doron Zeilberger [20].

Introduction. The problem for the non–expert in this field is in short, he needs a formula to reduce a certain sum, but he can't find exactly the required form in the tables. But to quote G. E. Andrews' statement about Gould's table from 1972, [12], from 1974 ([5], p. 470). "However, there still remains the difficulty of finding one's own identity among the vast number of others, and as Gould points out [12, p. viii], the situation may be further complicated through changes of variables and introduction of redundant factors."

What is needed is a way to write the sum in a *canonical* form, and to *classify* the formulas in the table according to the chosen form. In [5] G. E. Andrews claims that by using hypergeometric series one can reduce 450 of the 577 entries in Gould's table, [12], to 32 entries, but left to others to accomplish this reduction. We want to add that among the 122 leftovers in Gould's table are the Hagen–Rothe identity and some of the Abel identities which can <u>not</u> be written as a hypergeometric sum.

An important advance towards the solution is the algorithm of R. W. Gosper from 1978, [10]. This algorithm allows a systematic evaluation of indefinite hypergeometric sums analogous to the antiderivative in the calculus of integration. If the indefinite sum does not exist, the algorithm reports on this. For the more numerous and more important sums which are only summable between certain limits, this algorithm does not work.

Recently, in 1990, Herbert S. Wilf and Doron Zeilberger [19] published a very interesting method to attack definite sums by the help of computer algebra. Their method may in a large amount of cases be able to accomplish a proof of an identity, provided we already know or have guessed the result.

In this paper we present a different, more "human," approach. We shall introduce a canonical form avoiding the language of hypergeometric functions, and present the simplest and most useful formulas in this canonical form. More important, we give methods to recognize these identities, if we encounter them in one of their disguised forms. As examples we have chosen to show the most common and most useful identity, the Chu–Vandermonde convolution, in order to present the ideas as transparently as possible.

Historical introduction. The oldest known combinatorial identities are the *binomial formula,*

$$\sum_{k=0}^{n} \binom{n}{k} x^k y^{n-k} = (x+y)^n \tag{1}$$

and the *Chu–Vandermonde convolution,*

$$\sum_{k=0}^{n} \binom{x}{k} \binom{y}{n-k} = \binom{x+y}{n} \tag{2}$$

The first formula was discovered by Al–Karaǰi in Baghdad around 1008, and it was known in China as early as 1261 by Yang Hui [8]. The second formula has appeared in the famous Chinese algebra text by Shih–Chieh Chu, *Precious Mirror of the Four Elements,* in 1303, [7,11], while its discovery in Europe was much later. It was found in 1772 of A. T. Vandermonde, [18].

They are actually facets of the same formula, which is hardly recognized by their traditional appearance.

This may be surprising as the sign–alternating analogue of the binomial formula is just the same with the sign of x changed, while the analogue of the Chu–Vandermonde convolution is hardly known. Only for $x = y$ and for $x + y = 2n$ we have Kummer's formulas from 1836, [13],

$$\sum_{k=0}^{2n} (-1)^k \binom{x}{k} \binom{x}{2n-k} = (-1)^n \binom{x}{n} \tag{3}$$

$$\sum_{k=0}^{2n+1} (-1)^k \binom{x}{k} \binom{x}{2n+1-k} = 0 \tag{4}$$

$$\sum_{k=0}^{n} (-1)^k \binom{n+x}{k} \binom{n-x}{n-k} = 4^n \binom{\frac{n-x-1}{2}}{n} \tag{5}$$

These formulas do not explicitly appear in [13], but are specializations of evaluations of hypergeometric series.

In this paper we shall state a common generalization of the binomial and the Chu–Vandermonde formulas and some of the generalizations of Kummer's formulas, treated earlier in [1,2,3,4]. Besides this we shall give some of the more surprising variations, and a few hints to recognize the possible variations, when you meet them.

Motivation. It is a nice exercise to find the formula for the Laguerre polynomials as

$$L_m(x) = e^x \frac{d^m}{dx^m}\left(x^m e^{-x}\right) = m! \sum_{n=0}^{m} \binom{m}{n} \frac{(-x)^n}{n!} \tag{6}$$

The proof runs like the following

$$\begin{aligned}
e^x \frac{d^m}{dx^m}\left(x^m e^{-x}\right) &= \left(\sum_{n=0}^{\infty} \frac{x^n}{n!}\right) \frac{d^m}{dx^m}\left(x^m \sum_{k=0}^{\infty} \frac{(-x)^k}{k!}\right) \\
&= \left(\sum_{n=0}^{\infty} \frac{x^n}{n!}\right) \cdot \left(\sum_{k=0}^{\infty} \frac{(-1)^k [m+k]_m x^k}{k!}\right) \\
&= \sum_{n=0}^{\infty} x^n \sum_{k=0}^{n} \frac{(-1)^k [m+k]_m}{k!(n-k)!} \\
&= \sum_{n=0}^{\infty} \frac{x^n}{n!} \sum_{k=0}^{n} \binom{n}{k} (-1)^k [m+k]_m
\end{aligned}$$

where we have abbreviated the descending factorial as

$$[x]_m := x(x-1)\cdots(x-m+1) \tag{7}$$

In one form to be proven later, the Chu–Vandermonde convolution, (2), contains the special case

$$\sum_{k=0}^{n} \binom{n}{k} (-1)^k [m+k]_m = (-1)^n [m]_n [m]_{m-n} \tag{8}$$

which immediately gives the expression in (6).

There are two ways to extract the moral of this example. We encounter the Chu–Vandermonde convolution all the time, – also in this paper, e.g. twice in each of the proofs of (83), (87) and (95) and once in the proof of (91–93). And we need a way to recognize this event.

Definitions. To deal with formulas in some generality it is convenient to generalize the elementary concepts too. The **factorial** $[x,d]_n$ shall be defined for any **number,** $x \in \mathbb{C}$, any **stepsize,** $d \in \mathbb{C}$, and any **length,** $n \in \mathbb{Z}$, by

$$[x,d]_n := \begin{cases} \prod_{j=0}^{n-1} (x - jd) & n \in \mathbb{N} \\ 1 & n = 0 \\ \prod_{j=1}^{-n} \frac{1}{x + jd} & -n \in \mathbb{N} \end{cases} \tag{9}$$

As special cases we use the shorthands

$$[x]_n := [x,1]_n \qquad (x(x-1)\cdots(x-n+1) \text{ for } n>0) \tag{10}$$

$$(x)_n := [x,-1]_n \qquad (x(x+1)\cdots(x+n-1) \text{ for } n>0) \tag{11}$$

$$x^n = [x,0]_n \qquad n \in \mathbf{Z} \tag{12}$$

For the **nearest integers** to a real number, $x \in \mathbf{R}$, we use

$$\lceil x \rceil := min\{n \in \mathbf{Z} | n \geq x\} \tag{13}$$

$$\lfloor x \rfloor := max\{n \in \mathbf{Z} | n \leq x\} \tag{14}$$

And the **sign** of a real number, $x \in \mathbf{R}$, is defined as $\sigma(x)$ by

$$\sigma(x) := \begin{cases} 1 & x \geq 0 \\ -1 & x < 0 \end{cases} \tag{15}$$

Finally, we define the **maximum** and **minimum** of two numbers, $x, y \in \mathbf{R}$, as

$$x \vee y := max\{x, y\} \tag{16}$$

$$x \wedge y := min\{x, y\} \tag{17}$$

The **binomial coefficients**, $\binom{x}{n}$, are defined for $x \in \mathbf{C}$ and $n \in \mathbf{Z}$ by

$$\binom{x}{n} := \begin{cases} \dfrac{[x]_n}{[n]_n} & \text{for } n \in \mathbf{N}_0 \\ 0 & \text{for } -n \in \mathbf{N} \end{cases} \tag{18}$$

REMARK. *Does (2) also hold for $x, y \in \mathbf{C}$? Yes, as a matter of fact, but it holds even more generally!*

Elementary properties. The factorial satisfies some obvious, but very useful rules of computation. The most important are

$$[x,d]_n = [-x+(n-1)d, d]_n(-1)^n \qquad x,d \in \mathbf{C}, \quad n \in \mathbf{Z} \tag{19}$$

$$[x,d]_n = [x,d]_m[x-md,d]_{n-m} \qquad x,d \in \mathbf{C}, \quad n,m \in \mathbf{Z} \tag{20}$$

$$[x,d]_n = 1/[x-nd,d]_{-n} \qquad x,d \in \mathbf{C}, \quad n \in \mathbf{Z} \tag{21}$$

$$[x,d]_n = [x-d,d]_n + nd[x-d,d]_{n-1} \qquad x,d \in \mathbf{C}, \quad n \in \mathbf{Z} \tag{22}$$

$$[xd,d]_n = d^n[x]_n \qquad x,d \in \mathbf{C}, \quad n \in \mathbf{Z} \tag{23}$$

Similarly, the binomial coefficients obey a series of rules.

$$\binom{x}{n} = \binom{x-1}{n-1} + \binom{x-1}{n} \qquad x \in \mathbf{C}, \quad n \in \mathbf{N}_0 \tag{24}$$

$$\binom{x}{n}[n]_m = [x]_m \binom{x-m}{n-m} \qquad x \in \mathbf{C}, \quad m \in \mathbf{Z}, \quad n \in \mathbf{N}_0 \tag{25}$$

$$\binom{x}{n}\binom{n}{m} = \binom{x}{m}\binom{x-m}{n-m} \qquad x \in \mathbf{C}, \quad n, m \in \mathbf{N}_0 \tag{26}$$

$$\binom{x}{n}[y]_n = [x]_n \binom{y}{n} \qquad x, y \in \mathbf{C}, \quad n \in \mathbf{Z} \tag{27}$$

$$\binom{x}{n} = (-1)^n \binom{n-x-1}{n} \qquad x \in \mathbf{C}, \quad n \in \mathbf{Z} \tag{28}$$

$$\binom{m}{n} = \binom{m}{m-n} \qquad m \in \mathbf{N}_0, \quad n \in \mathbf{Z} \tag{29}$$

We omit the proofs, since the formulas (24–29) are easily proved using (18) and the elementary properties (19–23) of the factorials.

The Binomial–Chu–Vandermonde formula. Rather than considering the expression (2) with generalized binomial coefficients, it becomes nice if we multiply the equation with the integral denominator, $[n]_n$. Then (2) looks like:

$$\sum_{k=0}^{n} \binom{n}{k} [x]_k [y]_{n-k} = [x+y]_n \tag{30}$$

But this expression allows a generalization with arbitrary stepsize, i.e.,

THEOREM. *For $n \in \mathbf{N}_0$, $x, y \in \mathbf{C}$ and $d \in \mathbf{C}$, we have*

$$\sum_{k=0}^{n} \binom{n}{k} [x,d]_k [y,d]_{n-k} = [x+y,d]_n \tag{31}$$

PROOF: Let

$$S(n) := \sum_k \binom{n}{k} [x,d]_k [y,d]_{n-k}$$

We replace using (24) $\binom{n}{k}$ by $\binom{n-1}{k-1} + \binom{n-1}{k}$ and split the sum in two sums. In the sum with $\binom{n-1}{k-1}$ we substitute $k+1$ for k. Corresponding terms in the two sums now have $\binom{n-1}{k}[x,d]_k[y,d]_{n-k-1}$ as common factor. Using this we obtain

$$\sum_k \binom{n-1}{k} [x,d]_k [y,d]_{n-k-1} (x - k\cdot d + y - (n-k-1)\cdot d) = S(n-1)\cdot(x+y-(n-1)\cdot d)$$

Recursion now yields

$$S(n) = S(n-k)\cdot[x+y-(n-k)\cdot d, d]_k$$

Since $S(0) = 1$ we obtain (31).

Remark. *For $d = 1$ we have (30) and for $d = 0$ we have (1).*

Second remark. *The sum in (31) can not be written as a finite hypergeometric sum due to the arbitrary stepsize, d, in the factorial.*

Special cases of the Binomial–Chu–Vandermonde formula. The Binomial–Chu–Vandermonde formula appears in many disguises. In Gould's collection, [12], there are more than 150 forms, including special cases. Here we shall explore a few of the possibilities.

If we change the sign of all factors in one of the factorials, the form becomes a sum of sign–alternating products with a moving gap or overlap for suitable choices of x and y.

Let $x = -d - c$ and $y = c + (n-m)d$, then (31) becomes

$$\sum_{k=0}^{n} \binom{n}{k} [-d-c, d]_k [c + (n-m)d, d]_{n-k} = [(n-m-1)d, d]_n \tag{32}$$

Now, using (19) and (23), we get

$$\sum_{k=0}^{n} \binom{n}{k} (-1)^k [c + kd, d]_k [c + (n-m)d, d]_{n-k} = d^n [n-m-1]_n \tag{33}$$

We may then use (20) twice to write

$$\begin{aligned}
&[c + kd, d]_k [c + (n-m)d, d]_{n-k} \\
=&[c + kd, d]_m [c + (k-m)d, d]_{k-m} [c + (n-m)d, d]_{n-k} \\
=&[c + kd, d]_m [c + (n-m)d, d]_{n-m}
\end{aligned}$$

and then (21) on the last factorial

$$[c + (n-m)d, d]_{n-m} = 1/[c, d]_{m-n}$$

Multiplying and (19) applied to the right side gives

$$\sum_{k=0}^{n} \binom{n}{k} (-1)^{n-k} [c + kd, d]_m = [c, d]_{m-n} \cdot d^n [m]_n \tag{34}$$

which has (8) as the special case $c = m$ and $d = 1$.

Remark. *For $0 \leq m < n$ the right side becomes 0.*

Second remark. *For $c = 0$ the right side becomes 0, except for $m = n$. Hence*

$$\sum_{k=0}^{n} \binom{n}{k} (-1)^{n-k} [k]_m = \delta_{mn} \cdot [n]_n = \delta_{mn} \cdot [m]_m \tag{35}$$

THIRD REMARK. *For $m = -1$ we may apply (21) to get (with c replaced by $c - d$)*

$$\sum_{k=0}^{n} \binom{n}{k} (-1)^{n-k} \frac{1}{c+kd} = \frac{d^n[-1]_n}{[c+nd,d]_{n+1}} = \frac{d^n(-1)^n[n]_n}{[c+nd,d]_{n+1}} \tag{36}$$

or

$$\frac{1}{c(c+d)\cdots(c+nd)} = \frac{1}{d^n[n]_n} \sum_{k=0}^{n} \frac{(-1)^k\binom{n}{k}}{c+kd} \tag{37}$$

i.e., the partial fraction for a polynomial with equidistant roots; for $d = 1$ we even get

$$\frac{1}{(c)_{n+1}} = \frac{1}{[c+n]_{n+1}} = \frac{1}{[n]_n} \sum_{k=0}^{n} \frac{(-1)^k\binom{n}{k}}{c+k} \tag{38}$$

i.e., the partial fraction for a polynomial with consecutive integral roots.

Recognition of formulas. How could we have told from the outset that (8) was just a special form of the Chu–Vandermonde convolution?

Suppose we have an expression

$$T(a) = \sum_{k=m}^{n} t(a,k) \tag{39}$$

with $m, n, \in \mathbb{Z}$ and $n - m \in \mathbb{N}_0$ one less than the number of terms in the sum, $a \in \mathbb{C}^\ell$, a parameter–vector, and $k \in \mathbb{Z}$ the summation variable. The question is, do we know a formula, which by a trivial transformation gives us the value of the sum, (39)? By "trivial" we mean obtained by the following four operations:

1) Multiplying with a non–zero constant which may depend on a, m and n.
2) Translation of the interval of summation.
3) Special choice of the parameters and limits.
4) Reversing the direction of summation.

The way to treat the first triviality is to consider the *formal* quotient

$$q_t(a,k) := \frac{t(a,k+1)}{t(a,k)} \tag{40}$$

which characterizes the expressions up to proportionality, because of the formula:

$$T(a) = t(a,m) \sum_{k=m}^{n} \prod_{i=m}^{k-1} q_t(a,i) \tag{41}$$

E. g. the Binomial–Chu–Vandermonde formula is a typical example of (39):

$$[a+b,d]_n = \sum_{k=0}^{n} \binom{n}{k} [a,d]_k [b,d]_{n-k} \tag{42}$$

with parameter–vector (a,b,d,n), number of terms (minus one), n, and which has quotients which are rational functions of k:

$$q_t(a,b,d,n,k) = \frac{(n-k)(\frac{a}{d}-k)}{(-1-k)(n-1-\frac{b}{d}-k)} \tag{43}$$

If we substitute a for m in (8), we get

$$\sum_{k=0}^{n} \binom{n}{k} [a+k]_a (-1)^k \tag{44}$$

This sum has the quotient

$$q_s(a,n,k) = \frac{(n-k)[a+k+1]_a}{(-1-k)[a+k]_a} = \frac{(n-k)(-a-1-k)}{(-1-k)(-1-k)} = q_t(-a-1,n,1,n,k) \tag{45}$$

This means that we get a formula proportional to (44) by choosing in (42) $-a-1$ for a, n for b and $d=1$. I.e., using (19) on the right side, we get

$$\sum_{k=0}^{n} \binom{n}{k} [-a-1]_k [n]_{n-k} = [n-a-1]_n = (-1)^n [a]_n \tag{46}$$

To get what we want, we just apply (41), i.e., we shall multiply and divide by the two zero–terms, $(t(a,0))$:

$$\sum_{k=0}^{n} \binom{n}{k} [a+k]_a (-1)^k = \frac{[a]_a}{[n]_n} (-1)^n [a]_n = (-1)^n [a]_n [a]_{a-n} \tag{47}$$

Classification of formulas. In general, the quotient q_t may be any function of k, but for most sums occurring in combinatorial identities, this function will be a rational function of k. If we get a rational function of degrees (p,q) in numerator and denominator, respectively, and, having normalized the polynomials, with a constant factor χ, i.e.,

$$q_t(a,k) = \frac{(\alpha_1-k)\cdots(\alpha_p-k)}{(\beta_1-k)\cdots(\beta_q-k)} \cdot \chi \tag{48}$$

with $\chi \neq 0$ and $\alpha_i \neq \beta_j$ independent of k, we shall classify the formula by the degrees of the polynomials and the factor as parameters. In [1,2] we have considered different possibilities for the quotient with this rational form as the second. Hence it is classified as of **type**

$$II(p, q, \chi) \tag{49}$$

REMARK. *Only very few formulas are known with $p \neq q$.*

E.g. the binomial formula (1) has quotient

$$q_b = \frac{n-k}{-1-k}\left(-\frac{x}{y}\right) \tag{50}$$

and hence is of type $II(1, 1, \chi)$ with $\chi = -\frac{x}{y}$, while the Chu–Vandermonde for $d \neq 0$ is of type $II(2, 2, 1)$, cf. (43).

Given a formula of the form (39) with quotients of the form (48), then the formula (41) takes the form

$$T(a) = t(a, m) \sum_{k=m}^{n} \prod_{i=m}^{k-1} q_t(a, i) = t(a, m) \sum_{k=m}^{n} \frac{\prod_{j=1}^{p}[\alpha_j - m]_{k-m}}{\prod_{j=1}^{q}[\beta_j - m]_{k-m}} \chi^{k-m} \tag{51}$$

This is a reconstruction of (39), if we assume $t(a, m) \neq 0$ and $\beta_j \notin [m, n] \cap \mathbb{Z}$.

It is worth noting that in this standard form, the stepsize $d = 1$. As a matter of fact, it follows from (23) that $d \neq 1$ is only interesting, if $d = 0$ is included. But for the rest of the formulas, this case is trivial. So, we shall from now on assume $d = 1$.

The function of k in (51),

$$f(k) := \frac{\prod_{j=1}^{p}[\alpha_j - m]_{k-m}}{\prod_{j=1}^{q}[\beta_j - m]_{k-m}} \chi^{k-m} \tag{52}$$

may be defined for values of k outside the interval of summation, $[m, n]$, in which cases we may change the limits of summation.

If the function, $t(a, k)$, is defined for all $k \in \mathbb{Z}$, then it may happen, that it vanishes outside the limits of summation. In this case we can of course extend the limits arbitrarily without changing the value of the sum.

The function $f(k)$ in (52) may vanish, if $\alpha_j - m \in \mathbb{N}_0$ or $m - \beta_j \in \mathbb{N}$.

DEFINITION. *We shall call a sum (39) with quotient of the form (48) a* **sum of natural limits,** *if the function, $t(a, k)$, is defined for all $k \in [m, n] \cap \mathbb{Z}$ and $(m, n) = (\beta_j + 1, \alpha_i)$ for some (i, j).*

In this case we shall write

$$\frac{[\alpha_i - m]_{k-m}}{[\beta_j - m]_{k-m}} = \frac{[n - m]_{k-m}}{[-1]_{k-m}} = \binom{n-m}{k-m}(-1)^{k-m} \tag{53}$$

FIRST REMARK. *A sum may have several natural limits.*

SECOND REMARK. *If a sum has the form*

$$\sum_{k=m}^{n} \binom{n-m}{k-m} \cdots \tag{54}$$

then m and n are natural limits, because $n = \alpha_1$ and $m - 1 = \beta_1$, provided the terms are defined in the interval m, n.

Canonical form. The form (51) may be undefined if $\beta_j - m \in \mathbb{N}_0$ for $k > \beta_j$. This gives the restriction to the summation,

$$n \leq \beta_j \quad \vee \quad \beta_j < m \tag{55}$$

In this case we prefer to replace a denominator with a numerator. We apply the formula

$$\frac{1}{[\beta_j - m]_{k-m}} = \frac{[\beta_j - k]_{n-k}}{[\beta_j - m]_{n-m}} = \frac{[n-1-\beta_j]_{n-k}(-1)^{n-k}}{[\beta_j - m]_{n-m}} \tag{56}$$

to replace the denominator $[\beta_j - m]_{k-m}$ by the numerator $[n-1-\beta_j]_{n-k}(-1)^{n-k}$.

DEFINITION. *The* **canonical** *form of a sum (39) with quotient of the form (48) is*

(57)
$$\begin{aligned} S(a) &= \sum_{k=m}^{n} \binom{n-m}{k-m} \prod_{j=2}^{p} [\alpha_j - m]_{k-m} \prod_{j=2}^{q} [n-1-\beta_j]_{n-k} (-1)^{(k-m)q} \chi^{k-m} \\ &= T(a) \frac{s(a,m)}{t(a,m)} \end{aligned}$$

This form with $p = 2$, $q = 2$, $\alpha_2 = a$, $\beta_2 = n - 1 - b$ and $\chi = 1$ is the canonical Chu–Vandermonde.

Sums of arbitrary limits. We may try to reconstruct the sums of form (39) from the quotients (48) for all possible limits, $m \leq n \in \mathbb{Z}$. The problem is that we have difficulties summing past the integral roots of the rational function (48).

The obvious idea is to sum the products for the different situations of no roots in the interval, $[m, n[$. But does it really matter? Suppose we have two integral roots, say α and β, satisfying

$$m \leq \alpha < \beta < n \tag{58}$$

and that we have

$$t(\alpha, \beta, k) = [\beta - k]_{\beta - \alpha} \tag{59}$$

Then the term factor

$$\frac{t(\alpha,\beta,k)}{t(\alpha,\beta,m)} = f(k) = \frac{[\beta-k]_{\beta-\alpha}}{[\beta-m]_{\beta-\alpha}} \tag{60}$$

gives the quotient

$$q_f(k) = \frac{[\beta-k-1]_{\beta-\alpha}}{[\beta-k]_{\beta-\alpha}} = \frac{\alpha-k}{\beta-k} \tag{61}$$

But, we must admit that

$$f(k) = 0 \quad \text{for} \quad \alpha < k \leq \beta \tag{62}$$

The natural question is, are there other non–vanishing term factors giving the same quotient inside the interval $]\alpha,\beta]$?

The answer is yes, we may consider

$$g(k) = \frac{(-1)^k}{\binom{\beta-\alpha-1}{k-\alpha-1}} \tag{63}$$

which is defined for $\alpha < k \leq \beta$ and gives the quotient wanted,

$$q_g(k) = -\frac{\binom{\beta-\alpha-1}{k-\alpha-1}}{\binom{\beta-\alpha-1}{k-\alpha}} = -\frac{[\beta-\alpha-1]_{k-\alpha-1}[k-\alpha]_{k-\alpha}}{[\beta-\alpha-1]_{k-\alpha}[k-\alpha-1]_{k-\alpha-1}} = -\frac{k-\alpha}{\beta-k} = \frac{\alpha-k}{\beta-k} \tag{64}$$

As an example of this we have the formula due to Tor B. Staver, [17],

$$\sum_{k=0}^{n} \frac{1}{\binom{n}{k}} = \frac{n+1}{2^{-n-1}} \cdot \sum_{k=1}^{n+1} \frac{2^k}{k} \tag{65}$$

with the quotient $q(k) = \frac{1\ \ k}{n-k}(-1)$.

Similarly, if $\beta < \alpha$, the term factor

$$f(k) = \frac{1}{[\alpha-k]_{\alpha-\beta}} = [\beta-k]_{\beta-\alpha} \tag{66}$$

gives the quotient as above, and is defined for $k \leq \beta$ and for $\alpha < k$.

For the interval $]\beta,\alpha]$ we may choose the term factor

$$g(k) = (-1)^k \binom{\alpha-\beta-1}{k-\beta-1} \tag{67}$$

which has the quotient

$$q_g(k) = -\frac{\binom{\alpha-\beta-1}{k-\beta}}{\binom{\alpha-\beta-1}{k-\beta-1}} = \frac{\alpha-k}{\beta-k} \tag{68}$$

and is defined for $\beta < k \leq \alpha$.

EXAMPLE. *Consider the formula*

$$\sum_{k=m}^{n} \frac{[a]_k}{[b]_k} = \frac{1}{a-b-1}\left(\frac{[a]_{n+1}}{[b]_n} - \frac{[a]_m}{[b]_{m-1}}\right) \tag{69}$$

valid for $a \neq b+1$, and if $b \in \mathbb{N}_0$ for $n \leq b$, and if $-a \in \mathbb{N}$ for $a < m$.

Suppose $a, b \in \mathbb{Z}$ and that either $m \leq n \leq a \wedge b$ or $a \vee b < m \leq n$. Then we may compute the quotient

$$q_t(k) = \frac{[a]_{k+1}}{[b]_{k+1}} \cdot \frac{[b]_k}{[a]_k} = \frac{a-k}{b-k} \tag{70}$$

which is the same as the quotient of the terms

$$s(k) = [b-k]_{b-a} \tag{71}$$

Hence we may apply (51) twice to find

$$\sum_{k=m}^{n} \frac{[a]_k}{[b]_k} = \frac{[a]_m}{[b]_m} \cdot \frac{1}{[b-m]_{b-a}} \cdot \sum_{k=m}^{n} [b-k]_{b-a} \tag{72}$$

Now, from (22) follows

$$[b-k]_{b-a} = \frac{[b-k]_{b-a+1} - [b-k+1]_{b-a+1}}{a-b-1} \tag{73}$$

so that the sum (72) becomes by repeated use of (20)

$$\begin{aligned} &\frac{[a]_m}{[b]_{m+b-a}} \cdot \frac{1}{a-b-1}\left([b-n]_{b-a+1} - [b-m+1]_{b-a+1}\right) \\ =&\frac{1}{a-b-1}\left[\frac{[a]_m}{[b]_{m+b-a}} \cdot [b-k+1]_{b-a+1}\right]_m^{n+1} \\ =&\frac{1}{a-b-1}\left[\frac{[a]_m[b-k+1]_{b-a+1+m-k}[a-m]_{k-m}}{[b]_{k-1}[b-k+1]_{b-a+1+m-k}}\right]_m^{n+1} \\ =&\frac{1}{a-b-1}\left[\frac{[a]_k}{[b]_{k-1}}\right]_m^{n+1} \end{aligned} \tag{74}$$

The exception $b-a=-1$ corresponds to the case of harmonic numbers. The sum can be expressed as such, but the formula then looks different from (69).

This formula could of course also be found by the algorithm of Gosper, [10].

Recognition of classified formulas. The special choice of the parameters, a and n is often guided by considering the roots of the quotients. But sometimes this is not even necessary, e.g., the formula (35) as a special case of (34).

Reversal of the direction of summation changes any formula to a new one. The question is, whether this new formula may be obtained from the original by a suitable choice of parameters. It is necessary, that the type is $II(p, p, \pm 1)$.

Consider the formula (39) with quotient (40). Then we may define

$$S(a) = \sum_{k=m}^{n} t(a, n+m-k) \tag{75}$$

with quotient

$$q_s(a,k) = \frac{t(a, n+m-k-1)}{t(a, n+m-k)} = \frac{1}{q_t(a, n+m-1-k)} \tag{76}$$

This summation formula is independent of the direction of summation if there exists a parameter, b, such that

$$\frac{1}{q_t(a, n+m-1-k)} = q_t(b,k) \tag{77}$$

This means that

$$\frac{(\beta_1-(n+m-1-k))\cdots(\beta_q-(n+m-1-k))}{(\alpha_1-(n+m-1-k))\cdots(\alpha_p-(n+m-1-k))}\cdot\frac{1}{\chi} = \frac{(\hat{\alpha}_1-k)\cdots(\hat{\alpha}_p-k)}{(\hat{\beta}_1-k)\cdots(\hat{\beta}_q-k)}\cdot\hat{\chi} \tag{78}$$

In other words, the parameter b must be chosen such that

$$\begin{aligned}\hat{\alpha}_i &= n+m-1-\beta_i\\ \hat{\beta}_i &= n+m-1-\alpha_i\\ \hat{\chi} &= \frac{1}{\chi}\end{aligned} \tag{79}$$

For the Chu–Vandermonde convolution with $m = 0$ we have the roots $\alpha_1 = n$, $\alpha_2 = a$, $\beta_1 = -1$ and $\beta_2 = n-1-b$ and the factor $\chi = 1$. Hence the reversed Chu–Vandermonde convolution has the roots $\hat{\alpha}_1 = n-1-(-1) = n$, $\hat{\alpha}_2 = n-1-(n-1-b) = b$, $\hat{\beta}_1 = n-1-n = -1$ and $\hat{\beta}_2 = n-1-a$ and the factor $\hat{\chi} = \frac{1}{1} = 1$. Hence, reversing the Chu–Vandermonde convolution gives the same formula with a and b interchanged.

This means that we shall recognize a variation of the Chu–Vandermonde convolution by the characteristics, that the type is $II(2, 2, 1)$, it has natural limits, and the quotient as a rational function of k has one arbitrary root in the numerator besides the upper limit, and one arbitrary root in the denominator besides the lower limit -1.

The Kummer formulas. Consider a sum of the form

$$S(a,b,n) = \sum_{k=0}^{n} \binom{n}{k} [a]_k [b]_{n-k} (-1)^k \tag{80}$$

with the quotient

$$q_k = \frac{(n-k)(a-k)}{(-1-k)(n-1-b-k)} \cdot (-1) \tag{81}$$

which makes the sum alternating due to the factor (-1). Therefore, for $a = b$ and n odd we must have $S(a,a,2n+1) = -S(a,a,2n+1) = 0$.

THEOREM. *The general Kummer formula for* $a, b \in \mathbb{C}$ *is*

$$S(a,b,n) = \sum_{j=0}^{\lfloor \frac{n}{2} \rfloor} \binom{b-a}{n-2j} [n]_{n-j} [a]_j (-1)^j \tag{82}$$

PROOF: Let us define

$$S = \sum_{k=0}^{n} \binom{n}{k} [a]_k [b]_{n-k} (-1)^k$$

Then we apply (30) to the second factorial as

$$[b]_{n-k} = \sum_j \binom{n-k}{j-k} [a-k]_{j-k} [b-a+k]_{n-j}$$

Hence we obtain

$$S = \sum_k \sum_j \binom{n}{k} \binom{n-k}{j-k} (-1)^k [a]_k [a-k]_{j-k} [b-a+k]_{n-j}$$

We apply (26) and (20) and we interchange the order of summation to get

$$S = \sum_j \sum_k \binom{n}{j} \binom{j}{k} (-1)^k [a]_j [b-a+k]_{n-j} =$$

$$= \sum_j \binom{n}{j} [a]_j \sum_k \binom{j}{k} [b-a+k]_{n-j} (-1)^k$$

Now (34) implies that this is zero for $0 \le n-j < j$, but for $j \le \frac{n}{2}$ we get

$$S = \sum_j \binom{n}{j} [a]_j (-1)^j [b-a]_{n-2j} [n-j]_j$$

We apply (27) and (29) to write

$$\binom{n}{j}[n-j]_j = \binom{n-j}{j}[n]_j = \binom{n-j}{n-2j}[n]_j$$

and then (27) again to write

$$\binom{n-j}{n-2j}[b-a]_{n-2j} = \binom{b-a}{n-2j}[n-j]_{n-2j}$$

and finally (20) to write

$$[n]_j[n-j]_{n-2j} = [n]_{n-j}$$

The result is then

$$S = \sum_j \binom{b-a}{n-2j}[n]_{n-j}[a]_j(-1)^j$$

In general this theorem is the only formula known, but for two special cases there are improvements. If either $b-a$ or $b+a$ are integers, we can shorten the sum.

In the first case we have the quasi-symmetric Kummer formula with $p = b-a \in \mathbf{Z}$:

THEOREM. *The quasi-symmetric Kummer formula for $a \in \mathrm{C}$ and $p \in \mathbf{Z}$ is*

$$(83) \qquad S(a, a+p, n) = \sigma(p)^n \sum_{j=\lceil \frac{n-|p|}{2} \rceil}^{\lfloor \frac{n}{2} \rfloor} \binom{|p|}{n-2j}[n]_{n-j}[a+(p\wedge 0)]_j(-1)^j$$

PROOF: Let $p \geq 0$. Then the change to the natural limits gives the formula (83). The general formula is obtained by reversing the order of summation.

The special case of difference zero is the symmetric Kummer identity

COROLLARY. *The symmetric Kummer identity for $a \in \mathbf{C}$:*

$$(84) \qquad \sum_{k=0}^{n} \binom{n}{k}[a]_k[a]_{n-k}(-1)^k = \begin{cases} [n]_m[a]_m(-1)^m & \text{for } n = 2m \\ 0 & \text{for } n \text{ odd} \end{cases}$$

The formulas (3) and (4) follows from this corollary by division by $[n]_n = [n]_m[m]_m$. This procedure may be used on (83) as it stands. Hence we get

COROLLARY. *The quasi-symmetric Kummer formula analogues to (3) and (4)*

$$(85) \qquad \sum_{k=0}^{n} \binom{x}{k}\binom{x+p}{n-k}(-1)^k = \sigma(p)^n \sum_{j=\lceil \frac{n-|p|}{2} \rceil}^{\lfloor \frac{n}{2} \rfloor} \binom{|p|}{n-2j}\binom{x+(p\wedge 0)}{j}(-1)^j$$

In the second case, $a+b \in \mathbf{Z}$, we have the different nearly-poised and the well-poised Kummer formulas. To prove these we need the following

THEOREM. *The first nearly–poised Kummer formula for $a \in \mathbb{C}$ and $0 \leq n \geq p \in \mathbb{Z}$*

(86)

$$\sum_{k=0}^{n} \binom{n}{k} [n-p+a]_k [n-a]_{n-k} (-1)^k =$$

$$\frac{2^{n-(p\vee 0)}}{[n-p]_{-(p\wedge 0)}} \sum_{j=0}^{|p|} \binom{|p|}{j} (-\sigma(p))^{p+j} \cdot [n-a]_j [n-p+a]_{|p|-j} [n-j-1-a, 2]_{n-(p\vee 0)}$$

PROOF: We apply (19) to the first factorial to obtain

$$S = \sum_{k=0}^{n} \binom{n}{k} [n-p+a]_k [n-a]_{n-k} (-1)^k = \sum_{k} \binom{n}{k} [n-a]_{n-k} [p-n-a+k-1]_k$$

The product runs from $n-a$ to $p-n-a$ with the exception of the factors in the factorial $[k-a]_{n-p+1}$. Therefore we may write using (20) twice

$$S = \sum_{k} \binom{n}{k} \frac{[n-a]_{n-k} [k-a]_{n-p+1} [p-n-a+k-1]_k}{[k-a]_{n-p+1}}$$

$$= \sum_{k} \binom{n}{k} \frac{[n-a]_{2n-p+1}}{[k-a+p-n+(n-p)]_{n-p+1}}$$

We shall now apply (38) to the denominator with $c = k-a+p-n$ and $n = n-p$ to get

$$S = [n-a]_{2n-p+1} \sum_{k} \binom{n}{k} \frac{1}{[n-p]_{n-p}} \sum_{j} \binom{n-p}{j} \frac{(-1)^j}{k-a+p-n+j}$$

We substitute $i = k+j$ as summation variable in the second sum and get

$$S = \frac{[n-a]_{2n-p+1}}{[n-p]_{n-p}} \sum_{k} \binom{n}{k} \sum_{i} \binom{n-p}{i-k} \frac{(-1)^{i-k}}{p-n-a+i}$$

Then we interchange the order of summation and receive

$$S = \frac{[n-a]_{2n-p+1}}{[n-p]_{n-p}} \sum_{i} \frac{(-1)^i}{p-n-a+i} \sum_{k} \binom{n}{k} \binom{n-p}{i-k} (-1)^k$$

Now the integral quasi–symmetric Kummer formula (85) applies to the second sum (with the sign of p changed). Hence we get

$$S = \frac{[n-a]_{2n-p+1}}{[n-p]_{n-p}} \sum_{i} \frac{(-1)^i}{p-n-a+i} (-\sigma(p))^i \sum_{j} \binom{|p|}{i-2j} \binom{n-(p\vee 0)}{j} (-1)^j$$

Next we interchange the order of summation and get

$$S = \frac{[n-a]_{2n-p+1}}{[n-p]_{n-p}} \sum_j \binom{n-(p \vee 0)}{j} (-1)^j \sum_i \binom{|p|}{i-2j} \frac{(\sigma(p))^i}{p-n-a+i}$$

Now we substitute $k = i - 2j$ as summation variable in the second sum and get

$$S = \frac{[n-a]_{2n-p+1}}{[n-p]_{n-p}} \sum_j \binom{n-(p \vee 0)}{j} (-1)^j \sum_k \binom{|p|}{k} \frac{(\sigma(p))^k}{p-n-a+k+2j}$$

Again we interchange the order of summation while we divide the denominator by 2

$$S = \frac{[n-a]_{2n-p+1}}{[n-p]_{n-p}} \sum_k \binom{|p|}{k} \frac{(\sigma(p))^k}{2} \sum_j \binom{n-(p \vee 0)}{j} \frac{(-1)^j}{\frac{p-n-a+k}{2}+j}$$

Now we may again apply (38) with $c = \frac{p-n-a+k}{2}$, $n = n-(p \vee 0)$ on the inner sum. It becomes

$$S = \frac{[n-a]_{2n-p+1}}{[n-p]_{n-p}} \sum_k \binom{|p|}{k} \frac{(\sigma(p))^k}{2} \cdot \frac{[n-(p \vee 0)]_{n-(p\vee 0)}}{\left[\frac{p-n-a+k}{2}+n-(p \vee 0)\right]_{n-(p\vee 0)+1}}$$

Next step is to double each factor in the factorial of the denominator, which also doubles the stepsize, and we cancel common factors of the factorials before the sum, and finally we get

$$S = \frac{2^{n-(p\vee 0)}}{[n-p]_{-(p\wedge 0)}}[n-a]_{2n-p+1} \sum_k \binom{|p|}{k} \frac{(\sigma(p))^k}{[n-a+k-|p|,2]_{n-(p\vee 0)+1}}$$

We reverse the direction of summation and get

$$S = \frac{2^{n-(p\vee 0)}}{[n-p]_{-(p\wedge 0)}} \sum_k \binom{|p|}{k} (\sigma(p))^{p-k} \frac{[n-a]_{2n-p+1}}{[n-a-k,2]_{n-(p\vee 0)+1}}$$

We use (20) to split the numerator as

$$\begin{aligned}&[n-a]_{2n-p+1} = \\ &[n-a]_k[n-a-k]_{2n-2(p\vee 0)+1}[-a-n-k+2(p \vee 0)-1]_{2(p\vee 0)-p-k}\end{aligned}$$

and (24) to split

$$[n-a-k]_{2n-2(p\vee 0)+1} = [n-a-k,2]_{n-(p\vee 0)+1}[n-a-k-1,2]_{n-(p\vee 0)}$$

and finally (19) to write the last factorial as

$$[-a-n-k+2(p \vee 0)-1]_{2(p\vee 0)-p-k} = (-1)^{p-k}[a+n-p]_{|p|-k}$$

Then the result follows.

THEOREM. *The well–poised Kummer identity for $a \in \mathbb{C}$:*

$$\sum_{k=0}^{n} \binom{n}{k} [n+a]_k [n-a]_{n-k} (-1)^k = 2^n [n-1-a,2]_n \tag{87}$$

PROOF: We just have to apply (86) for $p = 0$.

THEOREM. *The second nearly–poised Kummer formula for $a \in \mathbb{C}$ and $p \in \mathbb{Z}$*

$$\sum_{k=0}^{n} \binom{n}{k} [n-p+a]_k [n-a]_{n-k} (-1)^k = \tag{88}$$

$$= \frac{2^{n-(p\vee 0)}}{[n-p]_{-(p\wedge 0)}} \sum_{j=0}^{|p|} \binom{|p|}{j} \sigma(p)^{p+j} [-a+n+j-1,2]_{n-(p\wedge 0)}$$

PROOF: We shall define

$$S(n,a,p) := \sum_{k=0}^{n} \binom{n}{k} [n-p+a]_k [n-a]_{n-k} (-1)^k$$

We shall apply (22) to the second factorial to split the sum in two, and then the formula (24) to the second sum to get

$$S(n,a,p) = \sum_{k=0}^{n} \binom{n}{k} [(n-(p+1))+(a+1)]_k [n-1-a]_{n-k} (-1)^k$$

$$+ n \sum_{k=0}^{n-1} \binom{n-1}{k} [((n-1)-(p-1))+a]_k [n-1-a]_{n-1-k} (-1)^k =$$

$$= S(n,a+1,p+1) + nS(n-1,a,p-1)$$

Next we shall apply (22) to the first factorial to split the sum in two, and then the formula (24) to the second sum to get

$$S(n,a,p) = \sum_{k=0}^{n} \binom{n}{k} [(n-(p+1))+a]_k [n-a]_{n-k} (-1)^k$$

$$+ n \sum_{k=1}^{n} \binom{n-1}{k-1} [((n-1)-(p-1))+(a-1)]_{k-1} \cdot$$

$$\cdot [n-1-(a-1)]_{n-1-(k-1)} (-1)^k =$$

$$= S(n,a,p+1) - nS(n-1,a-1,p-1)$$

By eliminating respectively the second and the first terms of the right sides of the two formulas, we obtain two useful recursions

$$2S(n,a,p) = S(n,a-1,p-1) + S(n,a,p-1)$$

$$2(n+1)S(n,a,p) = S(n+1,a,p+1) - S(n+1,a+1,p+1)$$

Repeating the first formula p times or the second $-p$ times and then using the formula (87) valid for $p = 0$ and canceling common powers of 2 yields the form (88).

COROLLARY 5. *For arbitrary c and arbitrary $n, p \in \mathbb{N}_0$ and $\delta = \pm 1$ we have*

$$
\sum_{j=0}^{p} \binom{p}{j} (\delta)^j [c+j, 2]_{n+p} =
\sum_{j=0}^{p} \binom{p}{j} (-\delta)^j [c-1]_j [2n+p-1-c]_{p-j} [c-j, 2]_n \tag{89}
$$

PROOF: We compare the right sides of (86) and (88) and cancel the common factors.

Other formulas. Besides the factors $\chi = \pm 1$ also the factor 2 appears in known formulas. E.g., the formulas of Gauß from 1813, [9], and Bailey from 1935, [6], which are the following formulas for the choice $p = 0$.

THEOREM. *The generalized Gauß identity for $a \in \mathbb{C}$, $p \in \mathbb{Z}$ and $n \in \mathbb{N}_0$*

$$
\sum_{k=0}^{n} \binom{n}{k} [a]_k [n-p-1-2a]_{n-k} 2^k =
= (-\sigma(p))^n \sum_{j=\lceil \frac{n-|p|}{2} \rceil}^{\lfloor \frac{n}{2} \rfloor} \binom{|p|}{n-2j} [n]_{n-j} [a+(p \wedge 0)]_j (-1)^j \tag{90}
$$

The quotient of Gauß is

$$
q_g = \frac{(n-k)(a-k)}{(-1-k)(2a-p-k)} \cdot 2 \tag{91}
$$

making it of type $II(2,2,2)$

THEOREM. *The generalized Bailey identity for $a \in \mathbb{C}$, $p \in \mathbb{Z}$ and $n \in \mathbb{N}_0$*

$$
2^{(p-n)\vee 0} [n-p]_{-(p \wedge 0)} \sum_{k=0}^{n} \binom{n}{k} [a]_k [(p-n-1)]_{n-k} 2^k =
(-1)^n 2^{(n-(p \vee 0)) \vee 0} \sum_{j=0}^{|p|} \binom{|p|}{j} \sigma(p)^{p+j} [-a+2n-p+j-1, 2]_{n-(p \wedge 0)} \tag{92}
$$

The quotient of Bailey is

$$
q_b = \frac{(n-k)(a-k)}{(-1-k)(2n-p-k)} \cdot 2 \tag{93}
$$

making it of type $II(2,2,2)$.

PROOF: We apply (19) and (31) to write

$$[b]_{n-k} = (-1)^{n-k}[-b+(n-k-1)]_{n-k} =$$
$$= (-1)^{n-k}\sum_{j=k}^{n}\binom{n-k}{j-k}[a-k]_{j-k}[n-1-a-b]_{n-j}$$

We substitute this in the sum (80) and obtain by using (26) and (20)

$$S(a,b,n) = \sum_{k=0}^{n}\sum_{j=k}^{n}\binom{n}{k}\binom{n-k}{j-k}(-1)^n[a]_k[a-k]_{j-k}[n-1-a-b]_{n-j}$$
$$= (-1)^n\sum_{j=0}^{n}\binom{n}{j}[a]_j[n-1-a-b]_{n-j}\sum_{k=0}^{j}\binom{j}{k}$$
$$= (-1)^n\sum_{j=0}^{n}\binom{n}{j}[a]_j[n-1-a-b]_{n-j}2^j$$

Now the substitution of $b = a + p$ yields the generalized Gauß formula [1,2,3], by the use of (83) and of $b = 2n - p - a$ yields the generalized Bailey formula, [1,2,3], by the use of (86).

The formulas due to Kummer, Gauß and Bailey are all special cases of evaluations of hypergeometric series by Γ–functions. These are generalized similarly to the generalizations presented here in [4].

There exist simple formulas of other types, e.g., $II(3,3,1)$. The oldest known is the Pfaff–Saalschütz formula from 1797, [14,15]:

THEOREM. *Let $m := a+x+b+y-n+1$ satisfy the Pfaff-Saalschütz condition, $m = 0$. Then we have*

$$\sum_{k=0}^{n}\binom{n}{k}[a]_k[x]_k[b]_{n-k}[y]_{n-k}(-1)^k = [a+b]_n[a+y]_n \tag{94}$$

The quotient for the left side is

$$q_{ps} = \frac{(n-k)(a-k)(x-k)}{(-1-k)(n-1-b-k)(n-1-y-k)} \tag{95}$$

REMARK. *This sum has natural limits. It is symmetric in the sense, that if we reverse the order of summation, the condition repeats itself.*

The formula (19) allows us to write the right side of (94) in four similar ways:

$$[a+b]_n[a+y]_n = [b+a]_n[b+x]_n(-1)^n = [x+y]_n[x+b]_n = [y+x]_n[y+a]_n(-1)^n \tag{96}$$

PROOF: We use (19) and (31) to write

$$(-1)^k[a]_k = [-a+k-1]_k = \sum_{j=0}^{k} \binom{k}{j} [b-(n-k)]_{k-j}[n-1-a-b]_j$$

We substitute this sum in the sum of (94) and apply (26) in order to interchange the order of summation

$$\begin{aligned} S &= \sum_{k=0}^{n} \binom{n}{k} [a]_k[x]_k[b]_{n-k}[y]_{n-k}(-1)^k \\ &= \sum_{k=0}^{n} \binom{n}{k} [x]_k[b]_{n-k}[y]_{n-k} \sum_{j=0}^{k} \binom{k}{j} [b-(n-k)]_{k-j}[n-1-a-b]_j \\ &= \sum_{j=0}^{n} \binom{n}{j} [n-1-a-b]_j[b]_{n-j} \sum_{k=j}^{n} \binom{n-j}{k-j} [x]_k[y]_{n-k} \end{aligned}$$

where we have used (20) on two factorials with b. We apply (20) again on the factorial with x, so we can apply (31) on the second sum

$$\begin{aligned} S &= \sum_{j=0}^{n} \binom{n}{j} [n-1-a-b]_j[b]_{n-j}[x]_j \sum_{k=j}^{n} \binom{n-j}{k-j} [x-j]_{k-j}[y]_{n-k} \\ &= \sum_{j=0}^{n} \binom{n}{j} [n-1-a-b]_j[b]_{n-j}[x]_j[x+y-j]_{n-j} \\ &= \sum_{j=0}^{n} \binom{n}{j} [n-1-a-b]_j[x]_j[n-1-x-y]_{n-j}[b]_{n-j}(-1)^{n-j} \end{aligned}$$

where we have used (19) on the factorial with $x+y$. This is similar to the expression S, so we may apply the formula once more to the final expression. Then we get the following

$$\begin{aligned} S &= \\ &\sum_{j=0}^{n} \binom{n}{j} [a+b+x+y-(n-1)]_j[x]_j[n-1-b-x]_{n-j}[n-1-x-y]_{n-j}(-1)^j \\ &= \sum_{j=0}^{n} \binom{n}{j} [m]_j[x]_j[n-1-b-x]_{n-j}[n-1-x-y]_{n-j}(-1)^j \end{aligned}$$

Until now, m has been arbitrary, but it is profitable to introduce the assumption, that $m \in \mathbf{N}_0$. Then we may apply (27) to write

$$\binom{n}{j}[m]_j = \binom{m}{j}[n]_j$$

Furthermore, $n-1-x-y=a+b-m$ and $n-1-b-x=a+y-m$. Applying (20) on the factorials yields

$$S=\sum_{j=0}^{m}\binom{m}{j}[n]_j[x]_j[a+b-m]_{n-j}[a+y-m]_{n-j}(-1)^j$$

For $m=0$ we get (94).

For $m\in\mathbb{N}_0$ we get the generalized Pfaff-Saalschütz formula due to H. M. Srivastava (1989), [16],

THEOREM. *Let $m := a+x+b+y-n+1$ satisfy the generalized Pfaff-Saalschütz condition, $m\in\mathbb{N}_0$. Then we have*

$$\begin{aligned}(97)\qquad &\sum_{k=0}^{n}\binom{n}{k}[a]_k[x]_k[b]_{n-k}[y]_{n-k}(-1)^k\\ &=\sum_{j=0}^{m}\binom{m}{j}[n]_j[x]_j[a+b-m]_{n-j}[a+y-m]_{n-j}(-1)^j\\ &=[a+b-m]_{n-m}[a+y-m]_{n-m}\cdot\\ &\sum_{j=0}^{m}\binom{m}{j}[n]_j[x]_j[a+b-n]_{m-j}[a+y-n]_{m-j}(-1)^j\end{aligned}$$

REMARK. *In applications the last form may be convenient.*

References.

[1] E. Sparre Andersen, *Classification of combinatorial identities*, KUMI Preprint Series 26 (1989) 21.

[2] E. Sparre Andersen and Mogens Esrom Larsen, *Algebraic Summation Identities I*, KUMI Preprint Series 5 (1991) 29.

[3] E. Sparre Andersen and Mogens Esrom Larsen, *Algebraic Summation Identities II*, KUMI Preprint Series 10 (1991) 11.

[4] E. Sparre Andersen and Mogens Esrom Larsen, *A generalization of Kummer's formula for hypergeometric series and its q-basic analogue*, KUMI Preprint Series 13 (1991) 7.

[5] George E. Andrews, *Applications of basic hypergeometric functions*, SIAM Review 16 (1974) 442–484.

[6] W. N. Bailey, *Generalized Hypergeometric Series,* Cambridge University Press London, 1935, p. 11.

[7] Shih–Chieh Chu, *Ssu Yuan Yü Chien (Precious Mirror of the Four Elements),* China, 1303.

[8] A. W. F. Edwards, *Pascal's Arithmetical Triangle,* Charles Griffin & Co., Ltd. London, 1987, p. 51–52.

[9] C. F. Gauß, *Disquisitiones generales circa seriem infinitam...*, Comm. soc. reg. sci. Gött. rec. II (1813) 123–162.

[10] R. William Gosper, Jr., *Decision procedure for indefinite hypergeometric summations*, Proc. Nat. Acad. Sci. U.S.A. 75 (1978) 40–42.

[11] John Hoe, *Les Systèmes d'Équations Polynômes dans le Siyuan Yujian (1303)*, Collège de France, Institut des hautes Études Chinoises Paris, 1977, p. 307–320.

[12] Henry W. Gould, *Combinatorial Identities*, Gould Publications, Morgantown, W. Va., 1972.

[13] E. E. Kummer, *Über die hypergeometrische Reihe* $F(\alpha, \beta, \gamma, x)$, J. reine u. ang. Math. 15 (1836) 39–83, 127–172.

[14] J. F. Pfaff, *Observationes analyticae ad L. Euler Institutiones Calculi Integralis*, Nova acta acad. sci. Petropolitanae 11 (1797) 38–57.

[15] L. Saalschütz, *Eine Summationsformel*, Zeitschr. Math. Phys. 35 (1890) 186–188.

[16] H. M. Srivastava, *An Extension of the q-Saalschützian Theorem*, Acta Math. Hung. 53 (1989) 115–118.

[17] Tor B. Staver, *Om summasjon av potenser av binomialkoeffisientene*, Norsk Matematisk Tidsskrift 29 (1947) 97–103.

[18] A. T. Vandermonde, *Mémoire sur des irrationnelles de différens ordres avec une application au cercle*, Mèm. Acad. Roy. Sci. Paris (1772) 489–498.

[19] Herbert S. Wilf and Doron Zeilberger, *Rational functions certify combinatorial identities*, J. Amer. Math. Soc. 3 (1990) 147–158.

[20] Doron Zeilberger, *The Method of Creative Telescoping*, J. Symbolic Computation 11 (1991) 195–204.

Local Variants of the Hausdorff–Young Inequality

MATS E. ANDERSSON

Matematiska institutionen
Uppsala universitet
S-751 06 Uppsala

ABSTRACT. The connection between the Banach valued Hausdorff–Young inequalities on $\mathbf{R}$, $\mathbf{Z}$ and $\mathbf{Z}_k$ is examined by developing the beginnings of a theory for transporting the inequality between the different groups. Secondly, the difference of operator norms is explained as being due to inherent properties of approximate identities on $\mathbf{T}$. This is done by examining the quotient appearing in HY for those identities. In some cases the heat kernel turn out to be extremal, while other cases are left unsettled.

PRELIMINARIES

One of the basic inequalities of Fourier analysis is the Hausdorff–Young inequality, which expresses the size of the transformed function as compared to the original one. For complex valued functions over compact or discrete groups it reads

$$\|\widehat{f}\|_{p'} \leq \|f\|_p.$$

Here and throughout the paper p obeys the restriction $1 \leq p \leq 2$. This form is not improvable due to the fact that at least one of the sides allow constant functions. In contrast to that stands the improvement due to Babenko and Beckner for $\mathbf{R}$, which says

$$\|\widehat{f}\|_{p'} \leq B_p \|f\|_p \qquad \text{where } B_p = \sqrt{\frac{p^{1/p}}{p'^{1/p'}}}.$$

This inequality is also best possible with maximizers being all Gaussian pulses. Thus there is an inherent difference between the two groups. When dealing with complex valued function it has been proved in [6]– [11] that the operator norm is strictly less than one if and only if the underlying group fails to have open compact subgroups. This is in fact proved for more general integral operators and in addition on nonunimodular groups.

This paper will examine to what extent it is possible to start with one instance of a Banach valued Hausdorff–Young inequality and from that conclude something about its presence on another group. The transition will mainly deal with $\mathbf{Z}_k \to \mathbf{T} \to \mathbf{R}$ or the other way around. We use the notation $\mathbf{Z}_k = \mathbf{Z}/k\mathbf{Z}$ from algebra. Closely related to this are two papers [4] and [5] by Bourgain where he deduces the presence of an HY-inquality from the type inequality satisfied by the space. It is not possible, however, to take the same exponent p in the resulting HY-inequality as the type of the space in question.

It turns out that the difference reflected above will persist and it will be explained as being due to a property shared by all approximate identities on $\mathbf{T}$. In fact we will see that each approximate identity $\{\phi_\alpha\}$ satisfies

$$\limsup_{\phi_\alpha \to \delta} \frac{\|\widehat{\phi_\alpha}\|_{p'}}{\|\phi_\alpha\|_p} \leq C_p < 1,$$

where the bound is independent of ϕ. This is established by proving an inequality looking like Beckner's valid for functions on $\mathbf{T}$ of small support.

NOTATIONS

This paper is concerned with different kinds of Fourier transforms depending on the choice of the underlying group. Here we deal with finite Abelian groups as well as $\mathbf{T}$, $\mathbf{Z}$ and $\mathbf{R}$. Finite groups and $\mathbf{T}$ are given Haar measures of unit mass, but their dual groups are given the counting measure in order to make Parseval's identity as clean as possible. The Fourier transform on $\mathbf{R}$ on the other hand is defined by the kernel $e^{ix\omega}$ and as a consequence the dual $\widehat{\mathbf{R}}$ gets the measure $\frac{d\omega}{2\pi}$. The large corpus of results will deal with vector valued functions. In all cases we use the Bochner integral and related conventions. The letter E will always denote a general Banach space and G a group. $\mathcal{F}_G$ and sometimes only ^ will be the Fourier transform for the group G.

We will use a number of tailored notations to fit closely to the Hausdorff–Young inequality:

Definition. (1) $M(G,E,p) = \sup\left\{ \frac{\|\widehat{g}\|_{L^{p'}(\widehat{G},E)}}{\|g\|_{L^p(G,E)}} \mid g : G \to E \right\}$

(2) $(G,E) \in HY(p)$ *etc. denotes that* $M(G,E,p) \leq 1$

(3) $\mathcal{N}_p(E) = \{n \mid (\mathbf{Z}_n, E) \in HY(p)\}$

(4) $\mathcal{O}_p(E) = \{G \mid G$ *finite*, $(G,E) \in HY(p)\}$

ENLARGING THE GROUP

This section establishes results that describe how one may move the HY inequality from a finite group to $\mathbf{T}$ and from $\mathbf{T}$ to $\mathbf{R}$. To pave our way we begin with some results, that are almost self-evident but whose detailed proofs are rather cumbersome.

Theorem 1. *For compact Abelian groups G and H we have*

(1) $M(G,E,p) \leq M(G \times H, E, p) \leq M(G,E,p)M(H,E,p)$.

(2) $M(\mathbf{Z}_n, E, p) \le M(\mathbf{Z}_{nm}, E, p) \le M(\mathbf{Z}_n, E, p) M(\mathbf{Z}_m, E, p)$.
(3) $(G, E) \in HY(p)$ *iff* $M(G, E, p) = 1$.
(4) $\mathcal{O}_p(E) = \{G : G \text{ finite}, |G| \in \mathcal{N}_p(E)\}$.

Proof. The parts (1) and (2) are done by splitting each character and switching the order of integration with the aid of Minkowski's integral inequality. In the first case the characters are factored into characters over the groups G and H respectively. In the second case consideration of each nth of the mnth roots with m applications of the HY-inequality on $\mathbf{Z}_m$ is the splitting. Part (3) is trivially due to the presence of the constant functions. Finally, (4) is accomplished by factoring the finite Abelian group G into its cyclic components. □

Theorem 2. *$\mathcal{O}_p(E) \neq \emptyset$ implies $(\mathbf{T}, E) \in HY(p)$.*

Proof. Since $\mathcal{O}_p(E) \neq \emptyset$, THEOREM 1 produces an integer $n \ge 2$ such that $\mathbf{Z}_n \in \mathcal{O}_p(E)$. The groups $\mathbf{Z}_{n^r}, r = 1, 2, \ldots$ are realized as $G_r = \{1, \rho, \rho^2, \ldots, \rho^{n^r - 1}\}$, where $\rho = \exp \frac{2\pi i}{n^r}$. The measures μ_r are taken as the Haar measures on G_r with unit mass and will be viewed as Borel measures on $\mathbf{T}$. The same theorem then proves $G_r \in \mathcal{O}_p(E)$ for each r.

Let $f : \mathbf{T} \to E$ be a step function, i.e., $f = \sum_{k=1}^{K} \alpha_k \chi_{I_k}$, where $\alpha_k \in E$ and χ_{I_k} is the characteristic function of $I_k = [e^{ia_k}, e^{ib_k}[$. Then

$$\|f\|_{L^p(\mu_r)}^p = \sum_k \|\alpha_k\|_E^p \cdot \frac{1}{n^r} |\{l : e^{\frac{2\pi i l}{n^r}} \in I_k\}| \longrightarrow$$

$$\underset{r\to\infty}{\longrightarrow} \sum_k \|\alpha_k\|_E^p \cdot \frac{|I_k|}{2\pi} = \|f\|_{L^p(\mathbf{T})}^p,$$

whence

$$\lim_{r\to\infty} \|\mathcal{F}_{G_r} f\|_{L^{p'}(\widehat{\mu_r})} \overset{HY}{\le} \lim_{r\to\infty} \|f\|_{L^p(\mu_r)} = \|f\|_{L^p(\mathbf{T})}.$$

Furthermore

$$\mathcal{F}_{G_r} f(j) = \sum_{l=0}^{n^r-1} f(e^{\frac{2\pi i l}{n^r}}) \cdot \frac{1}{n^r} e^{\frac{2\pi i l j}{n^r}} = \sum_{k=1}^{K} \alpha_k \sum_{a_k \le \frac{2\pi l}{n^r} < b_k} \frac{1}{n^r} e^{\frac{2\pi i l j}{n^r}} \longrightarrow$$

$$\underset{r\to\infty}{\longrightarrow} \sum_{k=1}^{K} \alpha_k \int_{a_k}^{b_k} e^{i\theta j} \frac{d\theta}{2\pi} = \int_{\mathbf{T}} f(e^{i\theta}) e^{i\theta j} \frac{d\theta}{2\pi} = \mathcal{F}_{\mathbf{T}} f(j).$$

Introducing $g_r : \mathbf{Z} \to E$, $\quad g_r(j) = \begin{cases} \mathcal{F}_{G_r} f(j) & -\frac{n^r}{2} < j \le \frac{n^r}{2} \\ 0 & \text{otherwise} \end{cases}$,
we have $\lim_{r\to\infty} g_r(j) = \mathcal{F}_{\mathbf{T}} f(j)$ for all $j \in \mathbf{Z}$, whence

$$\begin{aligned} \|\mathcal{F}_{\mathbf{Z}} f\|_{L^{p'}(\mathbf{Z})} &= \|\lim_{r\to\infty} g_r\|_{L^{p'}(\mathbf{Z})} \overset{Fatou}{\le} \varliminf_{r\to\infty} \|g_r\|_{L^{p'}(\mathbf{Z})} \\ &= \varliminf_{r\to\infty} \|\mathcal{F}_{G_r} f\|_{L^{p'}(\widehat{\mu_r})} \le \|f\|_{L^p(\mathbf{T})}. \end{aligned}$$

Finally the density of step functions among the Bochner integrable functions proves that the couple $(\mathbf{T}, E, p)$ fulfills the Hausdorff–Young inequality. □

Theorem 3. *$(\mathbf{T}, E) \in HY(p)$ implies $(\mathbf{R}, E) \in HY(p)$.*

Proof. It suffices to prove $\|\widehat{g}\|_{p',E,\widehat{\mathbf{R}}} \le \|g\|_{p,E,\mathbf{R}}$ for each C^∞, compactly supported $g : \mathbf{R} \to E$. By scaling we may suppose that $\operatorname{supp} g \subseteq [-\pi, \pi]$. The integral representation now says $\widehat{g}(\omega) = \int_{-\infty}^{\infty} g(\omega) e^{-i\omega x}\, dx = 2\pi \mathcal{F}_{\mathbf{T}} g(\omega)$ in a wide sense.

The assumption $(\mathbf{T}, E) \in HY(p)$ tells us that

$$\begin{aligned} \|g\|_{p,\mathbf{R}} &= (2\pi)^{1/p} \|g\|_{p,\mathbf{T}} \overset{HY}{\ge} (2\pi)^{1/p} \|\mathcal{F}_{\mathbf{T}} g\|_{p',\mathbf{Z}} = \\ &= (2\pi)^{-1/p'} \|\widehat{g}\|_{p',\mathbf{Z}} = \left[\frac{1}{2\pi} \sum_{k=-\infty}^{\infty} \|\widehat{g}(k)\|_E^{p'} \right]^{1/p'}. \end{aligned}$$

From $g \in C_c^\infty$ follows $\widehat{g} \in C_0(\widehat{\mathbf{R}})$ with $\int_{-\infty}^{\infty} \|\widehat{g}(\omega)\|_E^{p'}\, d\omega$ convergent, whence

$$\begin{aligned} \|\widehat{g}\|_{p',\widehat{\mathbf{R}}} &\underset{sum}{\overset{Riemann}{=}} \lim_{A\to\infty} \left[\frac{1}{2\pi} \sum_{k=-\infty}^{\infty} \frac{1}{A} \|\widehat{g}(k/A)\|_E^{p'} \right]^{1/p'} \le \\ &\le \varlimsup_{A\to\infty} \|A^{1/p} g(A\cdot)\|_{p,\mathbf{R}} = \|g\|_{p,\mathbf{R}}. \quad \square \end{aligned}$$

A closer examination of the necessary inequality yields

Theorem 4.

$$M(\mathbf{R}, E, p) \le \lim_{a\to 0+} \sup \left\{ \frac{\|\widehat{g}\|_{p'}}{\|g\|_p} \,\middle|\, g : \mathbf{T} \to E, \operatorname{supp} g \subseteq [-a, a] \right\}. \quad \square$$

This completes our results on transfering the HY-inequality to larger groups. We just notice that compactness and descreteness make results like THEOREM 4 uninteresting if one tries to replace $(\mathbf{R}, \mathbf{T})$ with something else from our present context: You always get a discrete group somewhere and hence the right-hand side is equal to one.

RESULTS FOR REDUCTION OF THE GROUP

Let us introduce the Gauss–Weierstrass kernel $W_t(\theta) = \sum_{-\infty}^{\infty} e^{-tm^2} e^{im\theta}$ as well as $\Psi_t = W_t/\|W_t\|_{L_p(\mathbf{T})}$.

Lemma 1.

(1)

$$t^{1/2p'}\left\|W_t\right\|_p \to \pi^{1/2p'} p^{-1/2p} \qquad as\ t \to 0.$$

(2)

$$t^{1/2p'}\left\|\widehat{W_t}\right\|_{p'} \to \pi^{1/2p'} p'^{-1/2p'} \qquad as\ t \to 0.$$

(3)

$$\lim_{t\to 0} \frac{\|\widehat{W_t}\|_{\ell^{p'}}}{\|W_t\|_{L^p(\mathbf{T})}} = B_p.$$

Proof. Properties proved by means of the Poisson summation formula. □

In the next theorem we need to study the norm of each kth Fourier coefficient of W_t. To that end we fix a positive integer k, which will be the order of a cyclic group, and we put $U_t(s) = \sum_{j=-\infty}^{\infty} |W_t(jk+s)|^{p'}$, $s \in \mathbf{Z}_+$. This notation improves readability in:

Lemma 2. $$k \sum_{j=-\infty}^{\infty} |\widehat{\Psi_t}(jk+s)|^{p'} \to {B_p}^{p'} \qquad as\ t \to 0.$$

Proof. Equivalently, according to LEMMA 1 relation (3), we show that

$$\frac{kU_t(s)}{\|\widehat{W_t}\|_{p'}^{p'}} \to 1.$$

In fact $\widehat{W_t} = e^{-tm^2}$, whence the identity

$$e^{-t(k\xi+s)^2} = \mathcal{F}\left[\frac{1}{2k\sqrt{\pi t}} e^{-s^2 t} e^{-(\frac{x}{2k}+ist)^2/t}\right],$$

due to integrability and Poisson's summation formula gives

$$kU_t(s) = k\sum_{j=-\infty}^{\infty} e^{-p't(jk+s)^2} = \frac{\sqrt{\pi}}{\sqrt{p't}} e^{-s^2 p't} \sum_{m=-\infty}^{\infty} \exp\left\{-\frac{(\frac{\pi m}{k}+isp't)^2}{p't}\right\}.$$

Comparison with the second relation in the previous lemma proves the sought limit. □

Theorem 5. *For all $k \in \mathbf{Z}_+$ we have $M(\mathbf{Z}_k, E, p) \leq B_p^{-1} M(\mathbf{T}, E, p)$.*

Proof. Let $g \in L^p(\mathbf{Z}_k, E)$ and consider the E-valued measure $\mu = \sum_{s=0}^{k-1} g(s)\delta_{2\pi\frac{s}{k}}$ on $\mathbf{T}$. We introduce the approximation

$$f_t(\theta) = k^{-1/p}\, \mu * \Psi_t(\theta) = k^{-1/p} \sum_{s=0}^{k-1} g(s)\Psi_t(\theta - 2\pi\frac{s}{k}).$$

The fact that Ψ_t is an approximate identity such that $\|\Psi_t\|_p \equiv 1$ yields in the limit when $t \to 0$

$$\|f_t\|_{L^p(\mathbf{T},E)} = \left[\frac{1}{k}\int \|\sum_s g(s)\Psi_t(\theta - 2\pi\frac{s}{k})\|_E^p \frac{d\theta}{2\pi}\right]^{1/p} \longrightarrow$$

$$\longrightarrow \left[\frac{1}{k}\sum_{s=0}^{k-1} \|g(s)\|_E^p\right]^{1/p} = \|g\|_{L^p(\mathbf{Z}_k,E)}.$$

Furthermore

$$\widehat{f_t}(j) = k^{-1/p}\, \widehat{\Psi_t}(j) \sum_{s=0}^{k-1} g(s) e^{-2\pi i \frac{sj}{k}} = k^{1-1/p}\, \widehat{\Psi_t}(j) \mathcal{F}_{\mathbf{Z}_k} g(j),$$

whence

$$\widehat{f_t}(jk+s) = k^{1/p'}\, \widehat{\Psi_t}(jk+s)\mathcal{F}_{\mathbf{Z}_k} g(s), \qquad \text{for } j \in \mathbf{Z},\ s \in \{0,1,\ldots,k-1\}$$

and

$$\|\widehat{f_t}\|_{\ell^{p'}(\mathbf{Z},E)}^{p'} = \sum_{s=0}^{k-1} \|\mathcal{F}_{\mathbf{Z}_k} g(s)\|_E^{p'} \cdot k \sum_{j=-\infty}^{\infty} |\widehat{\Psi_t}(jk+s)|^{p'} \longrightarrow$$

$$\underset{t\to 0}{\overset{Lemma}{\longrightarrow}} B_p^{p'} \sum_{s=0}^{k-1} \|\mathcal{F}_{\mathbf{Z}_k} g(s)\|_E^{p'} = B_p^{p'} \|\mathcal{F}_{\mathbf{Z}_k} g\|_{L^{p'}(\widehat{\mathbf{Z}_k},E)}.$$

Consequently

$$\begin{aligned}
\|\mathcal{F}_{\mathbf{Z}_k} g\|_{L^{p'}(\widehat{\mathbf{Z}_k},E)} &= \lim_{t\to 0} B_p^{-1} \|\widehat{f_t}\|_{\ell^{p'}(\mathbf{Z},E)} \\
&\overset{HY}{\leq} M(\mathbf{T},E,p) \lim_{t\to 0} B_p^{-1} \|f_t\|_{L^p(\mathbf{T},E)} \\
&= B_p^{-1} M(\mathbf{T},E,p)\, \|g\|_{L^p(\mathbf{Z}_k,E)}.
\end{aligned}$$

□

Theorem 6. $M(\mathbf{T}, E, p) \leq B_p^{-1} M(\mathbf{R}, E, p)$.

Proof. Suppose that $g(\theta) = \sum_{-n}^{n} e_k e^{ik\theta}$, where $e_k \in E$. Using a gadget suggested in [1], $h(x) = \lambda^{1/2p} g(x) e^{-\lambda x^2}$, we find as $\lambda \to 0$

$$\begin{aligned}
\|h\|_{L^p(\mathbf{R},E)} &= \left[\lambda^{1/2} \int_{-\infty}^{\infty} \|g(x)\|_E^p e^{-p\lambda x^2}\, dx\right]^{1/p} \\
&= p^{-1/2p} \left[(\lambda p)^{1/2} \int_{-\infty}^{\infty} \|g(x)\|_E^p e^{-p\lambda x^2}\, dx\right]^{1/p} \\
\|g(x)\| \ \xrightarrow{2\pi-periodic} &\ p^{-1/2p} \left[\sqrt{\pi} \int_0^{2\pi} \|g(\theta)\|_E^p \frac{d\theta}{2\pi}\right]^{1/p} \\
&= \pi^{1/2p} p^{-1/2p} \|g\|_{L^p(\mathbf{T},E)},
\end{aligned}$$

On the other hand

$$\widehat{h}(\omega) = \lambda^{1/2p} \sum_k e_k \mathcal{F}(e^{-\lambda x^2})(\omega - k) = \sqrt{\pi} \lambda^{-1/2p'} \sum e_k e^{-(\omega-k)^2/4\lambda},$$

whence

$$\begin{aligned}
\|\widehat{h}\|_{L^{p'}(\mathbf{R},E)} &= \sqrt{\pi} \left[\frac{1}{\sqrt{\lambda}} \int_{-\infty}^{\infty} \|\sum_{-n}^{n} e_k e^{-(\omega-k)^2/4\lambda}\|^{p'} \frac{d\omega}{2\pi}\right]^{1/p'} \longrightarrow \\
&\to \sqrt{\pi} \left[\frac{1}{2\pi} \sum_{-n}^{n} \|e_k\|_E^{p'} \cdot 2p'^{-1/2} \sqrt{\pi}\right]^{1/p'} \\
&= \pi^{1/2p} p'^{-1/2p'} \|\hat{g}\|_{\ell^{p'}(\mathbf{Z},E)}.
\end{aligned}$$

In conclusion the following inequality appears:

$$\begin{aligned}
\|\hat{g}\|_{\ell^{p'}(\mathbf{Z},E)} &= \lim_{\lambda \to 0} \pi^{-1/2p} p'^{1/2p'} \|\widehat{h_\lambda}\|_{L^{p'}(\widehat{\mathbf{R}},E)} \\
&\overset{HY}{\leq} M(\mathbf{R}, E, p) \lim_{\lambda \to 0} \pi^{-1/2p} p'^{1/2p'} \|h_\lambda\|_{L^p(\mathbf{R},E)} \\
&= M(\mathbf{R}, E, p)\, p^{-1/2p} p'^{1/2p'} \|g\|_{L^p(\mathbf{T},E)} \\
&= B_p^{-1} M(\mathbf{R}, E, p)\, \|g\|_{L^p(\mathbf{R},E)}.
\end{aligned}$$

The proof is complete once we recognize that the class of trigonometric polynomials is dense among all Bochner integrable functions. □

A local version of the HY-inequality

If we specialize to the case $E = \mathbf{C}$ then THEOREM 4 takes the form

$$B_p \leq \lim_{a \to 0+} \sup \left\{ \frac{\|\widehat{g}\|_{p'}}{\|g\|_p} \;\middle|\; g : \mathbf{T} \to \mathbf{C}, \;\; \operatorname{supp} g \subseteq [-a, a] \right\} \leq 1.$$

The right-hand bound stems from the the situation on $\mathbf{T}$. This could be viewed as a "microlocal" version of the Hausdorff–Young inequality hinting at the close relationship between $\mathbf{R}$ and $\mathbf{T}$. The largest obstacle to straightforward use of the relation is that it cannot be applied to any single function, quite irrespective of its support. Surprisingly enough there is a relation pertaining to every function that has sufficiently small support and which shows that the leftmost inequality is the typical answer. The true measure of sufficiently small depends only on the exponent p. The way to establish this goes via the connection between the Fourier transforms on $\mathbf{T}$ and $\mathbf{R}$ respectively.

To improve readability let us denote the d-fold convolution on $\mathbf{R}$ by f^{*d} and on $\mathbf{T}$ by $f^{*_T d}$.

Lemma. *For $d \in \mathbf{Z}_+$ and $p' = 2d$, each function $f \in L^p(\mathbf{R})$ fulfilling* $\operatorname{supp} f \subseteq [-\frac{\pi}{d}, \frac{\pi}{d}]$ *satisfies the identity*

$$\sum_n |\mathcal{F}_{\mathbf{R}} f(n)|^{p'} = \int_{\mathbf{R}} |\mathcal{F}_{\mathbf{R}} f(\omega)|^{p'} \, d\omega.$$

Proof. The assumption on the support provides

$$f^{*d}(x) = \begin{cases} 0 & |x| > \pi \\ (2\pi)^{d-1} f^{*_T d}(x), & |x| \leq \pi. \end{cases}$$

Hence

$$\begin{aligned} \sum_n |\mathcal{F}_{\mathbf{R}} f(n)|^{p'} &= (2\pi)^{2d} \sum_n |\mathcal{F}_{\mathbf{T}} f(n)^d|^2 = (2\pi)^{2d} \sum_n |\mathcal{F}_{\mathbf{T}}(f^{*_T d})(n)|^2 \\ &\overset{Parseval}{=} (2\pi)^{2d} \int_{\mathbf{T}} |f^{*_T d}(x)|^2 \frac{dx}{2\pi} = 2\pi \int_{\mathbf{R}} |f^{*d}(x)|^2 \, dx \\ &\overset{Plancherel}{=} \int_{\mathbf{R}} |\mathcal{F}_{\mathbf{R}}(f^{*d})(\omega)|^2 \, d\omega = \int_{\mathbf{R}} |\mathcal{F}_{\mathbf{R}} f(\omega)|^{p'} \, d\omega. \end{aligned}$$

□

Keeping the setting in the lemma we notice that is says

$$\frac{\|\mathcal{F}_{\mathbf{T}} f\|_{p',\mathbf{Z}}}{\|f\|_{p,\mathbf{T}}} = \frac{\|\mathcal{F}_{\mathbf{R}} f\|_{p',\widehat{\mathbf{R}}}}{\|f\|_{p,\mathbf{R}}},$$

simply by identifying the different normalizations. Now an application of Babenko–Beckner's inequality yields:

Theorem 7. *For $d \in \mathbf{Z}_+$ and $p' = 2d$ every function $f : \mathbf{T} \to \mathbf{C}$ with support in an interval of length $\frac{2\pi}{d}$ obeys*

$$\frac{\|\widehat{f}\|_{p',\mathbf{Z}}}{\|f\|_{p,\mathbf{T}}} \leq B_p.$$

□

This result will serve as endpoint estimates for an interpolation:

Theorem 8. *Given $1 \leq p \leq 2$ there are positive constants C_p and m_p such that every function $f : \mathbf{T} \to \mathbf{C}$ supported in an interval of length m_p satisfies*

$$\|\widehat{f}\|_{p'} \leq C_p \|f\|_p.$$

Furthermore they can be expressed as follows. Let p_0 and p_1 denote real numbers such that p_0' and p_1' are two consecutive even integers fulfilling $p_0' < p' \leq p_1'$. Then we may take $m_p = \frac{4\pi}{p_1}$ and $C_p = {B_{p_0}}^{\lambda} {B_{p_1}}^{1-\lambda}$, where $\frac{1}{p} = \frac{\lambda}{p_0} + \frac{1-\lambda}{p_1}$. There are two special values, viz. $C_2 = 1$ and $C_1 = 1$. As a consequence

$$(\dagger) \qquad 1 > C_p \geq B_p \quad \text{for} \quad 1 < p < 2.$$

Proof. Without restrictions we may suppose the functions to be supported in the interval $[-\frac{1}{2}m_p, \frac{1}{2}m_p]$; call that interval I. We next consider function spaces over I and view them as subspaces of their counterparts over $\mathbf{T}$.

The preceding theorem tells us, given the value of m_p above, that

$$\|\mathcal{F}\|_{L^{p_j}(I) \to L^{p_j'}(\mathbf{Z})} \leq B_{p_j} \quad \text{for } j = 0, 1.$$

An application of M. Riesz' interpolation theorem proves the claimed boundedness and gives the stated value for C_p. □

It must be pointed out that (†) has equality with the indicated value for C_p exactly for $p' = 4, 6, 8 \ldots$. This is due to the special kind of convexity arising from Riesz' method.

The preceding theorem gives immediate rise to an apparently new result for approximate identities on $\mathbf{T}$. On $\mathbf{R}$ already the result of Babenko-Beckner proves that

$$\frac{\|\widehat{\phi_\alpha}\|_{p'}}{\|\phi_\alpha\|_p} \leq B_p,$$

for each member ϕ_α of an approximate identity $\{\phi_\alpha\}$ or for that matter any nonzero function. If we however recognise that sufficiently late in the development of an approximate identity on $\mathbf{T}$ we may discard the contribution outside any fixed small interval around zero, then we have established

Theorem 9. *Any approximate identity on $\mathbf{T}$ conforms to*

$$\limsup_{\phi_\alpha \to \delta_0} \frac{\|\widehat{\phi_\alpha}\|_{p'}}{\|\phi_\alpha\|_p} \leq C_p.$$

□

Concluding remarks

It is clear that the gap exhibited in the upward respectively downward result for $\mathbf{T} \leftrightarrow \mathbf{R}$ cannot be eliminated; Beckner proved this. The same discrepancy has here been established for $\mathbf{Z}_k \leftrightarrow \mathbf{T}$. We hence ask ourselves whether there is a reason behind this.

To that end we return to the identification that was used in THEOREM 5. For our present purpose the function $g : \mathbf{Z}_n \to \mathbf{C}$, $g(0) = 1$ *and zero elsewhere,* is identified with the measure δ_0 = *the unit point mass at zero.* We then know $\widehat{g} \equiv \frac{1}{n}$ whence $\|g\|_p = n^{-1/p} = \|\widehat{g}\|_{p'}$. Since we want to approximate g or δ_0 in $L^1(\mathbf{T}) \subseteq L^p(\mathbf{T})$ we have to use an approximate identity $\{\varphi_n\}$. But the last of the theorems then says that

$$\frac{\|\widehat{\varphi_n}\|_{\ell^{p'}}}{\|\varphi_n\|_p} \leq C_p.$$

Hence we cannot arrive at $\frac{\|\widehat{g}\|_{p'}}{\|g\|_p} = 1$ unless we sacrifice something in the process; exactly as was done in THEOREM 5.

Finally it is well worth mentioning the bound C_p once more. Since B_p and C_p coincide when $p' = 2, 4, 6 \ldots$, the LEMMA 1 on the Gauss–Weierstrass kernel proves that W_t really is extremal in the sense of THEOREM 9. For all other values of p the used methods cannot decide whether the we have found the optimal C_p or not. It would be interesting from some aspects to be able to settle that. On grounds that the only maximizers over $\mathbf{R}$ are the kins of W_t – see [3] – it is not unreasonable to suspect that B_p and C_p coincide for all values of p.

Added in proof. Per Sjölin[12] has found an argument that improves THEOREM 9. He derives an upper bound that shows the limit to be the suspected $\leq B_p$. In fact his technique can be generalized to establish equality in THEOREM 4.

References

1. W. Beckner, Inequalities in Fourier analysis, *Annals of Mathematics* **102** (1975), 159–182.
2. E. M. Stein & G. Weiss, *Introduction to Fourier Analysis on Euclidean Spaces*, Princeton University Press, Princeton, New Jersey 1971.
3. E. H. Lieb, Gaussian kernels have only Gaussian maximizers, *Invent. Math.* **102** (1990), 179–208.
4. J. Bourgain, A Hausdorff–Young inequality for B-convex Banach spaces, *Pacific J. Math.* **101** (1982), 255–262.
5. J. Bourgain, Vector-valued Hausdorff–Young inequalities and applications, *Lecture Notes in Mathematics* **1317** (1988), 239–249.
6. J.J.F. Fournier, Sharpness in Young's inequality for convolution, *Pacific J. Math.* **72** (1978), 383–397.
7. B. Russo, The norm of the L^p-Fourier transform on unimodular groups, *Trans. Amer. Math. Soc.* **192** (1974), 293–305.
8. B. Russo, The norm of the L^p-Fourier transform II, *Can. J. Math* **28** (1976), 1121–1131.
9. B. Russo, Operator theory in harmonic analysis, *Lecture Notes in Mathematics* **604** (1977), 94–102.
10. B. Russo, On the Hausdorff–Young theorem for integral operators, *Pacific J. Math.* **68** (1977), 241–253.
11. B. Russo, The norm of the L^p-Fourier transform III, *J. Functional Anal.* **30** (1978), 162–178.
12. P. Sjölin, private communication.

How Integrals Can Be Used to Derive Matrix Perturbation Bounds

Anders Barrlund
Institute of Information Processing
University of Umeå
S-90187 Umeå, SWEDEN

Abstract
In recent years several papers have been written in which a technique based on integrals has been used to derive perturbation bounds for different matrix decompositions. In this paper we make a survey of the ideas behind this approach. We also derive a new perturbation bound for the symmetric eigenvalue problem.

1. Introduction

In this paper we will consider a technique, that we will call *the integral technique,* to solve problems of the following form: Let C be a matrix function of the elements in a matrix A,

$$C = C(A). \tag{1.1}$$

If we change A to $A + \Delta A$, then we (usually) get a change in C of the form

$$C + \Delta C = C(A + \Delta A). \tag{1.2}$$

We are interested in if we can find a bound of $\|\Delta C\|$ in terms of $\|\Delta A\|$, where $\|\cdot\|$ is an appropriate norm. We are in particular interested in the case where C is a factor in a matrix decomposition. The function C is often defined implicitly, and the hardest part of the process is often to check if C is differentiable and calculate (or find a bound of) the derivative. The main idea is to find a bound of an integral of the derivative. There also exists one case where we can use the integral technique when C is *not* differentiable.

The integral technique has in recent years been used to derive perturbation bounds for several matrix decompositions. The first published paper we know about, where an integral has been used to derive a matrix perturbation bound is from 1976. In [5] Ole H. Hald used an integral to derive a perturbation bound for the inverse eigenvalue problem. However, the technique has become popular in recent years and further developed by Anders Barrlund and Ji guang Sun. New or improved bounds have been derived for the inverse eigenvalue problem [11], the polar decomposition [2], the Cholesky factorization [12], the QR factorization [12], the LDL^H decomposition [3] and the generalized QR factorization [4]. These papers have in common that they are most interested in the bounds. In this paper we are most interested in making a survey of the ideas in the integral technique, since we assume that this technique can be applied to more problems then we know today. However, we also show (as an introductory example) how this technique can be used to derive a new proof for a known bound of the inverse matrix function, and we also derive a new bound for the symmetric eigenvalue problem.

It should also be mentioned that there exist examples where bounds derived with this technique, can be used in practice. In section 3 of [2], it is described how a bound for the polar decomposition can be modified to be used for rigid body movements. This bound has been used, in the form of a condition number, in a program package used in orthopedics [13]. The purpose was to find bounds for rigid body movements of different parts in the skeleton.

The outline of the paper is the following: In section 2, we describe how the integral technique can be used in the case where C is differentiable and in a case where C is not differentiable. In section 3 we show how the technique can be used on the inverse matrix function, and the symmetric eigenvalue problem. We will use the following notations: $\|\cdot\|$ indicates that the results are valid for any matrix norm, $\|\cdot\|_F$ denotes the Frobenius norm, $\|\cdot\|_2$ denotes the matrix norm induced by the Euclidean vector norm, σ_n denotes the smallest singular value of a matrix of order n, and λ_i denotes the ith eigenvalue of a matrix. A function $h(x)$ is of $o(x)$ if $\lim_{x \to 0} \frac{h(x)}{x} = 0$. The superscript H denotes the conjugate transpose of a matrix, and the superscript * denotes the conjugate of a number.

2. The integral technique

When we apply the integral technique we use the following parametrization of A,

$$A(t) = A + t\Delta A \ , \quad 0 \le t \le 1, \tag{2.1}$$

in particular $A = A(0)$ and $A + \Delta A = A(1)$. (One can also use other parametrizations of A, however (2.1) is convenient since it implies that $dA(t)/dt = \Delta A$). We also get a corresponding parametrization of C,

$$C(t) = C(A(t)) \ , \quad 0 \le t \le 1. \tag{2.2}$$

The integral technique is most simple to apply when $C(t)$ is differentiable with respect to t. In this case we will try to find a bound of $\|dC(t)/dt\|$ of the form

$$\|dC(t)/dt\| \le U(A(t))\|dA(t)/dt\|, \tag{2.3}$$

where U is a real function. It is also convenient to use the following notation with differentials

$$\|dC(t)\| \le U(A(t))\|dA(t)\| \tag{2.4}$$

introduced in [12], where $dC(t) = dC(t)/dt \cdot dt$ and $dA(t) = dA(t)/dt \cdot dt$. The inequality (2.4) means that (2.3) is satisfied for any differentiable parametrization of A.

In order to obtain something that can be integrated we also have to find a bound of U in terms of the original matrix $A = A(0)$, the perturbation ΔA, and t,

$$\|dC(t)\| \le U(A(t))\|dA(t)\| \le u(A(0), dA(t), t)\, \|dA(t)\| = u(A, \Delta A, t)\, \|\Delta A\| dt \tag{2.5}$$

Often U can be expressed in terms of the singular values or the eigenvalues of $A(t)$. In these cases we can use well-known inequalities for singular values or eigenvalues to get u. We will see examples of this in section 3.

Now we continue with the steps where we use integrals. We get

$$\|\Delta C\| = \|C(1) - C(0)\| = \|\int_0^1 dC(t)\| \le \int_0^1 \|dC(t)\| \le$$

$$\int_0^1 U(A(t))\|dA(t)\| \le \|\Delta A\| \int_0^1 u(A, \Delta A, t)\, dt. \tag{2.6}$$

By calculating, or finding an upper bound of, the right-hand side of (2.6) we get a perturbation bound for $\|\Delta C\|$.

Let us summarize the main steps

Algorithm 2.1

1. Check if C is a differentiable function of the elements of A.
2. Find a bound for the differential, (2.5), which can be integrated.
3. Calculate, or find a bound of, the integral, (2.6).

In the first step we can often use the implicit function theorem, see for instance [1, p. 187]. The second step is often the most technical part, where many different tricks have been used, see for instance [2,3,12] or section 3. In the third step the exact expression for the integral (2.6) is often very complicated, and therefore it is often better to use an upper bound [3,4,12]. We will look at two examples of how Algorithm 2.1 can be applied in section 3.

It is not absolutely necessary that C is a differentiable function. It is sufficient that the conditions in the following theorem are satisfied.

Theorem 2.2 *Assume that C is a matrix function of the elements of A*

$$C = C(A). \tag{2.7}$$

Also assume that there exist real functions $u(A, \Delta A, t)$ and $h(\|\Delta\tilde{A}\|)$, such that u is continuous, $h(\|\Delta\tilde{A}\|)$ is of $o(\|\Delta\tilde{A}\|)$ and

$$\|C(\tilde{A}+\Delta\tilde{A})-C(\tilde{A})\| \leq u(A, \Delta A, t)\|\Delta\tilde{A}\| + h(\|\Delta\tilde{A}\|) \tag{2.8}$$

for all $\tilde{A} = A+t\Delta A$, $0 \leq t \leq 1$.

Then ΔC has the bound

$$\|\Delta C\| \leq \|\Delta A\| \int_0^1 u(A, \Delta A, t)dt \tag{2.9}$$

Before we prove this theorem we make a few remarks. Note that it is necessary that the same $o(\|\Delta A\|)$-function can be used in the entire interval $0 \leq t \leq 1$. Also note that (2.9) is similar to (2.6). As a simple example where Algorithm 2.1 is unusable but Theorem 2.2 can be applied, we mention the absolute value function. It is not differentiable if we have zero elements. However, it is easy to verify that the conditions in Theorem 2.2 are satisfied for the absolute value function.

Proof of Theorem 2.2: We consider $C(A + \Delta A) - C(A)$. Divide ΔA in k equal pieces where k is a positive integer. Then from (2.8) we get

$$\|C(A+\frac{\Delta A}{k})-C(A)\| \leq u(A, \Delta A, 0)\frac{\|\Delta A\|}{k} + h(\frac{\|\Delta A\|}{k})$$

$$\|C(A+\frac{2\Delta A}{k})-C(A+\frac{\Delta A}{k})\| \leq u(A, \Delta A, \frac{1}{k})\frac{\|\Delta A\|}{k} + h(\frac{\|\Delta A\|}{k}) \tag{2.10}$$

$$\vdots$$

$$\|C(A+\frac{k\Delta A}{k})-C(A+\frac{(k-1)\Delta A}{k})\| \leq u(A, \Delta A, \frac{(k-1)}{k})\frac{\|\Delta A\|}{k} + h(\frac{\|\Delta A\|}{k}).$$

The triangle inequality gives

$$\|C(A+\Delta A)-C(A)\| \leq \sum_{i=0}^{k-1} u(A, \Delta A, \frac{i}{k})\frac{\|\Delta A\|}{k} + k \cdot h(\frac{\|\Delta A\|}{k}). \tag{2.11}$$

Let $k \to \infty$ then

$$\|\Delta C\| \le \|\Delta A\| \int_0^1 u(A, \Delta A, t)dt + \lim_{k \to \infty} k \cdot h(\frac{\|\Delta A\|}{k}). \tag{2.12}$$

But since

$$\lim_{k \to \infty} k \cdot h(\frac{\|\Delta A\|}{k}) = \|\Delta A\| \lim_{k \to \infty} \frac{h(\frac{\|\Delta A\|}{k})}{\frac{\|\Delta A\|}{k}} = 0 \tag{2.13}$$

the theorem is proven. Q.E.D.

3. Two examples

In this section we will first consider a simple example of how the integral technique can be used. The example is the inverse matrix function. Afterwards we consider the symmetric eigenvalue problem, which is a more advanced example. In both cases we use Algorithm 2.1.

3.1 The inverse

In this case we have

$$C(A) = A^{-1}. \tag{3.1}$$

It is well-known that C is differentiable, consequently we can omit step 1 of Algorithm 2.1. (However, it can be verified that C is differentiable by using $F(X,Y) = XY - I$, $X = A$ and $C = Y$ in the implicit function theorem). Let $A(t)$ and $C(t)$ be defined by (2.1) and (2.3), respectively. The derivative of $C(t)$ is

$$\frac{dC(t)}{dt} = -A^{-1}(t) \frac{dA(t)}{dt} A^{-1}(t) \tag{3.2}$$

which gives

$$\|dC(t)\|_F \le \|A^{-1}(t)\|_2^2 \|dA(t)\|_F = \frac{1}{\sigma_n^2(t)} \|dA(t)\|_F. \tag{3.3}$$

This is the function U in (2.3), in which $\sigma_n(t)$ is the smallest singular value of $A(t)$. A well-known inequality [6, p. 428] for singular values is

$$|\sigma_i(A+E) - \sigma_i(A)| \le \|E\|_2. \tag{3.4}$$

Consequently,

$$\sigma_n(t) \ge \sigma_n - t\|\Delta A\|_2, \tag{3.5}$$

where σ_n is the smallest singular value of $A = A(0)$. By inserting (3.5) and $\|dA(t)\|_F = \|\Delta A\|_F dt$ in (3.3) we get

$$\|dC(t)\|_F \le \frac{1}{(\sigma_n - t\|\Delta A\|_2)^2} \|\Delta A\|_F dt. \tag{3.6}$$

That is $u = 1/(\sigma_n - t\|\Delta A\|)^2$ in (2.5), and step 2 in Algorithm 2.1 is finished. The integral in (2.6) takes the form

$$\|\Delta A\|_F \int_0^1 \frac{dt}{(\sigma_n - t\|\Delta A\|_2)^2} = \frac{\|\Delta A\|_F}{\sigma_n(\sigma_n - \|\Delta A\|_2)} = \frac{\|A^{-1}\|_2^2 \|\Delta A\|_F}{1 - \|A^{-1}\|_2 \|\Delta A\|_2}. \tag{3.7}$$

This gives a new proof of an earlier known bound.

3.2 The symmetric eigenvalue problem

The symmetric eigenvalue problem is to find a diagonal matrix D and an orthogonal matrix Q such that

$$Q^H A Q = D \tag{3.8}$$

where A is a known Hermitian matrix. The diagonal elements of D are the eigenvalues of A, and the columns of Q are the corresponding eigenvectors. Here we should be aware that the solution of this problem is not unique. If we multiply any column in Q by a complex number of absolute value 1, we get a new solution. When we try to find a bound for ΔQ in the perturbed problem

$$(Q + \Delta Q)^H (A + \Delta A)(Q + \Delta Q) = D + \Delta D, \tag{3.9}$$

where ΔA is a Hermitian perturbation, we get a bound that is valid for *one* of the infinitely many solutions of (3.9). However the result is valid for a fixed solution of (3.8).

Before we apply Algorithm 2.1 we make some simplifications. (In a recent paper [9] the technique of reducing the problem to canonical form to calculate derivatives has been exploited. However, the idea in these simplifications is not new [2]). Let $\tilde{A} = Q^H A Q$, $\Delta\tilde{A} = Q^H \Delta A Q$, $\Delta\tilde{Q} = Q^H \Delta Q$ then (3.8) takes the form

$$\tilde{A} = D \tag{3.10}$$

and (3.9) takes the form

$$(I + \Delta\tilde{Q})^H (\tilde{A} + \Delta\tilde{A})(I + \Delta\tilde{Q}) = D + \Delta D \tag{3.11}$$

which is a simplified Hermitian eigenvalue problem. Since $\|\Delta Q\|_F = \|\Delta\tilde{Q}\|_F$ and $\|\Delta A\|_F = \|\Delta\tilde{A}\|_F$ we can without loss of generality study (3.10-11) instead of (3.8-9). When we study (3.10-3.11) we drop the tilde symbols, and (3.11) takes the form

$$(I + \Delta Q)^H (D + \Delta A)(I + \Delta Q) = D + \Delta D \tag{3.12}$$

where ΔA is Hermitian, ΔD is diagonal and

$$(I + \Delta Q)^H (I + \Delta Q) = I. \tag{3.13}$$

In step 1 of Algorithm 2.1, we will now select one of the infinitely many solutions and examine when ΔQ is differentiable. (The result is known [8], but we include it as an example of how step 1 can be verified). We will use the implicit function theorem [1, p.187]. When we select $F(X,Y)$ we will first rewrite (3.12). From (3.13) we get

$$\Delta Q^H = -\Delta Q - \Delta Q^H \Delta Q. \tag{3.14}$$

By replacing ΔQ^H in (3.12) by the right-hand side of (3.14) we get

$$(I - \Delta Q - \Delta Q^H \Delta Q)(D + \Delta A)(I + \Delta Q) - D + \Delta D = 0. \tag{3.15}$$

We define $F(X,Y)$ to be left-hand side of (3.15). The elements of ΔA will be the $n \cdot n$ known elements in X. The unknowns Y will be the diagonal elements of ΔD and the *off-diagonal* elements in ΔQ, hence $n \cdot n$ unknowns. We also replace all diagonal elements in ΔQ by

$$\Delta q_{ii} = \sqrt{1 - \sum_{i \neq j} |\Delta q_{ij}|^2} - 1. \tag{3.16}$$

in order to get $n \cdot n$ unknowns. This is also the step where we select one of the infinitely many solutions of (3.12). It is obvious that $F(X,Y)$ will be a continuously differentiable function, hence the first condition in the implicit function theorem is satisfied. To check when $\partial F / \partial Y$ is invertible we calculate the differential of (3.15). In the point where $\Delta A = 0$, $\Delta Q = 0$, $\Delta D = 0$ this differential is

$$- dQD + dA + DdQ - dD = 0. \tag{3.17}$$

The diagonal part of (3.17) gives

$$da_{ii} = d\lambda_i \tag{3.18a}$$

and the off-diagonal part of (3.17) gives

$$da_{ij} = (\lambda_j - \lambda_i)\, dq_{ij}. \tag{3.18b}$$

Note that from (3.16) we get

$$dq_{ii} = 0. \tag{3.19}$$

(in the point $\Delta Q = 0$), and hence the diagonal elements of ΔQ does not give any contribution in (3.18).

Note that we can solve for dY (i.e. dQ and dD) as a function of dX (i.e. dA) if and only if there are no multiple eigenvalues. Hence $\partial F / \partial Y$ is invertible, and hence from the implicit function theorem Y is a differentiable function of X if and only if no eigenvalues are equal. Consequently ΔQ given by (3.12-14) is a differentiable function of ΔA if and only if there are no multiple eigenvalues. Step 1 of algorithm 2.1 is finished.

Note that all steps from (3.8) to (3.19) can be performed for any $A(t)$ in the parametrization (2.1). From (3.18b) and (3.19) it is obvious that

$$\|dQ(t)\|_F \le \frac{\|dA(t)\|_F}{\min\limits_{i,j,i\neq j} |\lambda_i(A(t)) - \lambda_j(A(t))|} \tag{3.20}$$

and we have a bound of the form (2.3). From Wielandt-Hoffman's Theorem [6, p. 412] we have

$$\sum_{i=1}^{n} (\lambda_i(A+E) - \lambda_i(A))^2 \le \|E\|_F^2 \tag{3.21}$$

which implies that

$$|\lambda_i(A+E) - \lambda_j(A+E)| \ge |\lambda_i(A) - \lambda_j(A)| - \sqrt{2}\,\|E\|_F. \tag{3.22}$$

From Weyl's Theorem [10, p. 203] we have

$$|\lambda_i(A+E) - \lambda_i(A)| \le \|E\|_2, \tag{3.23}$$

which implies that

$$|\lambda_i(A+E) - \lambda_j(A+E)| \ge |\lambda_i(A) - \lambda_j(A)| - 2\,\|E\|_2. \tag{3.24}$$

Consequently, if $\min(2\,\|\Delta A\|_2, \sqrt{2}\,\|\Delta A\|_F) < \min\limits_{i,j,i\neq j} |\lambda_i(A) - \lambda_j(A)|$ then

$$\|dQ(t)\|_F \le \frac{\|\Delta A\|_F\, dt}{\min\limits_{i,j,i\neq j} |\lambda_i(A) - \lambda_j(A)| - t\min(2\,\|\Delta A\|_2, \sqrt{2}\,\|\Delta A\|_F)} \tag{3.25}$$

which is (2.5) and step 2 of Algorithm 2.1 is finished.

The integral in (2.6) takes the form

$$\|\Delta Q\|_F \le \|\Delta A\|_F \int_0^1 \frac{dt}{\min_{i,j,i\neq j} |\lambda_i(A) - \lambda_j(A)| - t \min(2\|\Delta A\|_2, \sqrt{2}\|\Delta A\|_F)} =$$

$$-\frac{\|\Delta A\|_F}{\min(2\|\Delta A\|_2, \sqrt{2}\|\Delta A\|_F)} \ln\left[1 - \frac{\min(2\|\Delta A\|_2, \sqrt{2}\|\Delta A\|_F)}{\min_{i,j,i\neq j} |\lambda_i(A) - \lambda_j(A)|}\right] \le \tag{3.26}$$

$$\frac{\|\Delta A\|_F}{\min_{i,j,i\neq j} |\lambda_i(A) - \lambda_j(A)| - \min(2\|\Delta A\|_2, \sqrt{2}\|\Delta A\|_F)}.$$

We can summarize the result in the following theorem.

Theorem 3.1 *Assume that* $A \in C^{n \times n}$ *is a Hermitian matrix with no multiple eigenvalues, and with the eigenvalue decomposition*

$$Q^H A Q = D. \tag{3.27}$$

Let $\Delta A \in C^{n \times n}$ *be a Hermitian matrix, different from the zero matrix, which satisfies*

$$\min(2\|\Delta A\|_2, \sqrt{2}\|\Delta A\|_F) < \min_{i,j,i\neq j} |\lambda_i(A) - \lambda_j(A)|. \tag{3.28}$$

Then there is an eigenvalue decomposition

$$(Q + \Delta Q)^H (A + \Delta A)(Q + \Delta Q) = D + \Delta D \tag{3.29}$$

where

$$\|\Delta Q\|_F \le -\frac{\|\Delta A\|_F}{\min(2\|\Delta A\|_2, \sqrt{2}\|\Delta A\|_F)} \ln\left[1 - \frac{\min(2\|\Delta A\|_2, \sqrt{2}\|\Delta A\|_F)}{\min_{i,j,i\neq j} |\lambda_i(A) - \lambda_j(A)|}\right] \tag{3.30}$$

$$\le \frac{\|\Delta A\|_F}{\min_{i,j,i\neq j} |\lambda_i(A) - \lambda_j(A)| - \min(2\|\Delta A\|_2, \sqrt{2}\|\Delta A\|_F)}.$$

The classical perturbation theory for eigenvalue problems is not given in terms of norms, instead it is given in terms of subspaces. An excellent survey is given in [10]. Since these results are of a different kind they are difficult to compare with (3.30). If we try to get results about subspaces from (3.30) they will not be better than the classical results. The bound (3.30) has disadvantages, note that (3.30) can only be used for sufficiently small perturbations when we have no multiple eigenvalues. (The case where we have multiple eigenvalues is however very sensitive [10], where discontinuous changes of the eigenvectors can occur). We should not over-estimate the scientific value of the result (3.30), certainly the classical results [10] are preferable in most cases.

However, there is no simple way to get bounds of $\|\Delta Q\|_F$ from the classical results in [10]. In a recent paper [7] perturbation bounds in terms of norms have been derived for the Schur decomposition. (The Schur decomposition is the same as the symmetric eigenvalue decomposition when A is Hermitian). The bounds in [7] have the advantage that they can be applied when A is not Hermitian. However, when A is Hermitian the bound (3.30) is sharper than the bounds in [7]. (In [7] it is proved that $\|\Delta Q\|_F \le (2c_2(\|\Delta A\|_F/\eta_A))^{1/2}$, see [7] for the definition of c_2 and η_A. However, for Hermitian matrices $(2c_2(\|\Delta A\|_F/\eta_A))^{1/2} \ge \sqrt{2}\|\Delta A\|_F/\eta_A \ge \sqrt{2}\|\Delta A\|_F / \min_{i,j,i\neq j} |\lambda_i(A) - \lambda_j(A)| >$ (3.30).) One advantage with (3.30) is that it has a simple form compared to the results in both [7] and [10]. Consequently, (3.30) must be of at least some interest.

In any case this is a good example of how integrals can be used to derive matrix perturbation bounds. It also illustrates that technical difficulties can occur when we apply Algorithm 2.1.

Acknowledgement

I am grateful to Ji-guang Sun who read the first version of this manuscript and gave me several corrections and suggestions of improvements. It was also Ji-guang Sun who informed me about the papers [5,7,8,11]. I would also like to thank Per Lindström for his comments and the referee who gave me significant comments and informed me about the paper [9].

References:

1. T. Apostol, *Mathematical Analysis, a Modern Approach to Advanced Calculus,* Addison-Wesley Publishing Company, Massachusetts, 1957.

2. A. Barrlund, "Perturbation bounds on the polar decomposition", *BIT 30* (1990), 101-113.

3. A Barrlund, "Perturbation bounds for the LDL^H and LU decompositions", *BIT 31* (1991), 358-363.

4. A Barrlund, "Perturbation bounds for the generalized QR factorization", *submitted to Linear Algebra Appl.*

5. O. H. Hald, "Inverse eigenvalue problems for jacobi matrices", *Linear Algebra Appl.,* (1976), 63-85.

6. G. H. Golub and C. F. Van Loan, *Matrix computations,* 2nd Edition, John Hopkins University Press, Baltimore, Maryland 1989.

7. M. M. Konstantinov, P. H. Petkov and N. D. Christov, "Non-local perturbation analysis of the schur system of a matrix", Department of Automatics, Sofia Technical University, Bulgaria. *Submitted to SIAM J, Matrix Anal Appl.*

8. F. Rellich, "Störungstheorie der Spektralzerlegung" *Mathematische Annalen 113* (1937), 601-619.

9. G. W. Stewart, "On the perturbation of LU, Cholesky and QR factorizations" Techical Report CS-TR-2848, Dept. of Computer Science, University of Maryland, College Park, 1992 *to appear in SIAM J, Matrix Anal Appl.*

10. G. W. Stewart and J. -G. Sun, *Matrix Perturbation Theory,* Academic Press, Inc., 1990.

11. J. -G. Sun, "The stability analysis of the solutions of inverse eigenvalue problems", *J. Comp. Math.,* (1986), 345-353.

12. J. -G. Sun, "Perturbation bounds for the Cholesky and QR factorizations", *BIT 31,* (1991), 341-352,

13. I. Söderkvist, "Some numerical methods for kinematical analysis", UMINF-186.90, Inst. of Information Processing of Umeå, Sweden, 1990.

Nonexistence of Periodic Solutions to Hyperbolic Partial Differential Equations

BJÖRN BIRNIR[1] U. CALIFORNIA, SANTA BARBARA, CA 93106, AND U. OF ICELAND, 107 REYKJAVIK, ICELAND.

There are many reasons why one is interested in periodic solutions to nonlinear hyperbolic PDE's. The first reason comes from physics where the nonlinear PDE's are models for classical field theories. We expect these fields to decay exponentially at infinity and a time–period solution represents a bound state. This is a state which stays in one place and does not travel. It represents a stationary classical particle. In quantum field theory the states are called bosons and fermions, they can be either localized or delocalized. The quantum field theories are obtained by quantizing the classical field theories. If the classical field theory does not have any time–periodic bound states, then the quantum theory may not have any localized quantum lumps that are fermions. This means that it cannot be used to describe elementary particles which are time–dependent bound states.

The second reason comes from modern dynamical systems theory. If we add dissipation to the nonlinear hyperbolic PDE's then they possess finite-dimensional attractors in many cases. This means that if we start with some initial data, then the resulting orbit can show some transient behavior but eventually gets closer and closer to a finite dimensional set, which is called the attractor. It turns out that periodic orbits are the components which give the interesting attractors. One can sometimes give a complete description of the possible attractors by describing the periodic orbit and all its bifurcations.

[1]This work was partiallly supported by the National Science Foundation, under Grants NSF-DMS 89-05770/03012/91-04532

The third reason for studying periodic solutions of nonlinear PDE's is that they are beautiful mathematical objects that may give insight leading towards a degree theory for nonlinear hyperbolic PDE's.

If you look through the literature you will discover how little is known about periodic solutions to nonlinear hyperbolic PDE's. This is a very different situation from that of linear PDE's or nonlinear ODE's where there is no shortage of periodic solutions. Periodic solutions to hyperbolic nonlinear PDE's are only known in two rather special cases. The first one is the separable case:

$$u(x,t) = u(x)e^{i\omega t}.$$

This requires the nonlinearities to be of special form; for example,

$$g(u) = |u|^p u - u.$$

Then the time–dependence can be factored out of the hyperbolic PDE and the problem becomes to solve an elliptic nonlinear PDE for $u(x)$. There is a well–known theory for doing this; see for example, Gilbarg and Trudinger [1977].

The second case is where the nonlinear periodic solution is a small variation of a periodic solution w to a linear PDE,

$$U(x,t) = w(x,t) + \epsilon u(x,t).$$

Then one can use the implicit function theorem to prove the existence of "small" periodic solutions u to the nonlinear PDE's. These methods go back to H. Lovicarovā [1969]. The problem was solved by P. Rabinowitz in a series of celebrated papers, CPAM 31 [1978]. H. Brézis, J.–M. Coron and L. Nirenberg found a simpler proof, CPAM 33 [1980].

I want to present a result which indicates that there is a good reason why so few periodic solutions are known for nonlinear PDE's. This is called the nonexistence of breathers theorem and is the result of a collaboration with H. McKean and A. Weinstein, CPAM [1992].

Theorem (BMW). *Suppose that the perturbed sine–Gordon equation*

$$u_{tt} - u_{xx} + \sin(u) = \epsilon g(u), \qquad (t,x) \in \mathbb{R} \times \mathbb{R},$$

where $g(u)$ is analytic and vanishes (H.I) at the origin, has a breather solution (H.II), close to the sine–Gordon breather. Then,

$$g(u) = a\sin(u) + bu\cos(u),$$

or

$$g(u) = c\left[1 + 3\cos(u) - 4\cos\left(\frac{u}{2}\right) + 4\cos(u)\log\left(\cos\left(\frac{u}{4}\right)\right)\right]$$

a,b and c constants.

We need to elaborate on the statement of the theorem and state the hypothesis (I and II) more precisely.

Firstly, a breather is a function which is periodic in t, period $T = 2\pi/\omega$, and vanishes exponentially at $x = \pm\infty$. There is exactly one quation of the above type which is known to have such a solution, namely the unperturbed sine–Gordon equation

$$u_{tt} - u_{xx} + \sin(u) = 0.$$

It has the sine–Gordon breather

$$\begin{aligned} u(x,t) &= 4\tan^{-1}\left[\frac{\alpha}{\omega}\frac{\sin(\omega t)}{\cosh(\alpha x)}\right], \qquad \alpha = \sqrt{1-\omega^2} \\ &= B \end{aligned}$$

as a solution. It is periodic in t, $T = 2\pi/\omega$, and decays exponentially in x,

$$|u(x,t)| = 4\frac{\alpha}{\omega}e^{-\alpha|x|} + \mathcal{O}(e^{-2\alpha|x|}) \quad \text{as} \quad |x| \to \infty.$$

Notice that we can scale t, x and u

$$t \to (1+\epsilon(b-a))^{1/2}t, \;\; x \to (1+\epsilon(b-a))^{1/2}x, \;\; u \to (1+\epsilon b)u$$

to get a breather solution, see Figure 1,

$$u(x,t) = \frac{4}{(1+\epsilon b)}\tan^{-1}\left[\frac{\alpha}{\omega}\frac{\sin(\omega(1+\epsilon(b-a))^{1/2}t)}{\cosh(\alpha(1+\epsilon(b-a))^{1/2}x)}\right]$$

of the perturbed equation

$$u_{tt} - u_{xx} + \sin(u) = \epsilon[a\sin(u) + bu\cos(u)] + \mathcal{O}(\epsilon^2).$$

Figure 1

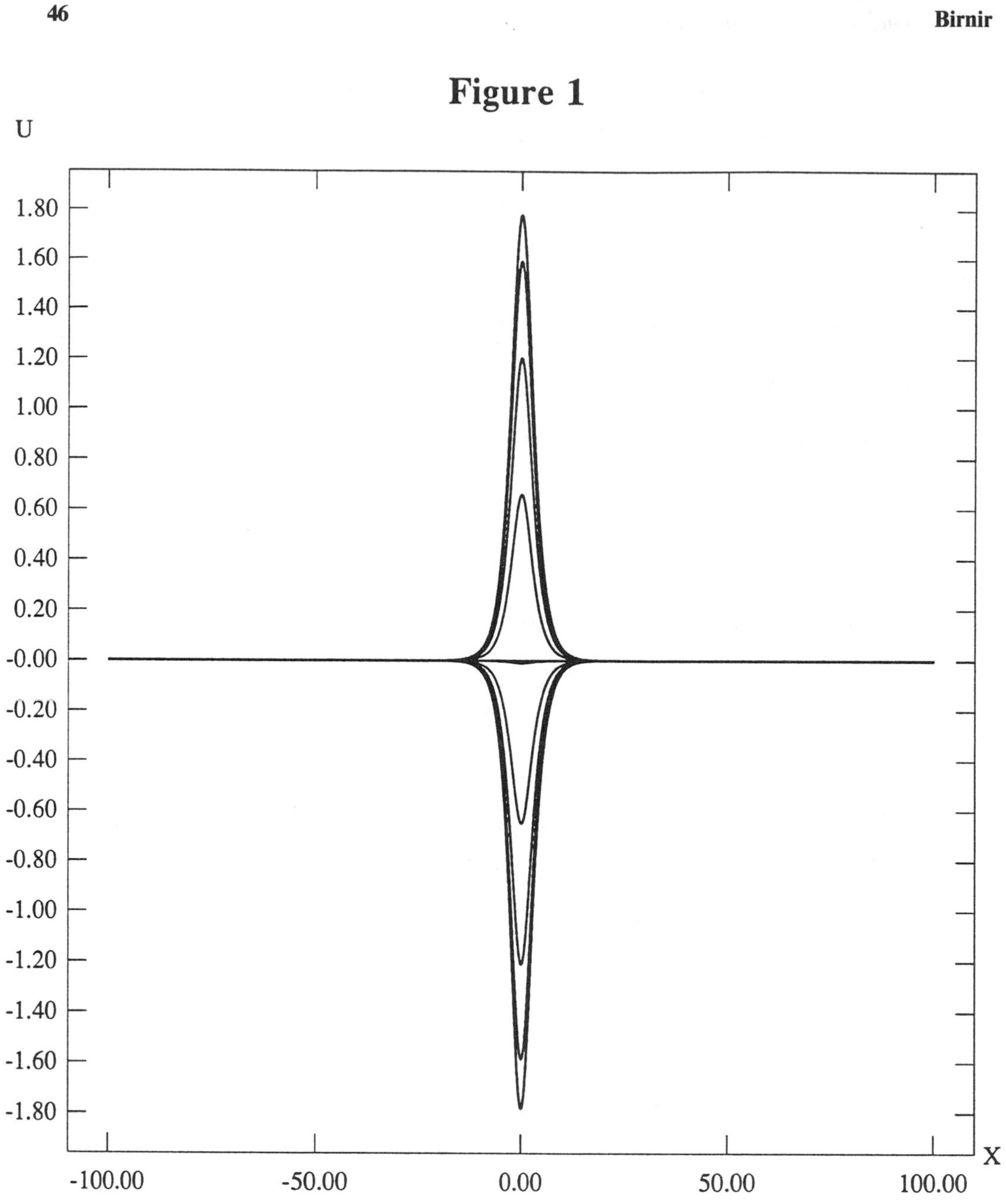
U
1.80
1.60
1.40
1.20
1.00
0.80
0.60
0.40
0.20
-0.00
-0.20
-0.40
-0.60
-0.80
-1.00
-1.20
-1.40
-1.60
-1.80
X
-100.00
-50.00
0.00
50.00
100.00

The theorem says that if hypothesis I and II are satisfied, then there are only two perturbed sine-Gordon equations that have breather solutions and this scaling produces one of them.

Hypothesis I. We require $g(u)$ to be analytic in an open strip $|u| < \pi$, and $g(0) = 0$.

Hypothesis II. We assume that the breather has the form

$$u(x,t,\omega,\epsilon) = B(x,t,\omega) + \epsilon u_1(x,t,\omega) + \mathcal{O}(\epsilon^2),$$

where B is the sine–Gordon breather, that u tends smoothly to B as $\epsilon \to 0$ and u_1 vanishes at $x = \pm\infty$. Moreover, we assume that there exists one such breather for every ω in an open subinterval of $[1/\sqrt{2}, 1)$

Comments: J. Denzler has shown in his Ph.D. dissertation at ETH that $g(u)$ only needs to be analytic in a small neighborhood of 0. We believe that it should suffice to require $g(u)$ to be smooth and assume that a breather exists only for finitely many values of ω. However, we have not been able to prove this.

Numerical experiments show that for $\omega < 1/\sqrt{2}$ the breather can separate into a kink–antikink pair which move to $\pm\infty$ respectively as $t \to \infty$ see Peyrard and Cambell [1983, 1986], and Birnir [1992]. The theorem strengthens the result of Kruskal and Segur [1987], that the ϕ^4 field theory does not have small breathers, (for small breathers ϕ^4 is a perturbation of sine–Gordon) and Kitchenassamy [1991] who showed the sine–Gordon is the only nonlinear wave equation having small analytic breathers. Compare these with Weinstein [1985] who proves the existence of breathers on the half line.

The theorem rules out all reasonable vector fields except a two–parameter scaling of the nonlinearity, and one additional function, as giving rise to breathers. In fact, it does not even require the solution u to be a breather. The first order correction u_1 can have any decay whatsoever, at $x = \pm\infty$! This indicates that nonlinear PDE's possessing periodic decaying solutions are isolated and very special indeed.

Now we present an outline of the proof. First notice that the period of the breather

$$T(\epsilon,\omega) = \frac{2\pi}{\omega} \quad \text{at} \quad \epsilon = 0,$$

so that

$$\frac{\partial T}{\partial \omega} = -\frac{2\pi}{\omega^2} \neq 0 \quad \text{at} \quad \epsilon = 0.$$

Therefore, there exists a function $\omega(\epsilon)$, by the implicit function theorem such that

$$T(\epsilon, \omega(\epsilon)) = 2\pi/\omega.$$

Then the breather

$$u(x, t, \epsilon, \omega) = B + \epsilon u_1 + \mathcal{O}(\epsilon^2)$$

is of a fixed period T.

Condition 1. *Let v be a C^2, T-periodic and bounded function of the 1st order equation*

$$(\Box + \cos(B))v = 0,$$

then

$$\int_0^T dt \int_{-\infty}^{\infty} vg(B)\, dx = 0. \tag{1}$$

Proof. u_1 is periodic of period T and

$$(\Box + \cos(B))u_1 = g(B),$$

i.e., u_1 is a solution of the inhomogeneous 1st (perturbation) order equation. Fix $0 < L < \infty$, then

$$\begin{aligned}
\int_0^T \int_{-L}^{L} vg(B) &= \int_0^T \int_{-L}^{L} v(\Box + \cos(B))u_1 \\
&= \int_0^T \int_{-L}^{L} (\Box + \cos(B))vu_1 \\
&+ \int_0^T (v'u_1 - vu_1')\Big|_{-L}^{L} dt
\end{aligned}$$

by integration by parts. We may have to mollify u_1 if it is not C^2. Then we let $L \to \infty$ and get the result by the boundedness of u_1. □

Remark. Condition 1 is just the Fredholm necessity condition once the space (of u_1 decay) is fixed.

The second step is to compute the radiation kernel of the operator $\Box + \cos(B)$. Numerical experiments, see Birnir [1992], show that the perturbed breather radiates, see Figure 2, and decays. That is, one gets small oscillations emanating from the breather and propagating towards $x = \pm\infty$. The kernel of $\Box + \cos(B)$ contains B_t and B_x, which satisfy Condition 1, but in addition we get infinitely many T–periodic solutions which are oscillatory in x at $x = \pm\infty$.

Lemma 1. *There exist infinitely many bounded, T–periodic solutions*

$$v_n^{\pm}(x,t) = \frac{1}{(1+y^2)}[d+ey^2+f\frac{\partial_x(\cosh^2(\alpha x))}{\cosh^2(\alpha x)}+h\partial_t y^2]e^{in\omega t \pm \sqrt{n^2\omega^2-1}x}, \qquad n \neq 0, \pm 1$$

where

$$d = \frac{1}{2}(n^2-1)\omega^2 - \alpha^2, \quad e = \frac{1}{2}(n^2+1)\alpha^2, \quad f = \frac{i}{2}(n^2\omega^2-1)^{1/2}, \quad h = i\frac{n\alpha^2}{\omega},$$

and $y = \frac{\alpha}{\omega}\frac{\sin(\omega t)}{\cosh(\alpha x)}$, *of the homogeneous 1st order equation*

$$(\Box + \cos(B))v = 0.$$

We compute the radiation about a breather by a method invented by McLaughlin and Scott [1979], their formulas contain minor errors which are corrected here. The sine–Gordon equation is a completely integrable PDE and one manifestation of this is that one can generate solutions by transformations called Bäcklund transformations:

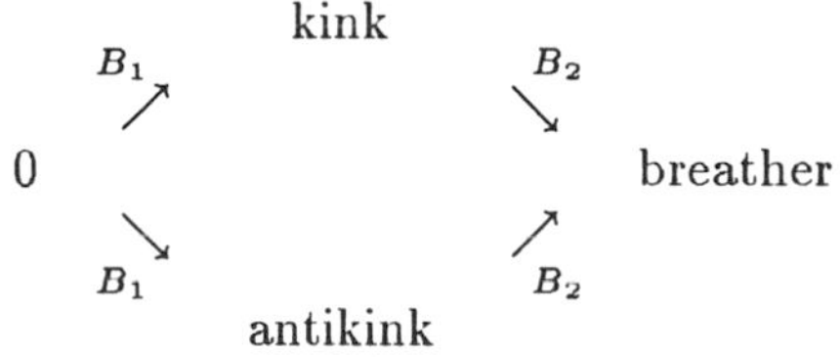

The idea of McLaughlin and Scott was to linearize the transformations about 0 and the kink and the anti–kink to get transformations that will create the radiation about the breather, from radiation about zero. We linearize the above transformation and get the diagram of linear "Bäcklund" transformations,

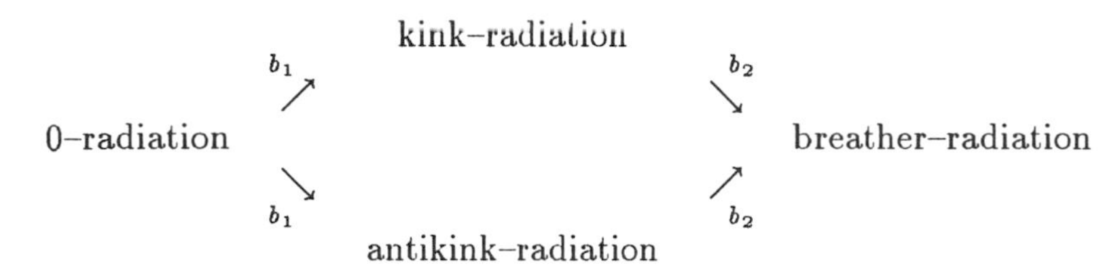

Figure 2

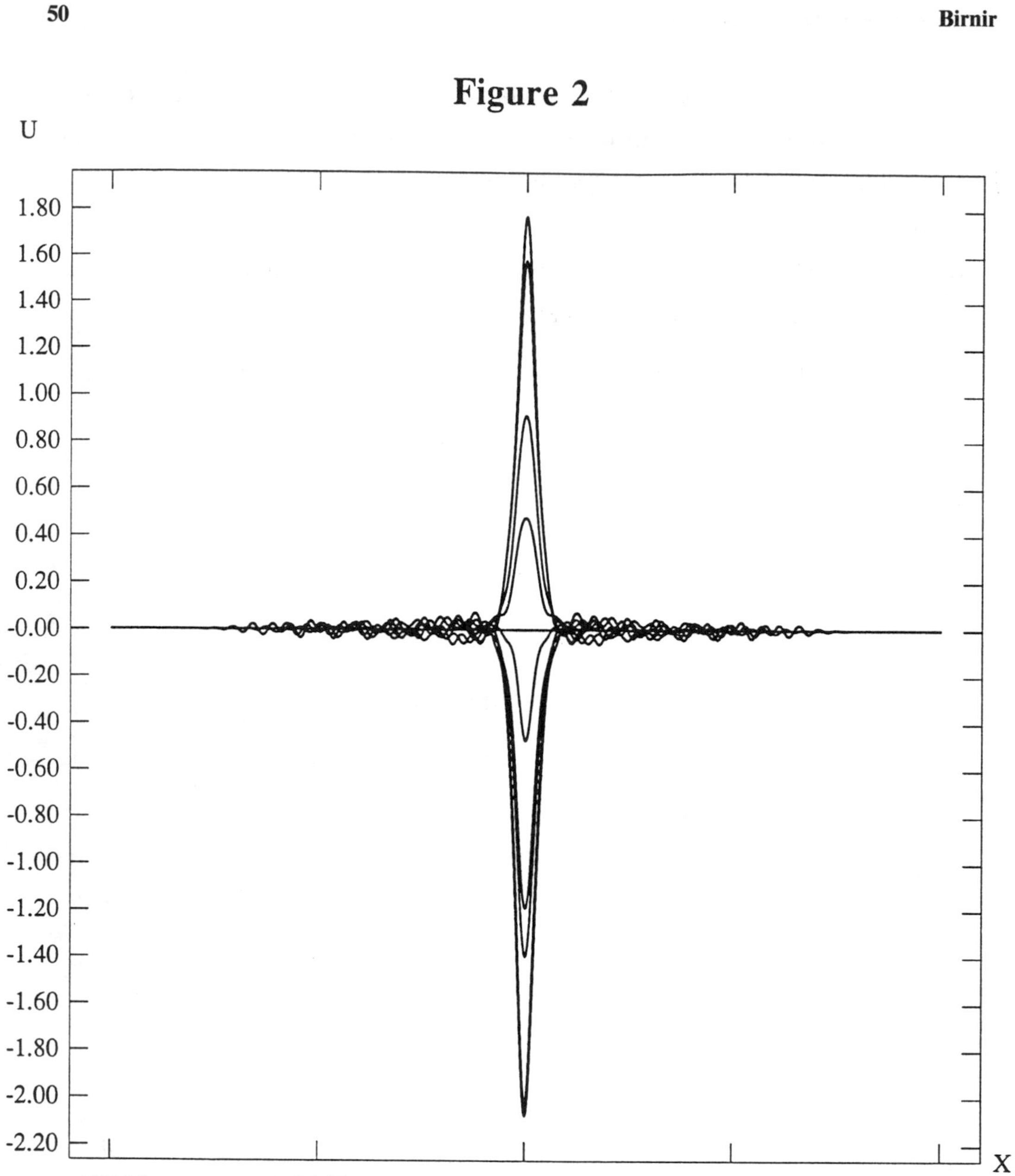
U
1.80
1.60
1.40
1.20
1.00
0.80
0.60
0.40
0.20
-0.00
-0.20
-0.40
-0.60
-0.80
-1.00
-1.20
-1.40
-1.60
-1.80
-2.00
-2.20
-100.00
-50.00
0.00
50.00
100.00
X

The $v_n^{\pm}$'s in Lemma 1 constitute the breather radiation and their formula is the result of the two transformations b_1 and b_2. The details of these computations are spelled out in Birnir, McKean and Weinstein [1992], they are inessential for the rest of the proof and we will omit them here.

The sine–Gordon breather can be written in the form

$$B = 4\tan^{-1}(y), \qquad y = \frac{\alpha}{\omega}\frac{\sin(\omega t)}{\cosh(\alpha x)}.$$

Thus for $|y| < 1$, $|B| < \pi$. Consequently, the function

$$\frac{g(4\tan^{-1}(y))}{1+y^2} = \sum_{k=1}^{\infty} d_k y^k$$

appearing in the Fredholm obstruction 1, to the existence of u_1 can be expanded in a series with coefficients d_k, provided $|y| < 1$, i.e., $\frac{\alpha}{\omega} < 1$, which means that $\omega > 1/\sqrt{2}$. This is how the lower limit on ω is obtained. We substitute this series into the Condition 1 and get,

Lemma 2. *A necessary condition for the existence of u_1, is, for $n \neq 0, 1$,*

$$\sum_{k=n}^{\infty} \left(\frac{\alpha}{\omega}\right)^k \binom{k}{\frac{k-n}{2}} [(k-1)!]^{-1} \left\{ d_k \left[\frac{1}{2}(n^2+1)\omega^2 - 1 - \frac{(n^2\omega^2-1)}{k}\right]\right.$$

$$\left. + d_{k-2}\left[\frac{1}{2}(n^2+1)\omega^2 + \frac{n^2\omega^2}{k}\right]\right\} \prod_{j}^{\frac{k}{2}-1} \left(j^2 + \frac{n^2\omega^2-1}{4\alpha^2}\right) = 0, \tag{2}$$

j is a half–integer $1/2, 3/2, \ldots$, if n is odd; j is an integer if n is even.

Proof.

$$\int_0^T \int_{-\infty}^{\infty} g(B)V_n = \sum_{k=1}^{\infty} \int_0^T \int_{-\infty}^{\infty} d_k y^k (1+y^2) v_n$$

$$= \sum_{k=1}^{\infty} d_k \left(\frac{\alpha}{\omega}\right)^k \int_0^T \int_{-\infty}^{\infty} \left(\frac{\sin(\omega t)}{\cosh(\alpha x)}\right)^k$$

$$\times \left[d + ey^2 + f\frac{\partial_x \cosh^2(\alpha x)}{\cosh^2(\alpha x)} + h\partial_t y^2\right] e^{in\omega + \sqrt{n^2\omega^2 - 1}x}.$$

This expression is then evaluated with the help of standard integrals. □

Lemma 4. *The obstruction 2 can be reduced to the recursion relation*

$$n(n+1)^2(n+3)^2 d_{n-2} + 2(n-1)^2(n+1)(n+3)^2 d_n + (n-1)^2(n+1)^2(n+2)d_{n+2} = 0, \tag{3}$$

for the coefficients of the perturbation

$$\frac{g(u)}{1+y^2} = \sum_{n=1}^{\infty} d_n y^n.$$

Proof. The obstruction 2 falls into an odd and an even part, namely $\frac{k-n}{2}$ is an integer only if both k and n are odd, or even, and $\binom{k}{\frac{k-n}{2}}$ is understood to vanish otherwise. The obstruction is an analytic function of ω in two regions symmetric about the positive and negative real axis and inside the hyparabola

$$\text{Real}^2\omega = \text{Im}^2\omega + 1/2.$$

(The analyticity region of α lies outside the hyparabola along the imaginary axis.) You see this by making $|\alpha/\omega| < q$ then the obstruction 2 is bounded by

$$\sum_{\substack{k=n \\ \text{is odd} \\ \text{or even}}} q^k d_k \prod^{k/2-1} \left(1 + \left(\frac{n^2}{4} + \frac{1}{2}\right) \Big/ j^2 q^2\right)$$

$$\leq \sinh\left[\left(\frac{n^2}{4} + \frac{1}{2}\right)^{1/2} \frac{\pi}{q}\right] \sum_{k}^{\infty} q^k d_k,$$

i.e., the products are bounded independently of k. This shows that they converge (uniformly) to an analytic function.

Now we can continue the obstruction 2 analytically to points on the real ω axis, these correspond to points on the pure imaginary α–axis, since in this region we maintain $|\alpha^2/\omega^2| < 1$. We choose ω so that the products in 2 vanish, i.e., let $j = n/2 + 1$ and choose ω such that

$$\left(\frac{n}{2} + 1\right)^2 + \frac{n^2\omega^2 - 1}{4\alpha^2} = 0, \quad \omega^2 = \frac{3+n}{4}, \quad \alpha^2 = \frac{1-n}{4}, \quad \frac{\alpha^2}{\omega^2} = \frac{1-n}{3+n}, \quad \left|\frac{\alpha^2}{\omega^2}\right| < 1.$$

Then 2 reduces to the 3–term recursion relation 3. □

The recursion relation 3 can be solved explicitly and has four solutions

n	d_n
odd	$(-1)^k(2k+2)^2$
$n = 2k+1$	$(-1)^k(2k+2)^2\sum_{j=0}^k 1/(2j+1)$
even	$(-1)^k(2k+1)^2$
$n = 2k$	$(-1)^k(2k+1)^2\sum_{j=1}^k 1/2j.$

Then we compare with the coefficients of $\sin(u)$ and $u\cos(u)$ for n odd, recalling that

$$\frac{sin(u)}{1+y^2} = \frac{4y(1-y^2)}{(1+y^2)^3} = \sum_{k=0}^{\infty}(-1)^k(2k+2)^2y^{2k+1}$$

$$\frac{u\cos(u)}{1+y^2} = u\frac{1-6y^2+y^4}{(1+y^2)^3} = \sum_{k=0}^{\infty}(-1)^k(2k+2)^2\left(\sum_{j=1}^{k}1/(2j+1)\right)y^{2k+1},$$

with $u = 4\tan^{-1}(y)$.

For n even we get

d_{2k}	$g(u)$
$(-1)^k(2k+1)^2$	$\cos(u)$
$(-1)^k(2k+1)^2\sum_{j=1}^k 1/2j$	$1+3\cos(u)-4\cos\left(\frac{u}{2}\right)+4\cos(u)\log\left(\cos\left(\frac{u}{4}\right)\right).$

The odd case is a nice check on the proof. $\cos(u)$ which corresponds to the scaling $u \to u + \epsilon \times$ constant, is ruled out because $\cos(0) \neq 0$. This leaves only the perturbation

$$g_4(u) = 1 + 3\cos(u) - 4\cos\left(\frac{u}{2}\right) + 4\cos(u)\log\left(\cos\left(\frac{u}{4}\right)\right).$$

It vanishes at the origin and so passes the test above. Numerical experiments, see Birnir [1992], indicate that this perturbation is real but its origin remains a mystery. This finishes the proof of the theorem.

I do not wish to leave you with the impression that sine–Gordon is the only nonlinear hyperbolic PDE that has a breather solution. There are in fact uncountably many such equations. Let $h(u)$ be any C^2 function with a positive derivative $h'(u) > 0,\ h(0) = 0$. Then the PDE

$$\Box u + \frac{h''(u)}{h'(u)}(u_t^2 - u_x^2) + \frac{1}{h'(u)}\sin(h(u)) = 0$$

has a breather solution $u = h^{-1}(B)$. In particular, if $h = u - \epsilon f(u)$ then we get a perturbed sine–Gordon equation

$$\Box u + \sin(u) = \epsilon[f''(u)(u_t^2 - u_x^2) - f'(u)\sin(u) + f(u)\cos(u)] + \mathcal{O}(\epsilon^2) \qquad (5)$$

which has a breather solution. There is an interesting connection here with the Christadoulou–Klainerman null condition [1986]. Namely, the quadratic terms $u_t^2 - u_x^2$ satisfy the null conditions. This may be the reason why this situation can be dealt with by the theorem above.

Proposition. *If the perturbed sine–Gordon equation*

$$\Box u + \sin(u) = \epsilon[f''(u)(u_t^2 - u_x^2) + g(u)]$$

has a breather solution satisfying Hypothesies I (f and g) and II, then

$$g(u) = -f'(u)\sin(u) + f(u)\cos(u) + a\sin(u) + bu\cos(u) + cg_4(u), \quad a, b, c \quad \text{constants.}$$

Proof. We substitute the perturbation into the Fredholm Condition 1 and integrate by parts,

$$\begin{aligned}
\int_0^T \int_{-\infty}^{\infty} vf''(B)(B_f^2 - B_x^2) + vg(B) &= \int_0^T \int_{-\infty}^{\infty} \Box v f(B) - vf'(B)\Box B + vg(B) \\
&= \int_0^T \int_{-\infty}^{\infty} (\Box v + \cos(B)v)f(B) - vf'(B)(\Box B + \sin B) \\
&+ \int_0^T \int_{-\infty}^{\infty} v(g(B) - f(B)\cos(B) + f'(B)\sin(B)) \\
&= \int_0^T \int_{-\infty}^{\infty} v(g(B) + f'(B)\sin(B) - f(B)\cos(B)).
\end{aligned}$$

Then we apply the BMW theorem to the perturbation

$$g(B) + f'(B)\sin(B) - f(B)\cos(B).$$

$\Box$

Results on the null condition [C,K] suggest that g as given by the Proposition may be the only perturbation, quadradic in u_t and u_x, that has breathers. But since sine–Gordon is really a nonlinear Klein–Gordon equation and these results on the nonlinear wave equation are open for the latter, this conjecture remains to be proven.

In closing I would like to come back to the finite dimensional attractors of dissipative nonlinear PDE's. You get the impression from the nonexistence of breathers theorem, that these attractors will not contain any periodic solutions and will not be very interesting. The surprising result, Birnir and Grauer [1992], is that there is a fairly large class of dissipative, forced nonlinear Klein–Gordon equations which have breather solutions. This is almost contradictory because that the dissipation and the forcing is small and when set to zero the breathers do not exist! But the reason is that you can put in a breather initial data, stabilize the breather by the right amount of forcing and kill its radiation by enough damping.

Finally, one should observe that the radiation decay of the perturbed breathers is exceedingly slow. Numerical experiments, see Birnir [1992], indicate that it takes 10^9 periods for the breather to loose its energy and even then it has not completely disappeared. It has become exponentially small and uncomputable. This means that the slowly radiating breathers may be fine as physical models for the lifetime of many experiments or that of the universe.

BIBLIOGRAPHY

BIRNIR, B., H. MCKEAN and A. WEINSTEIN: The isolated character of sine-Gordon breathers, to appear in *CPAM*.

BIRNIR, B.: Qualitative analysis of radiating breathers, to appear in *CPAM*.

BIRNIR, B. and R. GRAUER: An explicit description of the global attractor of the damped and driven sine-Gordon equation, to appear in *CMP*.

BRÉZIS, H., J-M. CORON and L. NIRENBERG: Free Vibrations for a nonlinear wave equation and a theorem of P. Rabinowitz, *CPAM* **33** (1980) 667–689.

CAMPBELL, D. K, and M. PEYRARD: Solitary wave collisions revisited, *Physica D* **18** (1986) 47-53.

GILBARG, D. AND N. D. TRUDINGER: Elliptic partial differential equations of second order, *Springer*, **Berlin** (1977).

CHRISTADOULOU, D.: Global solutions for nonlinear hyperbolic equations for small data, *CPAM* **39** (1986) 267–282.

R. GRAUER and B. BIRNIR: The center manifold and bifurcations of damped and driven sine-Gordon breathers, *Physica D* **56** (1992) 165-184.

KITCHENASSAMY, S.: Breather solutions of the nonlinear wave equation, *CPAM* **44** (1991) 789-818.

KLAINERMAN, S.: The null condition and global existence to nonlinear wave equations, *lect. Appl. Math.* **23** (1986) 293-326.

KRUSKAL, M., and H. SEGUR: Nonexistence of small amplitude breather solutions in ϕ^4 theory, *Phys. Rev. Lett.* **58** (1987) 747–750.

McLAUGHLIN, D. W., and A. C. SCOTT: Perturbation analysis of fluxon dynamics, *Phys. Rev. A* **18** (1978) 1652–1680.

PEYRARD, M. and D. K. CAMPBELL: Kink-antikink interactions in a modified sine-Gordon model, *Physica D* **18** (1986) 47-53.

RABIN0WITZ, P., Free vibrations for a semilinear wave equations, *CPAM* **31** (1978) 31–68.

WEINSTEIN, A.: Periodic nonlinear waves on a half-line, *Comm. Math. Phys.* **99** (1985), 385–388.

Methods to "Divide Out" Certain Solutions from Systems of Algebraic Equations, Applied to Find All Cyclic 8-Roots

Göran Björck and Ralf Fröberg

Department of Mathematics
Stockholm University
S-106 91 Stockholm, Sweden

Abstract

This paper presents some tricks which may be used when solving a system of algebraic equations which is too complex to be handled directly by a symbolic algebra system. These tricks are used to find all cyclic 8-roots. There are two types of infinite families of cyclic 8-roots and 1152 isolated ones.

0. Introduction.

The main purpose of this paper is to find all cyclic 8-roots (cf [4], where cyclic n-roots for general n are treated and related to Fourier transforms, and [5], where all cyclic n-roots are found for $n \leq 7$), i.e. all solutions $z = (a,b,c,d,e,f,g,h)$ of the system (over $\mathbf{C}$)

$$(S) \qquad C_1 = \ldots = C_7 = 0, \; C_8 = 1,$$

where

$$C_1 = a+b+c+d+e+f+g+h,$$

$$C_2 = ab+bc+cd+de+ef+fg+gh+ha,$$

$$C_3 = abc+bcd+cde+def+efg+fgh+gha+hab,$$

$$C_4 = abcd+bcde+cdef+defg+efgh+fgha+ghab+habc,$$

$$C_5 = abcde+bcdef+cdefg+defgh+efgha+fghab+ghabc+habcd,$$

$$C_6 = abcdef+bcdefg+cdefgh+defgha+efghab+fghabc+ghabcd+habcde,$$

$$C_7 = abcdefg+bcdefgh+cdefgha+defghab+efghabc+fghabcd+ghabcde+habcdef,$$

$$C_8 = abcdefgh.$$

This is well beyond what any symbolic algebra system known to the authors can handle by a straightforward attack. As in [5], our main idea is to *count* the solutions by using fast Gröbner algorithms for total reverse lexicographical ordering, applied to homogeneous systems, and then look for enough solutions of a simple kind to match the count.

The present case is complicated by the fact that there are infinitely many solutions (see [1]). This can be handled by methods to "divide out" certain solutions. In Section 1 we present two such methods in a form that may be of use in the treatment of other computationally hard systems. In Section 2 we present some of the ideas which we used to find simple (and in fact all) cyclic 8-roots. In Section 3, we present our list of cyclic 8-roots and formulate the main result that this list is complete. As a preparation for the proof of the main result, Section 4 contains some elementary properties of certain cyclic 8-roots. The proof (twice using one of the methods of Section 1) is in Sections 5 and 6.

1. A bag of tricks.

We begin by stating a weak form of the trick presented in [5]. For notation and basic facts in commutative algebra, we refer to [6] and [7].

Homogenization Lemma. *Let k be any field. Let (SI) be the (inhomogeneous) system of polynomial equations $f_1(x_1,\ldots,x_n) = \cdots = f_N(x_1,\ldots,x_n) = 0$, and let $I = (f_1,\ldots,f_N)$ in $k[x_1,\ldots,x_n]$. Let $f_{i,H}(x_1,\ldots,x_n,z) = z^{\deg f_i} f_i(x_1/z,\ldots,x_n/z)$, and let I_H be the ideal $(f_{1,H},\ldots,f_{N,H})$. For any positive integer m, let $H_m(t)$ be the power series $(\mathrm{Hilb}(P/I_H;t) - \mathrm{Hilb}(P/(I_H+(z^m));t)/t^m$. Then, if $H_m(t) = p(t)/(1-t)$ for some polynomial $p(t)$, the number of solutions of (SI) (counted with multiplicity) over k is $\leq p(1)$.*

The proof of this is contained in Section 1 of [5]. We remark that if $H_m(t) = H_{m+1}(t)$, then the inequality in the conclusion of the proposition is an equality. In particular this happens, if m is the maximum exponent of z in any initial term of a revlex order Gröbner basis for I_H. (This remark is not needed in the proof of our main result but explains our choices of exponents in Section 6.)

Next we present a trick that allows us to "divide out" the infinite families of solutions of a system having one-dimensional such families and get a one-variable equation satisfied by one component of all the isolated solutions.

The following lemma is well known.

GCD Lemma. *Let k be an arbitrary field, let (T) be the system of equations $h_1 = \cdots = h_t = 0$ in $k[x,y]$, and let $q = \mathrm{GCD}(h_1,\cdots,h_t)$. The solutions of (T) are (the possibly infinitely many) solutions of $q = 0$, together with the finitely many solutions of $h_1/q = \cdots = h_t/q = 0$.*

In the search for cyclic 8-roots, but not in the proof of the main result, we used the following corollary: Let (U) be the system of equations $f_1 = \cdots = f_u = 0$ in $k[x_1,\ldots,x_n]$, and choose a product ordering with $\{x_1,\ldots,x_{n-2}\}$ larger than $\{x_{n-1},x_n\}$. Suppose a Gröbner basis is $G = \{g_1,\ldots,g_N\}$ with $G \cap k[x_{n-1},x_n] = \{g_{k+1},\ldots,g_N\}$, and let $q = \mathrm{GCD}(g_{k+1},\ldots,g_N)$. Taking (T) to be $g_{k+1} = \cdots = g_N = 0$ in the GCD Lemma, we get:

Corollary. *In the situation just described, if the system* $(V): g_1 = \cdots = g_k = g_{k+1}/q = \cdots = g_N/q = 0$ *has finitely many solutions, then the system* $(W): g_1 = \cdots = g_k = q = 0$ *gives all infinite families of solutions.*

Finally, we present a trick that allows us to choose any homogeneous polynomial G and *count* only those solutions z of our system which satisfy $G(z) \neq 0$. This could be considered as "dividing out" all solutions where $G = 0$. Again we give the trick the form of a lemma dealing with slightly more general systems than (S):

Renormalization lemma. *Suppose we have two systems of equations*

$$(S_1) \qquad f_1 = \cdots = f_r = F - 1 = 0$$

and

$$(S_2) \qquad f_1 = \cdots = f_r = G - 1 = 0$$

where $f_1, \ldots, f_r, F,$ *and* G *are homogeneous. Then, if* (S_2) *has at most* $k \cdot \deg G$ *solutions, it follows that* (S_1) *has at most* $k \cdot \deg F$ *solutions* z *with* $G(z) \neq 0$.

Proof. Let us write $\deg F = f$ and $\deg G = g$. The only non-homogeneous equation in (S_1) is $F = 1$, where F is a form of degree f. Thus, in the set of solutions of S_1 there is an equivalence relation generated by multiplying a solution with any f:th root of unity. We will call a corresponding equivalence class an S_1-*cluster*. Clearly each S_1-cluster has f elements. Similarly, we define S_2-clusters (with g elements each).

Let z be a solution of (S_1) with $G(z) \neq 0$. Then $w = z/\rho$ will be a solution of (S_2) for every determination of $\rho = G(z)^{1/g}$. Denote by M_z the set of g values of w so obtained. Clearly, M_z is a S_2-cluster. Moreover, M_z does not change if z is replaced by another element in the same S_1-cluster. We have thus defined a map of the set of those S_1-clusters where $G \neq 0$ into the set of S_2-clusters. Since this map is injective, the result follows.

2. Special types of cyclic 8-roots, and methods to find them.

We introduce the notation $CR(8)$ for the set of cyclic 8-roots, that is the set of all solutions $z = (a, b, c, d, e, f, g, h)$ of (S) over $\mathbf{C}$. In this section we present some methods which can be used to find in $CR(8)$ elements with a simple structure. We have used these methods (among others) to prepare the list L in Section 3, using hand calculation and MATHEMATICA, MAPLE, and MACAULAY.

There is a large group acting on $CR(8)$: Without leaving $CR(8)$, one can conjugate all components of an element, invert them, multiply them by an eighth root of unity, cyclically permute the eight components, read them backwards, and "associate", i.e. if $z = (a, b, c, d, e, f, g, h) \in C(R)$, then $Az = (abc, def, gha, bcd, efg, hab, cde, fgh) \in C(R)$. We remark that "association" was not needed in the presentation of the list L.

The group action can be used in two different ways. The first one is simply to get many solutions from one, by letting the group act on it. A second way to use the group is to look for elements with short orbits under the group action. This leads to much simpler systems. As an example, if cyclic permutation 4 steps equals multiplication with -1, we get the pattern $(a, b, c, d, -a, -b, -c, -d)$, which gives the first parametric solution below (Section 3(a)). Many of the solutions in Section 3 have similar interpretations.

We next describe how the GCD Lemma could be used (using MACAULAY to identify all elements in $CR(8)$). In the corollary of the GCD Lemma, let $k = \mathbf{Z}_{31991}$, and let $U = S$. The system (W) gives exactly the two families (a) and (b) of Section 3. A new Gröbner base calculation for the system V gives a degree 368 equation for $x_n = h$. It would in principle be possible to add each factor of this polynomial to the original system and determine the solutions, in fact this is how we found some of them. It was not hard to identify corresponding solutions in characteristic 0 for these. But, luckily other methods described below saved us from carrying through this program.

Next, we describe two numerical methods. Both methods involve solving (S) numerically, starting from random points in $\mathbf{C}^8$ and applying an 8-dimensional Newton-Raphson method. The first proceeds just by inspecting the numerical solutions and looking for patterns like $a + e = 0$ or $a^8 = 1$. For example, with some luck, the patterns given in Section 3(d,e) will appear, and it is then a very simple calculation to find the single equation for a to which (S) is reduced.

The second numerical method uses a more sophisticated tool, namely the package "Recognize" in MATHEMATICA. We will describe how a successful application of this method made it possible for us to find the solutions given in Section 3(f) and at the same time feel convinced ("mod 31991"), that the list was thus made complete. We had at that time found all the patterns (a)-(e), and (e.g. by the corollary of the GCD Lemma just described) had good reason to believe that we still missed $64 \cdot 8$ isolated solutions. We then numerically found a cyclic 8-root $z = (a, b, c, d, e, f, g, h)$ with $a + e = 0$ which did not match any of then known patterns and which had all its components of modulus 1. Let k be any one of the components. We want to find the minimal equation for (the exact value of) k. Since $CR(8)$ is stable under multiplication with eighth roots of unity, it is clear that this will be an equation in $x = k^8$. Since $CR(8)$ is stable under inversion, it is enough to look for the minimal equation for $y = x + 1/x$, and we had thus reason to believe that there are at most 32 different values of y. We note that y is real, (since $|x| = 1$). After MATHEMATICA had calculated the numerical value of z and the various y to a few hundred decimal places, we could then use "Recognize" to get for each numerical y a low degree equation with integer coefficients, satisfied as well as possible by y. One equation obtained in this way led to equation (f_1) in Section 3(f). Adding (f_1) to our system (S) for the pattern discussed in Section 3(f), and calculating a Gröbner basis we found the $64 \cdot 8$ new solutions presented there.

Finally, we will discuss two types of particularly interesting solutions of (S). The first type are the *unimodular* solutions, i.e. those for which all components have absolute value one. These are important because their components are the *quotients* of the successive components of *bi-equimodular* sequences (cf [3] or [4]), that is sequences of modulus one, whose (finite) Fourier transforms also have constant modulus. We note that this is a particular instance of solutions with short orbits, namely those for which conjugation equals inversion.

The second type are those solutions whose components are all real. More generally, we call a solution $z = (z_1, \ldots, z_8)$ *essentially real* if there is a real ϕ such that $z = e^{i\phi}(r_1, \ldots, r_8)$ with r_i real (positive or negative). The equation $C_8 = 1$ gives that ϕ must be an integer multiple of $\pi/8$. Again, we have a case of solutions with short orbits, namely those for which conjugation equals multiplication with some eighth root of unity.

We finally remark that the group acts on the set of unimodular solutions, and on the set of essentially real solutions.

3. The main result.

In this section we will first give a description of the cyclic 8-roots, which is sufficiently precise for the interested reader to be able to check for herself that they actually are solutions of (S). Also the statements in the list about unimodularity and essential reality (for definitions, see Section 2) are easy to check, except perhaps in case (b) and (f), where proofs are given at the end of the section.

There are two types of infinite classes of solutions, and we begin by giving these.

(a) The first infinite class of solutions satisfy the pattern $(a, b, c, d, -a, -b, -c, -d)$. Parametrically these solutions can be given as $(s, t, is, it, -s, -t, -is, -it)$ with $(st)^4 = 1$. This can be considered as four one-dimensional families, or as one family (with e.g. $st = 1$) and its 1, 2 and 3 step cyclical permutations. The unimodular solutions constitute the families of real dimension one given by $s = e^{i\phi}$. There are no essentially real solutions of type (a).

(b) A second infinite class of solutions satisfy the pattern $(a, b, c, -a, -h, -c, -b, h)$, or the one obtained by permuting it cyclically one step. Letting $x = h - a$, $y = ha$, $u = b - c$, $v = bc$, we get from $C_2 = 0, C_4 = 0$ and $C_8 = 1$ that

$$\begin{cases} 2(y+v) - xu = 0 \\ x^2 v + u^2 y = 0 \\ y^2 v^2 = 1 \end{cases}$$

Letting $z = h + a$ and $w = b + c$, we get $h = \frac{1}{2}(z + x)$, $a = \frac{1}{2}(z - x)$, $b = \frac{1}{2}(w + u)$, $c = \frac{1}{2}(w - u)$, where

$$\theta^4 = 1, \quad \theta u^2 + 2\theta^2 - 2v^2 = 0, \quad xv - \theta u = 0, \quad yv + \theta^2 = 0, \quad z^2 = x^2 + 4y, \quad w^2 = u^2 + 4v.$$

We clearly get one-dimensional families of solutions, taking e.g. v as parameter. There are families of real dimension one of (essentially) real solutions, e.g those where $u = \sqrt{2v^2 - 2}$, $w = \sqrt{2v^2 + 4v - 2}$, $x = \sqrt{2v^2 - 2}/v$, $z = \sqrt{2v^2 - 4v - 2}/v$ (with all square roots *positive*), and v is real with $v > 1 + \sqrt{2}$.

We note that among the solutions of type (b), some are also of type (a), namely the set of

$$\omega \cdot (1, 1, i, i, -1, -1, -i, -i),$$

and their cyclic permutations, where ω is a (fixed) eighth root of unity. Only these are unimodular (proof at the end of the section).

Now we turn to isolated solutions.

(c) The pattern $(a, b, 1/b, 1/a, -1/a, -1/b, -b, -a)$, gives a solution if and only if

$$a^8 - 12a^6 + 6a^4 - 12a^2 + 1 = 0,$$

$$2a^2b^2 - ab(a^2 - 1)^2 - 2 = 0.$$

Following a suggestion by D. Lazard on how to present solutions, we remark that the values of a can be reached by the following sequence of extensions of degree two:

$$x^2 - 12x + 4 = 0, \;\; y^2 - xy + 1 = 0, \;\; a^2 = y.$$

This gives 16 solutions which can be cyclically permuted in 4 ways, and multiplied by an eighth root of unity (which gives for each solution only 4 new ones, since multiplication by -1 is already taken care of). In total we get $32 \cdot 8$ different solutions, and they are easily checked to be different from those in (a) and (b). Of these, $16 \cdot 8$, namely those coming from $x = 6 - \sqrt{32}$, are unimodular. The remaining $16 \cdot 8$ solutions, coming from $x = 6 + \sqrt{32}$, are essentially real.

(d) The pattern $(a, 1, -1/a, -1, -1, 1, -a, 1/a)$ gives a solution if and only if $3a^2 - 2a + 3 = 0$. This can be cyclically permuted, read backwards, inverted, and multiplied with an eighth root of unity to give $32 \cdot 8$ different solutions. As before it is easy to check that these are really new solutions. All are unimodular.

(e) The pattern $(a, 1, 1, 1, 1, 1, 1, 1/a)$ gives a solution if and only if $a^2 + 6a + 1 = 0$. This can be cyclically permuted, read backwards, and multiplied with an eighth root of unity to give $16 \cdot 8$ different new solutions. All are essentially real.

(f) The pattern $\kappa(1, b, c, d, -1, 1/d, 1/c, 1/b)$, gives a new solution (there are also some solution of type (a) with this pattern) if and only if

$$(f_1) \quad c^{16} - 184c^{14} + 1180c^{12} - 1800c^{10} + 1862c^8 - 1800c^6 + 1180c^4 - 184c^2 + 1 = 0,$$

$$(f_2) \quad 34816d = 17778c^{15} + 615c^{14} - 3271006c^{13} - 113175c^{12} + 20951162c^{11} + 728467c^{10} - 31825590c^9 - 1125955c^8 + 32831734c^7 + 1174357c^6 - 31749178c^5 - 1115621c^4 + 20720414c^3 + 726513c^2 - 3119538c - 127105,$$

$$(f_3) \qquad 34816b = 17778c^{15} - 615c^{14} - 3271006c^{13} + 113175c^{12} + 20951162c^{11} -$$
$$728467c^{10} - 31825590c^9 + 1125955c^8 + 32831734c^7 - 1174357c^6 - 31749178c^5 +$$
$$1115621c^4 + 20720414c^3 - 726513c^2 - 3119538c + 127105,$$

and κ is a primitive *sixteenth* root of unity. These can be cyclically permuted in 4 ways and read backwards, and there are 8 choices of κ, so we get $64 \cdot 8$ different solutions. As before one checks that these are new. Also here the values of c can be reached by a sequence of extensions of degree 2, namely by taking

$$(f_4) \qquad s^2 = 2, \ \ t^2 - 8ts - 41s - 26 = 0, \ \ x^2 - x(8t + 46) + 1 = 0, \ \ c^2 = x.$$

Of these solutions, those $32 \cdot 8$ which come from $s = -\sqrt{2}$, are unimodular, and the remaining $32 \cdot 8$ solutions, coming from $s = \sqrt{2}$, are essentially real (proof at the end of the section).

We note that in total we have found $144 \cdot 8 = 1152$ different isolated solutions, and that 640 of those are unimodular and the remaining 512 are essentially real.

Let us define L ("the list") to be the set of the ("infinitely many + 1152") solutions of (S) described in the above list. We now claim that there are no others:

Main theorem. *The point $z = (a, b, c, d, e, f, g, h) \in \mathbf{C}^8$ is a solution of the system S if and only if $z \in L$.*

The proof of the main theorem fills the remaining sections of the paper.

Proof of the claims about unimodularity and essential reality in cases (b) and (f). Let us consider a unimodular solution of type (b). If $x = 0$ or $u = 0$, the solution is clearly also of type (a). Let us therefore suppose that $x \neq 0$ and $u \neq 0$ and look for a contradiction. Since $|h| = |a| = 1$, we see that x and z must be orthogonal (as vectors in the complex plane). Thus $z = irx$ for some real number r, and we get $x^2(1+r^2) = -4y$. Introducing $x = \theta u/v$ and $yv = -\theta^2$, we get $u^2(1 + r^2) = 4v$. By the symmetric roles of h, a, x, y and b, c, u, v, we also find a real number s with $u^2(1 + s^2) = -4v$, and the absurdity $u^2(1 + r^2) = -u^2(1 + s^2)$ follows.

Finally, let us consider case (f). Checking in (f_4), we find immediately that $s = -\sqrt{2}$ gives eight c-values of modulus one, and that $s = \sqrt{2}$ gives eight other real values of c. By (f_2) and (f_3), real c implies real d and real b. If (f_2) is written as $d = D(c)$, where D is a certain polynomial, (f_3) will be $b = -D(-c)$. Hence it will be enough to show that there are sixteen values of d, eight of them real and the other eight of absolute value one. E.g. by eliminating c from (f_1) and (f_2), we get a symmetric equation of degree 16 for d, and taking $x = d + 1/d$, we get $F(x) = 0$, where $F(x) = x^8 + 16x^7 + 56x^6 - 96x^5 - 696x^4 - 896x^3 - 96x^2 + 256x - 16$. Since d is of absolute value one if and only if x is real with $|x| \leq 2$, and d is real if and only if x is real with $|x| \geq 2$, the result follows e. g. by noting that $F(x) < 0$ for $x =$ -11, -3, -1.4, 0.2, and 4, and $F(x) > 0$ for $x =$ -5, -2, 0, and 2.

4. Some simple lemmas.

Looking through our list L, we note that every element in the list has at least two components whose sum or difference is zero. Our proof of the main theorem will extensively use such relations. To prepare for this, we therefore collect in this section some simple consequences of our system S. We will start with the following result, which depends only on the equations $C_1 = C_7 = 0$ and $C_8 = 1$ and is thus independent of the order of the considered components:

Lemma 1. *If k, l, m, n, u, v, x, and y are the components of a cyclic 8-root* **in any order**, *and if $k + l = m + n = 0$, then $u + v = x + y = 0$, or $u + x = v + y = 0$, or $u + y = v + x = 0$.*

In the proof of Lemma 1, we will use the following lemma:

Lemma 2. *If k, l, m, n, u, v, x, and y are the components of a cyclic 8-root* **in any order**, *and if $k + l = m + n = 0$, but $u + v \neq 0$, then $uv = xy$.*

We remark that once Lemma 1 is proved, Lemma 2 will be of no further use, since Lemma 1 is trivially stronger.

Proof of Lemma 2. From $C_1 = 0$ we get

$$(u + v) + (x + y) = 0, \tag{1}$$

and from $C_8 = 1$ and $C_7 = 0$ we get $\frac{1}{u} + \frac{1}{v} + \frac{1}{x} + \frac{1}{y} = 0$, i.e.

$$\frac{1}{uv}(u + v) + \frac{1}{xy}(x + y) = 0. \tag{2}$$

Since $u \neq -v$, we have that (1) and (2), considered as a linear system of equations in the variables $(u + v)$ and $(x + y)$, has a non-trivial solution. Hence $\frac{1}{uv} = \frac{1}{xy}$, which proves the lemma.

Proof of Lemma 1. We first note that by $C_1 = 0$ and the hypothesis we have that $u + v = 0$ if and only if $x + y = 0$, and similarly for the other two pairings of u, v, x, and y. Suppose now that the conclusion of the lemma does not hold. In particular, $u + v \neq 0$, and $u + x \neq 0$, and Lemma 2 gives $uv = xy$ and $ux = vy$. These two relations lead to $uv^2 = xyv = ux^2$. Since $C_8 = 1$ implies $u \neq 0$, we get $v = \pm x$. By supposition, $v + x \neq 0$, and hence $v = x$. By the symmetric rolls of u, v, x, and y, we have thus $u = v = x = y$. But then $C_1 = 0$ gives the absurdity $u = v = x = y = 0$ which completes the proof.

We will now consider relations of the same kind as in Lemma 1, but where k, l etc are in specific positions. It is helpful to think of the components of a cyclic 8-root as arranged e.g. in the shape shown in the (four first) diagrams below and to draw an edge between two components whos sum is known to be zero. In particular we will consider cyclic 8-roots $z = (a, b, c, d, e, f, g, h)$, where $a + e = b + f = c + g = d + g$ (see Section 3(a)). Inspired by the diagram, we will call such a z a *star*. We will also consider the two cases $a + d = b + g = c + f = e + h = 0$ and $a + f = b + e = c + h = d + g = 0$, i.e. the cases considered in Section 3(b). We call solutions with any one of these two patterns a *sharp*.

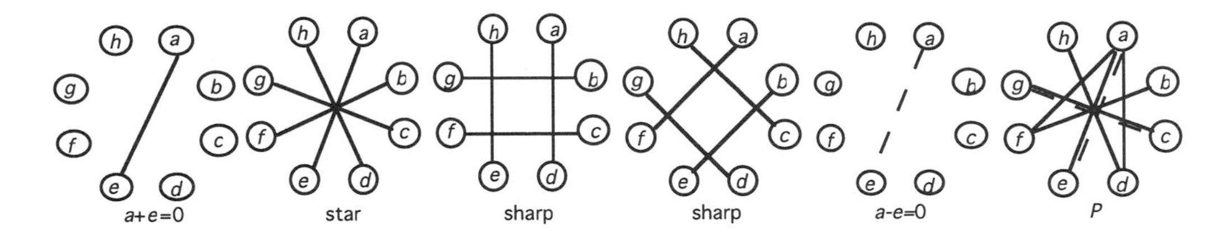

We recall the notation $CR(8)$ for the set of solutions over $\mathbf{C}$ of (S). In the following lemmas we denote the elements of $(CR(8))$ by $z = (z_1, \ldots, z_8)$ and count the indices modulo 8.

Lemma 3. *If $z \in CR(8)$, and for some i and j, with i and j noncongruent modulo 4, we have $z_i + z_{i+4} = 0$ and $z_j + z_{j+4} = 0$, then z is a star.*

Proof. Let $z = (a, b, c, d, e, f, g, h)$. By symmetry, it is enough to consider the two cases $a + e = c + g = 0$ and $a + e = b + f = 0$.

In the first case, either z is a star, or Lemma 1 gives that $b + h = 0$ or $b + d = 0$. Again by symmetry, it is enough to consider the case $b + h = 0$. We thus have $h = -b, g = -c, e = -a$, and $d = -f$. Now $C_2 = 2bc - 2cf = 0$, and hence $b = f$. Then $C_3 = 4acf = 0$, which is a contradicton.

In the second case, we have by Lemma 1 that z is a star, or that $c + d = g + h = 0$, or that $c + h = d + g = 0$. If $c + d = 0$, then $C_4 = -ab(c + g)^2$. Thus $c + g = 0$, which brings us back to the first case. The only possibility which remains to check is when $a + e = b + f = c + h = d + g = 0$. Then $C_4 = -ab(c - d)^2 = 0$ gives that $d = c$, and z is again a star (in fact a "sharp star" of the form described at the end of Section 3(b)).

Lemma 4. *If $z \in CR(8)$, and for every i at least one of the two relations $z_i + z_{i+3} = 0$ and $z_i + z_{i+5} = 0$ holds, then z is a sharp.*

Proof. In geometric language connected with the diagram, from every vertex there is an edge going three steps clockwise or counter-clockwise. Let us call such (undirected) edges "*3 edges*". We will first consider the case when there are two parallell 3-edges. Without loss of generality, we may thus suppose that $a + d = e + h = 0$ ("two vertical 3-edges"). Since the presence of one horisontal 3-edge would by $C_1 = 0$ imply the presence of the other one, the only way to avoid getting a vertical-horisontal sharp would be to have the 3-edges from b, c, f, and g all slanting. But this would lead to a slanting sharp. The only remaining case is then that we have exactly four 3-edges, one in each of the four possible directions. Without loss of generality, we may suppose that the vertical and horizontal 3-edges are given by $a+d=b+g=0$. Then considering the 3-edge from e, we see that it cannot go to h. Hence $e+b=0$. Similarly, starting from f, we see that $f+a=0$. But now we have two parallel slanting 3-edges, which is a contradiction. This completes the proof.

Finally we will prove a result similar to that in Lemma 4, but where we only know that at least from every second vertex there is a 3-edge but where we also know that at least two of the remaining components are equal:

Lemma 5. *If $z \in CR(8)$, and for every j at least one of the two relations $z_{2j}+z_{2j+3}=0$ and $z_{2j}+z_{2j+5}=0$ holds, and in addition $z_1-z_5=0$ or $z_3-z_7=0$, then z is a sharp.*

Proof. Without loss of generality we may assume that $z=(a,b,c,d,e,f,g,h)$ with $a-e=0$. One possibility might be that *all* the four known 3-edges go to a or to e. But then we would have $z=(a,-a,c,-a,a,-a,g,-a)$, and since $C_1=g+c-2a=0$ and $C_2=-2a(g+c+2a)=0$, we would get $a=0$, which is impossible. Another possibility might be that *none* of the four known 3-edges goes to a or to e. This would mean that we have $z=(a,b,c,b,a,-c,-b,-c)$, but then $C_4=4abc(b-c)$ would give $b=c$, and $C_1=2a+b-c=0$ would again give $a=0$. The only remaining case is that at least one of a and e receives exactly one of the four known 3-edges. Without loss of generality we may thus assume that $b+e=h+c=0$. The 3-edge from f must be either $f+a=0$ or $f+c=0$. The first alternative, (which by $C_1=0$ is equivalent to $g+d=0$), gives a sharp. Similarly, considering the 3-edge from d, we have either $d+g=0$, which gives a sharp, or $d+a=0$. Thus it only remains to consider the case $a-e=b+e=h+c=f+c=d+a=0$, which gives $z=(a,-a,c,-a,a,-c,g,-c)$. Now $C_1=0$ gives $g=c$, and then $C_2=-2(a+c)^2=0$ and $C_3=c(a-c)^2=0$. Thus $a=c=0$, which shows that this case is impossible, and the proof is complete.

5. Proof of the main theorem.

In addition to to our system (S), we will consider four auxiliary systems, namely

$$(S_P) \qquad C_1=\ldots=C_7=P-1=0,$$

$$(S_{Q+}) \qquad C_1=\ldots=C_7=Q-1=a+e=0,$$

$$(S_+) \qquad C_1=\ldots=C_7=C_8-1=a+e=0,$$

and

$$(S_{--}) \qquad C_1=\ldots=C_7=0=C_8-1=a-e=b-f=0,$$

where $P=bh(a+e)(b+f)(c+g)(d+h)(a-e)(c-g)(a+d)(a+f)$, and $Q=acdefgh(b+f)$.

We remark that the purpose of introducing P is to "divide out" at least all solutions in the infinite families. The precise choice of P reflects a delicate balance. Thus the "minimal" choice $P = (a+d)(a+e)(a+f)bcg$ worked well on a computer in characteristic 31991 but turned out to be too hard in characteristic 0. On the other hand, a "maximal" choice, "dividing out" *all* solutions, failed even in finite characteristic, because the degree became too high.

In Section 6, we will show how computer computations prove the following three propositions:

Proposition 1. *The system* (S_P) *has at most 240 solutions.*

Proposition 2. *The system* (S_{Q+}) *has at most 128 solutions.*

Proposition 3. *The system* (S_{--}) *has at most 64 solutions.*

(We could easily prove equality in Propositions 1 and 2, and it will follow from below that it is equality also in Proposition 3.)

We will now start the proof of the main theorem by claiming that it suffices to prove the following seemingly weaker result:

Proposition 4. *If* z *is a cyclic 8-root such that* $P(Tz) = 0$ *for every cyclic permutation* T*, then* $z \in L$.

Proof of the claim. We will start by exhibiting $24 \cdot 8$ elements in L for which $P \neq 0$. These will be $16 \cdot 8$ elements of type (c) (Section 3) and $8 \cdot 8$ elements of type (d). In fact, for type (c) we have

$$P(a, b, 1/b, 1/a, -1/a, -1/b, -b, -a) = a^{-3}b^{-3}(a^4 - 1)^2(b^4 - 1)(b^2 - 1)(1 - ab)$$

which is easily checked to be non-zero, using the equations defining a and b, and in the same way we see that $P(-a, a, b, 1/b, 1/a, -1/a, -1/b, -b) \neq 0$. Thus two of the four permutations mentioned in Section 3(c) will give what we want, and we have the desired $16 \cdot 8$ solutions of this type. For type (d), we get

$$P(-a, 1/a, a, 1, -1/a, -1, -1, 1) = 2a^{-4}(a^4 - 1)(a^2 - 1)(a - 1) \neq 0$$

and similarly for $P(a, 1, -1/a, -1, -1, 1, -a, 1/a)$. Thus two of the eight permutations work, and we have the desired $8 \cdot 8$ solutions.

By Proposition 1 and the Renormalization Lemma with $S_1 = S$ and $S_2 = S_P$ (thus $F = C_8$ and $G = P$) there are at most $24 \cdot 8$ cyclic 8-roots with $P \neq 0$. Thus any such cyclic 8-root must be among those we just exhibited, i.e. it must be in L.

Suppose now that Proposition 4 is true and that z is any cyclic 8-root. There is nothing to prove unless $P(Tz) \neq 0$ for some cyclic permutation T. Let us write $w = Tz$. Then w is a cyclic 8-root with $Pw \neq 0$. Hence, by what we have just proved, $w \in L$. Since L is stable under cyclic permutations, we have $z \in L$, which completes the proof.

Proof of Proposition 4. We will in analogy with "3-edges" introduce a terminology also for the two other types of relation considered in Lemmas 3 and 5: We will call any relation of the form $z_i + z_{i+4} = 0$ a *4-edge* and any relation of the form $z_i - z_{i+4} = 0$ a *4-minus-edge.*

We note that for $w = (a, b, c, d, e, f, g, h)$ with $bh \neq 0$ we have $Pw = 0$ if and only if w either contains a 3-edge from a or any 4-edge or at least one of the ("perpendicular") 4-minus-edges $a - e = 0$ and $c - g = 0$. (The sixth diagram above, with 4-minus-edges marked as in the fifth diagram, may help to visualize P.) Thus $P(Tz) = 0$ for every cyclic permutation T if and only if

(i) z contains at least one 4-edge, or
(ii) z contains two "adjacent" 4-minus-edges, i.e. we have some cyclic permutation of $a - e = b - f = 0$, or
(iii) z contains one 4-minus-edge and and at least one 3-edge from each of the vertices adjacent to the endpoints of the 4-minus-edge, i.e. we have either $a - e = 0$ together with a 3-edge from each of b, d, f and h or else a cyclic permutation of this pattern, or
(iv) z contains at least one 3-edge from each vertex.

Case (iv) is settled by Lemma 4.

Case (iii) is settled by Lemma 5 (possibly after a cyclic permutation one step).

In case (ii) we may without loss of generality suppose that z is a solution of (S_{--}). By Proposition 3 this case will be settled if we can exhibit 64 elements of L which are solutions of (S_{--}). We find at once 32 such elements by considering type (e) and noting that exactly one quarter of the possible permutations of $(a, 1, 1, 1, 1, 1, 1, 1/a)$ will have $a = b = e = f = 1$. The remaining 32 elements are (slanting) sharps with special values of the parameter. In fact, this corresponds to the specialization $w = 0$ for vertical-horisontal sharps treated in Section 3(b).

Finally, in case (i) we either have the situation covered by Lemma 3, or else z contains exactly one 4-edge. Without loss of generality we may then assume that $a + e = 0$ but $b + f \neq 0$, which means that z is a solution of (S_+) and such that $Q(z) \neq 0$. This case will therefore be settled, and Proposition 4 and thus the main theeorem will be proved, if we prove Proposition 5 below:

Proposition 5. *If z is a solution of (S_+) such that $Q(z) \neq 0$, then $z \in L$.*

Proof. We start by considering those elements of L which are of type (f) and as presented, without permutations. There are 128 such elements, and clearly they have $a+e = 0$ and a calculation shows that $b+f \neq 0$. In L we have thus found 128 solutions z of (S_+) with $Q(z) \neq 0$. But by Proposition 3 and the Renormalization Lemma, with $S_1 = S_+$ and $S_2 = Q_+$, there are at most 128 such z. Hence they must all lie in L, which completes the proof.

6. The computer computational parts of the proof.

Propositions 2 and 3 were proved by straightforward calculations of Gröbner bases in characteristic 0 in the program BERGMAN by Backelin which is described in [2]. Let

I_{S_+} and $I_{S_{--}}$ denote the homogenized ideals in $A = \mathbf{C}[a,b,c,d,e,f,g,h,z]$ of S_+ and S_{--} respectively. The calculations gave

$$\frac{\mathrm{Hilb}(A/I_{S_+};t) - \mathrm{Hilb}(A/(I_{S_+} + (z^{16}));t)}{t^{16}} = \frac{(1-t^8)(1+5t+9t^2+t^3)}{(1-t)^2}$$

$$= \frac{(1+t+t^2+t^3+t^4+t^5+t^6+t^7)(1+5t+9t^2+t^3)}{1-t}$$

and

$$\frac{\mathrm{Hilb}(A/I_{S_{--}};t) - \mathrm{Hilb}(A/(I_{S_{--}} + (z^{16}));t)}{t^{16}} = \frac{(1-t^8)(1+4t+3t^2)}{(1-t)^2}$$

$$= \frac{(1+t+t^2+t^3+t^4+t^5+t^6+t^7)(1+4t+3t^2)}{(1-t)}.$$

The Homogenization Lemma (Section 1) thus gives the estimates desired in Propositions 2 and 3. The computation times were a couple of minutes and a couple of seconds, respectively.

The system (S_P) is however much too hard to solve directly. To handle Proposition 1 in characteristic 0 we used the following method. (We have been informed that W. Trinks has used a similar idea.) We made a calculation of the Gröbner basis for the ideal $I_{S_P} = (f_1, \ldots, f_7, P - z^{10})$ in characteristic 31991, which besides the ideal J_{31991} of initial monomials (with 681 generators) also saved information about which S-polynomials were reduced to nonzero elements. We let this information govern a second run in characteristic 0. In this second run we just skipped all S-polynomial calculations which gave no new Gröbner basis elements in the first run. Thus we get a generating system for the ideal, whose initial monomials generate an ideal $J \subset J_0 = \mathrm{In}(f_1, \ldots, f_7, P - z^{10})$, the ideal of initial monomials in characteristic 0. It turned out that J had the same generators as J_{31991}, thus J_0 was as least as big as J_{31991}. Then we could estimate the number of solutions with the Homogenization Lemma. The result was

$$\frac{\mathrm{Hilb}(A/I_{S_P};t) - \mathrm{Hilb}(A/(I_{S_P} + (z^{20}));t))}{t^{20}} = \frac{(1-t^{10})(1+6t+17t^2)}{(1-t)^2}$$

$$= \frac{(1+t+t^2+t^3+t^4+t^5+t^6+t^7+t^8+t^9)(1+6t+17t^2)}{(1-t)}.$$

Since we used revlex, a Gröbner basis for $I_{S_P} + (z^{20})$ is obtained from a Gröbner basis for I_{S_P}, just by deleting all elements (in fact 60) whose initial monomial is divisible by z^{20}. Hence no new Gröbner basis calculation was needed, only a calculation of Hilbert series. The Homogenization Lemma and the reasoning above gives that the number of solutions of (S_P) is at most $10 \cdot 24$. This completes the proof of Proposition 1. The computation time for the first and second run were 6 and 15 hours CPU time on a SUN 490, respectively.

Acknowledgement.

We greatly appreciate that we had access to the fast Gröbner basis program BERGMAN, and its creator J. Backelin. At request he swiftly provided us with several special versions of his program. We also note that Backelin independently found the second type of infinite solutions by hand.

References.

[1] Backelin, J. Square multiples n give infinitely many cyclic n–roots, *Reports Matematiska Institutionen, Stockholms Universitet* (1989), No 8.

[2] Backelin, J. and Fröberg, R. How we proved that there are exactly 924 cyclic 7-roots, *Proc. ISSAC'91 (S. M. Watt, ed.)*, 103–111, ACM (1991).

[3] Björck, G. Functions of modulus one on $\mathbf{Z}_p$ whose Fourier transforms have constant modulus, *Proceedings of Alfred Haar Memorial Conference, Budapest, Colloquia Mathematica Societatis János Bolyai* **49** (1985), 193–197.

[4] Björck, G. Functions of modulus one on $\mathbf{Z}_n$ whose Fourier transforms have constant modulus, and "cyclic n-roots". *Recent Advances in Fourier Analysis and its Applications (J.S. Byrnes and J.F. Byrnes, eds), NATO Adv. Sci. Inst. Ser. C: Math. Phys. Sci.*, Kluwer **315** (1989), 131–140.

[5] Björck, G. and Fröberg R. A Faster Way to Count the Solutions of Inhomogeneous Systems of Algebraic Equations, with Applications to Cyclic n-roots , *J. Symb. Comp.* **12** (1991), 329–336.

[6] Zariski, O., Samuel, P. Commutative Algebra, van Nostrand, vol. I (1958).

[7] Zariski, O., Samuel, P. Commutative Algebra, van Nostrand, vol. II (1960).

The Exponential Function and Linearization of Quadratic Polynomials

Timo Erkama

Department of Mathematics
University of Joensuu
SF–80101 Joensuu, Finland

Abstract

The complex exponential function satisfies uncountably many functional equations of the form $f \circ \phi = P \circ f$ where P is a nonlinear polynomial and ϕ is an affine linear transformation. We show that this property is characteristic for the class of exponential functions.

Let D be a domain of the affine complex plane $\mathbf{C}$, and let $\mathrm{Aff}(D)$ be the group of affine conformal automorphisms of D. Every element $\phi \in \mathrm{Aff}(D)$ is a linear polynomial with a constant derivative $\phi' \neq 0$, and $\mathrm{Aff}(D)$ is a closed subgroup of the Lie group $\mathrm{Aff}(\mathbf{C})$. We say that a holomorphic function $f: D \to \mathbf{C}$ *linearizes* a polynomial P if there exists $\phi \in \mathrm{Aff}(D)$ such that

$$f \circ \phi = P \circ f. \tag{1}$$

If f linearizes P, then f also linearizes the nth iterate $P^{\circ n}$ of P because (1) implies

$$f \circ \phi^{\circ n} = P^{\circ n} \circ f$$

for each $n \geq 1$. More generally, the family $\Pi(f)$ of polynomials linearized by f is a semigroup under composition.

We say that two holomorphic functions $f: D \to \mathbf{C}$ and $g: D \to \mathbf{C}$ are *conjugate* if there exist $\sigma \in \mathrm{Aff}(D)$ and $\tau \in \mathrm{Aff}(\mathbf{C})$ such that

$$f \circ \sigma = \tau \circ g.$$

If f and g are conjugate, then $\Pi(f)$ and $\Pi(g)$ are isomorphic. We shall establish the following characterization of the exponential function.

Theorem 1. *Let $f: D \to \mathbf{C}$ be a nonconstant holomorphic function. Then the following conditions are equivalent:*

(i) *The number of quadratic polynomials linearized by f is uncountable.*

(ii) *D is either the whole plane $\mathbf{C}$ or a half plane and f is conjugate to the exponential function e^z.*

Note that e^z linearizes uncountably many quadratic polynomials because

$$e^{2z+\log A} = A(e^z)^2$$

for each $A \neq 0$. Hence (ii) implies (i).

Linearizations of polynomials play an important role in complex analytic dynamics. Entire functions linearizing simultaneously two different polynomials with no common iterates have been studied earlier by Fatou and Julia (see [4], [6]). However, they consider a more special situation where the given polynomials are permutable. Note that distinct quadratic polynomials never commute [5].

For entire functions the proof of Theorem 1 is simple and a stronger result is presented at the end of this article (Theorem 2).

In the proof of Theorem 1 we shall consider several sets of affine transformations associated with f. Given a subdomain Δ of D let $\text{Aff}(\Delta, D)$ be the set of all $\phi \in \text{Aff}(\mathbf{C})$ such that $\phi(\Delta) \subset D$. For each polynomial P let $\Phi(f, \Delta, P)$ be the set of all $\phi \in \text{Aff}(\Delta, D)$ such that $f(\phi(z)) = P(f(z))$ for each $z \in \Delta$. Note that elements of $\Phi(f, \Delta, P)$ need not map D into itself. The *stabilizer* Γ_ζ of a point $\zeta \in \mathbf{C}$ in $\text{Aff}(\mathbf{C})$ consists of all elements $\phi \in \text{Aff}(\mathbf{C})$ with $\phi(\zeta) = \zeta$.

Lemma 1. *Each Γ_ζ is a one-dimensional abelian subgroup of* $\text{Aff}(\mathbf{C})$, *and* $\bigcup_{\zeta \in \mathbf{C}} \Gamma_\zeta$ *is dense in* $\text{Aff}(\mathbf{C})$.

Proof. Let $\mathbf{C}^* = \mathbf{C} \setminus \{0\}$, and let $F: \mathbf{C}^* \times \mathbf{C} \to \text{Aff}(\mathbf{C})$ be the canonical homeomorphism such that $F(a,b)(z) = az + b$ for each $(a,b) \in \mathbf{C}^* \times \mathbf{C}$ and each $z \in \mathbf{C}$. The image $F(a,b)$ of a point $(a,b) \in \mathbf{C}^* \times \mathbf{C}$ is contained in Γ_ζ for some $\zeta \in \mathbf{C}$ if and only if $a \neq 1$, and it is clear that points (a,b) with this property are dense in $\mathbf{C}^* \times \mathbf{C}$. The first assertion of the lemma follows from the fact that each Γ_ζ is conjugate to Γ_0 and that the map $a \mapsto F(a,0)$ from $\mathbf{C}^*$ to Γ_0 is a group isomorphism.

Lemma 2. *Suppose that ϕ_1 and ϕ_2 are different elements of Γ_ζ. Then $\phi_1(z) \neq \phi_2(z)$ for every $z \neq \zeta$.*

Proof. Expanding ϕ_j in powers of $z - \zeta$ we have

$$\phi_j(z) = (z - \zeta)\phi_j{}' + \zeta \qquad (j = 1,2).$$

Since $\phi_1 \neq \phi_2$, we have $\phi_1{}' \neq \phi_2{}'$, and the assertion follows.

Lemma 3. *Let Δ be a subdomain of D, and let E be a subset of* $\mathrm{Aff}(\Delta, D)$. *Suppose that E is not discrete in* $\mathrm{Aff}(\Delta, D)$. *Then there exists at most one point $z_0 \in \Delta$ such that the set $E.z_0 = \{\,\phi(z_0);\ \phi \in E\,\}$ is discrete in$D$.*

Proof. Since E is not discrete in $\mathrm{Aff}(\Delta, D)$, there is a convergent sequence $\{\phi_n\}$ of distinct elements of E such that the limit $\phi = \lim_{n\to\infty} \phi_n$ is contained in E. We may assume that there exists $z_0 \in \Delta$ such that $E.z_0$ is discrete in D. Since $\lim_{n\to\infty} \phi_n(z_0) = \phi(z_0)$, it follows that $\phi_n(z_0) = \phi(z_0)$ for $n \geq n_0$. Thus for $n \geq n_0$ we may write $\phi_n = \chi_n \circ \phi$ where $\chi_n = \phi_n \circ \phi^{-1}$ is an element of $\Gamma_{\phi(z_0)}$. Let z be any point of Δ different from z_0, so that $\phi(z) \neq \phi(z_0)$. Then in view of Lemma 2 the points $\chi_n(\phi(z))$ are all distinct. On the other hand

$$\chi_n(\phi(z)) = \phi_n(z) \ \rightarrow\ \phi(z)$$

as $n \to \infty$, and we see that $\phi(z)$ is an accumulation point of $E.z$. This proves Lemma3.

Lemma 4. *Suppose that f is nonconstant. Then $\Phi(f, \Delta, P)$ is a discrete subset of* $\mathrm{Aff}(\Delta, D)$.

Proof. Assume that the set $E = \Phi(f, \Delta, P)$ is not discrete in $\mathrm{Aff}(\Delta, D)$. Then by Lemma 3 there exists $z \in \Delta$ such that $E.z$ is not discrete in D. Since $f(\phi(z)) = P(f(z))$ for each $\phi \in E$, it follows that the inverse image of $P(f(z))$ under f is not discrete. This is possible only if f is constant, because f is holomorphic.

Lemma 5. *Suppose that f is nonconstant, and let L be an abelian subgroup of* $\mathrm{Aff}(\mathbf{C})$. *Let ϕ_1 and ϕ_2 be elements of L with the following properties:*
(i) *$\phi_j \in \Phi(f, \Delta, P_j)$ for $j = 1, 2$.*
(ii) *$\phi_1 \circ \phi_2 \in \mathrm{Aff}(\Delta, D)$.*
Then $P_1 \circ P_2 = P_2 \circ P_1$.

Proof. Condition (i) implies for $j = 1, 2$

$$f(\phi_j(z)) = P_j(f(z)) \qquad (z \in \Delta).$$

By analytic continuation this equation holds at each point $z \in D$ such that $\phi_j(z) \in D$. It follows that

$$f((\phi_1 \circ \phi_2)(z)) = P_1(f(\phi_2(z))) = (P_1 \circ P_2)(f(z)) \qquad (z \in \Delta)$$

and similarly

$$f((\phi_2 \circ \phi_1)(z)) = (P_2 \circ P_1)(f(z)) \qquad (z \in \Delta).$$

Since $\phi_1 \circ \phi_2 = \phi_2 \circ \phi_1$, we conclude that

$$(P_1 \circ P_2)(f(z)) = (P_2 \circ P_1)(f(z)) \qquad (z \in \Delta).$$

Therefore $P_1 \circ P_2 = P_2 \circ P_1$ because f is nonconstant.

Lemma 6. *Suppose that f is nonconstant, and let L be an abelian subgroup of* $\mathrm{Aff}(\mathbf{C})$. *Let E be a subset of L with the following properties:*

(i) *Every element of E is in $\Phi(f,\Delta,P)$ for some nonlinear polynomial P.*
(ii) $\phi_1 \circ \phi_2 \in \mathrm{Aff}(\Delta, D)$ *for each* $\phi_1, \phi_2 \in E$.

Then E is at most countable.

Proof. Since each $\Phi(f,\Delta,P)$ is at most countable by Lemma4, it suffices to prove that the number of nonlinear polynomials P with $E \cap \Phi(f,\Delta,P) \neq \emptyset$ is at most countable. This will follow as soon as we show that all polynomials satisfying the above condition are permutable.

Let P_1 and P_2 be nonlinear polynomials such that $E \cap \Phi(f,\Delta,P_j) \neq \emptyset$ for $j = 1,2$. Then there exist $\phi_j \in E \cap \Phi(f,\Delta,P_j)$ such that $\phi_1 \circ \phi_2 = \phi_2 \circ \phi_1$ because L is abelian. By Lemma 5 it follows that $P_1 \circ P_2 = P_2 \circ P_1$. This proves Lemma 6 because the number of polynomials commuting with a given nonlinear polynomial is at most countable[5].

A domain D is called a *strip* if D is not a half plane but D is the intersection of two half planes with parallel boundaries. In this case the connected component of the identity of $\mathrm{Aff}(D)$ is an abelian subgroup L of index two in $\mathrm{Aff}(D)$, and each element of $\mathrm{Aff}(D)$ is either an element of L or an involution of order two. The following lemma gives information about maps defined in domains such as a strip, a punctured disk, or an annulus.

Lemma 7. *Suppose that f is nonconstant and that either D is a strip or* $\mathrm{Aff}(D)$ *is abelian. Then $\Pi(f)$ contains at most countably many nonlinear polynomials.*

Proof. For each nonlinear $P \in \Pi(f)$ choose $\phi_P \in \Phi(f,D,P)$, and let $E = \{\phi_P; P \in \Pi(f)\}$. It is clear that the map $P \mapsto \phi_P$ is one-to-one, because f is nonconstant. Therefore, it suffices to show that E is at most countable.

If ϕ_P is of order two for some nonlinear $P \in \Pi(f)$, iteration of the functional equation $f \circ \phi_P = P \circ f$ shows that $f = f \circ \phi_P \circ \phi_P = P \circ P \circ f$. This is impossible because f is nonconstant. Hence E is contained in an abelian subgroup L of $\mathrm{Aff}(D)$ even if D is a strip. By choosing $\Delta = D$ in Lemma 6 we see that E is at most countable, and the assertion follows.

Lemma 8. *Suppose that f is nonconstant and that $\Pi(f)$ contains uncountably many nonlinear polynomials. Then D is either the whole plane* $\mathbf{C}$ *or a half plane, and* $f'(z) \neq 0$ *for each* $z \in D$.

Proof. Let E be the set of all $\phi \in \mathrm{Aff}(D)$ such that $f \circ \phi = P \circ f$ for some nonlinear polynomial P. Then E is uncountable, and it follows that $\mathrm{Aff}(D)$ is not discrete. Domains with this property have been classified in [3] (see also [1]). The classification shows that if D is not the whole plane $\mathbf{C}$ or a half plane, then either D is a strip or $\mathrm{Aff}(D)$ is abelian. In view of Lemma 7 this proves the first assertion of Lemma 8.

To prove the second assertion let $X = \{ z \in D;\ f'(z) = 0 \}$. Differentiation of(1) yields

$$f'(\phi(z))\phi' = P'(f(z))f'(z) \qquad (z \in D),$$

and we conclude that

$$E.z \subset X \qquad \text{for each } z \in X. \tag{2}$$

It is clear that X is discrete, because f is nonconstant. Thus (2) shows that $E.z$ is discrete in D for each $z \in X$. Since E is uncountable, it follows by Lemma 3 that X contains at most one point. However, if $X = \{\zeta\}$, then (2) implies that $E \subset \Gamma_\zeta$. This is impossible by Lemma 6, because every element of E is contained in $\Phi(f, D, P)$ for some nonlinear polynomial P. Hence X is empty.

After these preparatory lemmas we begin our studies on the linearization of quadratic polynomials, that is, polynomials of the form

$$P(w) = Aw^2 + Bw + C$$

where A, B and C are complex constants and $A \neq 0$. If f linearizes P, there exists $\phi \in \mathrm{Aff}(D)$ such that

$$f \circ \phi = Af^2 + Bf + C. \tag{3}$$

For quadratic polynomials it is relatively easy to express the coefficients of P in terms of$f$ andϕ. Let us differentiate (3) twice with respect to z. Since ϕ' is a nonzero constant, this differentiation yields

$$(f' \circ \phi)\phi' = 2Aff' + Bf'; \tag{4}$$

$$(f'' \circ \phi)(\phi')^2 = 2A(f')^2 + 2Aff'' + Bf''. \tag{5}$$

At every point $z \in D$ such that $f'(z) \neq 0$ the system of equations (3), (4) and (5) is a nonsingular linear system with a unique solution (A, B, C). The explicit solution in terms of the Wronskian determinant $W(f', f' \circ \phi) = \phi' f'(f'' \circ \phi) - f''(f' \circ \phi)$ of f' and $f' \circ \phi$ is

$$\begin{aligned} A &= (\phi'/2)(f')^{-3} W(f', f' \circ \phi), \\ B &= \phi'(f' \circ \phi)/f' - \phi' f(f')^{-3} W(f', f' \circ \phi), \\ C &= f \circ \phi - \phi' f(f' \circ \phi)/f' + (\phi'/2) f^2 (f')^{-3} W(f', f' \circ \phi). \end{aligned} \tag{6}$$

Proof of Theorem 1. Suppose that f is nonconstant and that the number of quadratic polynomials linearized by f is uncountable. Then, by Lemma 8, D is either the whole plane $\mathbf{C}$ or a half plane, and $f'(z) \neq 0$ for each $z \in D$. Since the topology of $\mathrm{Aff}(D)$ has a countable basis, there is $\phi_0 \in \mathrm{Aff}(D)$ such that each neighborhood of

ϕ_0 contains uncountably many elements $\phi \in \mathrm{Aff}(D)$ satisfying (1) for some quadratic polynomial P.

The map $(\phi, z) \mapsto \phi(z)$ from $\mathrm{Aff}(\mathbf{C}) \times \mathbf{C}$ to $\mathbf{C}$ is continuous, because $\mathrm{Aff}(\mathbf{C})$ is a topological group. Therefore, given any point $z_0 \in D$ there exists an open neighborhoodΔ ofz_0 in$D$ and an open neighborhood N of ϕ_0 in $\mathrm{Aff}(\mathbf{C})$ such that for each $z \in \Delta$ and for each $(\phi_1, \phi_2, \phi_3, \phi_4) \in N^4$ the linear polynomials ϕ_1, $\phi_1 \circ \phi_2$ and $\phi_1 \circ \phi_2 \circ \phi_3 \circ \phi_4$ are in $\mathrm{Aff}(\Delta, D)$.

The set $N \times \Delta$ has a canonical complex analytic structure as an open subset of $\mathrm{Aff}(\mathbf{C}) \times \mathbf{C}$, and the three expressions on the right hand side of (6) are defined and holomorphic for $(\phi, z) \in N \times \Delta$. We shall denote these three holomorphic functions of $N \times \Delta$ by $A(\phi, z)$, $B(\phi, z)$ and $C(\phi, z)$, respectively.

If $\phi \in N$ satisfies (1) for some quadratic polynomial P, the expressions (6) are independent of z, and consequently their partial derivatives with respect to z satisfy

$$\frac{\partial^k A}{\partial z^k}(\phi, z_0) = \frac{\partial^k B}{\partial z^k}(\phi, z_0) = \frac{\partial^k C}{\partial z^k}(\phi, z_0) = 0 \qquad (k = 1, 2, \ldots). \tag{7}$$

Let M be the set of elements $\phi \in N$ satisfying (7). Then M is a complex analytic subset of N (see [2], p.62). Since N contains uncountably many elements of $\mathrm{Aff}(D)$ satisfying (7), the complex dimension of this subset is positive. Thus there exists a nonconstant holomorphic map $t \mapsto \phi_t$ from the open unit disk U into M. It follows that for $(t, z) \in U \times \Delta$

$$f(\phi_t(z)) = P_t(f(z)) \tag{8}$$

where P_t is a quadratic polynomial with coefficients $A(\phi_t, z_0)$, $B(\phi_t, z_0)$ and $C(\phi_t, z_0)$. By analytic continuation (8) holds for each $z \in D$ with $\phi_t(z) \in D$. Then it follows from (8) that for each $(t_1, t_2) \in U \times U$

$$f(\phi_{t_1} \circ \phi_{t_2}(z)) = P_{t_1} \circ P_{t_2}(f(z)) \qquad (z \in \Delta). \tag{9}$$

In other words, $\phi_{t_1} \circ \phi_{t_2} \in \Phi(f, \Delta, P_{t_1} \circ P_{t_2})$.

For each $t \in U$ we denote by $a(t)$ and $b(t)$ the coefficients of the linear polynomialϕ_t, so that

$$\phi_t(z) = a(t)z + b(t) \qquad (t \in U,\ z \in \mathbf{C}).$$

Clearly $a(t)$ and $b(t)$ are holomorphic functions of t, and $a(t) \neq 0$. There is a holomorphic map H from $U \times U$ to $\mathbf{C}^* \times \mathbf{C}$ such that

$$H(t_1, t_2) = \big(a(t_1)a(t_2),\ a(t_1)b(t_2) + b(t_1)\big);$$

the components of H are the coefficients of $\phi_{t_1} \circ \phi_{t_2}$. We show next that H is singular at each point $(t_1, t_2) \in U \times U$, so that the complex Jacobian

$$J_H(t_1, t_2) = \begin{vmatrix} a'(t_1)a(t_2) & a'(t_1)b(t_2) + b'(t_1) \\ a(t_1)a'(t_2) & a(t_1)b'(t_2) \end{vmatrix}$$

of H is identically zero.

Suppose that $J_H(t_1', t_2') \neq 0$ at some point $(t_1', t_2') \in U \times U$. Then the image of $U \times U$ under H contains a nonempty open subset Ω of $\mathbf{C}^* \times \mathbf{C}$ containing the coefficients of $\phi_{t_1'} \circ \phi_{t_2'}$. Let F be as in the proof of Lemma 1; then $F(H(t_1, t_2)) = \phi_{t_1} \circ \phi_{t_2}$ for each $(t_1, t_2) \in U \times U$, and by Lemma 1 there is $\zeta \in \mathbf{C}$ such that $F(\Omega)$ contains a nonempty open subset of Γ_ζ. Let S be the inverse image of $\Gamma_\zeta \cap F(\Omega)$ under $F \circ H$. Then S is uncountable, and (9) implies that $(F \circ H)(t_1, t_2) \in \Phi(f, \Delta, P_{t_1} \circ P_{t_2})$ for each $(t_1, t_2) \in S$. Thus the set $E = F(H(S))$ satisfies the hypotheses of Lemma 6 for $L = \Gamma_\zeta$. We conclude that E is at most countable. However, this is a contradiction because S is uncountable and $F \circ H$ is one-to-one in a neighborhood of (t_1', t_2'). Therefore $J_H \equiv 0$.

Given $t \in U$ the equation $J_H(t, t) = 0$ can be written

$$a'(t)a(t)^2 b'(t) = a'(t)a(t)[a'(t)b(t) + b'(t)]. \tag{10}$$

If $a'(t) \neq 0$, it follows from (10) that $a(t) \not\equiv 1$ and

$$\frac{d}{dt}[b(t)/(1 - a(t))] = 0.$$

Therefore $b(t)/[1 - a(t)]$ is independent of t while $a(t)$ is nonconstant. The corresponding linear polynomials $a(t)z + b(t)$ are all contained in the stabilizer of $b(t)/[1 - a(t)]$. As before this yields a contradiction because the set of these linear polynomials should be at most countable by Lemma 6. Hence $a'(t) = 0$, so that $\phi_t{}' = a(t)$ is constant with respect to t. Moreover, the constant value of $\phi_t{}'$ is different from one because all linear polynomials with leading coefficient one commute, and consequently an uncountable set of such polynomials could not satisfy the hypotheses of Lemma 6.

Given $t \in U$ it follows from the definition of M that

$$\frac{\partial A}{\partial z}(\phi_t, z) = 0 \qquad (z \in \Delta). \tag{11}$$

Since

$$A(\phi_t, z) = \phi_t{}' W_t(z)/2[f'(z)]^3$$

where W_t is the Wronskian of f' and $f' \circ \phi_t$, it follows from (11) that

$$(f')^3 W_t' - 3(f')^2 f'' W_t = 0. \tag{12}$$

If a denotes the constant value of $\phi_t{}'$, (12) can be written

$$\Lambda(b(t), z) = 3\frac{f''(z)}{f'(z)} \tag{13}$$

where

$$\Lambda(b, z) = \frac{a^2 f'(z) f'''(az + b) - f'''(z) f'(az + b)}{a f'(z) f''(az + b) - f''(z) f'(az + b)}$$

For a fixed $z \in \Delta$ the function $b \mapsto \Lambda(b, z)$ is holomorphic and assumes by (13) a constant value for $b = b(t)$. Since the function $t \mapsto b(t)$ is holomorphic and nonconstant, we conclude by analytic continuation that

$$\Lambda(b, z) = 3\frac{f''(z)}{f'(z)}$$

for each $b \in \mathbf{C}$ such that $az + b \in D$. The substitution $b = z - az$ yields

$$(a^2 - 1)f'''f' = 3(a-1)(f'')^2$$

and, because $a \neq 1$,

$$(a+1)f'''f' = 3(f'')^2.$$

Every nonconstant f' satisfying this differential equation is of the form

$$f' = \begin{cases} \sigma^\kappa, & \text{if } a \neq 2, \\ e^\sigma, & \text{if } a = 2, \end{cases}$$

where $\sigma \in \mathrm{Aff}(\mathbf{C})$ and $\kappa = (a+1)/(a-2)$.

Let us first consider the case $a \neq 2$. If $f' = \sigma^\kappa$ where $\sigma \in \mathrm{Aff}(\mathbf{C})$, there exists $\tau \in \mathrm{Aff}(\mathbf{C})$ such that either $f = \tau(\log \sigma)$ or $f = \tau(\sigma^{\kappa+1})$. Thus either f is the restriction of a polynomial or the analytic continuation of f has exactly one pole or branch point at $\zeta = \sigma^{-1}(0)$.

From (8) we see that f cannot be the restriction of a polynomial because the degrees of the left and right hand sides of (8) would not agree. Hence the analytic continuation of the left hand side of (8) has a singularity at ${\phi_t}^{-1}(\zeta)$. However, the only possible singularity of the right hand side occurs at ζ, and we conclude that ${\phi_t}^{-1}(\zeta) = \zeta$ for each $t \in U$. Hence $\phi_t \in \Gamma_\zeta$ for each $t \in U$ which again leads to a contradiction with Lemma6.

It follows that $a = 2$, so that f is conjugate to e^z. This completes the proof of Theorem 1.

Theorem 1 deals with linearizations of quadratic polynomials. It is natural to ask whether a similar result holds for cubic polynomials or more generally, for polynomials of any degree $n \geq 3$. For entire functions a positive answer is given by the following theorem.

Theorem 2. *Suppose that f is a nonconstant entire function which linearizes uncountably many nonlinear polynomials. Then f is conjugate to the exponential function e^z.*

Proof. Let L be the abelian group of translations of $\mathbf{C}$. By Lemma 6 at most countably many elements of L can be contained in $\Phi(f, \mathbf{C}, P)$ for some nonlinear $P \in \Pi(f)$. Since by hypothesis $\Phi(f, \mathbf{C}, P) \neq \emptyset$ for uncountably many nonlinear polynomials P, there exist a nonlinear polynomial $P \in \Pi(f)$ and a corresponding affine transformation $\phi \in \Phi(f, \mathbf{C}, P)$ such that $\phi \notin L$, i.e. $\phi \in \Gamma_\zeta$ for some $\zeta \in \mathbf{C}$.

The first step is to show that P has exactly one critical point and that this critical point is a Picard exceptional value for f. Since $\phi \in \Phi(f, \mathbf{C}, P)$, we have

$$f'(\phi(z))\phi' = P'(f(z))f'(z) \qquad (z \in \mathbf{C}). \tag{14}$$

By Lemma 8 the left hand side of this equation is nonzero for each $z \in \mathbf{C}$. Therefore $P'(f(z)) \neq 0$ for each $z \in \mathbf{C}$, and we conclude that the image of f does not contain any critical points of P. Thus in view of Picard's theorem P has exactly one critical pointω, and $\omega \neq f(z)$ for each $z \in \mathbf{C}$. In particular, $\omega \neq f(\phi(z)) = P(f(z))$ for each $z \in \mathbf{C}$, and it follows that the polynomial $P - \omega$ cannot have a root in the image of f. Hence $P(\omega) = \omega$, and because ω is the only critical point of P, P is of the form

$$P(w) = A(w - \omega)^n + \omega$$

where $A \in \mathbf{C}^*$.

Let $g = f - \omega$; then g is conjugate to f and satisfies the functional equation

$$g(\phi(z)) = P(f(z)) - \omega = Ag(z)^n \qquad (z \in \mathbf{C}). \tag{15}$$

Moreover, $g(z) \neq 0$ for each $z \in \mathbf{C}$. By dividing (14) by (15) we see that

$$\frac{g'(\phi(z))}{g(\phi(z))}\phi' = n\,\frac{g'(z)}{g(z)},$$

and the substitution $z = \zeta$ yields $\phi' = n$. Thus g'/g is automorphic with respect to the subgroup of Γ_ζ generated by ϕ. However, this is possible only if either g'/g is a constant or the order of ϕ is finite.

If ϕ is of finite order k, iteration of the functional equation (15) shows that

$$g(z) = g(\phi^{\circ k}(z)) = A^m g(z)^{n^k} \qquad (z \in \mathbf{C}),$$

where $m = (n^k - 1)/(n - 1)$. This is impossible because g is nonconstant. We conclude that g'/g is a nonzero constant, so that g is conjugate to the exponential function. This completes the proof of Theorem 2.

It is tempting to conjecture that Theorem 2 holds also for functions defined in a half plane. In order to generalize the proof of Theorem 1 for linearizations of polynomials of degree $n \geq 3$, the linear system of equations (3)–(5) should be replaced by a corresponding system of $n+1$ equations. During the Luleå congress, Larry Lambe kindly used the

Axiom computer algebra system to obtain the solution of the resulting equations in the cases $n = 3$ and $n = 4$. The analysis of these solutions is still unfinished.

References

[1] A. F. Beardon, *Conformal automorphisms of plane domains*, J. London Math. Soc.(2) 32 (1985), 245–253.
[2] E. M. Chirka, *Complex Analytic Sets*, Kluwer 1989.
[3] T. Erkama, *Möbius automorphisms of plane domains*, Ann. Acad. Sci. Fenn. Ser.AI Math. 10 (1985), 155–162.
[4] P. Fatou, *Sur les fonctions qui admettent plusieurs théorèmes de multiplication*, C.R. Acad. Sci. Paris Sér.I Math. 173 (1921), 571–573.
[5] E. Jacobsthal, *Über vertauschbare Polynome*, Math. Z. 63 (1955), 243–276.
[6] G. Julia, *Mémoire sur la permutabilité des fractions rationnelles*, Ann. Sci. École Norm. Sup. 39 (1922), 131–215.

From High-School Didactics to Curved Manifolds

JUHANI FISKAALI AND HEIKKI HAAHTI

University of Oulu
Department of Mathematics
90570-SF Linnanmaa

ABSTRACT. Geometries induced by commutative algebras are presented in an elementary and selfcontained manner. In the introduction on didactics, the algebras $A(q) = \{x + y\vec{f} \mid x, y \in \mathbf{R}\}$ are obtained as extensions of the reals $\mathbf{R}$ by giving the square $(\vec{f}\,)^2$ of a fixed element $\vec{f}$ prescribed values $(\vec{f}\,)^2 = q$. The equation $i^2 = -1$ has solutions $i(q) \in \mathbf{A}(q)$ if and only if q belongs to the parabola domain $\mathcal{C} : D(x,y) > 0$, where $D(x,y) = -(x + \frac{1}{4}y^2)$ and then $\mathbf{A}(q) \cong \mathbf{C} = \mathbf{A}(-1)$. This conclusion was made already by Hamilton [2]. The equation $j^2 = +1$ has non-real solutions $j(q) \in \mathbf{A}(q)$ if and only if $q \in \mathcal{L} : D(x,y) < 0$. Then $\mathbf{A}(q) \cong \mathbf{L} = \mathbf{A}(+1)$, $\mathbf{L}$ being the algebra of hyperbolic numbers.

The domains $\mathcal{C}$ and $\mathcal{I} = \{i(q) \mid q \in \mathcal{C}\}$ considered in the first two parts become, due to the algebras $\mathbf{A}(q)$, Riemannian manifolds with almost complex structures. Globally given diffeomorphisms are constructed, which transform the almost complex structures to the ordinary complex ones. The curvatures are determined; in particular, $\mathcal{I}$ appears to be an isometric representation of the Poincaré half plane.

Analogously, the domains $\mathcal{L}$ and $\mathcal{J} = \{j(q) \mid q \in \mathcal{L}\}$ become pseudo-Riemannian manifolds with involutive tensor fields.

In the third part of the paper the procedure to get extension algebras is repeated, the reals $\mathbf{R}$ being replaced by the algebra $\mathbf{A}(q)$, where $q \in \mathcal{D}^2 = \mathcal{C} \cup \mathcal{L}$ is fixed. Four dimensional pseudo-Riemannian manifolds $\mathcal{D}^4 - \mathcal{D}(q)$ with signature 2 arise in this case. For $q \in \mathcal{C}$ the manifolds are topologically non-trivial allowing loops, which change the orientations of time and space.

The extension procedure can be continued to get 2^n-dimensional pseudo-Riemannian manifolds ($n = 3, 4, \ldots$).

0. Introduction (didactics)

0.1. The following way to introduce complex numbers was given as early as 1834 by Hamilton in [3].[1] It is still today didactically useful in elementary courses in function theory. Surprisingly enough we were not able to find it in textbooks although we looked through several.

Instead of assuming directly the existence of a (necessarily non-real) solution to equation

$$i^2 = -1 \tag{1}$$

[1] We are indebted to Pertti Uolevi Hyyrynen for this knowledge.

suppose as a first step only, that there is at least one non-real element, say $\vec{f}$, which will be fixed. Supposing the familiar rules of addition and multiplication to be valid also for non-real elements we have

(I) -to find the smallest rings $\mathbf{A}$ for which $\vec{f} \in \mathbf{A}$, and where the reals $\mathbf{R}$ is a subring in the center of $\mathbf{A}$

and then

(II) - to select those rings, where (1) has a solution.

0.2. A smallest extension $\mathbf{A}$ is necessarily a real vector space $\mathbf{A} = \{x + y\vec{f} \mid x, y \in \mathbf{R}\}$ isomorphic to the vector space $\mathbf{R}^2$ by the correspondence $\mathbf{A} \ni x + y\vec{f} \leftrightarrow (x, y) \in \mathbf{R}^2$; in particular $\mathbf{A}$ can be identified in this way with $\mathbf{R}^2$, whereby $\vec{f}$ is represented by $(0,1)$ and 1 by $\vec{e} = (1,0)$. The sum and the product of $z = x + y\vec{f}$ and $\zeta = \xi + \eta\vec{f}$ are respectively given by

$$(2) \qquad z + \zeta = (x + \xi) + (y + \eta)\vec{f} \qquad \text{and} \qquad z\zeta = x\xi + (x\eta + y\xi)\vec{f} + (y\eta)(\vec{f})^2,$$

where $x, y, \xi, \eta \in \mathbf{R}$. On the other hand every value $p = u + v\vec{f} \in \mathbf{A}$ for $(\vec{f})^2$ gives via (2) a ring $\mathbf{A}$ of type (I). This ring is a real 2-dimensional and commutative algebra as well and will be denoted by $\mathbf{A}(p)$.

0.3. Defining the "discriminant"

$$(3) \qquad D(p) = D(u, v) = -(u + \frac{1}{4}v^2)$$

for the value $p = u + v\vec{f} = (\vec{f}\,)^2$, it follows by (2) that $i = x + y\vec{f}$ with $x, y \in \mathbf{R}$, is a solution of (1) if and only if

$$(1)' \qquad 2x = -yv \quad \text{and} \quad y^2 D(p) = 1.$$

Consequently, we get:

Proposition 1. *Equation (1) has solutions $i \in \mathbf{A}(p)$ if and only if the point $p = (\vec{f}\,)^2$ belongs to the open parabola domain*

$$(4) \qquad \mathcal{C} = \{p \in \mathbf{A} \mid D(p) > 0\}.$$

Then the solutions are

$$(1)'' \qquad i(p) = \frac{1}{\sqrt{D(p)}}(-\frac{v}{2} + \vec{f}\,)$$

and $-i(p)$.

0.4. For two values p and p' from the domain $\mathcal{C}$ the corresponding algebras $\mathbf{A}(p)$ and $\mathbf{A}(p')$ are isomorphic by the map $\mathbf{A} \ni x + yi(p) \to x + yi(p') \in \mathbf{A}(p')$. One chosen extension is $\mathbf{A}(-1) = \mathbf{C}$, the familiar representation of the complex numbers corresponding to the focal point $p = -1$ of the parabola $\partial\mathcal{C}$. In this algebra we hav thus by $(1)''$, $i(-1) = \vec{f}$ and $-1 = (\vec{f}\,)^2$. When we associate to the basis vector $\vec{f} = (0,1)$ this familiar law of multiplication, we denote $\vec{f}$ by i_o. The Corollary below gives an extra reason for the choice $p = -1$. First, we prove for any $p \in \mathcal{C}$:

Lemma 1. *When $Z = X + Y\vec{f}$ is represented in the basis $\{1, i(p)\}$ as $Z = X' + Y'i(p)$ and consequently its conjugate as $\bar{Z} = X' - Y'i(p)$, then*

$$\bar{Z}Z = (X + \frac{v}{2}Y)^2 + Y^2 D(p). \tag{5}$$

Proof. In coordinates X', Y' we have

$$\bar{Z}Z = \{X' - Y'i(p)\}\{X' + Y'i(p)\} = (X')^2 + (Y')^2. \tag{5$'$}$$

Since $\vec{f} = \sqrt{D(p)}\, i(p) + \frac{1}{2}v$ by $(1)''$, we get $Z = X + Y\vec{f} = \{X + \frac{v}{2}Y\} + Y\sqrt{D(p)}\, i(p)$, so

$$X' = X + \frac{v}{2}Y, \quad Y' = Y\sqrt{D(p)}. \tag{6}$$

By (5) the square of the canonical euclidean norm $|Z| = \sqrt{X^2 + Y^2}$ of $Z = X + Y\vec{f}$ equals $\bar{Z}Z$ if and only if $v = 0$ and $D(p) = 1$. Consequently by (3):

Corollary. *When $p \in \mathcal{C}$, the expression $\{ZZ\}^{\frac{1}{2}}$ defined in the algebra $\mathbf{A}(p)$ coincides with the canonical euclidean norm $|Z|$ if and only if $p = -1$.*

0.5. The question arises of what kind of extensions $\mathbf{A}(p)$ we get when the point $p \in \mathbf{A}$ is not in the domain $\mathcal{C}$. The proof of the following proposition is analogous to that of the former one.

Proposition 2. *The equation*

$$j^2 = +1 \tag{7}$$

has non-real solutions in $\mathbf{A}(p)$ if and only if $p = u + v\vec{f}$ belongs to the open domain

$$\mathcal{L} = \{p \in \mathbf{A} \mid D(p) < 0\}. \tag{8}$$

Then the only such solutions are

$$j(p) = \frac{1}{\sqrt{-D(p)}}(-\frac{1}{2}v + \vec{f}\,) \tag{7$'$}$$

and $-j(p)$.

Indeed, as one can readily see, condition (7) for $j = x + y\vec{f}$ is equivalent with

$$(7)'' \qquad 2x = -yv, \quad y^2 D(p) = -1.$$

Again the algebras $\mathbf{A}(p)$ and $\mathbf{A}(p')$ are isomorphic when here $p, p' \in \mathcal{L}$. In particular for all $p \in \mathcal{L}$ the extension $\mathbf{A}(p)$ is isomorphic to the "*Lorentz-algebra*" $\mathbf{L} = \mathbf{A}(+1)$ of hyperbolic numbers. Here by (7)', $j(+1) = \vec{f}$, this solution of (7) being denoted in the sequel by j_o. For $Z = X + Yj(p) \in \mathbf{A}(p)$ and $\bar{Z} = X - Yj(p)$ the form $\bar{Z}Z$ is the quadratic form corresponding to the canonical Lorentz scalar product in $\mathbf{R}^2$ if and only if $p = +1$.

Finally, the equation $z^2 = 0$ has non-real solutions in $\mathbf{A}(p)$ if and only if p belongs to the parabola $\partial\mathcal{C}$, the algebra $\mathbf{A}(p)$ being isomorphic to the algebra $\mathbf{A}_o = \mathbf{A}(0)$, where $(\vec{f}\ )^2 = 0$.

0.6. From the didactic point of view elementary arithmetic and analysis in the algebras $\mathbf{L}$ and $\mathbf{A}_o$, when compared to those in $\mathbf{C}$, offer students a variety of motivating problems. For instance, proving the following lemma might be a suitable exercise in the case $p = 1$.

Lemma 2. *For $p \in \mathcal{L}$ and $j = j(p)$ as above, the elements*

$$(8)' \qquad \vec{e}_o = \frac{1}{2}(1 - j), \quad \vec{e}_1 = \frac{1}{2}(1 + j)$$

define a basis of $\mathbf{A}(p)$ such that

$$(8)'' \qquad (\vec{e}_o\ )^2 = \vec{e}_o, \quad (\vec{e}_1\ )^2 = \vec{e}_1 \quad \text{and} \quad \vec{e}_o\vec{e}_1 = 0.$$

In particular, in this basis the multiplication in $\mathbf{A}(p)$ takes place coordinatewise, so for $z = z_o\vec{e}_o + z_1\vec{e}_1, w = w_o\vec{e}_o + w_1\vec{e}_1$ with real coordinates,

$$(8)''' \qquad zw = (z_o w_o)\vec{e}_o + (z_1 w_1)\vec{e}_1.$$

Corollary 1. *The product of two factors is zero, $zw = 0$, if and only if one factor is a real multiple of $\vec{e}_o$ and the other one that of $\vec{e}_1$.*

Corollary 2. *There exist zero divisors in the algebra $\mathbf{L}$.*

The picture below shows the domains $\mathcal{C}$ and $\mathcal{L} \subset \mathbf{R}^2$ and the focal point $p = -1$ of the parabola $\partial\mathcal{C}$.

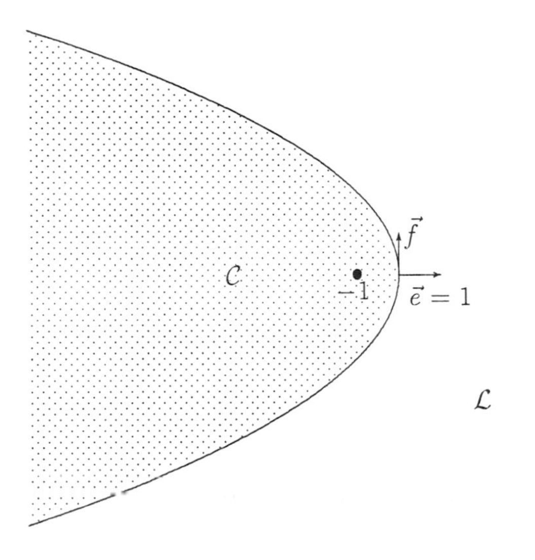

0.7. To prepare the geometric considerations to follow, we prove:

Lemma 3. *The mapping* $i : \mathcal{C} \ni p = u + v\vec{f} \to i(p) = x + y\vec{f}$ *defined by* $(1)''$ *is the only smooth function on* $\mathcal{C}$ *which satisfies* $i(-1) = \vec{f}$ *and whose value* $i(p)$ *solves (1) in* $\mathbf{A}(p)$. *Furthermore, the function* i *is a diffeomorphism from* $\mathcal{C}$ *onto the upper half plane*

$$\mathcal{I} = \{i(p) \mid p \in \mathcal{C}\} = \{x + y\vec{f} \mid x \in \mathbf{R}, y > 0\}. \tag{9}$$

Similarly equation $(7)'$ *defines the unique smooth function* j *on* $\mathcal{L}$ *with* $j(+1) = \vec{f}$, *whose value* $j(p)$ *solves (7) in* $\mathbf{A}(p)$. *The mapping* $-j$ *is a diffeomorphism from* $\mathcal{L}$ *onto the lower half plane*

$$\mathcal{J} = \{-j(p) \mid p \in \mathcal{L}\} = \{x + y\vec{f} \mid x \in \mathbf{R}, y < 0\}. \tag{10}$$

Proof. The mapping i defined by $(1)''$ is smooth, $i(-1) = \vec{f}$ and $i(p)$ solves (1) in $\mathbf{A}(p)$. Furthermore, since the y-coordinate of $z = i(p)$ is positive, $i(p) \in \mathcal{I}$. The inverse map of i appears to be

$$z = x + y\vec{f} \to -\frac{1}{y^2}\{(1 + x^2) + (2xy)\vec{f}\} = p(z) = u + v\vec{f}, \tag{9$'$}$$

so

$$u = -\frac{1 + x^2}{y^2}, \qquad v = -2\frac{x}{y}. \tag{9$''$}$$

To show the claimed uniqueness of the mapping i, suppose on the domain $\mathcal{C}$ there is defined a smooth function (denoted again by i) with the two given properties. So especially its y-coordinate satisfies $y(-1) = 1$. Given $p \in \mathcal{C}$, at every point $w(t) = u(t) + v(t)\vec{f}$ of a smooth curve $\ell : [0,1] \ni t \to w(t) \in \mathcal{C}$ from $-1 = w(0)$ to $p = w(1)$, the value $i(w(t))$ solves (1) and consequently its coordinates solve (1)'. In particular, the y-coordinate can't vanish. Differentiation of the second equation (1)' with respect to t gives

$$\frac{dy}{y} = -\frac{1}{2}\frac{dD}{D}, \tag{11}$$

where $D = D(w(t))$ is defined by (3) and $dD = -\{du + \frac{1}{2}vdv\}$. Now, since the domain $\mathcal{C}$ is simply connected, the y-coordinate necessarily has the unique value

$$y(p) = \exp\{-\frac{1}{2}\int_\ell \frac{dD}{D}\}, \tag{11)'$$

that is $y(p) = \frac{1}{\sqrt{D(p)}}$. Then by (1)' one gets the x-coordinate and consequently (1)'' gives the only possible value for $i(p)$. The latter part of the lemma can be proved analogously. In particular, the y-coordinate of the function j is given by (11)', where in this case ℓ is a curve in $\mathcal{L}$ from $+1$ to a given $p \in \mathcal{L}$.

1. Two-dimensional manifolds

1.1. As an open set in $\mathbf{A} \cong \mathbf{R}^2$ the parabola domain $\mathcal{C}$ given by (4) is a differentiable manifold. We have seen so far that to every value $p \in \mathcal{C}$ for the square $(\vec{f}\,)^2$ there is associated an algebra $\mathbf{A}(p)$ isomorphic to the complex numbers. It might be appropriate to regard the elements of $\mathbf{A}(p)$ as ones associated with p not merely algebraically but also geometrically. This is done by considering the elements $Z \in \mathbf{A}(p)$ as **tangent vectors** emerging from p. As is well known, the tangent space $\mathcal{C}_p$ at the point $p \in \mathcal{C}$ can be naturally identified with the vector space $\mathbf{R}^2 \cong \mathbf{A}(p)$ by letting the element $Z = X + Y\vec{f} \in \mathbf{A}(p)$ represent the tangent vector $\overrightarrow{p(p+Z)}$, (which again can be regarded as the differential operator $X\frac{\partial}{\partial u} + Y\frac{\partial}{\partial v}$ at $p = u + v\vec{f}$).

On the other hand, instead of defining $\mathbf{A}(p)$ by giving the value $p = (\vec{f}\,)^2$, the algebra is determined also by giving $z = i(p)$. Thus we may identify the algebra also with the tangent space $\mathcal{I}_z$ of the manifold $\mathcal{I}$ given by (9) regarding the element $Z \in \mathbf{A}(p)$ as the tangent vector $\overrightarrow{z(z+Z)}$.

1.2. The identifications above give $\mathcal{C}$ and $\mathcal{I}$ the structure of two-dimensional Riemannian manifolds by means of the scalar product

$$\langle Z, W\rangle_w = \Re\bar{Z}W, \tag{12}$$

where $w = p \in \mathcal{C}$ or $w = z = i(p) \in \mathcal{I}$ and $\Re\{\xi + \eta i(p)\} = \xi$ for all $\xi, \eta \in \mathbf{R}$. Here the conjugate $\bar{Z}$ of Z and the product $\bar{Z}W$ are given by the algebra $\mathbf{A}(p)$, the former as in lemma 1, and the elements Z and $W \in \mathbf{A}(p)$ are identified with the tangent vectors from $\mathcal{C}_p$ and from $\mathcal{I}_z$ respectively. The positive definiteness follows by (5)'. When these Riemannian metric tensors on $\mathcal{C}$ and $\mathcal{I}$ are denoted by G and H respectively, $G(p)$ and $H(z)$ are scalar products characterized by the fact that the tangent vector pair $\{1, i(p)\}$ is orthonormal. As in lemma 1, we get explicitly

$$(13) \qquad G(p)(Z,W) = (X + \frac{v}{2}Y)(U + \frac{v}{2}V) + D(p)YV$$

and by (9')

$$(13)' \qquad H(z)(Z,W) = \frac{1}{y^2}\{(yX - xY)(yU - xV) + (\operatorname{sign} y)YV\},$$

where $p = u+v\vec{f}$ and $z = i(p) = x+y\vec{f}$ are points on manifolds $\mathcal{C}$ and $\mathcal{I}$ respectively, $Z = X + Y\vec{f}$, $W = U + V\vec{f}$ are tangent vectors and $D(p) = -(u + \frac{1}{4}v^2)$. Here $\operatorname{sign} y = +1$ and later on, when $z \in \mathcal{J}$, $\operatorname{sign} y = -1$.

1.3. The tangent space $\mathcal{C}_p$ being identified with the algebra $\mathbf{A}(p)$ the multiplication of tangent vectors by $i(p)$ defines a smooth mixed tensor field I of degree two on $\mathcal{C}$ by $I(p)Z = i(p)Z$, where $p \in \mathcal{C}$ and $Z \in \mathcal{C}_p$. Hence for each $p \in \mathcal{C}$ we get a linear endomorphism $I(p) \in Hom_{\mathbf{R}}(\mathcal{C}_p; \mathcal{C}_p)$ with the property $\{I(p)\}^2 = -E(p)$, where $E(p)$ denotes the identity operator on the tangent space $\mathcal{C}_p$. This formula characterizes I as an **almost complex structure** on $\mathcal{C}$ and $(\mathcal{C}, I)$ as an almost complex manifold.

Besides $\mathcal{C}$, the manifold $\mathcal{I}$ gets an almost complex structure when the tangent space $\mathcal{I}_z$ at $z = i(p)$ is identified with the algebra $\mathbf{A}(p)$. In fact, in this case we get by (9)' explicitly

$$(14) \qquad I(z)Z = (xX - \frac{1+x^2}{y}Y) + (yX - xY)\vec{f},$$

where $z = x + y\vec{f} \in \mathcal{I}$ and $Z = X + Y\vec{f} \in \mathcal{I}_z$.

1.4. As for $\mathcal{C}$ and $\mathcal{I}$ above, the field of algebras $\{\mathbf{A}(p)\}$ gives for the manifolds $\mathcal{L}$ and $\mathcal{J}$ defined by (8) and (10), respectively, a metric structure. It is in this case a pseudo-Riemannian one. In fact, for $p \in \mathcal{L}$ and $z = -j(p) \in \mathcal{J}$, the tangent spaces $\mathcal{L}_p$ and $\mathcal{J}_z$ are identified with the algebra $\mathbf{A}(p)$, isomorphic to the algebra $\mathbf{L}$ of hyperbolic numbers. The scalar product at $p \in \mathcal{L}$ is introduced as the real part $G(p)(Z,W) = \Re\{\bar{Z}W\}$ of the product $\bar{Z}W \in \mathbf{A}(p)$, where $Z = X' + Y'j(p) = X + Y\vec{f}, \bar{Z} = X' - Y'j(p), W = U' + V'j(p) = U + V\vec{f}$ and "taking the real part" is defined by $\Re\{\xi + \eta j(p)\} = \xi$ for $\xi, \eta \in \mathbf{R}$. Since $\{j(p)\}^2 = +1$, we get $G(p)(Z,W) = X'U' - Y'V'$. Thus the scalar product is indefinite and consequently the manifold $(\mathcal{L}, G)$ is pseudo-Riemannian. Similarly, as in the proof of lemma 1 we get formula (5) and furthermore again expression (13). The indefiniteness of the scalar product is indicated in (5) by the inequality $D(p) < 0$ which characterizes the domain $\mathcal{L}$. In the same way, when the indefinite scalar product above is defined for tangent vectors at the point $z = -j(p) \in \mathcal{J}$, the lower half plane $\mathcal{I}$ becomes a pseudo- Riemannian manifold $(\mathcal{J}, H)$. Here the metric H is given by formula

(13)′, where in this case $z = x + y\vec{f} \in \mathcal{J}$, so $y < 0, \operatorname{sign} y = -1$ and $Z = X + Y\vec{f}$, $W = U + V\vec{f} \in \mathcal{J}_z$.

The multiplication of tangent vectors by $j(p)$ in the algebra $\mathbf{A}(p)$ defines a mixed tensor field J of degree two on $\mathcal{L}$ by $J(p)Z = j(p)Z$, where $p \in \mathcal{L}$ and $Z \in \mathcal{L}_p$. By (7) J is involutive: $\{J(p)\}^2 = E(p)$. A similar tensor field is obtained also on the manifold $\mathcal{J}$.

1.5. Recall that among all smooth real manifolds, a complex analytic one, $\mathcal{M}$ with the complex dimension n is characterized by an atlas, where each parametric transformation $\phi = \alpha^{-1} \circ \alpha'$ corresponding to two parametric representations α and α' is a local holomorphic map $\mathbf{C}^n \to \mathbf{C}^n$. This means that its derivative at the parametric point $z \in \mathbf{C}^n$, the $\mathbf{R}$–linear map $d\phi = d\phi(z) = (d\alpha)^{-1} \circ d\alpha' : \mathbf{C}^n \to \mathbf{C}^n$, commutes with $i_o = i(-1) \in \mathbf{C} = \mathbf{A}(-1)$, so

$$(d\phi)i_o = i_o(d\phi). \tag{15}$$

Thus $d\phi$ is not merely $\mathbf{R}$–linear but also $\mathbf{C}$–linear. Suppose that a tangent vector $Z = (d\alpha)h = (d\alpha')h'$ at $p = \alpha(z) = \alpha'(z') \in \mathcal{M}$ is represented at the parametric points z and z' by the $\mathbf{C}^n$–vectors h and h', respectively. Then $(d\alpha)h = (d\alpha')h' = d\alpha \circ \{(d\alpha)^{-1} \circ d\alpha'\}h' = d\alpha \circ (d\phi)h'$ yields the contravariant transformation law $h = (d\phi)h'$ of the representatives. The multiplication of h by i_o gives an $\mathbf{R}$-linear endomorphism $I(p) \in Hom_{\mathbf{R}}(\mathcal{M}_p; \mathcal{M}_p) : I(p)Z = (d\alpha)i_o h$, for $Z = (d\alpha)h$. By (15) this endomorphism does not depend on the choice of the chart used to define it. Namely, we have $i_o h = i_o(d\phi)h' = (d\phi)i_o h' = \{(d\alpha)^{-1} \circ d\alpha'\}i_o h'$, so $(d\alpha)i_o h = (d\alpha')\, i_o h'$. Since $\{I(p)\}^2 = -E(p)$, I defines an almost complex structure on $\mathcal{M}$. The definition of the tensor field I above can be expressed equivalently by

$$I(p)d\alpha = (d\alpha)\, i_o, \tag{16}$$

where $\mathbf{R}$–linear mappings operate on $\mathbf{C}^n$, $p = \alpha(z) \in \mathcal{M}$, $d\alpha = (d\alpha)(z) \in Hom_{\mathbf{R}}(\mathbf{C}^n; \mathcal{M}_p)$ and $z \in \mathbf{C}^n$.

Conversely, when $\mathcal{M}$ is a real smooth manifold having the real dimension $2n$, the n–dimensional complex vector space $\mathbf{C}^n$ – being also a $2n$–dimensional real vector space– can serve as the parametric space. An almost complex structure I on $\mathcal{M}$ is called **integrable** (or equivalently a complex structure)[2] if there exists such an atlas of $\mathcal{M}$ that in all its charts I is represented by the constant multiplier i_o in the sence of (16). Then the parametric transformations of the atlas are in fact holomorphic, the manifold $\mathcal{M}$ being thus a n–dimensional complex manifold. Indeed, since both α and α' satisfy (16), then for $\phi = \alpha^{-1} \circ \alpha'$, $(d\phi)i_o = (d\alpha)^{-1}\, d\alpha'\, i_o = (d\alpha)^{-1}\, I\, d\alpha' = i_o\, (d\alpha)^{-1}\, d\alpha'$, so $d\phi i_o = i_o d\phi$.[3]

If above, instead of $\mathbf{C}^n$, a real vector space V with $\dim_{\mathbf{R}} V = 2n$ is used as the parametric space of $\mathcal{M}$, then (16) can be replaced by

$$I(p)d\alpha = (d\alpha)I_o, \tag{16'}$$

[2] The vanishing of the **Nijenhuis tensor** [5, p.112], is a well known **tensorial condition** for the integrability of an almost complex structure.

[3] The smoothness condition above for the real $2n$-dimensional manifold $\mathcal{M}$ is not necessary and in fact, in our case it is sufficient to deal merely with C^1 manifolds $\mathcal{M}$. The commutativity $d\phi i_o = i_o d\phi$ implied by (16) expresses the Cauchy-Riemann equations and implies by the generalized Cauchy formula [4, p.28] the smoothness of ϕ.

where $I_o \in Hom_{\mathbf{R}}(V;V)$ and $-(I_o)^2$ is the identity on V.[4]

1.6. We will solve α from equation (16) in case $(\mathcal{C}, I)$ is the 2-dimensional manifold defined in section 1.3. Formula (16) can be read equivalently as

$$(16)'' \qquad I(p)(d\alpha)h = (d\alpha)(i_o h) \quad \forall h \in \mathbf{C}_z \cong \mathbf{C}.$$

Now, because the real dimension of $\mathbf{C}$ is two, we have:

Lemma 4. *If the equality in* $(16)''$ *holds for one non-zero vector* $h = h_o \neq 0$, *it holds for all* h, *and (16) is thus valid.*

Proof. The equation $I(p)(d\alpha)h_o = (d\alpha)i_o h_o$ gives, when multiplied by $I(p)$: $-d\alpha h_o = I(p)(d\alpha)(i_o h_o)$. Denoting $k_o = ih_o$, whereby $-(d\alpha)h_o = (d\alpha)(-h_o) = (d\alpha)i_o k_o$, it follows that $(16)''$ holds for $h = k_o$. Both sides of $(16)''$ being linear expressions of h, the lemma follows, since $\{h_o, k_o\}$ is a real basis of $\mathbf{C} = \mathbf{C}_z$.

To get $(16)''$ in form of partial differential equations, note that on the left hand side $I(p)(d\alpha)h = i(p)(d\alpha)h$ is the product of $i(p)$ and $Z = (d\alpha)h$ in the algebra $\mathbf{A}(p) = \mathcal{C}_p$, $i(p)$ being given by $(1)'$ for $p = \alpha(z) = u(z) + v(z)\vec{f}$. Furthermore using the subscript notation for partial derivatives we have for $h = h_1 + h_2\vec{f} = h_1\vec{e} + h_2\vec{f}$: $(d\alpha)h = h_1\alpha_x + h_2\alpha_y$.

Now by lemma 4 it is sufficient to solve $(16)''$ for α by choosing $h = \vec{e} = 1$. Then on the left hand side we get, remembering that $(\vec{f}\,)^2 = p = u + v\vec{f}$,

$$\begin{aligned} I(p)(d\alpha)\vec{e} = i(p)\alpha_x &= \frac{1}{\sqrt{D(p)}}(-\frac{v}{2} + \vec{f}\,)(u_x + uv_x\vec{f}\,) \\ &= \frac{1}{\sqrt{D(p)}}\{(-\frac{v}{2}u_x + uv_x) + (\frac{v}{2}v_x + u_x)\vec{f}\}. \end{aligned}$$

On the right hand side $i_o\vec{e} = \vec{f}$, so $(d\alpha)(i_o\vec{e}\,) = \alpha_y = u_y + v_y\vec{f}$. Thus (16) gets the equivalent form

$$(16)''' \qquad \begin{cases} u_y = \frac{1}{\sqrt{D(p)}}(-\frac{v}{2}u_x + uv_x), \\ v_y = \frac{1}{\sqrt{D(p)}}(\frac{v}{2}v_x + u_x), \end{cases}$$

where $D(p) = -\{u + \frac{1}{4}v^2\}$. These partial differential equations for u and v reduce at $p = -1$ to those of Cauchy-Riemann. By $(16)'''$ $u_y^2 + uv_y^2 = -(u_x^2 + uv_x^2)$, yielding

$$(17) \qquad u_x^2 + u_y^2 = -u(v_x^2 + v_y^2).$$

We have quite a lot freedom in selecting a solution. Namely, as was seen in section 1.5, if α is a solution to (16) and thus also to $(16)'''$, so is $\alpha' = \alpha \circ \phi$ for all holomorphic ϕ, too. When trying polynomial solutions u and v, v of degree one,

[4] This is seen when introducing the complex multipliers in V by $\gamma v = \alpha v + \beta I_o v$ for $\gamma = \alpha + \beta i_o \in \mathbf{C}$ to get a vector space $V_{\mathbf{C}} \cong \mathbf{C}^n$ over $\mathbf{C}$, where the multiplication by i_o corresponds to the operation by I_o on V.

u will be of degree two by (17). Demanding moreover that $\alpha = u + v\vec{f}$ maps the upper half plane bijectively onto the domain $\mathcal{C}$ and the border $\{z = x + yi_o \mid y = 0\}$ into the border $\partial\mathcal{C}$, we find the solution

$$\alpha(z) = -(x^2 + y^2) + (2x)\vec{f}. \tag{18}$$

When considering similarly the manifold $(\mathcal{I}, I)$, equation (16) reduces by (14) to partial differential equations

$$\begin{cases} u_y = uu_x - \frac{1+u^2}{v} v_x \\ v_y = vu_x - uv_x. \end{cases} \tag{19}$$

One solution β to (16) from upper half plane onto itself is then

$$\beta(z) = -\frac{x}{y} + \frac{1}{y} i_o. \tag{20}$$

1.7. The pull-back $g = \alpha^* G$ of the Riemannian metric G is the metric given by

$$g(z)(h, k) = G(p)(Z, W), \tag{21}$$

where h, k are tangent vectors at z and $Z = (d\alpha)h, W = (d\alpha)k$ those at $p = \alpha(z)$. We will calculate it using the lemma below, where $g_o = \langle \cdot, \cdot \rangle$ denotes the standard euclidean scalar product on $\mathbf{C} = \mathbf{R}^2$.

Lemma 5. *Any local solution $\alpha : \mathbf{C} \to \mathcal{C}$ to (16) defined in an open domain $\mathcal{U} \subset \mathbf{C}$ is a conformal map $(\mathcal{U}, g_o) \to (\mathcal{C}, G)$ that is, $\alpha^* G = \lambda g_o$, where λ is a positive function.*[5]

Proof. The vectors h and $i_o h$ are g_o-orthogonal. Since the real dimension of the tangent space is two, the vectors h, k at $z \in \mathcal{U}$ are orthogonal if and only if $\{k, i_o h\}$ are linearly dependent. In the same way the vectors Z and W at $p \in \mathcal{C}$ are $G(p)$-orthogonal if and only if $\{W, I(p)Z\}$ are linearly dependent. Suppose now that (h, k) is a g_o-orthogonal pair at z and $h \neq 0$. Thus $k = \mu i_o h$ for some $\mu \in \mathbf{R}$. For the corresponding tangent vectors $Z = (d\alpha)h$ and $W = (d\alpha)k$ at p, then $W = (d\alpha)(\mu i_o h) = \mu(d\alpha)(i_o h) = \mu I(p)(d\alpha)h = \mu I(p)Z$, so (W, Z) is $G(p)$-orthogonal. By (21) this means that (h, k) is orthogonal with respect to the pull-back $g(z)$, too. Thus the g_o-orthogonality of any vector pair (h, k) implies its $g(z)$-orthogonality. By a simple argument one concludes that the latter scalar product is a scalar multiple of the former, i.e. $g(z)(h, k) = \lambda(z)\langle h, k \rangle$ for all vectors h, k at z. Thus by (21), $\alpha^* G = \lambda g_o$.

Now, in our case where α is given by (18), we choose at z the unit length vector $h_o = \vec{f} = \frac{\partial}{\partial y}$. Then by (18) the corresponding vector $Z_o = (d\alpha)h_o$ at $p = \alpha(z)$ is the real element $Z_o = \alpha_y = -2y = \bar{Z}_o$ in the algebra $\mathbf{A}(p) = \mathcal{C}_p$. It follows by (12) that in the lemma now, $\lambda(z) = \lambda(z)\langle h_o, h_o \rangle = g(z)(h_o, h_o) = G(p)(Z_o, Z_o) = \bar{Z}_o Z_o = 4y^2$.

In the same way, choosing $h_o = 1 = \vec{e}$, one gets immediately $\beta^* H = \frac{1}{y^2} g_o$.

---

[5] In fact the conformality of α and the condition (16) are equivalent conditions. Since on the other hand the conformality of α means that α^{-1} as a map:$(\mathcal{C}, g_o) \to (\mathcal{U}, g_o)$ is a quasiconformal map with a given dilatation. One can conclude [7] that solutions α exist under weak smoothness condition for any almost complex sructure I, the latter being thus always integrable in the two-dimensional case.

1.8. The pseudo-Riemannian manifolds $\mathcal{L}$ and $\mathcal{J}$ given in section 1.4, can be treated similarly as the previous Riemannian ones. In this case the involutive tensor field J is analogous to the almost complex structure I above. In particular, J can be transformed to the constant multiplier j_o given in the algebra $\mathbf{L} = \mathbf{A}(-1)$ of hyperbolic numbers.

To formulate our results denote by $\mathcal{H}^+ = \{z = x + y\vec{f} \mid x \in \mathbf{R}, y > 0\}$ and $\mathcal{H}^- = \{z = x + y\vec{f} \mid x \in \mathbf{R}, y < 0\}$ the upper and lower half plane, respectively. Consider the conformally flat metric tensors g and h given for the tangent vectors $Z = X + Y\vec{f}$ and $W = U + V\vec{f}$ at $z = x + y\vec{f} \in \mathcal{H}^+ \cup \mathcal{H}^-$, by

$$g(z)(Z, W) = 4y^2\langle Z, W\rangle \tag{22}$$

and

$$h(z)(Z, W) = \frac{1}{y^2}\langle Z, W\rangle. \tag{23}$$

Here at $z \in \mathcal{H}^+, \langle Z, W\rangle = XU + YV$ is the euclidean scalar product, whereas at $z \in \mathcal{H}^-$, $\langle Z, W\rangle = XU - YV$ is the standard Lorentzian scalar product.

Define the transformations α and β :

$$\alpha(z) = \begin{cases} -(x^2 + y^2) + 2x\vec{f}, & \text{for } z = x + y\vec{f} \in \mathcal{H}^+ \\ x^2 - y^2 + 2x\vec{f}, & \text{for } z = x + y\vec{f} \in \mathcal{H}^- \end{cases} \tag{24}$$

and

$$\beta(z) = -\frac{x}{y} + \frac{1}{y}\vec{f}, \quad \text{for } z = x + y\vec{f} \in \mathcal{H}^+ \cup \mathcal{H}^-. \tag{25}$$

Then, denoting the Riemannian and pseudo-Riemannian manifolds $(\mathcal{C}, G)$, $(\mathcal{L}, G)$, $(\mathcal{I}, H)$ and $(\mathcal{J}, H)$ by $\mathcal{C}, \mathcal{L}, \mathcal{I}$ and $\mathcal{J}$ respectively,we get:

Proposition 3. *The transformation α is an isometry from $(\mathcal{H}^+, g)$ onto $\mathcal{C}$ and from $(\mathcal{H}^-, g)$ onto $\mathcal{L}$, when restricted to $\mathcal{H}^+$ and $\mathcal{H}^-$ respectively.*

Furthermore α transforms the almost complex structure I of $\mathcal{C}$ to the ordinary complex structure given on the complex plane $\mathbf{C} - \mathbf{A}(-1)$ by i_o. It also transforms the tensor field J on $\mathcal{L}$ to the constant multiplier j_o given in the algebra $\mathbf{L} = \mathbf{A}(+1)$ of the hyperbolic numbers.

The transformation β is an isometry from $(\mathcal{H}^+, h)$ onto $\mathcal{I}$ and from $(\mathcal{H}^-, h)$ onto $\mathcal{J}$, when resricted to $\mathcal{H}^+$ and $\mathcal{H}^-$ respectively.

Furthermore β transforms the almost complex structure I on $\mathcal{I}$ to the ordinary one i_o and the tensor field J on $\mathcal{J}$ to the constant multiplier j_o.

Now, the Gaussian curvature $\kappa = \kappa(z)$ of the conformally flat metric $g = \lambda g_o$ is $\kappa = -\frac{1}{2\lambda}(\mu_{xx} + \mu_{yy})$, $[\kappa = -\frac{1}{2\lambda}(\mu_{xx} - \mu_{yy})]$, where $\lambda = \lambda(z), \mu = \log\lambda$ and $g_o = \langle\ ,\ \rangle$ is the standard euclidean [Lorentzian] scalar product of $\mathbf{R}^2$ [1, p.112]

Corollary 1. *The Gaussian curvature κ of the Riemannian manifold $(\mathcal{C}, G)$ at the point $p = u + v\vec{f} \in \mathcal{C}$ is*

$$\kappa(p) = \frac{1}{4\{D(p)\}^2} = \frac{1}{4y^4}, \tag{26}$$

where $z = \alpha^{-1}(p) = x + y\vec{f} \in \mathcal{H}^+$. Similarly for the pseudo-Riemannian manifold $(\mathcal{L}, G)$ we have

$$\kappa(p) = -\frac{1}{4\{D(p)\}^2} = -\frac{1}{4y^4}, \tag{26$'$}$$

where $p = u + v\vec{f} \in \mathcal{L}$, $z = \alpha^{-1}(p) = x + y\vec{f} \in \mathcal{H}^-$ and in both formulas $D(p) = -\{u + \frac{1}{4}v^2\}$.

The Gaussian curvatures of $(\mathcal{I}, H)$ and $(\mathcal{J}, H)$ are the constants -1 and +1, respectively.

Corollary 2. *The half plane $(\mathcal{I}, H)$ of the imaginary units $\{i(p)\}$ given by $(1)''$ is an isometric representation of the Poincaré plane.*

2. Four-dimensional algebras and manifolds

2.1. In what follows we fix an element $q \in \mathcal{D}^2 = \mathcal{C} \cup \mathcal{L}$ and accordingly the extension $\mathbf{A}(q)$ of the reals. We will repeat the extension prosedure presented in sections 0.1-0.5, replacing the reals $\mathbf{R}$ by the algebra $\mathbf{A}(q)$ to get proper extensions of the latter. Recall that $\mathbf{A}(q) \cong \mathbf{A}(-1) = \mathbf{C}$ for $q \in \mathcal{C}$ and $\mathbf{A}(q) \cong \mathbf{A}(+1) = \mathbf{L}$ for $q \in \mathcal{L}$.

Given a proper extension $\mathbf{H}$ of $\mathbf{A}(q)$, with a fixed element $\vec{g} \in \mathbf{H}$, $\vec{g} \notin \mathbf{A}(q)$, the requirements (I) and (II) in section 0.1 lead to

$$\mathbf{H} = \{x + y\vec{g} \mid x, y \in \mathbf{A}(q)\},$$

a two dimensional module over $\mathbf{A}(q)$, isomorphic to the product $\mathbf{A}(q) \times \mathbf{A}(q)$. As an $\mathbf{R}-$vector space $\mathbf{A}(q) \cong \mathbf{R}^4$. The sum and product of $z = x + y\vec{g}$ and $\zeta = \xi + \eta\vec{g}$ are $z + \zeta = (x + \xi) + (y + \eta)\vec{g}$ and $z\zeta = x\xi + (x\eta + y\xi)\vec{g} + (y\eta)(\vec{g})^2$ respectively. In particular, $\mathbf{H}$ is a commutative algebra over $\mathbf{A}(q)$, thus also over $\mathbf{R}$. It is determined by giving the value $p = (\vec{g})^2 = u + v\vec{g} \in \mathbf{H}$, where $u, v \in \mathbf{A}(q)$ and it is denoted by $\mathbf{H}(p)$ or more accurately by $\mathbf{H}(p, q)$.

2.2. Suppose first that $q \in \mathcal{C}$ so $\mathbf{A}(q) \cong \mathbf{C}$ and denote by $i_1 = i(q) \in \mathbf{A}(q)$ the solution to (1) given by $(1)''$ (where p is replaced by q). The conditions

$$i^2 = -1, \qquad i \notin \mathbf{A}(q) \tag{27}$$

for $i = x + y\vec{g} \in \mathbf{H}(p)$ are equivalent with $(1)'$, where now x, y, u and v in equations $(1)'$ belong to $\mathbf{A}(q)$, thus also $D(p) = -(u + \frac{1}{4}v^2)$. By $(1)'$ there are in $\mathbf{H}(p)$ solutions to (27) if and only if $p = u + v\vec{g}$ belongs to

$$\mathcal{D}^4 = \mathcal{D}^4(q) = \{p \in \mathbf{H} \mid D(p) \neq 0\}. \tag{28}$$

This is an open arcwise connected domain in $\mathbf{H} \cong \mathbf{R}^4$. Its border

$$(28)' \qquad \partial\mathcal{D}^4 = \{x + y\vec{g} \mid x = -\frac{1}{4}y^2, y \in \mathbf{A}(q)\} \subset \mathbf{R}^4$$

is a connected surface of real dimension two. Taking into account the theorem of Alexander [6,Ch.6.sec.2] it follows that the first homology group of $\mathcal{D}^4$ is generated by the homology class $[\gamma]$ determined by the circle $\gamma(t) = \exp\{2\pi(t+\frac{1}{2})i_1\}$, $t \in [0,1]$ of the x-plane $y = 0$.

As in section 0.7, by choosing a smooth curve $\ell : [0,1] \ni t \to w(t) \in \mathcal{D}^4$ from $-1 = w(0)$ to $p = u + v\vec{g}$, where $u, v \in \mathbf{A}(q)$, we get for $i(w(t)) = x(t) + y(t)\vec{g}$ the differential equation (11). From this it follows, as in sections 0.2 and 0.7, that

$$(29) \qquad i(p) = i(p, \ell) = \frac{1}{\sqrt{D(p)}}\{-\frac{v}{2} + \vec{g}\ \}$$

is a solution to (27), where

$$(30) \qquad \sqrt{D(p)} = \exp\{\frac{1}{2}\int_\ell \frac{dD}{D}\}.$$

If besides ℓ also ℓ' is a path from -1 to p in $\mathcal{D}^4$ the homology classes are related as $[\ell^{-1}\ell'] = [\gamma]^m$, where m is an integer. By (30) and (29) we get then

$$(31) \qquad i(p, \ell') = (-1)^m i(p, \ell).$$

By commutativity, $j(p, \ell') = i_1 i(p, \ell')$ satisfies $j^2 = +1$ and is related to $j(p, l) = i_1 i(p, l)$ by

$$(31)' \qquad j(p, \ell') = (-1)^m j(p, \ell).$$

2.3. Let us fix a path ℓ from -1 to p in $\mathcal{D}^4$ and denote for short $i_2(p) = i(p, \ell)$ and $j(p) = j(p, \ell)$. It follows, that the algebra $\mathbf{H}(p)$ with the basis $\{1, i_1, i_2(p), j(p)\}$ is isomorphic to the direct product $\mathbf{A}(q) \times \mathbf{A}(q)$ (and hence also to $\mathbf{C} \times \mathbf{C}$), where the sum and the product are defined componentwise. In fact, an $\mathbf{R}$-linear map, which maps the basis vectors $(1,1), (i_1, i_1), (i_1, -i_1)$ and $(-1, +1)$ of $\mathbf{A}(q) \times \mathbf{A}(q)$ in that order to $1, i_1, i_2(p)$ and $j(p) \in \mathbf{H}(p)$, is an isomorphism.

2.4. As we did for $\mathcal{C}$ in section 1.1, we do now for $\mathcal{D}^4$; we identify the tangent space $\mathcal{D}^4_p$ at $p \in \mathcal{D}^4$ with the algebra $\mathbf{H}(p)$. Also, similarly as in section 1.2, define for $Z = X' + \vec{Y}' = X' + (\ Y_1' i_1 + Y_2' i_2(p) + Y_3' j(p)\ ) \in \mathcal{D}^4_p$ the conjugate vector $\bar{Z}$ by

$$(32) \qquad \bar{Z} = X' - \vec{Y}'$$

and the real part of Z by $\Re Z = X'$. If $W = U' + V_1' i_1 + V_2' i_2(p) + V_3' j(p) \in \mathcal{D}_p$ is another vector, then

$$(33) \qquad \Re\{\bar{Z}W\} = X'U' + Y_1'V_1' + Y_2'V_2' - Y_3'V_3'.$$

Since by (31) and (31)$'$, $i_2(p)$ and $j(p)$ are unique up to sign, formula (33) gives on $\mathcal{D}^4$ a well defined pseudo-Riemannian metric

$$(33)' \qquad G(p)(Z, W) = \Re\{\bar{Z}W\}$$

with signature 2. The basis $\{1, i_1, i_2(p), j(p)\}$ is orthonormal with respect to $G(p)$. Here $j(p)$ and $\{1, i_1, i_2(p)\}$ define orientations of time and space respectively; both of these will change when going around a loop in $\mathcal{D}^4$ homologous to the circle γ given in section 2.2.

2.5. As in sections 1.3 and 1.4, the formulas

$$(34) \qquad I(p)Z = i_2(p)Z, \qquad J(p)Z = j(p)Z, \quad \forall Z \in \mathcal{D}^4_p,$$

define mixed tensor fields I and J satisfying $(I)^2 = -E$ and $(J)^2 = E$, I being hence an almost complex structure and J an involutive field. However, because of the non-trivial topology, the tensor fields are now defined merely **locally** and by (31) and (31)$'$, on all of $\mathcal{D}^4$ only up to sign.

2.6. On the open domain

$$\mathcal{H}^4 = \{z = x + y\vec{g} \mid x, y \in \mathbf{A}(q), y \neq 0\} \subset \mathbf{R}^4$$

of real dimension four consider the tangent vectors $Z = X + Y\vec{g}$ and $W = U + V\vec{g} \in \mathcal{H}^4_z$ with the $A(q)$-valued coordinates $X = X_o + X_1 i_1, Y = Y_o + Y_1 i_1, U = U_o + U_1 i_1$ and $V = V_o + V_1 i_1$. Then at point $z = x + y\vec{g} \in \mathcal{H}^4$, where $x = x_o + x_1 i_1, y = y_o + y_1 i_1$, define by

$$(35) \qquad \frac{1}{4} g(z)(Z, W) = (y_o^2 + y_1^2)(Y_o V_o + Y_1 V_1) + (y_1^2 - y_o^2)(X_1 U_1 - X_o U_o) - 2y_o y_1 (X_o U_1 + X_1 U_o)$$

a pseudo-Riemannian metric g of $\mathcal{H}^4$. Analogously as in section 1.6, define for $z = x + y\vec{g} \in \mathcal{H}^4, (x, y \in \mathbf{A}(q))$,

$$(36) \qquad \alpha(z) = -(x^2 + y^2) + 2x\vec{g}.$$

By applying calculus as it is presented in [8] one easily verifies:

Proposition 4. *Let $q \in \mathcal{C}$ and accordingly $\mathbf{A}(q) \cong \mathbf{C}$. Formula (36) defines an isometric transformation from the pseudo-Riemannian manifold $(\mathcal{H}^4, g)$ onto $(\mathcal{D}^4, G)$. Furthermore it transforms the local almost complex structure I on $\mathcal{D}^4$ to the complex structure*

$$(37) \qquad I_o : Z = X + Y\vec{g} \to -Y + X\vec{g}$$

and the involutive tensor field J to the "constant" operator

$$(37)' \qquad J_o : Z = X + Y\vec{g} \to Y + X\vec{g}.$$

Here $Z = X + Y\vec{g} \in \mathcal{H}^4_z$ and $X, Y \in \mathbf{A}(q)$.

Corollary. *The locally given almost complex structure I on $\mathcal{D}^4$ is integrable.*

Note, that there is on $\mathcal{D}^4 = \mathcal{D}^4(q)$ also another almost complex structure given by $i_1(q)$, the solution of (1) in $\mathbf{A}(q) \subset \mathcal{D}^4_p = \mathbf{H}(p, q)$. For a fixed $q \in \mathcal{C}$ this structure is obviously integrable, thus a complex structure. In particular the subalgebra $\mathbf{A}(q)$ defines an integrable distribution of two-dimensional real vector spaces on $\mathcal{D}^4$, the associated submanifold through $p \in \mathcal{D}^4$ being $\mathcal{A}(p) = p + \mathbf{A}'(q) = \{p + X \mid X \in \mathbf{A}'(q)\}$, where $\mathbf{A}'(q)$ is the "punctured plane" $\mathbf{A}'(q) = \{X \mid X \in \mathbf{A}(q), X \neq -D(q)\}$.

2.7. Suppose next that $q \in \mathcal{L}$, so $\mathbf{A}(q) \cong \mathbf{L}$. Recall the basis $\{\vec{e}_o, \vec{e}_1\}$ of $\mathbf{A}(q)$ given by (8)′ in lemma 2 and denote $p = (\vec{g}\ )^2 \in \mathbf{H} \cong \mathbf{R}^4$. The multiplication on $\mathbf{H}(p)$ by the idempotent element $\vec{e}_k \in \mathbf{A}(q) \subset \mathbf{H}(p), (k = 0, 1)$ is a projection : $\mathbf{H}(p) \to \mathbf{H}_k = \vec{e}_k \mathbf{H}(p)$. Because $\vec{e}_o + \vec{e}_1 = 1$, these two projections are complementary,

$$\mathbf{H}(p) = \mathbf{H}_o \oplus \mathbf{H}_1. \tag{38}$$

Furthermore by the commutativity of the algebra $\mathbf{H}(p)$, $\vec{e}_k(zw) = (\vec{e}_k\ )^2(zw) = (\vec{e}_k z)(\vec{e}_k w)\ \forall z, w \in \mathbf{H}(p)$, consequently $\mathbf{H}_k$ is a subalgebra and the projection on it is an algebra homomorphism, $(k = 0, 1)$. For $z = x + y\vec{g} \in \mathbf{H}(p)$, where $x = x_o\vec{e}_o + x_1\vec{e}_1, y = y_o\vec{e}_o + y_1\vec{e}_1$, $(x_k, y_k \in \mathbf{R})$, the projection $z^{(k)}$ on $\mathbf{H}_k$ becomes: $z^{(k)} = \vec{e}_k z = x_k\vec{e}_k + y_k\vec{e}_k\vec{g} = x_k\vec{e}_k + y_k\vec{f}_k$, where

$$\vec{f}_k = \vec{e}_k\vec{g}, \quad (k = 0, 1). \tag{39}$$

In particular,

$$\mathbf{H}_k = \text{span}_{\mathbf{R}}\{\vec{e}_k, \vec{f}_k\} \tag{40}$$

and with

$$\begin{cases} (\vec{g}\ )^2 = p = p^{(0)} + p^{(1)} \\ p^{(k)} = \vec{e}_k p = u_k\vec{e}_k + v_k\vec{f}_k \quad (k = 0, 1), \end{cases} \tag{41}$$

we get the multiplication table for the basis $\{\vec{e}_k, \vec{f}_k\}$ of the algebra $\mathbf{H}_k$:

$$\vec{e}_k\vec{e}_k = \vec{e}_k, \quad \vec{e}_k\vec{f}_k = \vec{f}_k, \quad \vec{f}_k\vec{f}_k = p^{(k)}. \tag{42}$$

Comparing $\mathbf{H}_k$ to $\mathbf{A}$ considered in the introduction, we get the algebra isomorphism

$$\mathbf{H}_k \cong \mathbf{A}(w), \tag{42$'$}$$

where $w = u_k + v_k\vec{f}$, by the linear correspondence $(\vec{e}_k, \vec{f}_k) \leftrightarrow (1, \vec{f})$.

2.8. Pseudo-Riemannian manifolds $\mathcal{D}^4_m$, $(m = 1, 2, 3, 4)$ (se lemma 6 below) correspond again to extensions of $\mathbf{A}(q) \cong \mathbf{L}$, all manifolds being in this case topologically trivial. Indeed, by the previous section the extension procedures reduces to those in the first and second parts as follows. Besides of (41) consider the projections $\vec{e}_o z = z^{(0)} = x_o\vec{e}_o + y_o\vec{f}_o$ and $\vec{e}_1 z = z^{(1)} = x_1\vec{e}_1 + y_1\vec{f}_1$ of $z = x + y\vec{g}$, whereby $x = x_o\vec{e}_o + x_1\vec{e}_1$ and $y = y_o\vec{e}_o + y_1\vec{e}_1$ $(x_k, y_k \in \mathbf{R})$. Denote by $j = j(q) \in \mathbf{A}(q)$ the solution (7)′ of $j^2 = +1$. The solvability of the four equations

$$z^2 = c_m, \qquad z \notin \mathbf{A}(q) \tag{43}$$

with

$$\begin{cases} c_1 = -1 = -\vec{e}_o - \vec{e}_1, & c_2 = +1 = \vec{e}_o + \vec{e}_1, \\ c_3 = j = -\vec{e}_o + \vec{e}_1, & c_4 = -j = \vec{e}_o - \vec{e}_1, \end{cases}$$

for $z \in \mathbf{H}(p)$ leads now to four types of extensions respectively. By (43) we get for the projection on $\mathbf{H}_k$, $(k = 0, 1)$,

$$c_m^{(k)} = \vec{e}_k c_m = \vec{e}_k z^2 = (\vec{e}_k z)^2 = (z^{(k)})^2,$$

so the condition (43) can be written equivalently,

$$(43)' \qquad (z^{(k)})^2 = c_m^{(k)}, \quad k = 0, 1, \qquad z \notin \mathbf{A}(q).$$

Here

$$c_1^{(o)} = -\vec{e}_o, \qquad c_1^{(1)} = -\vec{e}_1, \qquad c_2^{(0)} = \vec{e}_o, \qquad c_2^{(1)} = \vec{e}_1$$
$$c_3^{(0)} = -\vec{e}_o, \qquad c_3^{(1)} = \vec{e}_1, \qquad c_4^{(0)} = \vec{e}_o, \qquad c_4^{(1)} = -\vec{e}_1.$$

Denote for points $w = \alpha\vec{e}_k + \beta\vec{f}_k$ in $\mathbf{H}_k : D(w) = -(\alpha + \frac{1}{4}\beta^2), k = 0, 1$, and define the plane domains $\mathcal{C}_k = \{w \mid w \in \mathbf{H}_k, D(w) < 0\}$, $\mathcal{L}_k = \{w \mid w \in \mathbf{H}_k, D(w) < 0\}$, $k = 0, 1$. By propositions 1 and 2, we get:

Lemma 6. *In the extension* $\mathbf{H}(p)$ *of* $\mathbf{A}(q) \cong \mathbf{L}$ *there exist solutions* z *of (43) if and only if* $p \in \mathcal{D}_m^4$, *where*

$$\mathcal{D}_1^4 = \mathcal{C}_o + \mathcal{C}_1 = \{w = w^{(0)} + w^{(1)} \mid w^{(k)} \in \mathcal{C}_k, k = 0, 1\}$$
$$\mathcal{D}_2^4 = \mathcal{L}_o + \mathcal{L}_1, \quad \mathcal{D}_3^4 = \mathcal{C}_o + \mathcal{L}_1, \quad \mathcal{D}_4^4 = \mathcal{L}_o + \mathcal{C}_1.$$

Since here $\mathcal{C}_k$ and $\mathcal{L}_k$ are simply connected, so are $\mathcal{D}_m^4$, $(k = 0, 1, \quad m = 1, ..., 4)$. In particular, we have:

Corollary. *There exists on* $\mathcal{D}_1^4$ $[\mathcal{D}_2^4]$ *a unique smooth function* $i = i_1$, $[j = j_2]$, *for which* $i(-1) = \vec{g}$, $[j(+1) = \vec{g}]$, *and for which the value* $i(p)$, $[j(p)]$, *solves the equation* $i^2 = -1$, $[j^2 = +1]$, *in* $\mathbf{H}(p)$.

As a matter of fact for both $i(p)$ and $j(p)$ we have the expression

$$\sum_{k=0}^{1} \frac{1}{\sqrt{|D(p^{(k)})|}} \{-\frac{v_k}{2}\vec{e}_k + \vec{f}_k\}$$

where for $p = p^{(0)} + p^{(1)}$, $p^{(k)} = u_k\vec{e}_k + v_k\vec{f}$, $k = 0, 1$.

2.9. Denote the solution $(7)'$ of $j^2 = 1$ in $\mathbf{A}(q)$ by j_1. For $p \in \mathcal{D}_1^4$ put $i_1 = i_1(p)$ and $i_2 = j_1 i_1$. Then $\{1, i_1, i_2, j_1\}$ is a basis of $\mathbf{H}(p)$. Identifying $\mathbf{H}(p)$ with the tangent space of $\mathcal{D}_1^4$ at the point p, the left hand side of the equation (33) gives a pseudo-Riemannian metric on $\mathcal{D}_1^4$ of signature 2. Also, for $p \in \mathcal{D}_2^4$, put $j_2 = j_2(p)$ and $j_3 = j_1 j_2$. Then $\{1, j_1, j_2, j_3\}$ is a basis of $\mathbf{H}(p)$. If for $Z = X_o + \sum_{k=1}^{3} Y_k j_k$,

$$(44) \qquad \bar{Z} = X_o - \sum_{k=1}^{3} Y_k j_k$$

and $\Re Z = X_o$, the expression $-\Re\bar{Z}W$ gives on $\mathcal{D}_2^4$ a pseudo-Riemannian metric with signature 2, the basis vector 1 giving now the time direction. In these pseudo-Riemannian manifolds $\mathcal{D}_m^4, (m = 1, 2)$ orientations of time and space are globally defined.

2.10. Finally we will add some remarks, which we will hopefully consider in more detail in forthcoming papers.
– In the study above, the parameter $q \in \mathcal{D}^2 = \mathcal{C} \cup \mathcal{L}$ and thus the algebra $\mathbf{A}(q)$ has been fixed. Considering $\mathcal{D}^2$ as a two-dimensional manifold (as in part 1) we get a "fiber space" $\{\mathcal{D}^4(q) \mid q \in \mathcal{D}^2\}$ over $\mathcal{D}^2$, where the fibers are four-dimensional pseudo-Riemannian manifolds. If we let the point $q \in \mathcal{C}$ move to the border $\partial\mathcal{C}$, the "world" $\mathcal{D}^4(q)$ collapses and then for $q \in \mathcal{L}$ splits in four parts.
– The "space of imaginary units" $\mathcal{I}^4 = \{i_2(p)\}$ and "units" $\mathcal{J}^4 = \{j(p)\}$ get naturally a pseudo-Riemannian metric in the cases $q \in \mathcal{C}$ and $q \in \mathcal{L}, m = 1$.
– The extension prosedure above can be continued to get N-dimensional commutative algebras and pseudo-Riemannian manifolds $\mathcal{D}^N$ with $N = 2^m, m = 3, 4, \dots$. When, for instance, at every extension step auxiliary solutions of $i^2 = -1$ are required, the signature of $\mathcal{D}^N$ will always be 2.
– The definition of the conjugate $\bar{Z}$ given by (32) is a direct copy of the two-dimensional case. However, in the four-dimensional algebras $\mathbf{H}(p)$ it is not compatible with the product, so in general $\overline{ZW} = \bar{Z}\bar{W}$ does'nt hold. On the contrary, by defining in the case $\mathbf{A}(q) \cong \mathbf{C}$ $[\mathbf{A}(q) \cong \mathbf{L}$ with $m = 1]$, $\bar{Z} = \bar{X} - \bar{Y}i_2$ for $Z = X + Yi_2$, the compatibility holds and the signature of the metric becomes 4.

We are indebted to Kalevi Suominen for discussions about the subject. We thank also Ari Heimonen for drawing the picture.

References

1. B.A. Dubrovin, A.T. Fomenko, S.P. Novikov, *Modern Geometry - Methods and Applications Part I*, Springer - Verlag, 1984.
2. –, *The mathematical papers of Sir William Rowan Hamilton Vol. III*, Cambridge at the University Press, 1967.
3. W.R. Hamilton, *On conjugate functions, or algebraic Couples*, Brit. Assoc. Report (1834), 519-523.
4. L. Hörmander, *An introduction to complex analysis in several variables*, D. van Nostrand Company, 1966.
5. S. Kobayashi, K. Nomizu, *Foundations of differential geometry Vol. II*, John Wiley et Sons, 1969.
6. A. Lichnerowicz, *Theorie globale des connexions et des groupes d'holonomie*, Consiglia nazionale delle ricerche, 1955.
7. O. Lehto und K. Virtanen, *Quasikonforme Abbildungen*, Springer, 1965.
8. F. und R. Nevanlinna, *Absolute Analysis*, Springer-Verlag, 1959.

Stable Iterated Function Systems

Erland Gadde

Department of Mathematics
University of Umeå
S-981 87 Umeå, Sweden

Abstract

This paper consists of the first two chapters in the author's doctoral thesis with the same title (Doctoral Thesis No 4, 1992, Dept. of Math., Univ. of Umeå).

The purpose of the thesis is to generalize the growing theory of *iterated function systems* (IFSs). Earlier, *hyperbolic* IFSs with finitely many functions had been studied extensively. Also, hyperbolic IFSs with infinitely many functions have been studied. In the thesis, more general IFSs are studied.

The *Hausdorff pseudometric* is studied. This is a generalization of the Hausdorff metric. *Wide and narrow limit sets* are studied. These are two types of limits of sequences of sets in a complete pseudometric space.

Stable Iterated Function Systems, a kind of generalization of hyperbolic IFSs, are defined. Some different, but closely related, types of stability for the IFSs are considered. It is proved that the IFSs with the most general type of stability have unique attractors. Also, invariant sets, addressing, and periodic points for stable IFSs are studied.

In the final two chapters of the thesis, which are not included in this paper, Stable IFSs with probabilities and appropriate metrics and norms on measure spaces are studied.

1991 Mathematics Subject Classification: 28A80.

0. SUMMARY.

This paper is mainly a study of *stable iterated function systems* (stable IFSs). There are some different, but related, kinds of stability for these IFSs, see Definitions 2.4.

Stable IFSs are a generalization of a type of IFSs introduced by Wallin and Karlsson ([26], p. 11), which they used to study continued fractions. All *hyperbolic* IFSs, studied by Hutchinson, Barnsley, Demko, Lewellen, and others ([14], [3], [4], [16]) are *boundedly stable* (Definitions 2.4).

This paper is mainly concerned with convergence properties of these IFSs and not with the dimensions and geometric properties of the *attractors* of the systems.

For these convergences, we use the *Hausdorff pseudometric* (see Definition 1.9), a generalization of the Hausdorff metric which is defined on *all* kinds of sets, not only closed and bounded ones. In Chapter 1, we first give a brief introduction to pseudometric spaces. Then, we use this knowledge to study the Hausdorff pseudometric. We conclude Chapter 1 by studying *wide and narrow limit sets*, which are two related types of limits of sequences of sets.

In Chapter 2, we define $(\mathfrak{T}, x_0)$–stable IFSs, and some special cases of these systems (Definitions 2.4). This is the most general kind of stability studied in this thesis. We show that all hyperbolic IFSs which satisfy a certain boundedness condition are boundedly stable (Proposition 2.9). Bounded stability is one of the special cases mentioned.

Then, we give a non–trivial example of a boundedly stable non–hyperbolic IFS (Example 2.10).

In the next section, we prove that the $(\mathfrak{T}, x_0)$–stable IFSs have unique attractors (Theorem 2.15). For the proof, we use a fixed point theorem (Theorem 2.12) in a similar way as Banach's Fixed Point Theorem is used for the proof of the corresponding theorem about hyperbolic IFSs ([3], p. 82 and [14], p. 728).

After illustrating with some examples and counterexamples, we introduce *addresses* for $(\mathfrak{T}, x_0)$–stable IFSs (Definitions 2.25), following the studies of Barnsley and Lewellen ([3], [16]). We give some relations between addresses, attractors, and invariant sets, which we then illustrate with some further examples. Then, we study periodic points (Definitions 2.36) and continuous addressing, giving further information about the attractor (Theorem 2.43 and Corollary 2.44). Finally, we make Ω^{∞} (see Definition 4.1) into a hyperbolic IFS (Definitions 2.47 and Remarks 2.48).

list of symbols

symbol	def no
$diam(A)$	1.2
$d(x, A)$	1.2
$B(x, r)$	1.2
$\bar{B}(x, r)$	1.2
A_r	1.2
$\langle X \rangle$	1.3
$\langle x \rangle$	1.3
$\langle d \rangle$	1.3
$\langle f \rangle$	1.3
$\mathfrak{S}(X)$	1.9
$\mathfrak{H}$	1.9
$\mathfrak{C}(X)$	1.9
$\mathfrak{H}'$	1.9
$\mathfrak{B}(X)$	1.9
$\mathfrak{C}\mathfrak{B}(X)$	1.9
$\mathfrak{K}(X)$	1.9
$\mathfrak{I}(X)$	1.9
$wl(A_n)$	1.14
$nl(A_n)$	1.14
Ω^n	2.2
Ω^∞	2.2
$\bar{\Omega}$	2.2
$\Delta_n(A)$	2.3
Adr	2.25
$Adr(\Omega^\omega)$	2.25
$\lvert \cdot, \cdot \rvert_B$	2.40
$\hat{s}_0$	2.47
$\hat{\Omega}$	2.47

1. PSEUDOMETRIC SPACES. HAUSDORFF PSEUDOMETRIC.

Pseudometric spaces

In the theory of iterated function systems, the Hausdorff metric plays an important role. However, in our study of *stable* IFSs it is convenient to generalize it to a pseudometric, defined on *all* nonempty sets in a space, instead of only closed, bounded sets. Also, we need a generalization of Banach's Fixed Point Theorem, which is formulated for a pseudometric space. Therefore, we start with a brief study of pseudometric spaces.

I should mention that most standard monographs do not treat pseudometric spaces, sometimes called *semimetric* spaces. Wilansky's book [27] is an exception. Therefore, much of the terminology below is my own invention, although the results stated are not new.

1.1 DEFINITION. A **pseudometric space** is an ordered pair (X, d), where X is a set and $d : X \times X \to [0, \infty]$ is a function, called a **pseudometric**, such that for any $x, y, z \in X$ the following conditions hold:

(1) $d(x, x) = 0$
(2) $d(x, y) = d(y, x)$
(3) $d(x, z) \leq d(x, y) + d(y, z)$.

In addition, if $d(x, y) = 0 \Longrightarrow x = y$, then d is a **metric** and (X, d) is a **metric space.**

If $d(x, y) < \infty$ for all $x, y \in X$, then d is a **finite** (pseudo)metric and (X, d) is a **finitely (pseudo)metric space**.

Note that we allow $d(x, y)$ to be infinite. *This is not standard*, but convenient here. Topologically, this convenience does not matter.

The following definitions are obvious generalizations from the metric case.

1.2 DEFINITIONS. Let (X, d) be a pseudometric space. For any subset $A \subset X$, $diam(A) = \sup_{x,y \in A} d(x, y)$. A is **bounded** if $diam(A) < \infty$. For any $x \in X$, $d(x, A) = \inf_{y \in A} d(x, y)$. For any $x \in X$ and any $r \in (0, \infty]$, $B(x, r) = \{y \in X | d(x, y) < r\}$, $\bar{B}(x, r) = \{y \in X | d(x, y) \leq r\}$ and for any nonempty $A \subset X$, $A_r = \{x \in X | d(x, A) < r\}$. For any sequence $\{x_n\}_{n=1}^{\infty}$ in X, $x_n \to x \in X$ as $n \to \infty$ if, for every $\varepsilon > 0$, $B(x, \varepsilon)$ contains x_n for all but finitely many n we say that x_n **converges** to x, x is then a **limit** of $\{x_n\}_{n=1}^{\infty}$. If $\{r_n\}_{n=1}^{\infty}$ is a sequence in $[0, \infty]$ converging to 0 such that $d(x_n, x) \leq r_n$ for all n, then we say that x_n converges to x with **speed** $\{r_n\}$. $\{x_n\}_{n=1}^{\infty}$ is **Cauchy** if, for every $\varepsilon > 0$, there is an N such that $d(x_m, x_n) < \varepsilon$ for all $m, n \geq N$. (X, d) is **complete** if every Cauchy sequence in X converges. (As in the metric case, a convergent sequence is always Cauchy.) For $\varepsilon > 0$, an ε**–net** in X is a set $S \subset X$ such that $S_\varepsilon = X$. X is **totally bounded** if it has a *finite* ε–net for *any* $\varepsilon > 0$. If (Y, d') is another pseudometric space and $f : X \to Y$ is a bijection such that $d'(f(x), f(y)) = d(x, y)$ for all $x, y \in X$, then f is an **isometry** and we say that the spaces (X, d) and (Y, d') are **isometric**.

As in the metric case, we observe that the family of balls $B(x,r)$ constitutes a basis for a topology on X. We always assume that X has this topology. This topology is *not* Hausdorff (in fact, not even $\mathcal{T}_0$) unless (X,d) is metric. Also, unless (X,d) is metric, some convergent sequences have no *unique* limits, because $d(x,y)$ may be 0 although $x \neq y$.

From this fact, it follows that for a non–metric pseudometric space (X,d), the following statements are *not* true:

(1) For every $S \subset X$, if every convergent sequence in S has a limit in S, then S is closed.
(2) Every complete subspace of X is closed.
(3) Every compact subspace of X is closed.

Instead of (1), the following is true:

(1') If *all* limits of all convergent subsequences in $S \subset X$ are in S, then S is closed.

The relation on X given by $d(x,y) = 0$ is an equivalence relation. Let $\langle X \rangle$ denote its quotient set, and let $\langle x \rangle$ denote the equivalence class containing $x \in X$. We now make $\langle X \rangle$ into a metric space by defining the metric $\langle d \rangle$ on X by $\langle d \rangle(\langle x \rangle, \langle y \rangle) = d(x,y)$ for any $\langle x \rangle, \langle y \rangle \subset \langle X \rangle$. It is easy to check that $\langle d \rangle$ will be a well defined metric on $\langle X \rangle$.

1.3 DEFINITION. The metric space $(\langle X \rangle, \langle d \rangle)$ just defined is called the **quotient space** of (X,d).

We observe that if a sequence $\{x_n\}$ in X converges to $x \in X$, then the equivalence class $\langle x \rangle$ is its set of limits. It follows that $x_n \to x$ in X if and only if $\langle x_n \rangle \to \langle x \rangle$ in $\langle X \rangle$ as $n \to \infty$ and that $\{x_n\}$ is Cauchy if and only if $\{\langle x_n \rangle\}$ is also Cauchy. Also, (X,d) is complete, compact, or totally bounded if and only if $(\langle X \rangle, \langle d \rangle)$ has the same property.

If (X,d) is metric, then (X,d) is isometric to $(\langle X \rangle, \langle d \rangle)$. The isometry is given by $x \mapsto \langle x \rangle$. Hence, in that case, $(\langle X \rangle, \langle d \rangle)$ can be identified with (X,d).

Now, let (Y,d') be another pseudometric space and let $f : X \to Y$ be a function such that $d(x,y) = 0 \Longrightarrow d'(f(x), f(y)) = 0$ for all $x,y \in X$. This condition is satisfied if f is continuous. This will induce a function $\langle f \rangle : \langle X \rangle \to \langle Y \rangle$ given by $\langle f \rangle(\langle x \rangle) = \langle f(x) \rangle$ for every $\langle x \rangle \in \langle X \rangle$. This function will be well defined.

1.4 DEFINITIONS. The function $\langle f \rangle$ just defined is called the **quotient function induced by** f.

f is **uniformly continuous** if $\langle f \rangle$ exists and is uniformly continuous.

A family Ω of functions from X to Y is **equicontinuous** if its family of quotient functions exists and is equicontinuous.

It is easy to see that the quotient function $\langle f \rangle$ exists and is continuous if f is continuous. If (X,d) and (Y,d') are metric and if we identify their quotient spaces with the original spaces, as mentioned above, then $\langle f \rangle$ will be identified with f.

1.5 EXAMPLES. Let X be the complex plane $\mathbb{C}$ and put $d(z,w) = |\text{Re } z - \text{Re } w|$, for all $z,w \in X$. Then, (X,d) is a pseudometric space. The eqivalence class $\langle z \rangle$ is

$\{w \in \mathbb{C} | \text{Im } w = \text{Im } z\}$, and its quotient space $\langle X \rangle$ can be identified with $\mathbf{R}$, if we identify $\langle z \rangle \in \langle X \rangle$ with Re z.

In this case, $d(\cdot, 0)$ is a seminorm on X. In general, *if p is any seminorm on a real vector space X, then $d(x,y) = p(x-y)$ is a pseudometric on X.*

Instead, let X be $\mathbb{C} \setminus \{0\}$ and put $d(z,w) = |Arg(z/w)|$ for $z, w \in X$. Again, (X,d) is a pseudometric space. For $z \in \langle X \rangle$, $\langle z \rangle$ is the ray $\{rz | r > 0\}$ and $\langle X \rangle$ is isometric to the unit circle, which is metrized by arc length.

We include a well-known result and the following definition.

1.6 DEFINITION. A sequence in a pseudometric space is called **relatively compact** if each one of its subsequences has a convergent (sub)subsequence.

As in the metric case, we observe that if X is complete, then it is compact if and only if it is totally bounded, and this holds if and only if every sequence in X is relatively compact.

1.7 LEMMA. *If (X_n, d_n) is a sequence of pseudometric spaces and if, for every n, $\{x_k^n\}_{k=1}^\infty$ is a relatively compact sequence in X_n, then there is a strictly increasing sequence of positive integers $\{k_j\}_{j=1}^\infty$ such that for every n, the sequence $\{x_{k_j}^n\}_{j=1}^\infty$ converges in X_n.*

PROOF: This is a standard diagonal argument.

By assumption, there is a strictly increasing sequence of positive integers $\{k_j^1\}_{j=1}^\infty$ such that $\{x_{k_j^1}^1\}_{j=1}^\infty$ converges in X_1. This sequence has a subsequence $\{k_j^2\}_{j=1}^\infty$ such that $\{x_{k_j^2}^2\}_{j=1}^\infty$ converges in X_2. Continuing in this manner, we obtain a sequence of sequences of positive integers, $\{\{k_j^n\}_{j=1}^\infty\}_{n=1}^\infty$, where each sequence is a subsequence of the preceding one and $\{x_{k_j^n}^n\}_{j=1}^\infty$ converges for every n. For every n, the *diagonal* sequence $\{k_j^j\}_{j=1}^\infty$ is, disregarding the first $n-1$ terms, a subsequence of $\{k_j^n\}_{j=1}^\infty$ and hence $\{x_{k_j^j}^n\}_{j=1}^\infty$ converges in X_n. Thus, we can take $k_j = k_j^j$ for all j.

We also include the following simple lemma.

1.8 LEMMA. *If $\{x_n\}$ is a relatively compact sequence in a pseudometric space (X,d) such that all convergent subsequences have a common limit x, then $x_n \to x$ as $n \to \infty$.*

PROOF: Assume that $\{x_n\}$ satisfies the conditions but does not converge to x. Then, there is an $\varepsilon > 0$ and a subsequence $\{x_{n_k}\}_{k=1}^\infty$ such that $d(x_{n_k}, x) \geq \varepsilon$ for all k. This subsequence must have a convergent subsequence, which must converge to x. But this is impossible. Hence, $x_n \to x$ as $n \to \infty$.

Hausdorff pseudometric

In this and the following section, (X,d) is always a complete pseudometric space.

1.9 DEFINITIONS. $\mathfrak{S}(X)$ denotes the family of all nonempty subsets of X. For $A, B \in \mathfrak{S}(X)$, we put $\mathfrak{H}(A,B) = \inf\{\varepsilon > 0 | A \subset B_\varepsilon \text{ and } B \subset A_\varepsilon\}$. $\mathfrak{H}$ is called the **Hausdorff pseudometric** on $\mathfrak{S}(X)$.

$\mathfrak{C}(X)$ denotes the family of nonempty *closed* subsets of X. $\mathfrak{H}'$ is the restriction of $\mathfrak{H}$ to $\mathfrak{C}(X)$. $\mathfrak{H}'$ is called the **Hausdorff metric** on $\mathfrak{C}(X)$.

$\mathfrak{B}(X)$ denotes the family of nonempty *bounded* subsets of X, we put $\mathfrak{C}\mathfrak{B}(X) = \mathfrak{C}(X) \cap \mathfrak{B}(X)$, and $\mathfrak{K}(X)$ denotes the family of nonempty *compact* subsets of X.

$\mathfrak{I}(X)$ denotes the family of *singleton* subsets of X.

1.10 REMARKS. It is easy to verify that $\mathfrak{H}$ indeed is a pseudometric on $\mathfrak{S}(X)$, so that $(\mathfrak{S}(X), \mathfrak{H})$ is a pseudometric space, and likewise that $\mathfrak{H}'$ metrizes $\mathfrak{C}(X)$. We will always assume that these spaces have these pseudometrics. When confusion is impossible, we write $\mathfrak{H}$ instead of $\mathfrak{H}'$.

Also, for any $A, A' \in \mathfrak{S}(X)$ it is easy to see that $\langle A \rangle = \{B \in \mathfrak{S}(X) | \overline{B} = \overline{A}\}$ and $\mathfrak{H}(A, A') = \mathfrak{H}(\langle A \rangle, \langle A' \rangle)$. The latter Hausdorff pseudometric is here defined on $(\langle X \rangle, \langle d \rangle)$ with $\langle A \rangle$ denoting $\{\langle x \rangle | x \in A\}$ and likewise for $\langle A' \rangle$. Notice that the notation is ambiguous at this point.

It follows that $(\langle \mathfrak{S}(X) \rangle, \langle \mathfrak{H} \rangle)$ can be identified with $(\mathfrak{C}(X), \mathfrak{H}')$ as well as $(\langle \mathfrak{S}(\langle X \rangle) \rangle, \langle \mathfrak{H} \rangle)$ (with the latter notation).

Also it is easy to see that $\mathfrak{B}(X)$ is a *closed* subspace of $\mathfrak{S}(X)$ and that $\mathfrak{C}\mathfrak{B}(X)$ is a *closed* subspace of $\mathfrak{C}(X)$.

If $\mathfrak{H}$ is restricted to $\mathfrak{I}(X)$, it is easy to see that (X, d) is isometric to $(\mathfrak{I}(X), \mathfrak{H})$. The isometry is given by $x \mapsto \{x\}$.

Let Y be a nonempty *closed* subspace of X. Then it is easy to see that $\mathfrak{S}(Y)$ is a *closed* subspace of $\mathfrak{S}(X)$.

The following result is well–known ([9], p. 37 and [3], p. 37), but we include a simple proof for it, which I have not seen before.

1.11 THEOREM. *The pseudometric space $\mathfrak{S}(X)$ and the metric space $\mathfrak{C}(X)$ are both complete.*

PROOF: Since $\mathfrak{H}(A, \overline{A}) = 0$ for every $A \in \mathfrak{S}(X)$, it suffices to prove the first statement.

So, let $\{A_n\}$ be any Cauchy sequence in $\mathfrak{S}(X)$. It has a subsequence $\{A_{n_k}\}_{k=1}^\infty$ such that $\mathfrak{H}(A_{n_k}, A_{n_{k+1}}) < 2^{-k}$ for all k. It suffices to show that this subsequence converges. Now, let $\mathfrak{A}$ be the family of all sequences $\{x_k\}$ in X such that (i) $x_k \in A_{n_k}$ and (ii) $d(x_k, x_{k+1}) < 2^{-k}$ for all k.

$\mathfrak{A}$ is nonempty, in fact we will show that *for every k and for every $y \in A_{n_k}$ there is a sequence $\{x_j\} \in \mathfrak{A}$ such that $x_k = y$.*

To show this, pick k and $y \in A_{n_k}$. Put $x_k = y$. Since $\mathfrak{H}(A_{n_k}, A_{n_{k+1}}) < 2^{-k}$, we can take $x_{k+1} \in A_{n_{k+1}}$ such that $d(x_k, x_{k+1}) < 2^{-k}$. Likewise, we can take $x_{k+2} \in A_{n_{k+2}}$, such that $d(x_{k+1}, x_{k+2}) < 2^{-(k+1)}$ etc. Similarly, if $k > 1$, we can "go backwards" and find $x_{k-1} \in A_{n_{k-1}}$ such that $d(x_{k-1}, x_k) < 2^{-(k-1)}$ and so on down to an $x_1 \in A_{n_1}$ such that $d(x_1, x_2) < 2^{-1}$. Thus, we have constructed a sequence satisfying our conditions.

By (ii) and the completeness of (X, d), every sequence in $\mathfrak{A}$ converges. Let A be the set of limits of sequences in $\mathfrak{A}$. Then, $A \neq \emptyset$. We will show that $\mathfrak{H}(A_{n_k}, A) < 2^{2-k}$ for all k, which will complete the proof.

Pick an arbitrary $y \in A_{n_k}$. Then, there is a sequence $\{x_j\} \in \mathfrak{A}$ with $x_k = y$. Let $x \in A$ be one of its limits. By (ii), $d(y, x) < 2^{1-k}$. Hence, $A_{n_k} \subset A_{2^{1-k}}$.

Conversely, pick an arbitrary $x \in A$. Then, there is a sequence $\{x_j\} \in \mathfrak{A}$ converging to x. By (i) and (ii), $x_k \in A_{n_k}$ and $d(x_k, x) < 2^{1-k}$. Hence, $A \subset (A_{n_k})_{2^{1-k}}$.

Hence, $\mathfrak{H}(A_{n_k}, A) \leq 2^{1-k} < 2^{2-k}$, which completes the proof.

I have not been able to decide whether or not the set A in the above proof must be closed.

We include two more well-known results with proofs.

1.12 THEOREM. *If X is metric, then $\mathfrak{K}(X)$ is a closed subspace of $\mathfrak{C}(X)$.*

PROOF: Let $\{A_n\}$ be a sequence in $\mathfrak{K}(X)$ which converges to a set $A \in \mathfrak{C}(X)$. We must prove that A is compact. To do this, let $\{x_k\}$ be an arbitrary sequence in A. We must prove that $\{x_k\}$ has a convergent subsequence, of which the limit will be in A since A is closed. For every n, we can select a sequence $\{x_k^n\}_{k=1}^{\infty}$ in A_n such that $d(x_k^n, x_k) \leq \mathfrak{H}(A_n, A)$ for all k. Since A_n is compact, $\{x_k^n\}_{k=1}^{\infty}$ is relatively compact. By Lemma 1.7, there is a strictly increasing sequence $\{k_j\}_{j=1}^{\infty}$ of positive integers such that for every n, $\{x_{k_j}^n\}_{j=1}^{\infty}$ converges and thus is Cauchy. We will show that $\{x_{k_j}\}_{j=1}^{\infty}$ is Cauchy and hence converges.

Pick $\varepsilon > 0$. Take n such that $\mathfrak{H}(A_n, A) < \varepsilon/3$ and take N such that $d(x_{k_i}^n, x_{k_j}^n) < \varepsilon/3$ for all $i, j \geq N$. Then, for $i, j \geq N$, $d(x_{k_i}, x_{k_j}) \leq d(x_{k_i}, x_{k_i}^n) + d(x_{k_i}^n, x_{k_j}^n) + d(x_{k_j}^n, x_{k_j}) < 3\varepsilon/3 = \varepsilon$.

It follows that $\{x_{k_j}\}_{j=1}^{\infty}$ is Cauchy, which completes the proof.

Our next result is the well-known *"Blaschke selection theorem"*, which we formulate for a compact space. We borrow the proof essentially from Falconer ([9], p. 37).

1.13 THEOREM. *If X is compact metric, then so is $\mathfrak{C}(X) = \mathfrak{K}(X)$.*

PROOF: Assume that X is compact metric, hence totally bounded. We will show that $\mathfrak{C}(X) = \mathfrak{K}(X)$ is totally bounded. Since it is complete, it is therefore compact (and of course metric).

Pick $\varepsilon > 0$. We must find a finite ε-net in $\mathfrak{K}(X)$. Let $S = \{x_1, \ldots, x_N\}$ be a finite $\varepsilon/2$-net in X. Then, $\mathfrak{S}(S)$ is a finite ε-net in $\mathfrak{K}(X)$. The reason is that for any $A \in \mathfrak{K}(X)$, $S \cap A_{\varepsilon/2} \in \mathfrak{S}(S)$. It is then easy to see that $\mathfrak{H}'(A, S \cap A_{\varepsilon/2}) \leq \varepsilon/2 < \varepsilon$.

Wide and narrow limit sets

Barnsley and Demko ([4], p. 245) define the *attractor* of an iterated function system as the set with the property that every neighbourhood of every point in the set intersects infinitely many sets in a certain sequence of sets. Two questions that arise are: What happens if "infinitely many" is changed to "all but finitely many"? What is the relation to convergence in the Hausdorff pseudometric?

In this section, we investigate this problem in a general setting.

1.14 DEFINITIONS. Let $\{A_n\}$ be a sequence in $\mathfrak{S}(X)$.

The **wide limit set** of $\{A_n\}$, denoted $wl\{A_n\}$, is the set of all $x \in X$ such that every neighbourhood of x intersects A_n for infinitely many n.

The **narrow limit set** of $\{A_n\}$, denoted $nl\{A_n\}$, is the set of all $x \in X$ such that every neighbourhood of x intersects A_n for all but finitely many n.

Notice that, by definition, these limit sets of a sequence must always exist, although one or both may be empty.

1.15 PROPOSITION. *If $\{A_n\}$ and $\{B_n\}$ are sequences in $\mathfrak{S}(X)$ such that $A_n \subset B_n$ for all n, and if $\{A_{n_k}\}_{k=1}^{\infty}$ is a subsequence of $\{A_n\}$, then:*

(1) $nl\{A_n\} \subset wl\{A_n\}$.
(2) $wl\{A_n\} \subset wl\{B_n\}$ *and* $nl\{A_n\} \subset nl\{B_n\}$.
(3) $wl\{A_{n_k}\}_{k=1}^{\infty} \subset wl\{A_n\}$ *and* $nl\{A_n\} \subset nl\{A_{n_k}\}_{k=1}^{\infty}$.
(4) $wl\{A_n\}$ *and* $nl\{A_n\}$ *are both closed.*
(5) $wl\{A_n\} = wl\{\overline{A}_n\}$ *and* $nl\{A_n\} = nl\{\overline{A}_n\}$.

PROOF: (1), (2), and (3) are obvious. (4) and (5) follow from the fact that if $A, U \subset X$ with U open, then $A \cap U = \varnothing \iff \overline{A} \cap U = \varnothing$.

1.16 EXAMPLES Let X be $\mathbb{C}$.
1) Let $A_n = \{re^{in} | 0 \leq r \leq 1\}$ for all n. Then, $wl\{A_n\}$ is the closed unit disc $\{z| \, |z| \leq 1\}$ and $nl\{A_n\} = \{0\}$.
2) Let $B_n = \bigcup_{k=1}^{n} A_k$ for all n. Then, both $wl\{B_n\}$ and $nl\{B_n\}$ are the closed unit disc.
3) Let $C_n = \{re^{in} | 1 \leq r \leq 2\}$ for all n. Then, $wl\{C_n\}$ is the annulus $\{z | 1 \leq |z| \leq 2\}$ and $nl\{C_n\} = \varnothing$.
4) Let $D_n = \{z| \, |z| = n\}$ for all n. Then, $wl\{D_n\} = nl\{D_n\} = \varnothing$.

We will now investigate the connection with the Hausdorff pseudometric.

1.17 THEOREM. *If $\{A_n\}$ is a relatively compact sequence in $\mathfrak{S}(X)$, then the following conditions are equivalent:*

(1) $\{A_n\}$ *converges in* $\mathfrak{S}(X)$.
(2) $wl\{A_n\} = nl\{A_n\}$.

Furthermore:

(3) *If* $A \in \mathfrak{C}(X) \subset \mathfrak{S}(X)$ *is a limit of* $\{A_n\}$, *then* $A = wl\{A_n\} = nl\{A_n\}$.

PROOF: (1)$\Longrightarrow$(2). Suppose that $\{A_n\}$ converges. By Proposition 1.15 (1), it suffices to prove $wl\{A_n\} \subset nl\{A_n\}$.

So, pick $x \in wl\{A_n\}$ and a neighbourhood U of x. Take $\varepsilon > 0$ such that $B(x,\varepsilon) \subset U$. Then, $A_n \cap B(x,\varepsilon/2) \neq \varnothing$ for infinitely many n. Since $\{A_n\}$ is Cauchy, there is an N' such that $\mathfrak{H}(A_m, A_n) < \varepsilon/2$ for all $m, n \geq N'$. Hence, there is an $N \geq N'$ such that A_N intersects $B(x,\varepsilon/2)$ at a point x_N. Hence, for any $n \geq N$, there is an $x_n \in A_n$ such that $d(x_n, x_N) < \varepsilon/2$. Thus, $x_n \in B(x,\varepsilon) \subset U$. It follows that $wl\{A_n\} \subset nl\{A_n\}$.

Next, we prove (3). Assume that the hypothesis in (3) holds. By the previous case, $wl\{A_n\} = nl\{A_n\}$. Hence, it suffices to show $A = nl\{A_n\}$. First, pick $x \in A$, a neighbourhood U of x, and an $\varepsilon > 0$ such that $B(x,\varepsilon) \subset U$. Since $\mathfrak{H}(A_n, A) \to 0$ as $n \to \infty$ and $B(x,\varepsilon) \subset U$, U intersects A_n for all but finitely many n. Hence, $A \subset nl\{A_n\}$.

Conversely, pick $x \in nl\{A_n\}$. Now, for any k, $A_n \cap B(x, 1/k) \neq \varnothing$ holds for all but finitely many n. Hence, we can find a strictly increasing sequence of positive integers $\{n_k\}_{k=1}^{\infty}$ such that, for all k, $A_n \cap B(x,1/k) \neq \varnothing$ for all $n \geq n_k$. Then, we take a sequence $\{x_k\}$ such that $x_k \in A_{n_k} \cap B(x, 1/k)$ for all k. Then, $x_k \to x$ as $k \to \infty$.

Now suppose that $x \notin A$. Then, $d(x, A) > 0$, since A is closed. Then there is an N such that, for all $k \geq N$, $\mathfrak{H}(A_{n_k}, A) < d(x,A)/2$, and hence $d(x_k, A) < d(x,A)/2$. But, if $k \geq \max(N, 2/d(x,A))$, then $d(x, x_k) < d(x,A)/2$, hence $d(x,A) \leq d(x,x_k) + d(x_k, A) < d(x,A)/2 + d(x,A)/2 = d(x,A)$ which is a contradiction. Hence, $x \in A$. It follows that $nl\{A_n\} \subset A$, which completes the proof of (3).

Finally, we prove (2)$\Longrightarrow$(1). So, assume that $wl\{A_n\} = nl\{A_n\}$. By Lemma 1.8, it suffices to prove that any convergent subsequence $\{A_{n_k}\}_{k=1}^{\infty}$ has the limit $wl\{A_n\} = nl\{A_n\}$. By (3), $\{A_{n_k}\}_{k=1}^{\infty}$ has the limit $wl\{A_{n_k}\}_{k=1}^{\infty} = nl\{A_{n_k}\}_{k=1}^{\infty}$. But, by the assumption and Proposition 1.15 (3), $wl\{A_{n_k}\}_{k=1}^{\infty} \subset wl\{A_n\} = nl\{A_n\} \subset nl\{A_{n_k}\}_{k=1}^{\infty} = wl\{A_{n_k}\}_{k=1}^{\infty}$. It follows that these inclusions must be equalities, and the conclusion follows.

1.18 COROLLARY. *If X is compact, then the conclusions in Theorem 1.17 hold for any sequence in $\mathfrak{S}(X)$.*

PROOF: Since X is compact, $\langle X \rangle$ is compact metric. By Theorem 1.13 and Remarks 1.10, $\mathfrak{C}(\langle X \rangle) = \langle \mathfrak{S}(\langle X \rangle)\rangle = \langle \mathfrak{S}(X) \rangle$ is compact, so every sequence in $\langle \mathfrak{S}(X)\rangle$ is relatively compact. Hence, every sequence in $\mathfrak{S}(X)$ is relatively compact. Thus, Theorem 1.17 applies.

From the above theorem and corollary, the following result can be derived:

1.19 COROLLARY. *Suppose that $\{A_n\}$ is a monotonic (i.e monotonically increasing ($A_n \subset A_{n+1}$ for all n) or monotonically decreasing ($A_{n+1} \subset A_n$ for all n)) and relatively compact sequence in $\mathfrak{S}(X)$.*

In particular, if X is compact, $\{A_n\}$ could be any monotonic sequence in $\mathfrak{S}(X)$.

Then, $\{A_n\}$ converges in $\mathfrak{S}(X)$ to $A = \overline{\bigcup_n A_n}$ if it is monotonically increasing, and it converges to $A = \bigcap_n \overline{A_n}$ if it is monotonically decreasing.

PROOF: It is easy to see that, in both cases $A = wl\{A_n\} = nl\{A_n\}$. Hence, Theorem 1.17 or Corollary 1.18 applies.

2. STABLE ITERATED FUNCTION SYSTEMS.

Definitions and examples

2.1 DEFINITION. An **iterated function system** (IFS) is an ordered triple (X, d, Ω), where (X, d) is a complete pseudometric space and Ω is a nonempty family of functions $s : X \to X$.

If Ω consists of one single function w, we write (X, d, w) instead of $(X, d, \{w\})$. We say that (X, d, w) is a **simple** IFS.

The word "iterated" indicates that we are interested in how these functions behave when they are applied several times.

There is a growing theory of IFSs today, and our investigation is a contribution to this. We will generalize the theory of *hyperbolic* IFSs developed by Hutchinson, Barnsley and Demko, Lewellen, and others ([14], [4], [16], [13]). In particular, much of the investigation in this chapter will overlap with Lewellen's work [16], since he too studied attractors, addresses etc. for IFSs with infinitely many functions, although he only studied *hyperbolic* IFSs.

2.2 DEFINITIONS. Let (X, d, Ω) be an IFS. For any n, Ω^n denotes the family of ordered n-tuples $(s_1, \ldots, s_n)$ of functions in Ω. Ω^∞ denotes the family of infinite sequences $\{s_n\}_{n=1}^{\infty}$ of functions in Ω. $\bar{\Omega} : \mathfrak{S}(X) \to \mathfrak{S}(X)$ denotes the map given by $\bar{\Omega}(B) = \cup_{s \in \Omega} s(B)$ for all $B \in \mathfrak{S}(X)$.

Note, in these definitions, s, s_1, s_2, s_n, etc. denote function *variables*. We choose this way of notation rather than indexing the functions.

2.3 DEFINITION. Let (X, d, Ω) be an IFS and let $A \in \mathfrak{S}(X)$. For any $n \geq 0$, the number $\sup\{diam(\{s_1 \circ \cdots \circ s_n(A) | (s_1, \ldots, s_n) \in \Omega^n\})\} \in [0, \infty]$ is denoted $\Delta_n(A)$.

We use the convention that the case $n = 0$ in this definition corresponds to the set A. Thus, $\Delta_0(A) = diam(A)$.

2.4 DEFINITIONS. Let (X, d, Ω) be an IFS. Let $x_0 \in X$ and let $\mathfrak{T}$ be a family of nonempty, closed subspaces of X, which all contain x_0, such that:

(1) For any $B \in \mathfrak{B}(X)$ such that $x_0 \in B$, $B \subset A$ for some $A \in \mathfrak{T}$.
(2) For any $A \in \mathfrak{T}$ and any $s \in \Omega$, $s(A) \subset A$.
(3) For any $A \in \mathfrak{T}$, $\lim_{n \to \infty} \Delta_n(A) = 0$.

Then, (X, d, Ω) is called $(\mathfrak{T}, x_0)$**–stable**.

(X, d, Ω) is called $\mathfrak{T}$**–stable** if, for *every* $x \in X$, it is $(\mathfrak{T}, y)$–stable for some $y \in B(x, \infty)$.

If it is $\{X\}$–stable, it is called **totally stable** and if it is $\mathfrak{T}$–stable for some $\mathfrak{T} \subset \mathfrak{B}(X)$, it is called **boundedly stable**.

(2) above is equivalent to the following condition: *$\bar{\Omega}(\mathfrak{S}(A)) \subset \mathfrak{S}(A)$ for all $A \in \mathfrak{T}$.*

Our definition is admittedly technical. Actually, the only cases which we are really interested in are totally and boundedly stable IFSs. But, in the proof of the fundamental Theorem 2.15, we need some auxiliary spaces which may be neither. Since we want a theory as general as possible, we choose the above definition.

The notion of a totally stable IFS was introduced by Wallin and Karlsson ([26], p. 11). They used it to reformulate a classical theorem about continued fractions.

2.5 PROPOSITION. *If (X, d, Ω) is a $(\mathfrak{T}, x_0)$–stable IFS, then the following statements hold:*

(1) *If d is finite, then (X, d, Ω) is $\mathfrak{T}$–stable.*
(2) *If $X \in \mathfrak{T}$, then (X, d, Ω) is totally stable.*
(3) *If $A \in \mathfrak{T}$ and if d and the functions in Ω are resticted to A, then (A, d, Ω) is totally stable.*
(4) *If $B \in \mathfrak{C}(A)$ for some $A \in \mathfrak{T}$, $x_0 \in B$ and $\bar{\Omega}(B) \subset B$, then (X, d, Ω) is $(\mathfrak{T} \cup \{B\}, x_0)$–stable.*
(5) *If B is as in (4) and if d and the functions in Ω are restricted to B, then (B, d, Ω) is totally stable.*

PROOF: (1), (2), (3), and (4) are obvious, and (5) is a consequence of (3) and (4).

In the sequel, with B as in (5) above, we write (B, d, Ω) and tacitly understand that d and the functions in Ω should be restricted to B.

The simplest case of all is the following:

2.6 COROLLARY. *Let (X, d, Ω) be an IFS. Let $x_0 \in X$ be any point, and let $\mathfrak{T}$ be a family of nonempty, closed subspaces of X satisfying (2) of Definitions 2.4.*

Consider the following conditions:

(1) *(X, d, Ω) is $(\mathfrak{T}, x_0)$–stable.*
(2) *(X, d, Ω) is $\mathfrak{T}$ stable.*
(3) *(X, d, Ω) is totally stable.*
(4) *(X, d, Ω) is boundedly stable.*

If X is bounded, then (1) $\Longrightarrow$ (2) $\Longrightarrow$ (3) $\Longleftrightarrow$ (4). In addition, if $x_0 \in A$ for all $A \in \mathfrak{T}$, then (1) $\Longleftrightarrow$ (2), and if also $\{X\} \in \mathfrak{T}$, then the four conditions are all equivalent.

PROOF: These are easy consequences of Definitions 2.4 and Proposition 2.5.

Our next proposition gives an important example, which has been extensively studied for finitely many functions ([14], [4], [3]), but also for infinitely many functions [16].

2.7 DEFINITION. A **hyperbolic** IFS is an IFS (X, d, Ω) with a *finite* metric d, such that there is a constant $c \in [0, 1)$ such that $d(s(x), s(y)) \leq cd(x, y)$ for all $s \in \Omega$ and for all $x, y \in X$.

2.8 REMARK. By induction, it follows that $\Delta_n(B) \leq c^n diam(B)$ for all $B \in \mathfrak{B}(X)$ and all $n \geq 0$.

2.9 PROPOSITION. *Let (X, d, Ω) be a hyperbolic IFS. Assume that, for some $x_0 \in X$, $\cup_{s \in \Omega}\{s(x_0)\}$ is bounded. Then, (X, d, Ω) is boundedly stable.*

PROOF: Let c be as in Definition 2.7 and let $x_0 \in X$ satisfy the assumption. Put $d = \sup_{s \in \Omega} d(s(x_0), x_0) < \infty$. Put $\mathfrak{T} = \{\bar{B}(x_0, R) | d/(1-c) \le R < \infty\}$. Then, $\mathfrak{T}$ is a family of closed bounded subspaces of X, all containing x_0. We will show that (X, d, Ω) is $(\mathfrak{T}, x_0)$–stable, hence it is boundedly stable by Proposition 2.5 (1).

In Definitions 2.4, (1) is obvious and (3) follows from Remark 2.8.

To prove (2), take $R \in [d/(1-c), \infty)$. Then, $d + cR \le R$. Hence, if $x \in \bar{B}(x_0, R)$ and $s \in \Omega$, then $s(x) \in \bar{B}(s(x_0), cR) \subset \bar{B}(x_0, R)$. Hence, $s(\bar{B}(x_0, R)) \subset \bar{B}(x_0, R)$ for every $s \in \Omega$, so (2) holds.

Of course, if Ω above is finite, then the boundedness condition above is satisfied, but if Ω is infinite, it may fail.

As an example of the latter, take $X = \mathbb{R}$ with the usual metric, and put $\Omega = \{\frac{x+n}{2} | n > 0\}$. This IFS is obviously hyperbolic, but since, for any $x \in X$, $\cup_{n=1}^{\infty}\{\frac{x+n}{2}\}$ is unbounded, the condition is not satisfied, and it is easy to see that (X, d, Ω) is not boundedly stable.

The theory in the following sections is developed for stable IFSs (of different kinds), but for hyperbolic systems, the theory becomes simpler, as we will see. In order to justify our study of stable systems, we should therefore find examples of stable non–hyperbolic IFSs. As we have remarked, an example has been studied in the theory of continued fractions ([26], p. 15). Also, it is not hard to find simple stable systems, some of these are studied in the following sections.

Here, we give another nontrivial class of examples.

2.10 EXAMPLE. We will need some prerequisites.

We define a function $f : [0, \infty) \to [0, \infty)$ as

$$f(t) = \max\left(t\frac{1+\frac{t}{3}}{1+\frac{2t}{3}}, t\left(1 - \frac{(\frac{t}{3})^2}{(1+\frac{t}{3})^2}\right)\right).$$

$f(t)$ is continuous and strictly increasing on $[0, \infty)$, which could be verified by differentiation, for example. Also, for all $t > 0$, $0 < f(t) < t$.

It follows that, for any $t_0 > 0$, the sequence $\{f^n(t_0)\}_{n=0}^{\infty}$ decreases strictly to a limit $s \ge 0$. But if $s > 0$, then $f(s) < s$, and by the continuity of f, $f(t) < s$ for all t in some interval $[s, s+\varepsilon)$ for some $\varepsilon > 0$. But this interval contains $f^n(t_0)$ for some n. Hence $f^{n+1}(t_0) < s$, which is impossible. Hence, $s = 0$.

Now, we turn to our example.

Let X be a Banach space and take any $A \in \mathfrak{B}(X)$. To each $a \in A$, we assign an arbitrary norm–preserving map T_a on X (i.e. an isometry on X fixing 0, which is linear by Mazur–Ulam's Theorem ([2], p. 166)), and a map w_a on X defined by

$$w_a(x) = T_a\left(\frac{x-a}{1+\|x-a\|}\right) + a$$

for all $x \in X$. Put $\Omega = \{w_a | a \in A\}$. Pick $x_0 \in A$. Put $\mathfrak{T} = \{\bar{B}(x_0, R) | 3diam(A) + 1 \leq R < \infty\}$.

We will show that $(X, \| \cdot - \cdot \|, \Omega)$ is a $(\mathfrak{T}, x_0)$–stable non–hyperbolic IFS. Thus, it is boundedly stable by Proposition 2.5 (1).

Since, for any $a \in A$ and any $x \in X \backslash \{a\}$, $\|w_a(x) - w_a(a)\| / \|x - a\| = 1/(1 + \|x - a\|) \to 1$ as $x \to a$, the system is not hyperbolic.

$\mathfrak{T}$ is obviously a family of nonempty, closed subspaces of X, of which all contain x_0 and satisfy (1) of Definitions 2.4.

To prove (2), take any $R \in [3diam(A) + 1, \infty)$. Pick $a \in A$ and $x \in \bar{B}(x_0, R)$ arbitrarily. We will show that $w_a(x) \in \bar{B}(x_0, R)$. Hence, (2) will hold.

If $\|x - a\| \leq 1$, then

$$\|w_a(x) - x_0\| \leq \|w_a(x) - a\| + \|a - x_0\| \leq \|w_a(x) - w_a(a)\| + diam(A)$$

$$\leq \|x - a\| + diam(A) \leq 1 + diam(A) \leq R.$$

If, instead, $\|x - a\| \geq 1$, then

$$\|w_a(x) - x_0\| \leq \|w_a(x) - a\| + \|a - x_0\| \leq \frac{\|x - a\|}{2} + diam(A)$$

$$\leq \frac{\|x - x_0\| + \|x_0 - a\|}{2} + diam(A) \leq \frac{R + 3diam(A)}{2} \leq R.$$

Thus, in both cases, $w_a(x) \in \bar{B}(x_0, R)$.

Finally, we will show that, for any $\bar{B}(x_0, R) \in \mathfrak{T}$, $\Delta_n(\bar{B}(x_0, R)) \leq f^n(2R)$ for all n, with f as above. Hence, (3) will hold.

By induction, using that f is strictly increasing on $[0, \infty)$, it suffices to show that for any $a \in A$ and any $x, y \in X$, $\|w_a(x) - w_a(y)\| \leq f(\|x - y\|)$.

So, pick $a \in A$, $x, y \in X$, put $d = \|x - y\|$, and assume, without loss of generality, that $\|x - a\| \geq \|y - a\|$. Then,

$$\|w_a(x) - w_a(y)\| = \left\| \frac{x - a}{1 + \|x - a\|} - \frac{y - a}{1 + \|y - a\|} \right\|$$

$$= \frac{\|(1 + \|y - a\|)(x - y) - (\|x - a\| - \|y - a\|)(y - a)\|}{(1 + \|x - a\|)(1 + \|y - a\|)}$$

$$\leq d \frac{1 + 2\|y - a\|}{(1 + \|x - a\|)(1 + \|y - a\|)}$$

$$= d \left(\frac{1 + \|y - a\|}{1 + \|x - a\|} - \frac{\|y - a\|^2}{(1 + \|x - a\|)(1 + \|y - a\|)} \right)$$

$$\leq \min \left(d \frac{1 + \|y - a\|}{1 + \|x - a\|}, d \left(1 - \frac{\|y - a\|^2}{(1 + \|y - a\|)^2} \right) \right).$$

If $\|y-a\| \leq d/3$, then $\|x-a\| \geq 2d/3$. Hence, in this case

$$d\frac{1+\|y-a\|}{1+\|x-a\|} \leq d\frac{1+\frac{d}{3}}{1+\frac{2d}{3}} \leq f(d).$$

On the other hand, if $\|y-a\| \geq d/3$, then

$$d\left(1-\frac{\|y-a\|^2}{(1+\|y-a\|)^2}\right) \leq d\left(1-\frac{(\frac{d}{3})^2}{(1+\frac{d}{3})^2}\right) \leq f(d)$$

since the function $1-t^2/(1+t)^2$ decreases on $[0,\infty)$. Hence, in both cases, $\|w_a(x)-w_a(y)\| \leq f(d)$, and we are finished.

Attractors for stable IFSs

In this section, we prove a generalized version of Banach's Fixed Point Theorem, and then use this version to prove the existence of an *attractor* for a stable IFS.

2.11 LEMMA. *Let (X,d,Ω) be a totally stable IFS and let $\{s_n\} \in \Omega^\infty$. For any n and any $y \in X$, put $S_n(y) = s_1 \circ \cdots \circ s_n(y)$, $(S_0(y) = y)$.*
Then, there is a point $x \in X$ such that:

1. *For any $B \in \mathfrak{S}(X)$, the sequence $\{S_n(B)\}$ converges to $\{x\}$ (in $\mathfrak{S}(X)$) with speed $\{\Delta_n(X)\}$.*
2. *For any $y \in X$, the sequence $\{S_n(y)\}$ converges to x (in X) with speed $\{\Delta_n(X)\}$.*

PROOF: The sequence $\{S_n(X)\}$ is monotonically decreasing. It is then easy to see that for any N and any $m,n \geq N$ $\mathfrak{H}(S_m(X), S_n(X)) \leq diam(S_N(X)) \leq \Delta_N(X)$. By (3) of Definitions 2.4, $\{S_n(X)\}$ is Cauchy. Hence, it converges in $\mathfrak{S}(X)$, by Theorem 1.11. By Corollary 1.19, a limit is $C = \cap_{n=0}^{\infty}\overline{S_n(X)}$. By (3) of Definitions 2.4, $diam(C) = 0$. $C \neq \varnothing$, so pick $x \in C$.

Now, for any $B \in \mathfrak{S}(X)$ and any n, $S_n(B) \subset \overline{S_n(X)}$. Since $x \in \overline{S_n(X)}$, we get $\mathfrak{H}(S_n(B),\{x\}) \leq diam(S_n(X)) \leq \Delta_n(X)$. Hence, (1) follows.

For any $y \in X$ and any n, $\{S_n(y)\} \in \mathfrak{I}(X)$. By (1) and Remarks 1.10, (2) follows.

2.12 THEOREM. *Let (X,d,w) be a simple $(\mathfrak{T},x_0)$-stable IFS. Then, there is an $x \in \cap_{A\in\mathfrak{T}}A$, such that the following statements hold:*

1. *For any $y_0 \in \cup_{A\in\mathfrak{T}}A$, the sequence $\{w^n(y_0)\}$ converges to x with speed $\{\Delta_n(A)\}$ if $y_0 \in A \in \mathfrak{T}$.*
2. *If $w(y) = y$ for some $y \in \cup_{A\in\mathfrak{T}}A$, then $y \in \langle x\rangle$.*
3. *If, for some $A \in \mathfrak{T}$, $d|_{A\times A}$ is a metric and $w|_A$ is continuous, then x is a unique fixed point of w in $\cup_{A\in\mathfrak{T}}A$.*
4. *If d is finite or if (X,d,w) is totally stable, then in (1)—(3), X may be substituted for $\cup_{A\in\mathfrak{T}}A$.*

PROOF: By Proposition 2.5 (3), (A, d, w) is totally stable for every $A \in \mathfrak{T}$. Hence, for every $A \in \mathfrak{T}$, there is an $x_A \in A$ such that for any $y \in A$, the sequence $\{w^n(y)\}$ converges to x_A with speed $\{\Delta_n(A)\}$, by Lemma 2.11 (2). This holds in particular for $y = x_0$, since by definition, $x_0 \in A$ for all $A \in \mathfrak{T}$. Hence, $d(x_A, x_{A'}) = 0$ for all $A, A' \in \mathfrak{T}$. By definition, each $A \in \mathfrak{T}$ is closed. Hence, we can take a common x_A for all A. Put x as this common x_A. This gives (1).

To prove (2), assume that $w(y) = y$ for some $y \in \cup_{A\in\mathfrak{T}} A$. Then, by induction, $w^n(y) = y$ for all n. Hence, $w^n(y) \to y$ as $n \to \infty$. By (1), $y \in \langle x \rangle$.

If the assumptions of (3) hold, then (1) gives $w(x) = \lim_{n\to\infty} w \circ w^n(x) = \lim_{n\to\infty} w^n(x) = x$, and by (2), the fixed point x is unique in $\cup_{A\in\mathfrak{T}} A$. So, (3) holds.

If d is finite, then (1) of Definitions 2.4 gives $\cup_{A\in\mathfrak{T}} A = X$. Also, the same is obviously true if (X, d, w) is totally stable. Hence, (4) holds.

2.13 REMARK. Using Proposition 2.9, we see that the classical Banach's Fixed Point Theorem is a special case of Theorem 2.12.

In the same way as Banach's Fixed Point Theorem is used to prove the corresponding result for hyperbolic IFSs with finitely many functions ([3], p. 82 and [14], p. 728), we use Theorem 2.12 to prove our next theorem.

2.14 DEFINITION. Let (X, d, Ω) be an IFS. A set $B \in \mathfrak{S}(X)$ is called **Ω–invariant** if $\bar{\Omega}(B) = B$.

If it is clear from the context which Ω we mean, B is just called **invariant.**

2.15 THEOREM. *Let (X, d, Ω) be a $(\mathfrak{T}, x_0)$–stable IFS. Then, there is a set $S \in \mathfrak{C}(\cap_{A\in\mathfrak{T}} A)$ such that:*

(1) *For any $B \in \cup_{A\in\mathfrak{T}} \mathfrak{S}(A)$, $\{\bar{\Omega}^n(B)\}$ converges to S (in $\mathfrak{S}(X)$) with speed $\{\Delta_n(A)\}$ if $B \in \mathfrak{S}(A)$.*

(2) *If, for some $B \in \cup_{A\in\mathfrak{T}} \mathfrak{S}(A)$, $\bar{\Omega}(B) = B$, then $\overline{B} \subset S$.*

(3) *If, for some $A \in \mathfrak{T}$, $\bar{\Omega}(\mathfrak{C}(A)) \subset \mathfrak{C}(A)$ and $\bar{\Omega}|_{\mathfrak{S}(A)}$ is continuous, then S is a unique closed Ω–invariant set in $\cup_{A\in\mathfrak{T}} \mathfrak{S}(A)$.*

(4) *If (X, d, Ω) is totally stable, then, in (1)—(3), $\mathfrak{S}(X)$ may be substituted for $\cup_{A\in\mathfrak{T}} \mathfrak{S}(A)$.*

PROOF: Put $\overline{\mathfrak{T}} = \{\mathfrak{S}(A) | A \in \mathfrak{T}\}$. We will show that the *simple* IFS $(\mathfrak{S}(X), \mathfrak{H}, \bar{\Omega})$ is $(\overline{\mathfrak{T}}, \{x_0\})$–stable with $\Delta_n(\mathfrak{S}(A)) \leq \Delta_n(A)$ for all n and all $A \in \mathfrak{T}$.

By Remarks 1.10, each $\mathfrak{S}(A) \in \overline{\mathfrak{T}}$ is a closed subspace of $\mathfrak{S}(X)$ containing $\{x_0\}$.

Now, let $\mathfrak{A}$ be a *bounded* subset $\mathfrak{S}(X)$ containing $\{x_0\}$. Then, for any $B \in \mathfrak{A}$, $\mathfrak{H}(B, \{x_0\}) \leq diam(\mathfrak{A}) < \infty$. Hence, $B \subset \bar{B}(x_0, diam(\mathfrak{A}))$, which is bounded. Since $\bar{B}(x_0, diam(\mathfrak{A}))$ is contained in some $A \in \mathfrak{T}$, $B \in \mathfrak{S}(A)$. It follows that $\mathfrak{A} \subset \mathfrak{S}(A) \in \overline{\mathfrak{T}}$. Hence, (1) of Definitions 2.4 holds.

(2) of Definitions 2.4 follows easily from the corresponding property of the $(\mathfrak{T}, x_0)$–stability of (X, d, Ω).

Next, pick $A \in \mathfrak{T}$, an n, and two sets $B, B' \in \bar{\Omega}^n(\mathfrak{S}(A))$. Then, there are two sets $E, E' \in \mathfrak{S}(A)$ such that $\bar{\Omega}^n(E) = B$ and $\bar{\Omega}^n(E') = B'$. Now, pick $y \in B$. Then, there is a $z \in E$ and an n-tuple $(s_1, \ldots, s_n) \in \Omega^n$ such that $s_1 \circ \cdots \circ s_n(z) = y$. Pick $z' \in E'$ and put $y' = s_1 \circ \cdots \circ s_n(z')$. Then, $y' \in B'$ and $d(y, y') \leq \Delta_n(A)$. It follows that $B \subset \overline{B'_{\Delta_n(A)}}$. By the same argument, $B' \subset \overline{B_{\Delta_n(A)}}$. Hence, $\mathfrak{H}(B, B') \leq \Delta_n(A)$. It follows that $diam(\bar{\Omega}^n(\mathfrak{S}(A))) \leq \Delta_n(A)$ and (3) of Definitions 2.4 holds.

We have now proved the statement above.

We can now apply Theorem 2.12 to the simple $(\overline{\mathfrak{T}}, \{x_0\})$-stable IFS $(\mathfrak{S}(X)), \mathfrak{H}, \bar{\Omega})$. Since to every $B \in \mathfrak{S}(X)$ there is a unique *closed* $C \in \mathfrak{S}(X)$ such that $\mathfrak{H}(B, C) = 0$, namely $C = \overline{B}$, and since all $\mathfrak{S}(A) \in \overline{\mathfrak{T}}$ are *closed* subspaces of $\mathfrak{S}(X)$, (1) follows from (1) of Theorem 2.12 as does (2).

If the assumptions of (3) hold, then on the *metric* quotient space $(\mathfrak{C}(A), \mathfrak{H})$ (see Remarks 1.10), the quotient function $\langle \Omega \rangle$ exists, is continuous, and can be identified with $\bar{\Omega}|_{\mathfrak{C}(A)}$.

Then, it is easy to see that $(\mathfrak{C}(A), \mathfrak{H}, \bar{\Omega}|_{\mathfrak{C}(A)})$ is a simple, totally stable IFS.

Now, (3) follows from (3) of Theorem 2.12 and (2).

Finally, (4) is obvious.

2.16 PROPOSITION. *Let (X, d, Ω) be a $(\mathfrak{T}, x_0)$-stable IFS. If, for some $A \in \mathfrak{T}$, the family $\Omega|_A = \{s|_A | s \in \Omega\}$ is equicontinuous, then $\bar{\Omega}|_{\mathfrak{S}(A)}$ is uniformly continuous.*

PROOF: Assume that $\Omega|_A$ is equicontinuous for some $A \in \mathfrak{T}$. Pick $\varepsilon > 0$. Then, there is a $\delta > 0$ such that, for any $y, z \in A$ with $d(y, z) < \delta$, $d(s(y), s(z)) < \varepsilon/2$ for all $s \in \Omega$. Hence, for any $s \in \Omega$ and any $B, B' \in \mathfrak{S}(A)$ with $\mathfrak{H}(B, B') < \delta$, we get $s(B) \subset s(B')_{\varepsilon/2}$ and $s(B') \subset s(B)_{\varepsilon/2}$. It follows that $\mathfrak{H}(\bar{\Omega}(B), \bar{\Omega}(B')) \leq \varepsilon/2 < \varepsilon$. Hence, $\bar{\Omega}|_{\mathfrak{S}(A)}$ is uniformly continuous.

We now get the corresponding result for hyperbolic IFSs as a corollary (see [3], p. 82 and [14], p. 728): :

2.17 COROLLARY. *Let (X, d, Ω) be a hyperbolic IFS, with c as in Definition 2.7, which satisfies the assumption of Proposition 2.9. Assume that for some $C \in \mathfrak{C}(X)$, $\bar{\Omega}(\mathfrak{C}(C)) \subset \mathfrak{C}(C)$ (this holds if Ω is finite, C is compact, and $s(C) \subset C$ for all $s \in \Omega$, for example).*

Then there is a unique $S \in \mathfrak{CB}(X)$ such that $\bar{\Omega}(S) = S$. Furthermore, for this S, all sequences $\{\bar{\Omega}^n(B)\}$ with $B \in \mathfrak{S}(X)$ converge to S in $\mathfrak{S}(X)$ with speed $\{c^n D\}$, for some D depending upon B.

PROOF: By Proposition 2.9, (X, d, Ω) is boundedly stable, that is, it is $\mathfrak{T}$-stable for some $\mathfrak{T} \subset \mathfrak{B}(X)$. For each $A \in \mathfrak{T}$, $\Omega|_A$ is equicontinuous (with $\Omega|_A$ as in Proposition 2.16). Hence, $\bar{\Omega}|_{\mathfrak{S}(A)}$ is uniformly continuous by Proposition 2.16.

Now, with C as above, pick $x_0 \in C$, remove from $\mathfrak{T}$ all sets A such that $x_0 \notin A$, and for some A still in $\mathfrak{T}$ (such an A exists, since $\{x_0\}$ is bounded), adjoin $C \cap A$ to $\mathfrak{T}$, using Proposition 2.5 (4). Then, (X, d, Ω) is $(\mathfrak{T}, x_0)$–stable with this new $\mathfrak{T}$.

Now the assumptions of (3) of Theorem 2.15 hold. Hence, using that theorem and Remark 2.8, we obtain the desired results.

2.18 DEFINITION. The set S in Theorem 2.15 is called the **attractor** of the $(\mathfrak{T}, x_0)$–stable IFS (X, d, Ω).

In the next section, we will investigate attractors and sets with Hausdorff–pseudo-distance 0 from them.

Addressing and examples

First, we take a closer look at the Theorems 2.12 and 2.15, and point out what they say and don't say. This will also show the relevance of the technical Definitions 2.4.

2.19 EXAMPLE. Let (X, d, w) and (X', d', w') be two simple totally stable IFSs such that $X \cap X' = \varnothing$, d and d' are metrics, and w and w' are continuous.

We now define d'' as the only pseudometric on $X'' = X \cup X'$ such that $d''|_{X \times X} = d$, $d''|_{X' \times X'} = d'$ and $d''(X \times X') = \{\infty\}$. We also define w'' as the unique function on X'' such that $w''|_X = w$ and $w''|_{X'} = w'$. w'' is then continuous.

Then, (X'', d'', w'') is a simple IFS which is *both* $\{X\}$–stable and $\{X'\}$–stable, but not totally stable. Theorem 2.12 gives *two* different fixed points in X'', $x \in X$ and $x' \in X'$, with $d''(x, x') = \infty$, such that $\{w''^n(y_0)\}$ converges to x for all $y_0 \in X$ and $\{w''^n(y_0')\}$ converges to x' for all $y_0' \in X$.

This is a reason why we need a family $\mathfrak{T}$ and a point x_0 in Definitions 2.4. We need a point to specify in which finitely–metric subspace we are (compare Proposition 2.5 (1)).

Now, the reader might think: Infinite pseudometrics just seem to complicate the theory. Why allow them in the first place?

The answer is that we want to define the Hausdorff pseudometric not only on $\mathfrak{B}(X)$ but also on $\mathfrak{S}(X)$ for unbounded spaces X. Otherwise we cannot use Theorem 2.12 in the proof of Theorem 2.15, for example in the case of a totally stable IFS on an unbounded space X.

2.20 EXAMPLE. This example is perhaps more natural than the preceding one.

Consider $(\mathbf{R}, |\quad|, w)$, where $|\quad|$ is the usual metric on $\mathbf{R}$ and $w(x) = x/2$ for all $x \in \mathbf{R}$. This is a simple hyperbolic IFS. Hence, it is boundedly stable, by Proposition 2.9 and the remark succeding it. $\{0\}$ is its attractor and it is invariant. But $\mathbf{R}$ is closed and invariant too. This does not contradict (2) and (3) of Theorem 2.15 since $\mathbf{R}$ is not bounded, and therefore, $\mathbf{R} \notin \cup_{A \in \mathfrak{T}} \mathfrak{S}(A)$ if $\mathfrak{T} \subset \mathfrak{B}(X)$. Again, this shows the necessity of considering a family $\mathfrak{T}$.

2.21 EXAMPLE. Consider the first pseudometric space $(\mathbf{C}, d)$ of Examples 1.5. Put $w(x + iy) = \frac{1}{2}x + i(y + 1)$ for all $x, y \in \mathbf{R}$. Then, w is continuous and $(\mathbf{C}, d, w)$ is a simple, boundedly stable IFS with the imaginary axis as the (invariant) attractor. But, w has no fixed point in $\mathbb{C}$. This does not contradict (3) of Theorem 2.12 since d is no metric.

2.22 EXAMPLE. Let X be $[0,1]$ with d as the usual metric. Put $w(0) = 1$ and $w(x) = x/2$ for $x \in (0,1]$. Then, (X, d, w) is a simple totally stable IFS with $\{0\}$ as the attractor. But, neither $\{0\}$ nor any other set is invariant. The reason is that w is neither continous at 0, nor does it take closed sets to closed sets in any neighbourhood of 0, so the hypotheses of (3) of Theorem 2.15 are not satisfied.

I have not been able to decide whether or not the continuity assumption of (3) of Theorem 2.15 can be dispensed with.

However, the assumption of taking closed sets to closed sets cannot:

2.23 EXAMPLE. Let X be the closed unit disk in $\mathbb{C}$ with d as the usual metric. Put $\Omega = \{(z + e^{2i\pi\theta})/2 | \theta \in \mathbb{Q}\}$. Then, (X, d, Ω) is a hyperbolic and totally stable IFS.

Put $Y = X^\circ \cup \{e^{2i\pi\theta} | \theta \in \mathbb{Q}\}$. We will show that $\bar{\Omega}(X) = Y = \bar{\Omega}(Y)$. Then, Y is invariant and, by (2) of Theorem 2.15, X is the attractor although X is not invariant. $\bar{\Omega}$ is uniformly continuous, but it maps the closed set X onto the non–closed set Y.

First, put $z = e^{2i\pi\theta}$, with $\theta \notin \mathbb{Q}$. Then, $z \notin s(X)$ for all $s \in \Omega$. Hence, $\bar{\Omega}(X) \subset Y$.

Hence, it suffices to show $Y \subset \bar{\Omega}(Y)$. Take $z = e^{2i\pi\theta}$ with $\theta \in \mathbb{Q}$. Then, $(z + e^{2i\pi\theta})/2 = z$. Hence, $z \in \bar{\Omega}(Y)$.

Next, since $(1 + e^{i\pi})/2 = 0$, then $0 \in \bar{\Omega}(Y)$.

Finally, take $z = re^{2i\pi\theta}$ with $r \in (0,1)$ and $\theta \in \mathbb{R}$. If $r \geq 1/2$, put $\theta' = \theta$ and $r' = 2r - 1$, and if $r < 1/2$, put $\theta' = \theta + 1/2$ and $r' = 1 - 2r$. Put $z' = r'e^{2i\pi\theta'} \in X^\circ$. Put $s'(z'') = (z'' + e^{2i\pi\theta})/2$ for all $z'' \in X^\circ$. Then, $z = s'(z') \in s'(X^\circ)$. Since $s'(X^\circ)$ is open, it contains $re^{2i\pi\phi}$ for some $\phi \in \mathbb{Q}$. Put $\phi' = \phi$ if $r \geq 1/2$ and put $\phi' = \phi + 1/2$ if $r < 1/2$. Put $s(z'') = (z'' + e^{2i\pi\phi})/2$ for all $z'' \in X$. Then, $s \in \Omega$, $s(r'e^{2i\pi\phi'}) = re^{2i\pi\phi}$ and by symmetry, $z \in s(X^\circ) \subset s(Y)$. It follows that $Y \subset \bar{\Omega}(Y)$.

2.24 EXAMPLE. Let X be $[0,1]$ with d as the usual metric. Put $\Omega = \{\frac{x}{2}, \frac{x+1}{2}\}$. Then, (X, d, Ω) is a hyperbolic and totally stable IFS with X as invariant attractor. However, $[0,1)$ and $(0,1]$ are invariant too (their closures are the attractor [0,1], of course). Hence, there can be several invariant sets, but at most one is *closed*, namely, the attractor.

Instead of finding the attractor by constructing a simple IFS and then using Theorem 2.12, as we did in the proof of Theorem 2.15, we can use Lemma 2.11 directly.

We will generalize the *addresses* studied by Barnsley and Lewellen ([3], ch. 4, [16], ch. 2) for hyperbolic systems.

2.25 DEFINITIONS. Let (X, d, Ω) be a $(\mathfrak{T}, x_0)$–stable IFS. We define the **address map** $Adr : \Omega^\infty \to \langle X \rangle$ by putting $Adr(\{s_n\}) = \lim_{n\to\infty} \langle s_1 \circ \cdots \circ s_n(x_0) \rangle$ for all $\{s_n\} \in \Omega^\infty$. The convergence of this sequence is guaranteed by (2) of Lemma 2.11 and (3) of Proposition 2.5. If $x \in Adr(\{s_n\})$ for some $\{s_n\} \in \Omega^\infty$, we say that $\{s_n\}$ is an **address** of x and that x is an **addressate** of $\{s_n\}$. The set of *all* addressates, denoted $Adr(\Omega^\infty)$ (somewhat improperly), is called the **addressate set**.

If d is a *metric*, then we consider X instead of $\langle X\rangle$ as the target of the address map. In this case, every $\{s_n\} \in \Omega^\infty$ has a *unique* addressate.

2.26 REMARK. In this definition, we chose x_0 as the starting point for the sequence above. But by Lemma 2.11, we could have chosen any $y \in \cup_{A\in\mathfrak{T}} A$ without changing the result since each (A, d, Ω) is totally stable. In fact, by that lemma, we could have started with any $B \in \cup_{A\in\mathfrak{T}}\mathfrak{S}(A)$ and then taken a corresponding sequence in $\mathfrak{S}(X)$ instead and then arrived at an equivalence class $\langle x\rangle$ of addressates (a singleton set if d is a metric).

2.27 PROPOSITION. *Let (X, d, Ω) be a $(\mathfrak{T}, x_0)$-stable IFS with attractor S. Then,* $\overline{Adr(\Omega^\infty)} = S$.

PROOF: Pick $A \in \mathfrak{T}$. Then (A, d, Ω) is totally stable. We will show that $nl\{\bar{\Omega}^n(A)\} = \overline{Adr(\Omega^\infty)}$. Since the convergent sequence $\{\bar{\Omega}^n(A)\}$ in $\mathfrak{S}(A)$ is relatively compact, the conclusion then follows from (3) of Theorem 1.17.

So, pick $x \in nl\{\bar{\Omega}^n(A)\}$ and pick $\varepsilon > 0$. Take N such that $\Delta_n(A) < \varepsilon/2$ and $B(x, \varepsilon/2)$ intersects $\bar{\Omega}^n(A)$ for all $n \geq N$. Then, for any $n \geq N$, there is an $(s_1, \ldots, s_n) \in \Omega^n$ and a $y \in A$ such that $s_1 \circ \cdots \circ s_n(y) \in B(x, \varepsilon/2)$. Now, lengthen $(s_1, \ldots, s_n)$ to a sequence $\{s_k\} \in \Omega^\infty$. By Lemma 2.11 (2), $\langle d\rangle(\langle s_1 \circ \cdots \circ s_n(y)\rangle, Adr(\{s_k\})) \leq \Delta_n(A) < \varepsilon/2$. Hence, $B(x, \varepsilon) \cap Adr(\Omega^\infty) \neq \varnothing$.

It follows that $nl\{\bar{\Omega}^n(A)\} \subset \overline{Adr(\Omega^\infty)}$.

Conversely, pick $x \in Adr(\Omega^\infty)$ and $\varepsilon > 0$. Then, there is an $\{s_n\} \in \Omega^\infty$ such that $\{s_1 \circ \cdots \circ s_n(x_0)\}$ converges to x with speed $\{\Delta_n(A)\}$. Take N such that $\Delta_n(A) < \varepsilon$ for all $n \geq N$. Then, for any $n \geq N$, $B(x, \varepsilon) \cap \bar{\Omega}^n(A) \neq \varnothing$.

Using Proposition 1.15 (4), $\overline{Adr(\Omega^\infty)} \subset nl\{\bar{\Omega}^n(A)\}$ follows, which completes the proof.

This proposition can be established more directly without using Theorem 1.17. This was done by Wallin and Karlsson for totally stable systems ([26], Th. 2, p. 13).

The attractor, the addressate set, and the invariant sets in $\cup_{A\in\mathfrak{T}}\mathfrak{S}(A)$, if there are any, have all Hausdorff pseudodistance 0 from each other. Therefore they represent the same equivalence class in the quotient space $\langle\mathfrak{S}(X)\rangle$. The following propositions give other relations between these sets.

2.28 PROPOSITION. *Let (X, d, Ω) be a $(\mathfrak{T}, x_0)$-stable IFS, and let $B \in \cup_{A\in\mathfrak{T}}\mathfrak{S}(A)$ satisfy $B \subset \bar{\Omega}(B)$.*

Then, $B \subset Adr(\Omega^\infty)$.

In particular, the conclusion holds for every invariant $B \in \cup_{A\in\mathfrak{T}}\mathfrak{S}(A)$.

PROOF: Pick $y_0 \in B$. Since $B \subset \bar{\Omega}(B)$, $y_0 = s_1(y_1)$ for some $s_1 \in \Omega$ and for some $y_1 \in B$. Also, $y_1 = s_2(y_2)$ for some $s_2 \in \Omega$ and for some $y_2 \in B$. Hence, $y_0 = s_1 \circ s_2(y_2)$. Continuing in this manner, we obtain a sequence $\{s_n\} \in \Omega^\infty$ such that

$y_0 \in s_1 \circ \cdots \circ s_n(B)$ for all n. It follows that $\mathfrak{H}(s_1 \circ \cdots \circ s_n(B), \{y_0\}) \le \Delta_n(A)$ for all n and all $A \in \mathfrak{T}$ such that $B \subset A$. Hence, the sequence $\{s_1 \circ \cdots \circ s_n(B)\}$ converges to $\{y_0\}$ in $\mathfrak{S}(X)$. By Remark 2.26, $\{s_n\}$ is an address of y_0.

2.29 COROLLARY. *If (X, d, Ω) is a $(\mathfrak{T}, x_0)$–stable IFS with the attractor S such that $S \subset \bar{\Omega}(S)$, then $Adr(\Omega^\infty) = S$.*

In particular, the conclusion holds if S is invariant.

PROOF: This follows immediately from Propositions 2.27 and 2.28.

2.30 PROPOSITION. *Let (X, d, Ω) be a $(\mathfrak{T}, x_0)$–stable IFS such that all $s \in \Omega$ are continuous on $Adr(\Omega^\infty)$. Then, $\bar{\Omega}(Adr(\Omega^\infty)) \subset Adr(\Omega^\infty)$.*

In addition, if $d|_{Adr(\Omega^\infty)\times Adr(\Omega^\infty)}$ is a metric, then $\bar{\Omega}(Adr(\Omega^\infty)) = Adr(\Omega^\infty)$.

PROOF: Pick $y \in \bar{\Omega}(Adr(\Omega^\infty))$. Then, $y = s_0(z)$ for some $s_0 \in \Omega$ and for some $z \in Adr(\Omega^\infty)$. Let $\{s_n\}_{n=1}^\infty \in \Omega^\infty$ be an address of z. Since s_0 is continuous at z, it follows that $s_0 \circ s_1 \circ \cdots \circ s_n(x_0) \to s_0(z) = y$ as $n \to \infty$. Hence, $\{s_{n-1}\}_{n=1}^\infty$ is an address of y. It follows that $\bar{\Omega}(Adr(\Omega^\infty)) \subset Adr(\Omega^\infty)$.

Conversely, if $d|_{Adr(\Omega^\infty)\times Adr(\Omega^\infty)}$ is a metric, pick $y \in Adr(\Omega^\infty)$ and let $\{s_n\}_{n=1}^\infty$ be an address of y. Put $\langle z\rangle = Adr(\{s_{n+1}\}_{n=1}^\infty)$. Then, s_1 is continuous at z and $y = \lim_{n\to\infty} s_1 \circ s_2 \circ \cdots \circ s_n(y) = s_1(z)$, by Remark 2.26, Proposition 2.27, the previous case, and the metricity assumption. Hence, $y \in \bar{\Omega}(Adr(\Omega^\infty))$. It follows that $Adr(\Omega^\infty) \subset \bar{\Omega}(Adr(\Omega^\infty))$.

As is seen by Example 2.22, the continuity assumption cannot be dispensed with. The metricity assumption for the identity case is also essential, since otherwise there might be problems with the different points in the equivalence classes in $\langle X\rangle$.

Another question that arises is: Must the addressate set and the attractor coincide, i.e. must the addressate set be closed? The answer is no, as the following example shows:

2.31 EXAMPLE. Let (X, d, Ω) be the totally stable IFS of Example 2.23. Since $\bar{\Omega}(B) = Y$ for all B such that $Y \subset B \subset X$, Propositions 2.28 and 2.30 give $Adr(\Omega^\infty) = Y$.

We will now discuss another question. Consider a complete pseudometric space (X, d), and consider two families Ω and Ω_0 of functions on X such that $\bar{\Omega} = \bar{\Omega}_0$. What can we say about the relation between the IFSs (X, d, Ω) and (X, d, Ω_0)?

Not much, unfortunately.

2.32 EXAMPLE. Let (X, d, Ω) be the totally stable IFS of Example 2.24. Define $w_1 : [0,1] \to [0,1]$ by $w_1(x) = x/2$ if $x \in \mathbb{Q}$ and $w_1(x) = (x+1)/2$ if $x \notin \mathbb{Q}$. Define $w_2 : [0,1] \to [0,1]$ by $w_2(x) = (x+1)/2$ if $x \in \mathbb{Q}$ and by $w_2(x) = x/2$ if $x \notin \mathbb{Q}$. Put $\Omega_0 = \{w_1, w_2\}$, and consider the IFS (X, d, Ω_0). This IFS is not totally stable, and hence is not stable in any way by Corollary 2.6 since it is easy to see that $w_1^n([0,1])$ converges to $\{0,1\}$ in $\mathfrak{S}(X)$, contradicting (3) of Definitions 2.4.

But, it is obvious that $\bar{\Omega} = \bar{\Omega}_0$, and yet (X,d,Ω) is totally stable, but (X,d,Ω_0) is not stable in any way!

If we assume that both systems are $(\mathfrak{T},x_0)$–stable, then the invariant sets must be the same, of course, and by (1) of Theorem 2.15, the attractors must coincide too.

But the addressate sets need not coincide:

2.33 EXAMPLE. Let (X,d) and w be as in Example 2.22. Let Ω be the family of all constant functions mapping $[0,1]$ into $(0,1]$ and put $\Omega_0 = \Omega \cup \{w\}$. Then, it is easy to see that (X,d,Ω) and (X,d,Ω_0) are both totally stable and that $\bar{\Omega} = \bar{\Omega}_0$. $[0,1]$ is the attractor to both systems.

But, $Adr(\Omega^\infty) = (0,1]$ and $Adr(\Omega_0^\infty) = [0,1]$.

However:

2.34 PROPOSITION. *Let (X,d,Ω) and (X,d,Ω_0) both satisfy the assumptions of Proposition 2.30, including the metricity assumption, with the same $\mathfrak{T}$ and x_0.*

Assume that $\bar{\Omega}$ and $\bar{\Omega}_0$ agree on the set $\{Adr(\Omega^\infty), Adr(\Omega_0^\infty)\}$.

Then, $Adr(\Omega^\infty) = Adr(\Omega_0^\infty)$.

PROOF: The assumptions and Propositions 2.30 and 2.28 give $Adr(\Omega^\infty) \subset Adr(\Omega_0^\infty)$ and likewise $Adr(\Omega_0^\infty) \subset Adr(\Omega^\infty)$.

2.35 REMARKS. Hitherto, two different "continuity" conditions have occurred. In (3) of Theorem 2.15 and Proposition 2.16, we considered continuity of $\bar{\Omega}$, while in Proposition 2.30 we considered continuity of *every* $s \in \Omega$ on a set. How are these related to each other?

First, in the totally stable IFS (X,d,Ω_0) in Example 2.33, $\bar{\Omega}_0$ is continuous (in fact, constant), although $w \in \Omega_0$ is not continuous at 0.

Second, without any assumption of stability, it is easy to find IFSs (X,d,Ω) with all $s \in \Omega$ continuous but with $\bar{\Omega}$ discontinuous.

However, I have not been able to answer the following question: *Can the equicontinuity assumption of Proposition 2.16 be weakened to the assumption of continuity on A for every $s \in \Omega$?*

Periodic points and continuous addressing

We saw in Example 2.24 that there can be invariant proper subsets of the attractor other than the addressate set. We will now see that in a large class of cases, there are such sets.

To do this we generalize the notion of *periodic points*, which were studied by Barnsley ([3], p. 132 ff.), Lewellen ([16], sect. 2.4), and others, from hyperbolic to stable IFSs.

2.36 DEFINITIONS. Let (X,d,Ω) be a $(\mathfrak{T},x_0)$–stable IFS.

A point $x \in Adr(\Omega^\infty)$ is called **eventually periodic** if it has an address $\{s_n\}$ such that for some $k \geq 0$ and some $N > 0$, $s_{n+N} = s_n$ for all $n > k$. If $k = 0$, x is called **periodic**.

2.37 PROPOSITION. *Let (X, d, Ω) be a $(\mathfrak{T}, x_0)$–stable IFS with attractor S. If P is its set of periodic points, then $\overline{P} = S$.*

PROOF: By Proposition 2.27, it suffices to prove that P is dense in $Adr(\Omega^\infty)$.

So, pick $y \in Adr(\Omega^\infty)$ with address $\{s_n\}$, and pick $\varepsilon > 0$. Take an N such that $\Delta_N(A) < \varepsilon$ for some $A \in \mathfrak{T}$. Take $\{s'_n\}$ such that $s'_{mN+j} = s_j$ for all $m \geq 0$ and all $j : 1 \leq j \leq N$. Then, $\langle d\rangle(Adr(\{s'_n\}), Adr(\{s_n\})) \leq \Delta_N(A) < \varepsilon$. Since every $x \in Adr(\{s'_n\})$ is periodic, the conclusion follows.

2.38 PROPOSITION. *The complete statement of Proposition 2.30 is still true if the set of all eventually periodic points, denoted Q, everywhere in that statement is substituted for $Adr(\Omega^\infty)$.*

PROOF: The proof is very similar to the proof of Proposition 2.30 and is left to the reader.

In Example 2.22, $Q = \{0\}$, but it is not invariant, so the continuity condition taken from Proposition 2.30 is essential.

Of course, in most cases, $Q \neq Adr(\Omega^\infty)$ so that we have at least two different invariant sets, but in some "degenerate" cases, they may coincide.

2.39 EXAMPLE. Let X be $[0, 1]$ with d as the usual metric, and let Ω be the set of all constant functions on $[0, 1]$. Then, (X, d, Ω) is stable in all ways (Corollary 2.6), and (with S, P, and Q as in the above propositions) $P = Q = Adr(\Omega^\infty) = S = [0, 1] = X$.

We will now pseudometrize Ω and Ω^∞.

2.40 DEFINITIONS. Let (X, d, Ω) be an IFS.

For any $B \in \mathfrak{S}(X)$, we define a pseudometric $|\cdot,\cdot|_B$ on Ω by: $|s, s'|_B = \min(\sup_{x \in B} d(s(x), s'(x)), 1)$ for all $s, s' \in \Omega$.

We define a pseudometric $|\cdot,\cdot|_B$ on Ω^∞ by $|\{s_n\}, \{s'_n\}|_B = \sum_{n=1}^{\infty} 2^{-n} |s_n, s'_n|_B$ for all $\{s_n\}, \{s'_n\} \in \Omega^\infty$.

If it is clear which B we mean, we drop the subscripts. If nothing else is stated, we assume that $B = X$.

2.41 REMARK. It is easy to see that if $B' \subset B$ ($B' \neq \emptyset$), then the identity map from $(\Omega, |\cdot,\cdot|_B)$ to $(\Omega, |\cdot,\cdot|_{B'})$ is uniformly continuous. The corresponding statement for Ω^∞ is also true.

2.42 REMARKS. $|\cdot,\cdot|_B$ will give Ω^∞ the product topology given by the $|\cdot,\cdot|_B$–topology on Ω.

$(\Omega, |\cdot,\cdot|_B)$ is metric if and only if $(\Omega^\infty, |\cdot,\cdot|_B)$ is metric.

Also, it is not difficult to verify that, with these pseudometrics, Ω is complete if and only if Ω^∞ is complete. (Take componentwise limits.)

2.43 THEOREM. *Let (X, d, Ω) be a $(\mathfrak{T}, x_0)$–stable IFS such that all $s \in \Omega$ are continuous on $Adr(\Omega)$ and such that the family $\Omega|_{Adr(\Omega^\infty)} = \{s|_{Adr(\Omega^\infty)} | s \in \Omega\}$ is equicontinuous.*

Suppose also that Ω^∞ has the pseudometric $|\cdot,\cdot|_B$ for some B such that $Adr(\Omega^\infty) \subset B \subset X$.

Then, Adr is uniformly continuous.

PROOF: By Remark 2.41, we may assume that $B = Adr(\Omega^\infty)$. By repeated applications of Proposition 2.30, $\bar{\Omega}^n(Adr(\Omega^\infty)) \subset Adr(\Omega^\infty)$ for all n.

It follows that for any n, the family $\{s_1|_{Adr(\Omega^\infty)} \circ \cdots \circ s_n|_{Adr(\Omega^\infty)} | (s_1, \ldots, s_n) \in \Omega^n\}$ is equicontinuous. Pick $\varepsilon > 0$. Take an N such that $\Delta_N(Adr(\Omega^\infty)) < \varepsilon/3$. For $1 \le k \le N$, take $\delta_k \in (0, 1]$ such that $d(s_1 \circ \cdots \circ s_{k-1}(x), s_1 \circ \cdots \circ s_{k-1}(y)) < \varepsilon/(3N)$ for all $x, y \in Adr(\Omega^\infty)$ such that $d(x, y) < \delta_k$ and all $s_1, \ldots, s_{k-1} \in \Omega$. Put $\delta = \min_{1 \le k \le N} 2^{-k}\delta_k$.

Now, take $\{s_n\}, \{s'_n\} \in \Omega^\infty$ such that $|\{s_n\}, \{s'_n\}| < \delta$. Then, $|s_k, s'_k| < \delta_k$ holds for all k such that $1 \le k \le N$. Hence, for any $n \ge N$ and any $x \in Adr(\Omega^\infty)$,

$$d(s_1 \circ \cdots \circ s_n(x), s'_1 \circ \cdots \circ s'_n(x))$$

$$\le d(s_1 \circ \cdots \circ s_n(x), s_1 \circ \cdots \circ s_N \circ s'_{N+1} \circ \cdots \circ s'_n(x)) +$$

$$\sum_{k=1}^{N} d(s_1 \circ \cdots \circ s_{k-1} \circ s'_k \circ \cdots \circ s'_n(x), s_1 \circ \cdots \circ s_k \circ s'_{k+1} \circ \cdots \circ s'_n(x)) < 2\frac{\varepsilon}{3}.$$

If we let n tend to ∞, we get $\langle d \rangle(Adr(\{s_n\}), Adr(\{s'_n\})) \le 2\varepsilon/3 < \varepsilon$.

It follows that Adr is uniformly continuous.

2.44 COROLLARY. *Let (X, d, Ω) be an IFS which satisfies the assumptions of Theorem 2.43 with the attractor S.*

Also, assume that the pseudometric space $(\Omega, |\cdot,\cdot|_B)$ (with B as in Theorem 2.43) is compact.

Then $Adr(\Omega^\infty) = S$, and this set is compact.

In addition, if $d|_{Adr(\Omega^\infty) \times Adr(\Omega^\infty)}$ is a metric, then this set is invariant.

PROOF: Using the compactness assumption, Remarks 2.42, and Tychonov's Theorem ([27], p. 134), $(\Omega^\infty, |\cdot,\cdot|_B)$ is compact too. By Theorem 2.43, Adr is continuous, hence $\langle Adr(\Omega^\infty) \rangle$ is compact. Hence, so is $Adr(\Omega^\infty)$.

Also, $\langle x \rangle \subset Adr(\Omega^\infty)$ for all $x \in Adr(\Omega^\infty)$, so the compactness implies that $Adr(\Omega^\infty)$ is closed. Hence, by Proposition 2.27, $Adr(\Omega^\infty) = S$.

If $d|_{Adr(\Omega^\infty)\times Adr(\Omega^\infty)}$ is a metric, then Proposition 2.30 gives that $Adr(\Omega^\infty) = S$ is invariant.

Lewellen proved variants of Theorem 2.43 and Corollary 2.44 for hyperbolic systems ([16], ch:s 6 and 7).

2.45 EXAMPLE. Let (X, d, Ω) be a hyperbolic IFS satisfying the assumption of Proposition 2.9. Then, it is boundedly stable, i.e. it is $\mathfrak{T}$–stable for some $\mathfrak{T} \subset \mathfrak{B}(X)$. Assume that for some $A \in \mathfrak{T}$, $(\Omega, |\cdot,\cdot|_A)$ is compact. Since Ω is equicontinuous on A, Corollary 2.44 gives that the addressate set and the attractor coincide and that this set is invariant and compact.

2.46 REMARKS. Examples 2.23 and 2.31 show that the compactness assumption in Corollary 2.44 cannot be dispensed with, not even for hyperbolic systems.

If Ω is finite, the $|\cdot,\cdot|_B$–compactness is obvious for *any* $B \in \mathfrak{S}(X)$, so the conclusions of Corollary 2.44 hold in this case.

We will now define an IFS on Ω^∞ with connections to Barnsley ([3], p. 155 ff.) and Lewellen ([16], ch:s 6 and 7).

2.47 DEFINITIONS. Let (X, d, Ω) be a $(\mathfrak{T}, x_0)$–stable IFS.

For any $s_0 \in \Omega$, we define $\hat{s}_0 : \Omega^\infty \to \Omega^\infty$ by $\hat{s}_0(\{s_n\}_{n=1}^\infty) = \{s_{n-1}\}_{n=1}^\infty$. We put $\hat{\Omega} = \{\hat{s} | s \in \Omega\}$.

2.48 REMARKS. Let $B \in \mathfrak{S}(X)$ and assume that $(\Omega, |\cdot,\cdot|_B)$ is complete. Then, Ω^∞ is complete too by Remarks 2.42. Also, it is easy to see that $(\Omega^\infty, |\cdot,\cdot|_B, \hat{\Omega})$ is a hyperbolic IFS (with $c = 1/2$ in Definition 2.7) which is stable in all ways since Ω^∞ is bounded (Corollary 2.6).

Ω^∞ is its attractor and its addressate set, and Ω^∞ is invariant.

The address map on $\hat{\Omega}^\infty$ is given by $\{\hat{s}_n\} \mapsto \langle\{s_n\}\rangle$. Hence, $\{\hat{t}_n\} \in \hat{\Omega}^\infty$ is an address of $\{s_n\} \in \Omega^\infty$ if and only if $|\{s_n\}, \{t_n\}|_B = 0$. By Remarks 2.42, this address map is bijective if $(\Omega, |\cdot,\cdot|_B)$ is metric.

The set of (eventually) periodic points, P (Q), is the set of all sequences with $|\cdot,\cdot|_B$–pseudodistance 0 to (eventually) periodic sequences in Ω^∞. These sets have the same closure: Ω^∞, Q is invariant, and the inclusions $P \subset Q \subset \Omega^\infty$ are proper if Ω contains two functions with positive $|\cdot,\cdot|_B$–pseudodistance from each other.

This is actually a special case of the following:

2.49 PROPOSITION. *Let (X, d, Ω) be a $(\mathfrak{T}, x_0)$–stable IFS such that, for some $A \in \mathfrak{T}$, $s|_A$ is injective, $s|_A$ takes closed sets to closed sets, and $s(A) \cap s'(A) = \varnothing$ for all $s, s' \in \Omega$ such that $s \neq s'$.*

Then, each $x \in Adr(\Omega^\infty)$ has a unique address, and the inclusions $P \subset Q \subset Adr(\Omega^\infty)$ (with P and Q as above) are proper unless Ω contains only one function.

PROOF: It is easy to see that for all n and for any n–tuples $(s_1, \ldots, s_n), (s'_1, \ldots, s'_n) \in \Omega^n$, $s_1 \circ \cdots \circ s_n|_A$ is injective, it takes closed sets to closed sets and $s_1 \circ \cdots \circ s_n(A) \cap s'_1 \circ \cdots \circ s'_n(A) = \varnothing$. Hence, if $\{s_n\}$ is an address of $x \in Adr(\Omega\infty)$, then $x \in s_1 \circ \cdots \circ s_n(A)$ for all n from which the conclusions follow.

2.50 EXAMPLE. The middle–third Cantor set.

Let X be [0,1] with d as the usual metric. We define w_0 and w_2 on [0,1] by $w_0(x) = x/3$ and $w_2(x) = (x+2)/3$ for all $x \in [0,1]$. We put $\Omega = \{w_0, w_2\}$. Then, (X, d, Ω) is a hyperbolic IFS with the invariant attractor S, which is defined as the set of all points in [0,1] which have a ternary expansion without 1s.

For any $\{s_n\} \in \Omega^\infty$ and any k, the k:th digit in the ternary expansion of $Adr(\{s_n\})$ is i if $s_k = w_i$, $(i = 0, 2)$.

It is not difficult to verify that with Ω^∞ metrized by $|\cdot,\cdot|$, Adr is a homeomorphism onto S.

The addresses are indeed unique. The set of eventually periodic points is $S \cap \mathbb{Q}$, and the set of periodic points is the subset thereof of numbers which could be written in the form $m/(3^n - 1)$ with m and n as nonnegative integers $(n > 0)$. The ternary period will be m written ternary.

So, the inclusions in Proposition 2.49 are indeed proper.

We conclude this chapter by giving an example of a weaker kind of stability, where the "attractor" depends upon the starting point.

2.51 EXAMPLE. Let (X, d, Ω) be a totally stable IFS. For any $s \in \Omega$, we define a map $s \times I : X \times X \to X \times X$ by putting $s \times I(x, y) = (s(x), y)$ for all $x, y \in X$. We put $\Omega \times I = \{s \times I | s \in \Omega\}$. We pseudometrize $X \times X$ by d' such that $d'((x,y),(u,v)) = d(x,u) + d(y,v)$. Then, $(X \times X, d', \Omega \times I)$ is an IFS.

For any $\{s_n \times I\} \in (\Omega \times I)^\infty$ and any $x, y \in X$, the sequence $\{s_1 \times I \circ \cdots \circ s_n \times I(x,y)\}$ converges to a point (z, y) with z independent of (x, y), but y will be the same as in the starting point. Hence, if we try to define the "attractor" and the "addressate set" here, it is possible, but they will depend upon the starting points.

REFERENCES.

[1] F.J. Almgren, Jr., Existence and regularity almost everywhere of solutions to elliptic variational problems among surfaces of varying topological type and singularity structure, *Ann. of Math.* **87** (1968), 321–391.
[2] S. Banach, "Théorie des opérations linéaires", Monografje Matematyczne, Warzaw, 1932.
[3] M.F. Barnsley, "Fractals everywhere", Academic Press, Inc., New York, 1988.
[4] M.F. Barnsley and S. G. Demko, Iterated function systems and the global construction of fractals, *Proc. Royal Soc. London*,**A 399** (1985), 243–275.
[5] M.F. Barnsley, S.G. Demko, J.H. Elton and J.S. Geronimo, Invariant measures for Markov processes arising from iterated function systems with place–dependent probabilities, *Ann. Inst. Henri Poincaré*, **24** (1988), 367–394.
[6] R.L. Dobrushin, Prescribing a System of Random Variables by Conditional Distributions, *Theory Prob. Applications* **15**, **3** (1970), 458–486.
[7] J.H. Elton, An ergodic theorem for iterated maps, *Ergod. Th. & Dynam. Sys.* (1987), **7**, 481–486.
[8] J.H. Elton and Z. Yan, Approximation of Measures, *Constructive Approximation*, **5**, **1** (1989), 69–87.
[9] K.J. Falconer, "The Geometry of Fractal Sets", Cambridge University Press, New York, 1985.
[10] H. Federer, "Geometric Measure Theory", Springer-Verlag, Berlin, 1969.
[11] A.A. Fraenkel, Y. Bar–Hillel and A. Levy, "Foundations of Set Theory", 2nd ed, North–Holland Publishing Company, Amsterdam, 1973.
[12] P.R. Halmos, Measure Theory, D. Van Nostrand Company, Inc., Princeton, New Jersey, 1950.
[13] M. Hata, On the Structure of Self–Similar Sets, *Japan J. of Math.*, **2**, (1985), 381–414.
[14] J. Hutchinson, Fractals and self–similarity, *Indiana Univ. J. Math.* **30** (1981), 713–747.
[15] G. Letac, A contraction principle for certain Markov chains and its applications, *Contemp. Math.* **50** (1986), 263–273.
[16] G.B. Lewellen, "Topological investigations of Self–Similarity", Ph.D. dissertation, Georgia Inst. of Tech., June, 1989.
[17] E.J. McShane, Extension of Range of Functions, *Bull. Amer. Math. Soc.* **40** (1934), 837–842.
[18] Yu. V. Prokhorov, Convergence of Random Processes and Limit Theorems in Probability Theory, *Theory of Prob. Applications*, **1**, **2** (1956).
[19] H.L. Royden, "Real Analysis", 2nd ed, The Macmillan Company Collier–Macmillan Limited, London, 1968.
[20] W. Rudin, "Functional Analysis", McGraw–Hill Book Company, New York, 1973.
[21] W. Rudin, "Real and Complex Analysis", 3rd ed, McGraw–Hill Book Company,

New York, 1987.

[22] D.W. Stroock and S.R.S Varadhan, "Multidimensional Diffusion Processes". Springer–Verlag, Berlin, 1979.

[23] A. Szulga, On the Wasserstein metric, *Trans. 8th Prague Conference*, **B**, D. Reidel Publishing Company, Dordrecht, Holland, 1978.

[24] G. Takeuti and W.M. Zaring, "Introduction to Axiomatic Set Theory", Springer–Verlag, Berlin, 1971.

[25] L.N. Vaserhstein, Markov Processes over Denumerable Products of Spaces, describing large Systems of Automata, *Problems of information transmission*, **5**, **3**, (1969).

[26] H. Wallin and J. Karlsson, "Continued Fractions and Iterated Function Systems", Preprint No. 2, Dpt. of Math, Univ. of Umeå, 1991.

[27] A. Wilansky, "Topology for Analysis", Ginn & Co, Waltham, Massachusetts, 1970.

[28] S. Willard, "General Topology", Addison–Wesley Publishing Company, Reading, Massachusetts, 1970.

Rational Supremum Norm Approximation of Transfer Functions by the cf Method

Christer Glader

Department of Mathematics
Åbo Akademi
Fänriksgatan 3, 20500 Åbo, Finland

Abstract

A method for using the cf approximation technique on the unit disk on rational supremum norm approximation of continuous time transfer functions is presented.

1. Introduction

We will look at the following approximation problem on the imaginary axis. Let R_{mn}^{im} denote the space of rational functions with real coefficients and with numerator degree m, denominator degree n, $m < n$, and with all poles in the open left half-plane. Let $G(s)$ be a function that is analytic in the open right half-plane and continuous in the closed right half-plane and vanishes at $\pm\infty i$ on the extended imaginary axis. We wish to find an approximation $r^* \in R_{mn}^{im}$ so that

$$\|G - r^*\| = \inf_{r \in R_{mn}^{im}} \{ \max_{-\infty<\omega<\infty} |G(\omega i) - r(\omega i)| \}. \tag{1.1}$$

$G(s)$ could for instance be a transfer function for a high order continuous time system which we want to approximate with a low order real-rational strictly proper transfer function corresponding to a continuous time linear system.

Define D, $\overline{D}$ and S by $D = \{z : |z| < 1\}$, $\overline{D} = \{z : |z| \leq 1\}$ and $S = \{z : |z| = 1\}$. Denote by $\mathcal{L}_2(\overline{D})$ the inner product space of functions that are analytic in D, continuous in $\overline{D}$ and zero in the point $z = 1$. The inner product and $\|\cdot\|_2$ for $f, g \in \mathcal{L}_2(\overline{D})$ are defined by

$$\begin{aligned}(f,g) &= \int_0^{2\pi} f(e^{i\theta})\overline{g(e^{i\theta})}d\theta,\\ \|f\|_2 &= \sqrt{(f,f)}.\end{aligned} \tag{1.2}$$

The approximationproblem on the imaginary axis can be transformed to the unit disk using the following conformal bilinear mapping,

$$z = \gamma_a(s) = \frac{s-a}{s+a} \tag{1.3}$$
$$s = \gamma_a^{-1}(z) = a\frac{1+z}{1-z},$$

which maps the open right half-plane into D, the extended imaginary axis onto S and the point at infinity onto $z = 1$. The positive parameter a is free to choose, resulting in the mapping of the intervall $[-ai, ai]$ onto the left half circle of S.

Now the transformed function, $\tilde{G}(z) := G(\gamma_a^{-1}(z))$, is in $\mathcal{L}_2(\overline{D})$, and our approximation problem on the unit disk is to find a rational function r^* of the form

$$r^*(z) = (z-1)^k \frac{a_0 + a_1 z + \ldots + a_m z^m}{b_0 + b_1 z + \ldots + b_n z^n}, \quad k = n - m, \tag{1.4}$$

so that the supremum norm on the unit circle,

$$\|\tilde{G} - r^*\| = \max_{z \in S} |\tilde{G}(z) - r^*(z)|, \tag{1.5}$$

is minimized over such rational forms with all poles outside $\overline{D}$. If we had not the interpolation condition in the point $z = 1$ this would be a rational Chebyshev approximation problem on the unit disk and we could use the cf method of Trefethen to get a near-best approximation. For a description of the cf method the reader may consult the reference [4]. To make use of the cf method we expand the function $\tilde{G}$ in an orthogonal expansion,

$$\tilde{G}(z) \sim (z-1)^k \sum_{j=0}^{\infty} c_j^* z^j, \tag{1.6}$$

using for each k, $k = n - m$, a different family, $\{\varphi_j^k\}_{j=0}^{\infty}$, of polynomials which satisfy the interpolation condition. Each family is a complete orthogonal system in $\mathcal{L}_2(\overline{D})$. Under certain conditions on $\tilde{G}$ uniform convergence of the expansion can be obtained, and the truncated sum $\sum_{j=0}^{M} c_j^* z^j$ from (1.6) is then approximated by a rational form with numerator degree m and denominator degree n using the cf method. Finally by appending the interpolation condition, $(z-1)^k$, to the approximant we get an approximant of the desired form in (1.4). The expansion in the φ_n^k polynomials is treated in section 2. In section 3 we perform a numerical case study by approximating a delay system using the described method and comparing the results with the classical Padé approximation method and with lower bounds for the achievable errors.

2. Orthogonal expansion of functions in $\mathcal{L}_2(\overline{D})$

Define families of polynomials for $k = 1, 2, \ldots$ by

$$\varphi_n^k(z) = \frac{n!}{(k+n)!}(z-1)^k \sum_{j=0}^{n} \frac{(k+j)!}{j!} \binom{k+n-j-1}{n-j} z^j, \quad n = 0, 1, \ldots. \tag{2.1}$$

And introduce the notation

$$(a)_n = a \cdot (a+1) \cdot \ \ldots \ \cdot (a+n-1) \ , \ (a)_0 = 1, \tag{2.2}$$
$${}_nF_m[a_1, \ldots, a_n; b_1, \ldots, b_m; z] = \sum_{i=0}^{\infty} \frac{(a_1)_i \cdot \ \ldots \ \cdot (a_n)_i}{(b_1)_i \cdot \ \ldots \ \cdot (b_m)_i \cdot i!} z^i \ .$$

for the upper factorial and for the generalized hypergeometric series. Then the $\varphi_n^k(z)$ polynomials can be written as follows,

$$\varphi_n^1(z) = \frac{1}{n+1}(z-1)\sum_{j=0}^{n}(j+1)z^j = z^{n+1} - \frac{1}{n+1}\sum_{j=0}^{n} z^j \ , \tag{2.3}$$
$$\varphi_n^k(z) = \frac{k}{n+k}(z-1)^k \ {}_2F_1[k+1, -n; 1-n-k; z] \ , \ k > 1.$$

We note that for every k and n, $\varphi_n^k(z) \in \mathcal{L}_2(\overline{D})$.

Proposition 2.1. *The sequences* $(\varphi_n^k)_{n=0}^{\infty}$ *are orthogonal in* $\mathcal{L}_2(\overline{D})$ *for every* k, $k = 1, 2, \ldots$, *and*

$$\|\varphi_n^k\|_2^2 = (\varphi_n^k, \varphi_n^k) = 2\pi \frac{n!(2k+n)!}{((k+n)!)^2}. \tag{2.4}$$

Proof. Let $0 \le n \le m$. We write

$$\varphi_n^k(z) = (z-1)^k \sum_{i=0}^{n} a_i^n z^i, \text{ where } a_i^n = \frac{n!}{(k+n)!}\frac{(k+i)!}{i!}\binom{k+n-i-1}{n-i}, \tag{2.5}$$
$$\varphi_m^k(z) = (z-1)^k \sum_{i=0}^{m} b_i^m z^i, \text{ where } b_i^m = \frac{m!}{(k+m)!}\frac{(k+i)!}{i!}\binom{k+m-i-1}{m-i}.$$

The inner product is then

$$(\varphi_n^k, \varphi_m^k) = \int_0^{2\pi} (e^{i\theta} - 1)^k (e^{-i\theta} - 1)^k (\sum_{r=0}^{n} a_r^n e^{r\theta i})(\sum_{j=0}^{m} b_j^m e^{-j\theta i}) d\theta \tag{2.6}$$
$$= \ \ldots \ =$$
$$= 4^k \sum_{r=0}^{n}\sum_{j=0}^{m} a_r^n b_j^m \int_0^{2\pi} sin^{2k}(\frac{\theta}{2}) e^{(r-j)\theta i} d\theta$$
$$= \ \ldots \ =$$
$$= \sum_{r=0}^{n} \sum_{j=max(0,r-k)}^{min(m,k+r)} a_r^n b_j^m I(r,j),$$
$$I(r,j) = \begin{cases} 2\pi(-1)^{(j-r)}\binom{2k}{k+j-r}, & \text{if } |j-r| \le k; \\ 0, & \text{otherwise.} \end{cases}$$

A. Let us first investigate the case $k = 1$. Then by (2.5) we have $a_r^n = \frac{r+1}{n+1}$ and $b_j^m = \frac{j+1}{m+1}$, and by (2.6)

$$(\varphi_n^1, \varphi_m^1) = \frac{2\pi}{(n+1)(m+1)} \sum_{r=0}^{n}(r+1) \sum_{j=max(0,r-1)}^{min(m,r+1)} (-1)^{(j-r)}(j+1)\binom{2}{j-r+1}. \tag{2.7}$$

If we assume that $0 \le r \le n < m$ then it is easily verified that the inner sum in (2.7) is zero. Thus we have $(\varphi_n^1, \varphi_m^1) = 0$ when $n < m$. Because $(\varphi_m^1, \varphi_n^1) = \overline{(\varphi_n^1, \varphi_m^1)}$ we have $(\varphi_n^1, \varphi_m^1) = 0$ when $n \neq m$. If $n = m$ the inner sum in (2.7) is zero if $r < n$. If $r = n = m$ we get from (2.7) that $(\varphi_n^1, \varphi_n^1) = 2\pi\frac{(n+2)}{(n+1)}$, and then the proposition is true for $k = 1$.

B. Now we turn to the case $k > 1$. The inner product in (2.6) written in full detail is

$$(2.8)\qquad \begin{aligned} (\varphi_n^k, \varphi_m^k) &= \frac{2\pi(2k)!n!m!}{(k+n)!(k+m)!((k-1)!)^2} \sum_{r=0}^{n} \frac{(k+n-r-1)!(k+r)!}{(n-r)!r!} s(k,m,r), \\ s(k,m,r) &= \sum_{j=max(0,r-k)}^{min(m,k+r)} (-1)^{(j-r)} \frac{(k+m-j-1)!(k+j)!}{(k+j-r)!(k-j+r)!(m-j)!j!}. \end{aligned}$$

The aim is to prove that the inner sum, $s(k,m,r)$, in (2.8) is zero when $r < m$. Let us assume that $0 < r \le n < m$. We have to investigate 4 cases depending on the summation limits of $s(k,m,r)$. Suppose that $max(0, r-k) = 0$ and $min(m, k+r) = m$. Then we have

$$(2.9)\qquad \begin{aligned} s(k,m,r) &= \sum_{j=0}^{m} (-1)^{(j-r)} \frac{(k+m-j-1)!(k+j)!}{(k+j-r)!(k-j+r)!(m-j)!j!} \\ &= (-1)^r \frac{(k+m-1)!k!}{(k-r)!(k+r)!m!} \sum_{j=0}^{m} \frac{(k+1)_j(-(k+r))_j(-m)_j}{(k-r+1)_j(1-m-k)_j j!} \\ &= (-1)^r \frac{(k+m-1)!k!}{(k-r)!(k+r)!m!} {}_3F_2[(k+1), -(k+r), -m; (k-r+1), (1-m-k); 1]. \end{aligned}$$

By Saalschutz's theorem, see Henrici [3], we have that ${}_3F_2[a,b,-n;c,d;1] = \frac{(c-a)_n(c-b)_n}{(c)_n(c-a-b)_n}$, if $c + d = a + b - n + 1$. Applying the theorem to (2.9) gives us

$$(2.10)\qquad s(k,m,r) = (-1)^r \frac{(k+m-1)!k!}{(k-r)!(k+r)!m!} \frac{(-r)_m(2k+1)_m}{(k-r+1)_m(k)_m}.$$

But since $r < m$ we have that $(-r)_m = 0$, and thus $s(k,m,r) = 0$. The 3 remaining cases are treated in a similar manner, and the result is that the inner sum, $s(k,m,r)$, is always zero when $r < m$. Then we get that $(\varphi_n^k, \varphi_m^k) = 0$ when $n < m$. By the fact that $(\varphi_m^k, \varphi_n^k) = \overline{(\varphi_n^k, \varphi_m^k)}$ we get that $(\varphi_n^k, \varphi_m^k) = 0$ when $n \neq m$.

Suppose now that $n = m$. Then (2.8) takes the form

$$(2.11)\qquad \begin{aligned} (\varphi_n^k, \varphi_n^k) &= \frac{2\pi(2k)!(n!)^2}{((k+n)!)^2((k-1)!)^2} \sum_{r=0}^{n} \frac{(k+n-r-1)!(k+r)!}{(n-r)!r!} s(k,n,r), \\ s(k,n,r) &= \sum_{j=max(0,r-k)}^{min(n,k+r)} (-1)^{(j-r)} \frac{(k+n-j-1)!(k+j)!}{(k+j-r)!(k-j+r)!(n-j)!j!}. \end{aligned}$$

By our earlier investigations we know that $s(k,n,r) = 0$ when $r < n$. Putting $r = n$ in (2.11) we obtain

$$(2.12)\qquad (\varphi_n^k, \varphi_n^k) = \frac{2\pi(2k)!n!}{(k+n)!(k-1)!} \sum_{j=max(0,n-k)}^{n} (-1)^{(j-n)} \frac{(k+n-j-1)!(k+j)!}{(k-n+j)!(k+n-j)!(n-j)!j!}.$$

Now we have to consider 2 cases depending on the lower summation limit in the above formula. Suppose that $max(0, n-k) = 0$. Then (2.12) can be put into the following form

$$
\begin{aligned}
(\varphi_n^k, \varphi_n^k) &= \frac{2\pi(2k)!k(-1)^n}{(k+n)(k+n)!(k-n)!}{}_3F_2[(k+1), (-(k+n)), (-n); (1-k-n), (k-n+1); 1] \\
&= \frac{2\pi(2k)!k(-1)^n}{(k+n)(k+n)!(k-n)!}\frac{(-n)_n(2k+1)_n}{(k-n+1)_n(k)_n} = 2\pi\frac{n!(2k+n)!}{((k+n)!)^2}.
\end{aligned}
\tag{2.13}
$$

The case $max(0, n-k) = n-k$ yields the same result as in (2.13). The proof is then completed.

Proposition 2.2. *The sequences $(\varphi_n^k)_{n=0}^\infty$ are complete orthogonal systems in $\mathcal{L}_2(\overline{D})$ for every k, $k = 1, 2, \ldots$.*

Proof. A theorem by Runge-Walsh, see Walsh [5], says that if f is analytic in the interior of a Jordan curve C and continuous in the closed domain bounded by C, then f can be approximated on C with arbitrary accuracy in the supremum norm by polynomials. We will first show that $f \in \mathcal{L}_2(\overline{D})$ can be approximated arbitrarily well, for every $k = 1, 2, \ldots$, by a polynomial of the form $(z-1)^k(a_0 + a_1 z + \ \ldots \ + a_m z^m)$.

Let $\varepsilon > 0$ and $f \in \mathcal{L}_2(\overline{D})$. We proceed with a proof by induction.

1. $k = 1$. By the Runge-Walsh theorem there is a polynomial p so that $\max_{z\in S}|f(z) - p(z)| \le \frac{\varepsilon}{2}$. Since $f(1) = 0$ we have $|p(1)| \le \frac{\varepsilon}{2}$. Define $g(z) = p(z) - p(1)$. Then we have $g(z) = (z-1)p_1(z)$, where p_1 is a polynomial. Thus $\max_{z\in S}|f(z) - g(z)| \le \max_{z\in S}|f(z) - p(z)| + |p(1)| \le \varepsilon$. We conclude that $f \in \mathcal{L}_2(\overline{D})$ can be approximated arbitrarily well in the supremum norm by a polynomial of the form $(z-1)p(z)$.
2. Suppose that there is a polynomial p_{k-1}, for a certain value on k, $k \ge 2$, so that $\max_{z\in S}|f(z) - (z-1)^{k-1}p_{k-1}(z)| \le \frac{\varepsilon}{2}$. Define the positive constant p by $p = \max_{z\in S}|p_{k-1}(z)|$ and the function $g \in C_0(\overline{D})$ by $g(z) = (z-1)p_{k-1}(z)/(z - 1 - \frac{\varepsilon}{p2^k})$. We can choose the polynomial $p_k(z)$ so that $\max_{z\in S}|g(z) - (z-1)p_k(z)| \le \frac{\varepsilon}{2^{k+1}}$. Then we have

$$
\max_{z\in S}|(z-1)^{k-1}g(z) - (z-1)^k p_k(z)| \le 2^{k-1}\cdot\frac{\varepsilon}{2^{k+1}} = \frac{\varepsilon}{4}. \tag{2.14}
$$

Furthermore

$$
\begin{aligned}
&\max_{z\in S}|(z-1)^{k-1}(p_{k-1}(z) - g(z))| = \max_{z\in S}|(z-1)^{k-1}p_{k-1}(z)\frac{-\frac{\varepsilon}{p2^k}}{z - 1 - \frac{\varepsilon}{p2^k}}| \\
&< \max_{z\in S}|\frac{p_{k-1}(z)}{p}|\max_{z\in S}\frac{|z-1|^{k-1}}{|z - 1 - \frac{\varepsilon}{p2^k}|}\frac{\varepsilon}{2^k} \le 1\cdot 2^{k-2}\cdot\frac{\varepsilon}{2^k} = \frac{\varepsilon}{4}.
\end{aligned}
\tag{2.15}
$$

Then by (2.14) and (2.15) we get that

$$
\begin{aligned}
&\max_{z\in S}|f(z) - (z-1)^k p_k(z)| \le \max_{z\in S}|f(z) - (z-1)^{k-1}p_{k-1}(z)| + \max_{z\in S}|(z-1)^{k-1}p_{k-1}(z) - (z-1)^k p_k(z)| \\
&\le \frac{\varepsilon}{2} + \max_{z\in S}|(z-1)^{k-1}(p_{k-1}(z) - g(z))| + \max_{z\in S}|(z-1)^{k-1}g(z) - (z-1)^k p_k(z)| \le \frac{\varepsilon}{2} + \frac{\varepsilon}{4} + \frac{\varepsilon}{4} = \varepsilon.
\end{aligned}
\tag{2.16}
$$

The induction proof is completed and we have that $f \in \mathcal{L}_2(\overline{D})$ can be approximated arbitrarily well in the supremum norm by a polynomial of the form $(z-1)^k p_k(z)$, $k = 1, 2, \dots$. Every polynomial of the form $(z-1)^k p_k(z)$ can be expressed as a finite linear combination of polynomials from the sequence $(\varphi_n^k)_{n=0}^\infty$. But then every $f \in \mathcal{L}_2(\overline{D})$ can be expanded in an orthogonal expansion, by polynomials in the sequence $(\varphi_n^k)_{n=0}^\infty$, that converges to f in the mean. The proof is completed.

Proposition 2.3. *The following recursion formula holds for the polynomials* $(\varphi_n^k)_{n=0}^\infty$.

$$\varphi_{n+1}^k(z) = z\varphi_n^k(z) + (-1)^k \frac{k}{k+n+1} z^{k+n} \varphi_n^k(\frac{1}{z}). \tag{2.17}$$

Proof. The formula is easily verified using the expression (2.1).

Proposition 2.4. *The following inequalities hold for every* k, $k = 1, 2, \dots$

$$2\pi < \|\varphi_n^k\|_2^2 \le 2\pi \binom{2k}{k}, \quad n = 0, 1, \dots . \tag{2.18}$$

Proof. From proposition 2.1 we have that $\|\varphi_n^k\|_2^2 = (\varphi_n^k, \varphi_n^k) = 2\pi \frac{n!(2k+n)!}{((k+n)!)^2}$. Then we have $\|\varphi_0^k\|_2^2 = 2\pi\binom{2k}{k}$ and $\lim_{n\to\infty} \|\varphi_n^k\|_2^2 = 2\pi$. On the other hand we have that $\|\varphi_{n+1}^k\|_2^2 / \|\varphi_n^k\|_2^2 = (n+1)(2k+n+1)/(k+n+1)^2 < 1$, and the proof is completed.

The next proposition gives us a lower and upper bounds for the supremumnorm on the unit circle in the case $k = 1$.

Proposition 2.5. *For the supremumnorm,* $\|\cdot\|$, *on* S *we have when* $n = 0, 1, \dots$

$$1 < \|\varphi_n^1\| \le 2. \tag{2.19}$$

Proof. Making use of (2.3) we get $\|\varphi_n^1\| \le \|z^{n+1}\| + \frac{1}{n+1}\sum_{i=0}^n \|z^i\| = 2$. From (2.18) we get that $\|\varphi_n^k\|_2 > \sqrt{2\pi}$. On the other hand $\|\varphi_n^k\|_2 = (\int_0^{2\pi} \varphi_n^k(e^{i\theta})\varphi_n^k(e^{-i\theta})d\theta)^{\frac{1}{2}} \le \sqrt{2\pi}\|\varphi_n^k\|$. So $\|\varphi_n^k\| > 1$. The proof is completed.

Numerical experimentation suggests that in the cases $k > 1$ we might have $1 < \|\varphi_n^k\| \le 2^k$.

Let $f \in \mathcal{L}_2(\overline{D})$. Denote the Fourier coefficients of f by c_j,

$$c_j = \frac{1}{2\pi}\int_0^{2\pi} f(e^{i\theta})e^{-ij\theta}d\theta \, . \tag{2.20}$$

Proposition 2.6. *Let* $f \in \mathcal{L}_2(\overline{D})$. *The Fourier coefficients* c_j^* *in the orthogonal expansion* $f \sim \sum_{j=0}^\infty c_j^* \varphi_j^k$ *are given by*

$$\begin{aligned} c_j^* &= \frac{1}{j+2}[(j+1)c_{j+1} - \sum_{p=0}^j c_p] \, , \quad k = 1, \\ c_j^* &= \frac{(k+j)!}{(2k+j)!}\sum_{p=0}^j [\frac{(k+p)!}{p!}\binom{k+j-p-1}{j-p}\sum_{l=0}^k (-1)^l \binom{k}{l} c_{k-l+p}], \quad k > 1. \end{aligned} \tag{2.21}$$

Proof. We have that $c_j^* = \frac{(f,\varphi_j^k)}{(\varphi_j^k,\varphi_j^k)}$. By proposition 2.1 $(\varphi_j^k,\varphi_j^k) = 2\pi\frac{n!(2k+n)!}{((k+n)!)^2}$. So we have to calculate (f,φ_j^k). If $k=1$ we get

$$(2.22)\quad \begin{aligned}(f,\varphi_j^1) &= \frac{1}{j+1}\int_0^{2\pi}[(e^{-i\theta}-1)f(e^{i\theta})\sum_{p=0}^{j}(p+1)e^{-ip\theta}]d\theta \\ &= \frac{1}{j+1}\sum_{p=0}^{j}(p+1)[\int_0^{2\pi} f(e^{i\theta})(e^{-i(p+1)\theta}-e^{-ip\theta})d\theta] \\ &= \frac{2\pi}{j+1}\sum_{p=0}^{j}(p+1)(c_{p+1}-c_p) = \frac{2\pi}{j+1}((j+1)c_{j+1}-\sum_{p=0}^{j}c_p).\end{aligned}$$

If $k>1$ we write $\varphi_j^k(z) = (z-1)^k\sum_{p=0}^{j}a_pz^p$, where $a_p = \frac{j!}{(k+j)!}\frac{(k+p)!}{p!}\binom{k+j-p-1}{j-p}$. Then
(2.23)

$$\begin{aligned}(f,\varphi_j^k) &= \int_0^{2\pi}[(e^{-i\theta}-1)^kf(e^{i\theta})\sum_{p=0}^{j}a_pe^{-ip\theta}]d\theta = \sum_{p=0}^{j}a_p\int_0^{2\pi}[f(e^{i\theta})e^{-ip\theta}\sum_{l=0}^{k}(-1)^l\binom{k}{l}e^{-i(k-l)\theta}]d\theta \\ &= \sum_{p=0}^{j}a_p\sum_{l=0}^{k}(-1)^l\binom{k}{l}[\int_0^{2\pi}f(e^{i\theta})e^{-i(k-l+p)\theta}d\theta] = 2\pi\sum_{p=0}^{j}a_p\sum_{l=0}^{k}(-1)^l\binom{k}{l}c_{k-l+p}.\end{aligned}$$

Inserting a_p in (2.23) and dividing by $(\varphi_j^k,\varphi_j^k)$ gives the proposition for $k>1$. The proof is completed.

Let $f\in\mathcal{L}_2(\overline{D})$. Define $P_k(f,n)$ by

$$(2.24)\quad P_k(f,n) = \sum_{j=0}^{n}c_j^*\varphi_j^k(z).$$

Proposition 2.7. *If $f\in\mathcal{L}_2(\overline{D})$ and f has the Fourier coefficients c_j we have*

$$(2.25)\quad \begin{aligned}P_1(f,n) &= (z-1)\sum_{j=0}^{n}[\frac{1}{(j+1)(j+2)}[(j+1)c_{j+1}-\sum_{p=0}^{j}c_p]\sum_{p=0}^{j}(p+1)z^p] \\ &= \sum_{j=0}^{n+1}c_jz^j-\frac{1}{n+2}(\sum_{j=0}^{n+1}c_j)(\sum_{j=0}^{n+1}z^j).\end{aligned}$$

Proof. The result is obtained by forming $\sum_{j=0}^{n}c_j^*\varphi_j^1(z)$ using (2.3) and (2.21).

Proposition 2.8. *If the partial sums, $S(f,n)$, of the Fourier series of $f\in\mathcal{L}_2(\overline{D})$ converge uniformly to f as $n\to\infty$, then $P_1(f,n)$ converges uniformly to f as $n\to\infty$, and*

$$(2.26)\quad \|f-P_1(f,n)\|\le 2\|f-S(f,n+1)\|.$$

Proof. Because $f(1)=0$ and the Fourier series converges uniformly to f there is for every $\varepsilon>0$ an index n_ε so that $|\sum_{j=0}^{n+1} c_j|<\varepsilon$ if $n>n_\varepsilon$. Using (2.25) we get

$$
\begin{aligned}
|S(f,n+1)-P_1(f,n)| &= |\sum_{j=0}^{n+1} c_j z^j - (\sum_{j=0}^{n+1} c_j z^j - \frac{1}{n+2}(\sum_{p=0}^{n+1} c_p)(\sum_{j=0}^{n+1} z^j))| \\
&= |\frac{1}{n+2}(\sum_{p=0}^{n+1} c_p)(\sum_{j=0}^{n+1} z^j)| \le \frac{1}{n+2}|\sum_{p=0}^{n+1} c_p|\sum_{j=0}^{n+1}|z|^j \\
&= |\sum_{p=0}^{n+1} c_p| < \varepsilon \text{ if } n>n_\varepsilon.
\end{aligned} \tag{2.27}
$$

We conclude that $P_1(f,n)$ converges uniformly to f when $n\to\infty$. Furthermore we have $\|f-P_1(f,n)\| \le \|f-S(f,n+1)\| + \|S(f,n+1)-P_1(f,n)\| \le \|f-S(f,n+1)\| + |\sum_{p=0}^{n+1} c_p| \le 2\|f-S(f,n+1)\|$. The proof is completed.

3. Approximation of a delay system

Consider a delay system with the transfer function

$$G(s)=e^{-hs}\frac{1}{Ts+1}, \tag{3.1}$$

where $h,T>0$, and h is the time delay. Note that $G(\pm i\infty)=0$. After the bilinear mapping (1.3) to the unit disk we get

$$\tilde{G}_a(z)=G(a\frac{1+z}{1-z})=e^{-ah\frac{1+z}{1-z}}\frac{1-z}{1+aT+(aT-1)z}. \tag{3.2}$$

Now $\tilde{G}_a\in\mathcal{L}_2(\overline{D})$, so the cf method cannot be applied directly to get an approximant of the desired form (1.4). We approximate the delay system with a finite dimensional linear system of low order. Approximants with numerator degree $(n-1)$ and denominator degree n, $n=1,\ldots,5$, are computed for different ratios h/T for the parameters in the transfer function. We will use the cf approximation method on the expansion $P_1(\tilde{G}_a,100)$, from which we have dropped the $(z-1)$ factor. The results are compared with the classical Padé approximation method, where the delay term e^{-hs} is approximated by its Padé approximant of the form $(n-1)/(n-1)$, and with lower bounds for the achievable supremum norm errors. Lower bounds for the achievable supremum norm errors can be computed as the eigenvalues of a Hankel operator, see [2]table 1.

The Padé approximant of e^{-hs} can be expressed as $\frac{Q_n(-hs)}{Q_n(hs)}$, where

$$Q_n(hs)=\sum_{j=0}^{n}\frac{(n+j)!}{j!(n-j)!}(hs)^{n-j}, \tag{3.3}$$

and is thus easily computed. The Fourier series of the transfer function $\tilde{G}_a(z)$ is absolutely convergent, but the convergence rate is very slow, see [2].

Table 3.1. *Table of supremum norm errors for approximants of the form* $(n-1)/n$, $n = 1, \dots, 5$. *The upper value is for the Padé approximant, the middle value for the cf approximant and the lower value is a lower bound, not necessarily tight, for the achievable error.*

h/T	0/1	1/2	2/3	3/4	4/5
	0.0972	0.0396	0.0267	0.0204	0.0166
0.1	0.0876	0.0468	0.0309	0.0292	0.0275
	0.0573	0.0209	0.0127	0.0090	0.0071
	0.1889	0.0791	0.0534	0.0408	0.0332
0.2	0.1576	0.0825	0.0427	0.0402	0.0325
	0.1054	0.0413	0.0252	0.0181	0.0141
	0.4346	0.1966	0.1332	0.1018	0.0828
0.5	0.3187	0.1491	0.0851	0.0840	0.0508
	0.2180	0.0993	0.0621	0.0449	0.0351
	0.7618	0.3862	0.2644	0.2028	0.1652
1.0	0.4944	0.2633	0.1715	0.1328	0.1064
	0.3528	0.1863	0.1208	0.0884	0.0695
	1.1990	0.7246	0.5138	0.3990	0.3269
2.0	0.6919	0.4496	0.3157	0.2351	0.1942
	0.5261	0.3300	0.2279	0.1712	0.1362
	1.45881	0.0060	0.7379	0.5832	0.4820
3.0	0.7942	0.5817	0.4332	0.3412	0.2798
	0.6361	0.4421	0.3215	0.2475	0.1994
	1.7196	1.3671	1.0959	0.9024	0.7630
5.0	0.8997	0.7478	0.6124	0.5111	0.4331
	0.7664	0.6011	0.4724	0.3801	0.3144

As can be seen from the table the cf method performs well for all h/T values. The Padé approximant is good when h/T ratios are small, corresponding to $\tilde{G}_a$ having its biggest fluctuations in modulus close to the origin.

4. Conclusions

Certain families of orthogonal polynomials on the unit circle have been studied and results that makes it possible to expand functions in these orthogonal bases are presented. A question that remains open is if proposition 2.8 may be generalized to cover the cases when $k > 1$.

A method for using the cf approximation technique on an approximation problem on the imaginary axis, by transforming the problem to the unit disk and expanding the function using orthogonal polynomials, has been presented and shown to perform well in a numerical case study.

Acknowledgments

The author wishes to thank Göran Högnäs, Pertti Mäkilä and Hannu Toivonen for helpful discussions.

References

[1] L. Ya. Geronimus,*Orthogonal polynomials.* Consultants bureau, New York, 1961.

[2] C. Glader, G. Högnäs, P.M. Mäkilä and H.T. Toivonen, *Approximation of delay systems - a case study*, Int. J. Control 53 (1991), 369-390.

[3] P. Henrici, *Applied and computational complex analysis, vol 1*, John Wiley & Sons, New York, 1974.

[4] L.N. Trefethen, *Rational Chebyshev approximation on the unit disk*, Numer. Math. 37 (1981), 297-320.

[5] J.L. Walsh, *Interpolation and approximation.* Amer. Math. Soc. Coll. Publs, XX.1935.

On the Colength Function in One-dimensional Noetherian Domains

Christian Gottlieb

Department of Mathematics, University of Stockholm
S-106 91 Stockholm, Sweden

Abstract

The subject of this paper is the colength function $L(x) = \ell(R/(x))$, where R is a Noetherian domain of dimension one. The main result is that the sets $\{x;\ L(x) > k\}$ are ideals for all k if and only if the integral closure $\overline{R}$ of R is a local ring. We also consider the question whether or not L is bounded in the sets $\overline{R} \setminus R$ and $R \setminus I$, where I is an ideal.

1. Introductory remarks on the colength function

We assume throughout that R is a *Noetherian domain of dimension one*. Thus, for any nonzero element x, the ring $R/(x)$ is Artinian and hence of finite length. This length, the colength of x, will be denoted by $\ell(R/(x))$ or, most frequently, by $L(x)$. In case we want to call special attention to which ring is under consideration, we write $\ell_R(R/(x))$ or $L_R(x)$. The subject of this paper is the study of some problems concerning the colength function L which, so far, takes nonzero elements of R to nonnegative integers. We extend the domain of L to the whole of R by putting $L(0) = \infty$. It is convenient also to consider the colengths of ideals in R, i.e. $L(I) = \ell((R/I))$.

Our first proposition is quite elementary but nevertheless a starting-point for our discussion.

Proposition 1. *$L(xI) = L(x) + L(I)$ for all $x \in R$ and all ideals I in R. In particular, $L(xy) = L(x) + L(y)$ for all $x, y \in R$.*

Proof. This is obvious if $x = 0$ or $I = 0$, so suppose $x \neq 0$, $I \neq 0$. Multiplication by x yields an isomorphism of R-modules $R/I \cong (x)/xI$. Thus $L(xI) = L(x) + \ell((x)/xI) = L(x) + L(I)$.

In other words the colength function L defines a monoid homomorphism from the multiplicative monoid $R\backslash\{0\}$ into the additive monoid of nonnegative integers. It is, however, a more difficult task to say anything about $L(x+y)$, the colength of the sum of two elements. A natural question is the following

When is $L(x+y) \geq \min(L(x), L(y))$?

Another way of formulating this is

When is it true that $\{x; L(x) \geq k\}$ is an ideal?

Our main result on the subject appears in Section 2, where we characterize (see Proposition 9) those rings for which $\{x; L(x) \geq k\}$ is an ideal for all k. A few further results appear in Section 3. It should be noted at once that if R is not local then $\{x; L(x) \geq 1\}$ is certainly not an ideal. Indeed the questions are of a local nature. Still we prefer not to restrict ourselves entirely to local rings, since all our introductory considerations work out equally well in the nonlocal case.

In attempting to show that $\{x; L(x) \geq k\}$ is an ideal we have come across with another question, namely

Is $L(x)$ bounded when x lies outside a given ideal?

We shall also be able to say a few things on this point. See especially Proposition 12.

It proves very useful to extend the domain of the colength function L from R to K, the field of fractions of R. In order to do so in an unambiguous way we prove the following lemma.

Lemma 2. *Let I be a fractional ideal and let x, y be nonzero elements in R such that $xI, yI \subseteq R$. Then $L(xI) - L(x) = L(yI) - L(y)$.*

Proof. We have $xyI = yxI$ and hence from Proposition 1 we deduce $L(x) + L(yI) = L(xyI) = L(yxI) = L(y) + L(xI)$.

Definition. Let I be a fractional ideal and let x be a nonzero element in R such that $xI \subseteq R$. Then we define $L(I) = L(xI) - L(x)$. In particular we let $L(\frac{a}{b}) = L(a) - L(b)$ for $a, b \in R$.

It is a fact that the formula $L(\frac{a}{b}) = L(a) - L(b)$ holds even if a, b lie outside R. This follows from the next proposition.

Proposition 3. *Let I be a fractional ideal and let $x \in K$. Then $L(xI) = L(x) + L(I)$. In particular, $L(xy) = L(x) + L(y)$ for all $x, y \in K$.*

Proof. This is obvious if $x = 0$ or $I = 0$ so we suppose this is not the case. We can write $x = \frac{a}{b}$, where $a, b \in R$ and $aI \subseteq R$. Then $L(xI) - L(I) = L(\frac{aI}{b}) - L(I) = L(aI) - L(b) - (L(aI) - L(a)) = L(a) - L(b) = L(x)$.

We have thus extended the colength function L to a homomorphism of groups $K\backslash\{0\} \mapsto Z$. The colength function, extended as it is, is still closely related to the concept of R-module length.

Proposition 4. *Let I and J be nonzero fractional ideals with $I \supseteq J$. Then $\ell_R(I/J) = L(J) - L(I)$.*

Proof. Take a nonzero $a \in R$, such that $aI \subseteq R$. Multiplication by a yields an R-module isomorphism $I/J \cong aI/aJ$ and hence $\ell_R(I/J) = \ell_R(aI/aJ) = L(aJ) - L(aI) = L(a) + L(J) - L(a) - L(I) = L(J) - L(I)$.

We now introduce the sets K_n.

Definition. For every integer n we put $K_n = \{x \in K; L(x) \geq n\}$.

We have thus a chain of subsets of K, namely $\cdots \supseteq K_{-2} \supseteq K_{-1} \supseteq K_0 \supseteq K_1 \supseteq K_2 \supseteq \cdots$. Clearly $K_0 \supseteq R$, but it is not true in general that $K_0 = R$. It is also possible that $K_n = K_{n+1}$ for some n so that colength n does not occur.

Lemma 5. *Let $a \in K$ with $L(a) = m$. Then $K_{n+m} = aK_n$.*

Proof. $x \in K_{m+n} \Leftrightarrow L(x) \geq m + n \Leftrightarrow L(xa^{-1}) \geq n \Leftrightarrow xa^{-1} \in K_n \Leftrightarrow x \in aK_n$.

Proposition 6. *If K_n is an R-module for some n then K_n is an R-module for all n.*

Proof. Suppose K_m is an R-module. There is evidently no restriction in presuming that there is an x such that $L(x) = m$. Take any K_n. Again we may presume that colength n actually occurs, say $L(y) = n$. Then, by Lemma 5, $K_n = \frac{y}{x}K_m$ and hence K_n is an R-module.

2. When I_n is an ideal for all n and on the boundedness of L outside an ideal.

We denote by $\overline{R}$ the integral closure of R in K.

Proposition 7. *For any $x \in \overline{R}$ we have $L(x) \geq 0$. Equality $L(x) = 0$ holds if and only if x is a unit in $\overline{R}$.*

Proof. Take $x \in \overline{R}$ and put $T = R[x]$. Then T is a finitely generated R-module and hence a fractional ideal. Using Propositions 3 and 4 we obtain $L(x) = L(xT) - L(T) = \ell_R(T/xT)$. Thus $L(x) \geq 0$ with equality if and only if x is a unit in T. From the integrality, however, it follows that x is a unit in T if and only if x is a unit in $\overline{R}$.

It is thus clear that $K_0 \supseteq \overline{R}$ but it is, as we shall see, far from always the truth that $K_0 = \overline{R}$. For its own interest we interfoliate the following related and well-known result.

Proposition 8. *Let $x \in K$ and suppose there exists a nonzero fractional ideal I such that $xI \subseteq I$. Then $x \in \overline{R}$.*

Proof. Since $I \neq 0$ we have canonically embeddings $R \hookrightarrow R[x] \hookrightarrow \mathrm{End}_R I$. But I is finitely generated and hence $\mathrm{End}_R I$ is integral over R by the Cayley-Hamilton theorem.

We shall henceforth use the notation I_n for $K_n \cap R$, i.e. $I_n = \{x \in R; L(x) \geq n\}$. What follows next is the main result of the paper.

Proposition 9. *The following conditions are equivalent*
(1) $\overline{R}$ *is local*
(2) $K_0 = \overline{R}$
(3) K_n *is an* R*-module for some* n
(4) K_n *is an* R*-module for all* n
(5) I_n *is an ideal for all* n.

Proof. The equivalence of (3) and (4) was already shown in Proposition 6 and (2) $\Rightarrow$ (3) and (4) $\Rightarrow$ (5) are obvious, so what remains to show is (1) $\Rightarrow$ (2) and (5) $\Rightarrow$ (1). Assume first that $\overline{R}$ is local, i.e. that $\overline{R}$ is a discrete valuation domain and let $t\overline{R}$ be the maximal ideal in $\overline{R}$. We showed in Proposition 7 that $L(t) > 0$, say $L(t) = c$. Each element in K is of the form ut^k where u is a unit in $\overline{R}$ and where k is an integer. We obtain $L(ut^k) = L(u) + kL(t) = kc$. Thus $L(ut^k) \geq 0$ if and only if $k \geq 0$ and hence $K_0 = \overline{R}$. This proves (1) $\Rightarrow$ (2). Assume next that $\overline{R}$ is not local. Then there exist nonunits $x, y \in \overline{R}$ with $1 = x + y$. Write x and y with a common denominator in R, say $x = \frac{a}{s}, y = \frac{b}{s}$, where $a, b, s \in R$. Then, by Proposition 7, $L(a), L(b) > L(s)$. Put $n = L(s)$. Then $L(a), L(b) \geq n + 1$ or in other words $a, b \in I_{n+1}$. However $s \notin I_{n+1}$ although $s = a + b$. This proves that I_{n+1} is not an ideal. Thus (5) $\Rightarrow$ (1) is proved which completes the proof of the proposition.

Can we add the condition "I_n is an ideal for some n" to the five conditions of Proposition 9? The answer to this is no. It may very well happen that I_n is an ideal for some n, although not for all. We shall return to this question in Section 3.

We now turn our attention to the question whether or not $L(x)$ is bounded when x is outside a given ideal. We start our inquiry by two lemmas. The first will later be generalized as Proposition 18.

Lemma 10. *Assume that* R *is local but that* $\overline{R}$ *is not local. Then* $\overline{R} \setminus R$ *possesses elements of arbitrary high colength.*

Proof. Let $\overline{\mathfrak{m}}$ and $\overline{\mathfrak{n}}$ be two different maximal ideals in $\overline{R}$ and let $\mathfrak{m} = \overline{\mathfrak{m}} \cap R = \overline{\mathfrak{n}} \cap R$ be the maximal ideal in R. Take $x \in \overline{\mathfrak{m}} \setminus \overline{\mathfrak{n}}$. Then $L(x) > 0$ so $L(x^n)$ will attain arbitrarily high values if n is chosen high enough. Since $x^n \notin \overline{\mathfrak{n}}$ we have $x^n \notin \mathfrak{m}$. But, having positive colength, x^n is not a unit in R. Thus $x^n \notin R$ for all n.

Lemma 11. *Assume that* $\overline{R}$ *is finitely generated as an* R*-module and that* $L(x)$ *is bounded in* $R \setminus I$ *for all nonzero ideals* I. *Then* $L(x)$ *is bounded in* $\overline{R} \setminus R$.

Proof. Since $\overline{R}$ is finite over R there is a nonzero element $a \in R$ such that $a\overline{R} \subseteq R$. Take any $x \in \overline{R} \setminus R$. Then $ax \in R \setminus (a)$. But $L(x)$ is bounded in $R \setminus (a)$, say by N. Thus $L(ax) \leq N$ and hence $L(x) \leq N - L(a)$.

Proposition 12. *Assume that $\overline{R}$ is finitely generated as an R-module. Then the following conditions are equivalent.*

(1) *$\overline{R}$ is local.*
(2) *For every nonzero ideal I the colength function $L(x)$ is bounded in $R \setminus I$.*

Proof. Assume first that $\overline{R}$ is local. Using the same notation as in the proof of Proposition 9, the maximal ideal in $\overline{R}$ is $t\overline{R} = \overline{\mathfrak{m}}$, say, $L(t) = c > 0$ and we have $K_{kc} = t^k\overline{R} = \overline{\mathfrak{m}}^k$ for all $k \geq 0$. Since $\overline{R}$ is finite over R the conductor $(R : \overline{R}) \neq 0$. Thus, for some n, we have $\overline{\mathfrak{m}}^n \subseteq (R : \overline{R})$ and hence $\overline{\mathfrak{m}}^n$ is an R-ideal. But $1 \notin \overline{\mathfrak{m}}^n$ so $\overline{\mathfrak{m}}^n$ is a proper ideal and hence, for each nonzero ideal I in R, there is a k such that $\overline{\mathfrak{m}}^{nk} \subseteq I$. Thus $K_{nkc} \subseteq I$ and hence $L(x) < nkc$ for all $x \in R \setminus I$. This proves (1) $\Rightarrow$ (2). To prove (2) $\Rightarrow$ (1) assume that $\overline{R}$ is not local. If R too is nonlocal there are two distinct maximal ideals $\mathfrak{m}$ and $\mathfrak{n}$ in R. Take $x \in \mathfrak{m} \setminus \mathfrak{n}$. Then $L(x^n) \geq n$ but $x^n \notin \mathfrak{n}$ for all n, so in this case L is unbounded in $R \setminus \mathfrak{n}$. On the other hand if R is local but $\overline{R}$ nonlocal it is immediate from Lemmas 10 and 11 that L is unbounded in $R \setminus I$ for some nonzero I. This proves (2) $\Rightarrow$ (1).

Remark. Consider again the proof of the implication (2) $\Rightarrow$ (1) of Proposition 12. We assumed that $\overline{R}$ was nonlocal and then distinguished the case when R was local from the case when R was nonlocal. In the nonlocal case it appears from the proof that there is an infinite strictly decreasing sequence $(x) \supset (x^2) \supset (x^3) \supset \cdots$ such that x^n avoids $\mathfrak{n}$ for all n. In the local case there is indeed an ideal $I \neq 0$ such that L is unbounded outside I. However if $(x_0) \supset (x_1) \supset (x_2) \supset \cdots$ we have for each n that $x_n \in \mathfrak{m}x_{n-1}$ so that $x_n \in \mathfrak{m}^n x_0$. But I is $\mathfrak{m}$-primary, whence it follows that $\mathfrak{m}^n \subseteq I$ and hence $x_n \in I$ for some n.

Example. Let $S = Z \setminus 5Z$, i.e the set of integers not divisible by 5, and consider the rings $Z[5i]$ and $Z[i]$, where $i^2 = -1$. Localize at S to get $Z_S[5i]$ and $Z_S[i]$. Then $Z_S[i]$ is the integral closure of $Z_S[5i]$, so we put $R = Z_S[5i]$ and $\overline{R} = Z_S[i]$. In $\overline{R}$ there are two maximal ideals $\overline{\mathfrak{m}} = (1+2i)$ and $\overline{\mathfrak{n}} = (1-2i)$. R, however, is local with the only maximal ideal $\mathfrak{m} = \overline{\mathfrak{m}} \cap R = \overline{\mathfrak{n}} \cap R = (5, 5i)$. We have $\mathfrak{m}^2 = 5\mathfrak{m}$, whence $L(5) = \ell_R(\mathfrak{m}/\mathfrak{m}^2) = 2$. Thus $I_1 = I_2 = \mathfrak{m}$ is an ideal. We have $(1+2i)(1-2i) = 5$ and $L(1 \pm 2i) > 0$. Thus $L(1 \pm 2i) = 1$. It follows that $L(5 \pm 10i) = 3$. But $L(5 + 10i + (5 - 10i)) = L(10) = L(2) + L(5) = 2$ since 2 is a unit in R. This shows that I_3 is not an ideal. Let $I = (5)$. Then L is unbounded in $R \setminus I$ because $5(1+2i)^n \in R \setminus I$ for all n.

Assume that R is local with maximal ideal $\mathfrak{m}$ and define for any $x \neq 0$ the number $d(x)$ by the relation $x \in \mathfrak{m}^{d(x)} \setminus \mathfrak{m}^{d(x)+1}$. As an application of Proposition 12 we shall, when $\overline{R}$ is local, prove a formula which relates $L(x)$ to the asymptotic behaviour of $d(x^k)$.

Since the proof makes use of the concept of superficial element we recall that $t \in \mathfrak{m}^r$ is said to be superficial of degree r if $d(xt) = d(x) + r$ for all x such that $d(x)$ is large enough. In our case, R being a one-dimensional domain, this amounts to $t\mathfrak{m}^n = \mathfrak{m}^{n+r}$ or, equivalently (cf Proposition 1), $L(t) = \ell(\mathfrak{m}^n/\mathfrak{m}^{n+r})$ for large enough n. But for large n we have $\ell(\mathfrak{m}^n/\mathfrak{m}^{n+r}) = re(R)$, where $e(R)$ is the multiplicity of R. Thus $t \in \mathfrak{m}^r$ is superficial if and only if $L(t) = re(R)$. There is always an integer r such that a superficial element of degree r exists in R. For this and other basic facts on superficial elements we refer to [1].

Proposition 13. *Assume that $\overline{R}$ is local and finitely generated as an R-module. Let $x \neq 0$. Then $L(x) = e(R) \lim_{k\to\infty} \frac{d(x^k)}{k}$.*

Proof. Let t be superficial of degree r and suppose $t\mathfrak{m}^n = \mathfrak{m}^{n+r}$. Choose k so large that $d(x^k) \geq n + r$, say $n + h_k r \leq d(x^k) < n + (h_k + 1)r$, where $h_k \geq 1$ is an integer. Then we may write $x^k = t^{h_k} a_k$ where $a_k \in \mathfrak{m}^n \setminus \mathfrak{m}^{n+r}$. Thus $L(x^k) = h_k L(t) + L(a_k) = h_k re(R) + L(a_k)$ and hence $L(x) = \frac{h_k}{k} re(R) + \frac{L(a_k)}{k}$. Now, since L is bounded outside $\mathfrak{m}^{n+r}$ and $a_k \notin \mathfrak{m}^{n+r}$ for all k, we have $\lim_{k\to\infty} \frac{L(a_k)}{k} = 0$. Further we have $n \leq d(x^k) - h_k r < n + r$, whence $\lim_{k\to\infty} \frac{d(x^k)}{k} = \lim_{k\to\infty} \frac{h_k r}{k}$. Putting things together we obtain $L(x) = e(R) \lim_{k\to\infty} \frac{d(x^k)}{k}$.

For example it follows that, in the situation of Proposition 13, an element x of $\mathfrak{m}^r$ is superficial of degree r if and only if $\lim_{k\to\infty} \frac{d(x^k)}{k} = r$.

3. Some further results.

If $\overline{R}$ is nonlocal we know that, for some n, I_n is not an ideal in R. On the other hand it should be clear for instance from the example in Section 2 that I_n might be an ideal for some other values of n. We shall prove another two propositions on the matter and then end the paper with some further remarks concerning the boundedness of L in $\overline{R} \setminus R$.

Proposition 14. *Assume that $\overline{R}$ is not local. Then I_n is an ideal only for a finite number of n.*

Proof. Let x, y be nonunits in $\overline{R}$ with $1 = x + y$. We have, say, $x = \frac{a}{s}$, $y = \frac{b}{s}$, where $a, b, s \in R$. Thus $L(a), L(b) > L(s)$ and $s = a + b$. We obtain $s^k = \sum_{i=0}^{k} \binom{k}{i} a^i b^{k-i}$. We propose to show that, for n large enough, we can always choose k such that $s^k \notin I_n$, whereas $a^i b^{k-i} \in I_n$ for $i = 0, \ldots, n$. This will show that I_n is not an ideal. Consider the intervals of integers $[kL(s) + 1, kL(s) + k] = S_k$. It is obvious that only a finite number of positive integers avoid all S_k, i.e. there is an integer N such that for all $n \geq N$ there is a k such that $n \in S_k$. Choose arbitrarily such an n and a corresponding k. Then $L(a^i b^{k-i}) \geq k(L(s)+1) \geq n$, i.e. $a^i b^i \in I_n$ for $i = 0, \ldots, n$. But $L(s^k) = kL(s) < n$, so $s^k \notin I_n$.

Now consider the subsets $(R :_{\overline{R}} I_n)$ of $\overline{R}$. It is a nice fact that these are actually subrings of $\overline{R}$, as we show next.

Lemma 15. *$(R :_{\overline{R}} I_n) = (I_n :_{\overline{R}} I_n)$ and this is a subring of $\overline{R}$ for all n.*

Proof. Let $x \in (R :_{\overline{R}} I_n)$. Then $xa \in R$ for every $a \in I_n$. But, since $x \in \overline{R}$, $L(x) \geq 0$ and hence $L(xa) \geq L(a)$. Thus $xa \in I_n$. Thus $(R :_{\overline{R}} I_n) = (I_n :_{\overline{R}} I_n)$. That this is a subring of $\overline{R}$ is now elementary.

Proposition 16. *Suppose that $I_n \neq I_{n+1}$ and that I_{n+1} is an ideal. Then $(R :_{\overline{R}} I_n)$ is a local ring.*

Proof. Let x be a nonunit in the ring $(R :_{\overline{R}} I_n)$. Since $\overline{R}$ is integral over $(R :_{\overline{R}} I_n)$ we know that x is also a nonunit in $\overline{R}$, i.e. that $L(x) > 0$. Thus $xI_n \subseteq I_{n+1}$, i.e. $x \in (I_{n+1} :_{\overline{R}} I_n)$. Conversely let $x \in (I_{n+1} :_{\overline{R}} I_n)$. Since $I_n \neq I_{n+1}$, there is an element $a \in R$ of colength exactly n. But $xa \in I_{n+1}$, whence $L(x) \geq 1$. Thus x is a nonunit in $\overline{R}$ and hence in $(R :_{\overline{R}} I_n)$. The set of nonunits in $(R :_{\overline{R}} I_n)$ is therefore $(I_{n+1} :_{\overline{R}} I_n)$. We omit the last and elementary part of the proof which is to show that $(I_{n+1} :_{\overline{R}} I_n)$ is an ideal in $(R :_{\overline{R}} I_n)$ when I_{n+1} is an ideal in R.

Corollary 17. *Assume that R is integrally closed. Then the following conditions are equivalent.*

(1) *I_n is an ideal for some $n > 0$.*
(2) *I_n is an ideal for all n.*

Proof. Suppose condition (1) is satisfied. Then choose n such that $I_n \neq I_{n+1}$ and I_{n+1} is an ideal. From Proposition 16 we know that $(R :_{\overline{R}} I_n)$ is local. But, since $R = \overline{R}$, we have $(R :_{\overline{R}} I_n) = \overline{R}$. Now use Proposition 9.

Example. Let, as in the example of Section 2, $S = Z \setminus 5Z$, but put $R = Z_S[625i]$ this time. Again we get $\overline{R} = Z_S[i]$ with the two maximal ideals $\overline{\mathfrak{m}} = (1+2i)$ and $\overline{\mathfrak{n}} = (1-2i)$. R is local with maximal ideal $\mathfrak{m} = \overline{\mathfrak{m}} \cap R = \overline{\mathfrak{n}} \cap R = (5, 625i)$. We obtain $L(5) = 2$ and $L(1 \pm 2i) = 1$. We can write any $z \in R$ in the form $z = \dfrac{a5^n + b625i}{c}$, where 5 does not divide either a or c. A few minutes of pondering over this will reveal the following $L(z) = 0$ if $n = 0$, $L(z) = 2$ if $n = 1$, $L(z) = 4$ if $n = 2$, $L(z) = 6$ if $n = 3$ and $L(z) \geq 8$ if $n \geq 4$. It follows that $I_0 = R$, $I_1 = I_2 = \mathfrak{m}$, $I_3 = I_4 = (25, 625i)$, $I_5 = I_6 = (125, 625i)$ and $I_7 = I_8 = (625, 625i)$. Thus I_n is an ideal for $n \leq 8$. Note also that $(R : I_0) = R$, $(R : I_2) = Z_S[125i]$, $(R : I_4) = Z_S[25i]$ and $(R : I_6) = Z_S[5i]$ are all local rings. However $(R : I_8) = \overline{R}$ is not local. Thus we conclude from Proposition 16 that I_9 is not an ideal.

If $\overline{R}$ is local and finitely generated as an R-module, we know from Lemma 11 and Proposition 12 that L is bounded in $\overline{R} \setminus R$. The converse of this is not true, but we have the following generalization of Lemma 10.

Proposition 18. *If the colength function is bounded in $\overline{R} \setminus R$ then $\overline{R}_{\mathfrak{m}}$ is local for every maximal ideal $\mathfrak{m}$ in R.*

Proof. Assume there is an $\mathfrak{m}$ such that $\overline{R}_{\mathfrak{m}}$ is not local and let n be an arbitrary integer. Then Lemma 10 yields an $x \in \overline{R}_{\mathfrak{m}} \setminus R_{\mathfrak{m}}$ such that $L_R\ (x) \geq n$. We can write $x = \frac{a}{b}$, where $a \in \overline{R} \setminus R$ and $b \in R \setminus \mathfrak{m}$. Note that b is a unit in $R_{\mathfrak{m}}$ so that $L_R\ (x) = L_R\ (a)$. Put $T = R[a]$. Then $T_{\mathfrak{m}} = R_{\mathfrak{m}}[a]$. Using the facts that T is a finitely generated R-module and that $T_{\mathfrak{m}}$ is a finitely generated $R_{\mathfrak{m}}$-module we obtain $n \leq L_R\ (a) = \ell_R\ (T_{\mathfrak{m}}/aT_{\mathfrak{m}}) \leq \ell_R(T/aT) = L_R(a)$. But n is arbitrary, which shows that L is unbounded in $\overline{R} \setminus R$.

Note that Proposition 18 does not require $\overline{R}$ to be finite over R. A partial converse terminates the paper.

Proposition 19. *Assume that $\overline{R}$ is finitely generated as an R-module, that R is semilocal and that $\overline{R}_{\mathfrak{m}}$ is local for every maximal ideal $\mathfrak{m}$ in R. Then the colength function is bounded in $\overline{R} \setminus R$.*

Proof. $\overline{R}_{\mathfrak{m}}$ is finite over $R_{\mathfrak{m}}$ for all $\mathfrak{m}$, so the colength function is (by Lemma 11 and Proposition 12) bounded in each $\overline{R}_{\mathfrak{m}} \setminus R_{\mathfrak{m}}$ and thus in each $\overline{R} \setminus R_{\mathfrak{m}}$. However $R = \bigcap_{\mathfrak{m}} R_{\mathfrak{m}}$ so $\overline{R} \setminus R = \bigcup_{\mathfrak{m}} (\overline{R} \setminus R_{\mathfrak{m}})$. The result now follows from the fact that the $\mathfrak{m}$:s are finite in number.

That $\overline{R}_{\mathfrak{m}}$ is local can, of course, be rephrased as follows: there is exactly one maximal ideal in $\overline{R}$ lying over $\mathfrak{m}$.

REFERENCE

[1] O. Zariski and P. Samuel: *Commutative Algebra* Vol. 2, Springer (1976).

Null Recurrence in a Stochastic Ricker Model

Mats Gyllenberg*, **Göran Högnäs**†, **Timo Koski***

* Department of Applied Mathematics
Luleå University of Technology
S-95187 Luleå
SWEDEN

† Department of Mathematics
Åbo Akademi
SF-20500 Åbo
FINLAND

Abstract

We consider a nonlinear first order stochastic difference equation which may be viewed as a stochastic perturbation of Ricker's deterministic model of population growth. Numerical experiments seem to suggest that the corresponding Markov process has a stationary probability distribution but this is shown to be false by proving that the process is in fact null recurrent.

Subject Classification: AMS 60J05, 92D25, 92D40

Key words: Markov processes, Population Dynamics, Non-linear Stochastic Difference Equations, Stochastic Lyapunov functions

1 Introduction

In this paper we consider the long term behaviour of the Markov process (chain) $\mathbf{N} = \{N_t | t = 0, 1, 2, \cdots\}$ on the state space $(0, +\infty)$ generated by the first order stochastic difference equation

$$N_{t+1} = N_t e^{r(1-\xi_t \cdot N_t)}, \tag{1.1}$$

where $r > 0$ is constant and $\{\xi_t | t = 0, 1, 2, \cdots\}$ is a sequence of independent identically distributed real random variables having a density which is positive on all of $(-\infty, 0)$.

Equation (1.1) can be regarded as a stochastic perturbation of the widely studied deterministic model

$$N_{t+1} = N_t e^{r(1-\alpha N_t)} \tag{1.2}$$

of single-species population growth. Here N_t is the population density at time t, $r > 0$ is the per capita growth rate at low densities and α is the inverse of the equilibrium density or *carrying capacity*. The parameter α is a measure of intraspecific competition and its biological interpretation requires it to be non-negative. In the case $\alpha = 0$ there is no competition among individuals and the resulting linear difference equation models density independent growth. Note that in the biologically irrelevant case $\alpha < 0$ the difference equation (1.2) still defines a dynamical system on $(0, +\infty)$. In this case $1/\alpha$ cannot of course be interpreted as a carrying capacity and N_t tends rapidly (superexponentially) to infinity as $t \longrightarrow +\infty$. As r increases the deterministic model (1.2) exhibits convergence, period doubling, and chaos in the same way as the standard logistic difference equation $x_{t+1} = rx_t(1-x_t)$. The model (1.2) is known in ecology as the *Ricker model* [15] (c.f. [12]) and has since its introduction been comprehensively treated by several authors, see e.g. [11].

In the stochastic model (1.1) the carrying capacity of the environment is not assumed to be constant but to vary randomly and independently of population density N_t. Equation (1.1) is therefore a model of population growth under *environmental stochasticity*. Again the biological interpretation of carrying capacity requires the ξ_t to have a distribution with support on $[0, +\infty)$. This case has been thoroughly treated in [7]. In this paper we concentrate on the case where ξ_t is assumed to have a density which is positive on all of $(-\infty, 0)$. This case is not, however, completely void of biological interest. For instance Hanski and Woiwod [9] succesfully used equation (1.1) with normally distributed ξ_t to explain observed relationships between population mean density and variance.

Let us start our investigation by performing a numerical experiment. We simulated the process in (1.1) with Gaussian ξ_t (mean $\mu = 2$, variance $\sigma^2 = 0.1$). The resulting realization is depicted in Fig. 1. The process N looks very much like having a stationary distribution. In fact, if there were a stationary probability distribution, then its mean would be given by

$$E[N_t] = \frac{1}{\mu}, \tag{1.3}$$

as is easily seen. Computing the arithmetic mean of the time series in Fig. 1 one obtains the value ≈ 0.4961 which agrees extremely well with the value

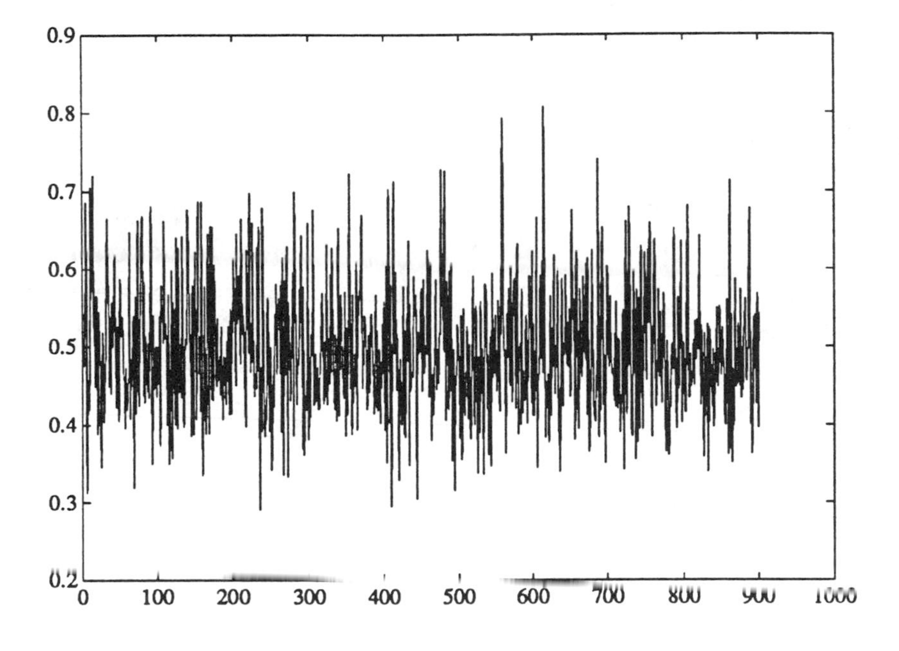

Figure 1: *A Simulation of* N_t, $r = 1$, ξ_t *is* $N(2, 0.1)$- *distributed.*

0.5 given by (1.3).

But appearances are deceptive. Our main result shows that if ξ_t has a density, which is positive on (all of) $(-\infty, 0)$ then $\mathbf{N}$ is a null recurrent process and therefore does not possess a stationary (probability) distribution. We conclude that mathematics is not an empirical science.

The stochastic difference equation (1.1) has been previously investigated by the present authors in [7], where a weaker version of Proposition 3.1 of the present paper was proved. The strenghtened result is obtained with the aid of a new stochastic Lyapunov function which is a kind of inverse to a particular superexponential sequence.

Other related types of stochastic first order difference equations with applications to ecology have been studied by a number of authors, the interested reader is in particular referred to the work of Ellner [4] and the references therein as well as to [7]. The main difference between Ellner's models and ours is that Ellner [4] imposes certain monotonicity conditions not satisfied by the Ricker model.

Environmental stochasticity is an important issue in theoretical ecology. It has been studied using a number of different probabilistic models, see e.g. [14, chap. 8].

2 Classification of States and a Recurrence Criterion for Markov Chains

We recall first, for ease of reference and for clarity of exposition, a few salient facts and definitions about Markov processes. For more extensive accounts and surveys of the statements below we refer to [10, 13, 18, 19]. Let now $\mathbf{X} = \{X_t | t = 0, 1, 2, \cdots, \}$ be a Markov process with the *state space* generically denoted by $\mathcal{X}$. In the subsequent study the state space $\mathcal{X}$ will always be the real numbers $\mathcal{R}$. We refer hence to the measurable subsets of $\mathcal{X}$ as Borel sets. $\mathbf{P}_x$ will designate the probability measure governing the process starting at $x \in \mathcal{X}$ and $\mathbf{P}$ stands for probability expressions which are true for all $x \in \mathcal{X}$ simultaneously. Similarly, E_x and E denote the corresponding expectations. $P(x; A)$, $x \in \mathcal{X}$, A a Borel set, is the temporally homogeneous transition probability of the process, i.e. $P(x; A) = \mathbf{P}_x(X_1 \in A)$. We let the t *-step (temporally homogeneous) transition probability* be $\mathbf{P}^t(x; A) := \mathbf{P}_x(X_t \in$

A), where A is a Borel set. For the Markov process under consideration here this can be expressed as $\mathbf{P}^t(x; A) = \int_A p_t(v \mid x)dv$. In other words there exists the *t-step transition density*, $p_t(v \mid x)$, of the process $\mathbf{X}$ and in addition for all x and v we have $p_t(v \mid x) > 0$. Some significant properties of the process $\mathbf{X}$ follow readily from these facts by the results in [19] or [10, 13, 18]. In fact, the preceding implies that we have for every $x \in \mathcal{X}$ that $\sum_{t=1}^{\infty} 2^{-t}\mathbf{P}^t(x; A) > 0$, if the set A has a positive measure w.r.t. some σ-finite measure ϕ on the state space $\mathcal{X}$. Here the measure ϕ can be taken as the Lebesgue measure. This fact recapitulates one of many similar definitions of *(ϕ-) irreducibility*, see [4, 10, 18, 19]. This means that every set in $\mathcal{X}$ with positive ϕ -measure is attainable by the process $\mathbf{X}$ from any $x \in \mathcal{X}$. Furthermore, these facts imply that $\mathbf{X}$ is an *aperiodic* process, see [13, pp.20-21], i.e. there are no cyclic subsets in the state space. For proofs of the indicated general facts we refer e.g. to [13, p.11] or [18, p. 741].

An irreducible Markov process is *recurrent*, if

$$\sum_{t=1}^{\infty} \mathbf{P}^t(x; A) = +\infty \tag{2.1}$$

for all sets A with positive ϕ-measure. Otherwise the process is called *transient.* For a full treatment of these topics the reader is referred to [19].

Every recurrent process has a σ -finite invariant measure, denoted by π, a measure satisfying

$$\pi(A) = \int_{\mathcal{X}} \pi(dx)\mathbf{P}_x(A) \tag{2.2}$$

for any Borel set A. If the measure π is finite (and thus can be normalized to a probability), then we say that $\mathbf{X}$ is a *positively recurrent* process. Otherwise the process is called *null-recurrent.* For a null-recurrent process there exists at least one set A with positive ϕ-measure such that $\lim_{n \to +\infty} \frac{1}{n} \sum_{t=1}^{n} \mathbf{P}^t(x; A) = 0$. This means that the process returns to such an A very rarely, in fact so rarely that the expected time for first return to A will be infinite.

In the theory of dynamical systems questions of fundamental importance deal with the existence and stability of equilibria and periodic orbits. Stability analysis of Ricker's model and certain other difference equation models encountered in population biology is found in [5, 6]. For Markov processes problems concerning the existence of stationary distributions (invariant probability measures), recurrence and first return time play a similar part.

Positive recurrence of a Markov process, in particular when determined by a stochastic difference equation, is often conveniently established by checking a *recurrence criterion*, see [7, 10, 16, 17, 18, 19] and also [20] for a brief and clear statement, or by finding what is sometimes known as a *stochastic Lyapunov function* (see [8]) for the given process. The phrase stochastic Lyapunov function is used for a function prescribed in the recurrence criterion for the reason that the technique is analogous with the Lyapunov functions used in the stability analysis of deterministic difference equations, see e.g. [1, ch. 5] or [3]. The Lyapunov functions for stochastic difference equations discussed in [2] have a different purpose than that of deriving recurrence properties.

However, we wish to establish null-recurrence. We are in a position to do this by using a first return time argument to rule out the possibility of positive recurrence for the Ricker model and then by showing that the model is recurrent. In the second step we use the following recurrence criterion, given in [18, thm. 4.3].

Proposition 2.1 *If there exists a non-negative function* V*, a* stochastic Lyapunov function, *defined on* $\mathcal{X}$*, a topological space, such that there exists a compact set* $K \subseteq \mathcal{X}$ *with*

$$E_x\left[V(X_1)\right] - V(x) \leq 0 \text{ as } x \notin K, \tag{2.3}$$

and that for every sufficiently large M*, there is a compact set* K_M *of positive* ϕ*- measure such that*

$$V(x) \geq M \quad \text{on the complement of } K_M, \tag{2.4}$$

then the process $\mathbf{X}$ *is recurrent.*

The proof can be done by checking an equivalent form of (2.1). In the proof in [18, loc.cit] it is required that the process is irreducible and that there is Feller (or strong) continuity. The latter requirement is in our case automatically satisfied because of the continuity of the pertinent transition probability $\mathbf{P}(x; A)$ in x.

3 The Main Results

In this section we show that under certain conditions the process $\mathbf{N}$ generated by equation (1.1) is null-recurrent and hence has no invariant probability measure. These conditions do not involve the parameter r and therefore the behaviour of the stochastically perturbed equation does not depend on r as in the deterministic case. In figure 2 we display a simulated time-series of $X_t = \log(N_t)$ which shows that the process $\mathbf{N}$ will once in a while grow fast for a few steps to a large value v, say, and then suffer a tremendous setback (or a population catastrophe in biological terms) to $\exp(-v)$. After a catastrophe the population grows almost deterministically and it takes a very long time until the population density has been restored to the levels preceding the catastrophe.

Our formal proof will be derived from this insight. We shall show that the *expected return time* to a neighborhood of 1 in a logarithmic scale turns out to be infinite. For recurrence we construct a Lyapunov function V that grows extremely slowly to $+\infty$ as $|x| \uparrow +\infty$ thus emulating the essential feature of null-recurrence.

Proposition 3.1 *Assume that ξ_t are independent, identically distributed random variables on R with a density that is positive on all of $(-\infty, 0)$. Assume in addition that*

1. $E[|\xi_t|] < +\infty$
2. $E[|\xi_t|^2] < +\infty$
3. $\mathbf{P}(\xi_t > 0) > 0$
4. $N_0 > 0$.

Then $\mathbf{N}$ *is a null recurrent Markov process on the state space* $(0, +\infty)$.

The result holds true even in the case where ξ_t has a density that is positive on subsets of the form $(-\beta, 0)$, $0 < \beta < +\infty$. This case has been completely treated in [7].

The proof of the assertion is presented after we have provided some notational preliminaries and two lemmas that exclude the positive recurrence.

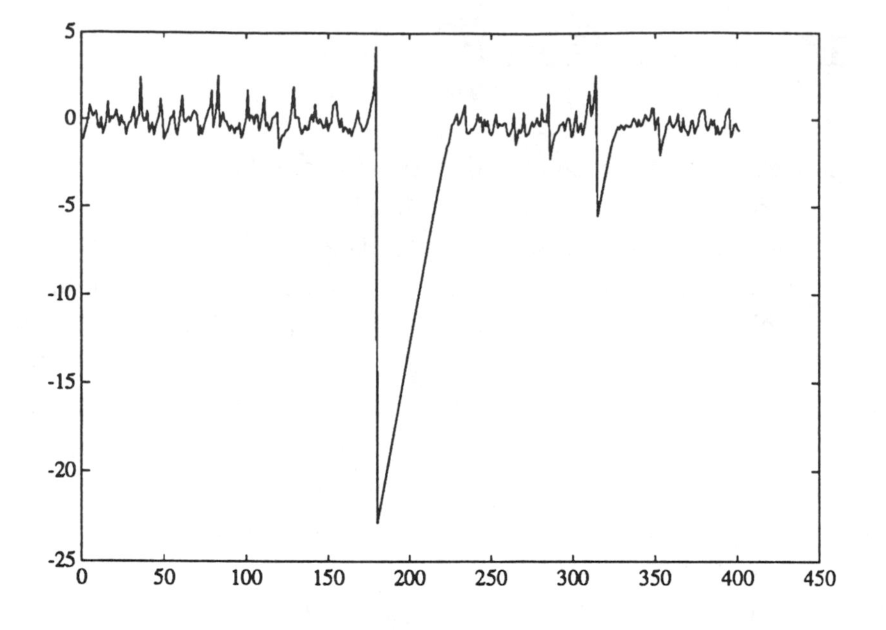

Figure 2: *A Simulation of* $X_t = \log(N_t)$, $r = 1$, ξ_t *is* $N(1,1)$*- distributed.*

Taking logarithms of equation (1.1) one obtains

$$X_{t+1} = X_t + r\left(1 - \xi_t e^{X_t}\right), \quad t = 0, 1, 2, \cdots, \tag{3.1}$$

where $X_t := \log(N_t)$. Equation (3.1) generates the Markov process $\mathbf{X} = \{X_t \mid t = 0, 1, \ldots\}$ with state space $\mathcal{R}$.

It is obvious that this process is irreducible (with respect to the Lebesgue measure). It is also obvious that the recurrence properties are inherited by $\mathbf{X}$ from $\mathbf{N}$ and vice versa. We therefore prove proposition 3.1. by investigating the process $\mathbf{X}$ and proving that it is null-recurrent. We note in passing that (3.1) is not one of the standard explicit processes encountered in the theory of nonlinear stochastic difference equations or nonlinear autoregressions, c.f. [16, 17].

Let T_A be the *first return time* to the set A, given $X_0 \in A$,

$$T_A := \min\{t \geq 1 | X_t \in A\}. \tag{3.2}$$

We take $A = [0, 1]$. The following lemma has been proved in [7].

Lemma 3.2 *For any c and $\epsilon \in (0,1)$, there is a negative number $-L$ such that if $x < -L$, then*

$$\mathbf{P}_x\left(T_A \geq (1-c)\frac{|x|}{r+1}\right) > 1-\epsilon. \tag{3.3}$$

■

Let us now recall that if an invariant probability measure π exists for a process, then the first return time T_A to any set with $\pi(A) > 0$ satisfies

$$1 = \int_A E_x(T_A)\pi(dx).$$

For this reason the next lemma suffices to show that the process $\mathbf{X}$ cannot be positively recurrent.

Lemma 3.3 $E_y(T_A) = +\infty$ *for all* $y \in A$.

Proof: Let first $\{f_k\}_{k=0}^{+\infty}$ be the sequence defined as follows:

$$f_0 = 0,\ f_{k+1} = f_k + r(1+e^{f_k}),\ (k = 0,1,2,\cdots). \tag{3.4}$$

Thus the numbers f_k grow "superexponentially". Let

$$p := \min\left[\mathbf{P}(\xi_t < -1), \mathbf{P}(\xi_t > 2)\right].$$

Then the event $\{\xi_0 < -1, \xi_1 < -1, \cdots, \xi_{k-1} < -1, \xi_k > 2\}$ has probability $\geq p^{k+1}$. Hence $\mathbf{P}_y(X_k > f_k, X_{k+1} < -f_k) \geq p^{k+1}$ for $y \geq 0$. When this observation is combined, via a conditioning argument, with the result in lemma 3.2, we obtain that

$$\mathbf{P}_y\left(T_A \geq (1-c)\frac{f_k}{r+1} + k + 1\right) > (1-\epsilon)p^{k+1}.$$

Checking now the expected first return time we have

$$E_y(T_A) = \sum_{i=0}^{+\infty} \mathbf{P}_y(T_A > i) > \sum_{i=0}^{\lceil c' f_k \rceil} \mathbf{P}_y(T_A > i),$$

where c' is a suitable constant ($> (1-c)/(r+1)$) and k is a large enough integer. The last sum is larger than $c' f_k p^{k+1}$. This term increases superexponentially fast. Hence the series is divergent and the expected return time to A is infinite for all starting points in A. ∎

Before proving proposition 3.1 we introduce some notations and definitions. Since ξ_t are i.i.d. we shall in the rest of this proof omit, for ease of writing, the subscript t. Hence we introduce

$$p(x) := \mathbf{P}_x (X_1 > 0) = \mathbf{P}\left(x + r\left(1 - \xi e^x\right) > 0\right)$$

or, equivalently,

$$p(x) = \mathbf{P}\left(\xi < \frac{e^{-x}(x+r)}{r}\right) \tag{3.5}$$

and

$$0 \le p_o := \lim_{x \longrightarrow +\infty} p(x). \tag{3.6}$$

The case $p_o = 0$ is covered in [7], hence we shall hereafter take

$$p_o > 0. \tag{3.7}$$

Let us set

$$m_- := E\left(\mid \xi \mid I_{\{\xi \le 0\}}\right), \quad m_+ := E\left(\mid \xi \mid I_{\{\xi > 0\}}\right),$$

where I_A is the indicator function of the Borel set A. We note that $\mu \equiv E(\xi) = m_+ - m_-$ and that

$$E_x\left(\xi \mid X_1 > 0\right) = E\left(\xi I_{\{x + r(1-\xi e^x) > 0\}}\right) \cdot \frac{1}{p(x)}. \tag{3.8}$$

It follows that

$$E\left(\xi I_{\{x + r(1-\xi e^x) > 0\}}\right) \cdot \frac{1}{p(x)} \longrightarrow \frac{m_-}{p_o} < +\infty, \text{ as } x \longrightarrow +\infty. \tag{3.9}$$

Similar formulas link the event $\{X_1 \le 0\}$ (conditioned on $X_0 = x$) and m_+. It turns out to be useful for the subsequent analysis to introduce an $\overline{m}$ chosen such that

$$\overline{m} > \max\left\{\frac{m_-}{p_o}, \frac{m_+}{(1-p_o)}\right\}. \tag{3.10}$$

Note that $p_o < 1$ by assumption *3.* above. Furthermore we define another superexponential sequence, without risk of confusion with (3.4) again denoted by $\{f_n\}_{n=0}^{+\infty}$, in terms of the iteration

$$f_0 = 1, f_{n+1} = g(f_n) = g^n(f_0), \tag{3.11}$$

where g^n denotes the n:th successive iteration of the map

$$g(x) := x + r(1 + \overline{m}e^x). \tag{3.12}$$

Clearly $f_{n+1} > r\overline{m}e^{f_n}$, so $f_n \uparrow +\infty$ as $n \longrightarrow +\infty$. We have also $f_{n+1} - f_n < f_{n+2} - f_{n+1}$, since

$$f_{n+1} - f_n = g(f_n) - f_n = r\left(1 + \overline{m}e^{f_n}\right), \tag{3.13}$$

which by definition means that $\{f_n\}_{n=0}^{+\infty}$ is a *convex sequence*. We put

$$W(f_n) := n, \quad n = 0, 1, 2, \ldots, \tag{3.14}$$

and define $W(x)$ for $x > 0$ by linear interpolation i.e.

$$W(x) := \frac{x}{f_{n+1} - f_n} - \frac{f_n}{f_{n+1} - f_n} + n \text{ for } f_n \le x \le f_{n+1} \tag{3.15}$$

or inserting from the above

$$W(x) = \frac{x}{r(1 + \overline{m}e^{f_n})} - \frac{f_n}{r(1 + \overline{m}e^{f_n})} + n \text{ for } f_n \le x \le f_{n+1}. \tag{3.16}$$

Evidently W is indeed a very slowly increasing function. Now we can introduce our candidate for a *stochastic Lyapunov function* V in the recurrence critorion outlined above by

$$V(x) := \begin{cases} W(x) & x \ge 1 \\ 0 & -f_1 \le x \le 1 \\ W(|x|) - 1 & x < -f_1. \end{cases} \tag{3.17}$$

Proof of Proposition 3.1: The idea is to use the fact that V is concave on $(-\infty, -f_1)$ and on $(1, +\infty)$ to prove that

$$E_x\{V(X_1)\} - V(x) \le 0 \tag{3.18}$$

for $| x |$ large enough. It is suitable for our purposes to split up the demonstration into two cases.

Case 1: $x > 1$.

Here as in *Case 2* below we write the one-step conditional expectation of the Lyapunov function as

$$\begin{aligned} E_x \{V(X_1)\} &= E_x \left[V(X_1) I_{\{X_1 \geq 0\}} \right] \\ &+ \\ &E_x \left[V(X_1) I_{\{X_1 < 0\}} \right] \end{aligned} \tag{3.19}$$

and then deal with the two terms in the right hand side separately.

We shall first decompose the event $\{X_1 \geq 0\}$ conditioned on $X_0 = x$. For $n \geq 2$ we set

$$A_n(x) := \left\{ \frac{g^{n-1}(x) - x - r}{re^x} \leq | \xi | \leq \frac{g^n(x) - x - r}{re^x} \right\}. \tag{3.20}$$

i.e.

$$A_n(x) = \left\{ g^{n-1}(x) \leq x + r(1 - \xi e^x) \leq g^n(x) \right\}. \tag{3.21}$$

For ξ assuming values in $A_n(x)$ it is easily seen using the given definitions that

$$W(x) + (n-2) \leq W(x + r(1 - \xi e^x)) \leq W(x) + (n+1). \tag{3.22}$$

In addition we introduce the sets

$$A_{x,M} := \left\{ 0 \leq x + r(1 - \xi e^x) \leq \frac{g(x)}{M} \right\} \tag{3.23}$$

and

$$A^c_{x,M} := \left\{ \frac{g(x)}{M} \leq x + r(1 - \xi e^x) \leq g(x) \right\}. \tag{3.24}$$

Using the same reasoning as above we obtain that for $\xi \in A^c_{x,M}$

$$W(x + r(1 - \xi e^x)) \leq W(x) + 2. \tag{3.25}$$

Let us next consider the case, where ξ hits $A_{x,M}$. For any given $M > 1$ it holds for x large enough that

$$W(x + r(1 - \xi e^x)) \leq W\left(\frac{g(x)}{M}\right) < W(x) + \frac{1}{M} \tag{3.26}$$

in view of the fact that the function W behaves almost like a constant for very large values of the argument.

The event $\{X_1 \geq 0\}$ (conditioned on $X_0 = x$) can be written as a disjoint union of $A_n(x)$'s, $A_{x,M}$ and $A^c_{x,M}$. Hence in view of the preceding estimates and using (3.1), the expectation of $E_x\left[V(X_1) I_{\{X_1 \geq 0\}}\right]$ is bounded above by

$$E_x\left[W(x + r(1 - \xi e^x)) I_{\{X_1 \geq 0\}}\right] \leq W(x)p(x) \quad + \\ + \frac{1}{M}\mathbf{P}(A_{x,M}) + 2\mathbf{P}\left(A^c_{x,M}\right) \\ + \sum_{n=2}^{+\infty}(n+1)\mathbf{P}\left(\frac{g^{n-1}(x) - x - r}{re^x} \leq \mid \xi \mid \leq \frac{g^n(x) - x - r}{re^x}\right).$$

The sum is convergent since $W(x) < \mid x \mid$ and $E \mid \xi \mid < +\infty$. As $x \longrightarrow +\infty$ the terms in the sum approach zero. Also $\mathbf{P}(A_{x,M}) \longrightarrow 1$ and $\mathbf{P}\left(A^c_{x,M}\right) \longrightarrow 0$ as $x \longrightarrow +\infty$. In other words we have obtained

$$E_x\left[W(x + r(1 - \xi e^x)) I_{\{X_1 \geq 0\}}\right] < W(x)p(x) + 3\epsilon_1 \tag{3.27}$$

for x large enough and for an arbitrary $\epsilon_1 > 0$.

Next we consider $E_x\left[V(X_1) I_{\{X_1 \leq 0\}}\right]$ for x large enough. By construction this is

$$E_x\left[\left(V(X_1) I_{\{X_1 \leq 0\}}\right] = E_x\left[(W(r\xi e^x - x - r) - 1)\right) I_{\{X_1 \leq -f_1\}}\right].$$

Since $W(|x|)$ is a concave function for $x \in (-\infty, f_1)$, an iterated conditional expectation and Jensen's inequality yield the upper bound for the first term as

$$E_x\left[W(r\xi e^x - x - r) I_{\{X_1 \leq -f_1\}}\right] \leq \\ E_x\left[W(rE_x(\xi\{X_1 \leq 0\}) e^x - x - r) I_{\{X_1 \leq 0\}}\right].$$

since $\{X_1 \leq -f_1\} \subseteq \{X_1 \leq 0\}$. As pointed out earlier

$$E_x(\xi \mid X_1 \leq 0) \longrightarrow \frac{m_+}{1 - p_o}$$

as x increases. Here we note that since $\frac{m_+}{1-p_o} < \overline{m}$ by our choice in (3.10) we have for large x

$$W(rE_x(\xi \mid X_1 \leq 0) e^x - x - r) < W(g(x)) \leq W(x) + 1. \tag{3.28}$$

To verify the right hand inequality in (3.28) we take $f_n \leq x \leq f_{n+1}$ so that $x = \alpha f_n + (1-\alpha) f_{n+1}$. Clearly $f_{n+1} \leq g(x) \leq f_{n+2}$. Then, since g is a convex map,

$$\begin{aligned} g(x) &\leq \alpha g(f_n) + (1-\alpha) g(f_{n+1}) \\ &= \alpha f_{n+1} + (1-\alpha) f_{n+2}. \end{aligned} \tag{3.29}$$

Hence we obtain by definition of W that

$$\begin{aligned} W(\alpha f_{n+1} + (1-\alpha) f_{n+2}) &= \alpha(n+1) + (1-\alpha)(n+2) \\ &= \alpha n + (1-\alpha)(n+1) + 1 \\ &= W(\alpha f_n + (1-\alpha) f_{n+1}) + 1, \end{aligned} \tag{3.30}$$

which proves the asserted inequality.

Since $0 < I_{\{x+r(1-\xi e^x)\leq 0\}} - I_{\{x+r(1-\xi e^x)\leq -f_1\}} < \epsilon_2$ for arbitrary $\epsilon_2 > 0$ as x becomes sufficiently large we have by the preceding shown that

$$E_x\left[(W(r\xi e^x - x - r) - 1) I_{\{X_1 \leq -f_1\}}\right] < W(x)(1-p(x)) + \epsilon_2 \tag{3.31}$$

for large x. The inequalities (3.27) and (3.31) constitute the desired result in *Case 1*.

Case 2: $x < -f_1$.

As above we consider, using (3.1),

$$\begin{aligned} E_x[V(X_1)] &= E_x\left[V(x + r(1-\xi e^x)) I_{\{X_1 \geq 0\}}\right] \\ &\quad + \\ &\quad E_x\left[V(x + r(1-\xi e^x)) I_{\{X_1 < 0\}}\right]. \end{aligned} \tag{3.32}$$

The first term in the right hand side of (3.32) can be handled by a different kind of bounding technique. By force of the assumptions made in this proposition Chebyshev's inequality yields

$$\mathbf{P}\left(-\xi > e^{-x}\frac{(|x| - r)}{r}\right) \leq \frac{c_1 e^{-2|x|}}{|x|^2} \tag{3.33}$$

for a constant c_1 depending on the second moment of $\xi I_{\{\xi\leq 0\}}$. In the same way we obtain, with another constant c_2,

$$\mathbf{P}_x(X_1 > a) \leq \frac{c_2 e^{-2|x|}}{(|x| + a)^2}. \tag{3.34}$$

Hence we have, because $V(x) < x$,

$$\begin{aligned} E_x\left[V(X_1)\, I_{\{X_1\geq 0\}}\right] &\leq E_x\left[X_1 I_{\{X_1\geq 0\}}\right] \\ &= \int_0^{+\infty} \mathbf{P}_x(X_1 \geq a)\, da \\ &\leq \int_0^{+\infty} \frac{c_2 e^{-2|x|}}{(|\, x\,| + a)^2} da \\ &= \frac{c_3 e^{-2|x|}}{|\, x\,|}. \end{aligned} \tag{3.35}$$

Finally we consider the remaining term of (3.32). V is piecewise linear and concave on $(-\infty, -f_1)$ so for $x << 0$ with $-f_{n+1} \leq x \leq -f_n$ and $X_1 < -f_1$ we obtain again by iterated conditional expectation, Jensen's inequality and setting $m(x) := E_x[\xi \mid X_1 < 0]$ that

$$\begin{aligned} &E_x\left[V(x + r(1 - \xi e^x))\, I_{\{X_1<0\}}\right] \leq \\ &V(x + r(1 - m(x)e^x))\, \mathbf{P}_x(X_1 < -f_1). \end{aligned} \tag{3.36}$$

As $x \longrightarrow -\infty$, $\mathbf{P}_x(X_1 < -f_1) \longrightarrow 1$ and $m(x) \longrightarrow m$. Since $\{f_n\}$ is a convex sequence that grows superexponentially we may assume that both x and $x + r(1 - m(x)e^x)$ are included in one and the same interval of the type $[-f_{n+1}, -f_n[$ for some n.

Then we have by construction of V in the negative axis and by an application of (3.15) for $x << 0$ that

$$V(x + r(1 - m(x)e^x)) - V(x) = \frac{-(1 - m(x)e^x)}{1 + \overline{m}e^{f_n}} < 0, \tag{3.37}$$

where we have also used (3.13).

In view of (3.35) and (3.37) we have obtained bounds that together with the result in *Case 1* show that

$$E_x\{V(X_1)\} - V(x) \leq 0 \tag{3.38}$$

for $|\, x\,|$ large enough.

Since lemma 3.3 eliminates positive recurrence, we have in view of (3.38) proved the claim in proposition 3.1 as asserted i.e. that V given in (3.17) satisfies the recurrence criterion stated in [18, thm. 4.3] and recapitulated in (2.3) - (2.4). ■

It is has been observed, see [3], that in many situations there is a systematic approach to prove ergodicity of a stochastic difference equation via a Lyapunov function corresponding to its deterministic part. In the present case, where we are concerned with just recurrence, this approach would not seem to be particularly useful in a direct sense, if checked against the deterministic Lyapunov functions for Ricker's model established in [6]. Rather our choice of Lyapunov function is motivated by a careful analysis of the growth-catastrophe behaviour of the stochastic Ricker model.

Acknowledgements: The research of Mats Gyllenberg is partially supported by the Bank of Sweden Tercentenary Foundation, The Swedish Council for Forestry and Agricultural Research and the Swedish Cancer Foundation. The research of Timo Koski is partially supported by the Bank of Sweden Tercentenary Foundation.

References

[1] R.P. Agarwal: *Difference Equations and Inequalities. Theory, Methods, and Applications.* Marcel Dekker Inc. New York, Basel, Hong Kong, 1992.

[2] R.S. Bucy: Stability and Positive Supermartingales. *J. Diff. Eq.*, 1, 1965, pp. 151 - 155.

[3] K.S. Chan and H. Tong: On the Use of the Deterministic Lyapunov Function for the Ergodicity of Stochastic Difference Equations. *Adv. Appl. Prob.*, 17, 1985, pp. 666 - 678.

[4] S.P. Ellner: Asymptotic behaviour of some stochastic difference equation population models. *J. Math. Biol.*, 19, 1984, pp. 169 - 200.

[5] M.E. Fisher, B.S. Goh and T.L. Vincent: Some Stability Conditions for Discrete Time Single Species Models. *Bull. Math. Biol.*, 41, 1979, pp. 861 - 875.

[6] B.S. Goh: Stability in a Stock Recruitment Model of Exploited Fishery. *Math. Biosci.*, 33, 1977, pp. 359 - 372.

[7] M.Gyllenberg, G. Högnäs and T. Koski: Population Models with Environmental Stochasticity. *J. Math. Biol.*, to appear, 1992.

[8] H.J. Kushner: *Introduction to Stochastic Control.* Holt, Rinehart and Winston, New York, 1971.

[9] I. Hanski and I.P. Woiwod: Mean-related Stochasticity and Population Variability. *Oikos*, 1993, in the press.

[10] G.M. Laslett, D.B. Pollard and R.L. Tweedie: Techniques for Establishing Ergodic and Recurrence Properties of Continuous-Valued Markov Chains. *Nav. Res. Log. Quart.*, 25, 1978, pp. 455 - 472.

[11] R.M. May: Simple mathematical models with very complicated dynamics. *Nature*, Vol. 261, 1976, pp. 459 - 467.

[12] P.A.P. Moran: Some remarks on animal population dynamics. *Biometrics*, 6, 1950, pp. 250 - 258.

[13] E. Nummelin: *General Irreducible Markov Chains and Nonnegative Operators.* Cambridge University Press, Cambridge, 1984.

[14] E. Renshaw: *Modelling Biological Populations in Space and Time.* Cambridge Studies in Mathematical Biology, Vol. 11. Cambridge University Press, Cambridge, 1991.

[15] W.E. Ricker: Stock and Recruitment. *J. Fish. Res. Bd. Can.*, 11, 1954, pp. 559- 623.

[16] D. Tjøstheim: Non-linear Time Series and Markov Chains. *Adv. in Appl. Prob.*, 22, 1990, pp. 587 - 611.

[17] H. Tong: *Non-Linear Time Series. A Dynamical System Approach.* Oxford Science Publ. Clarendon Press, Oxford, 1990.

[18] R.L. Tweedie: Sufficient Conditions for Ergodicity and Recurrence of Markov Chains on a General State Space. *Stoch. Proc. Appl.*, 3, 1975, pp. 385 - 403.

[19] R.L. Tweedie: Criteria for Classifying General Markov Chains. *Adv. Appl. Prob.*, 8, 1976, pp. 737 - 771.

[20] R.L. Tweedie: Recurrence Criterion. *Encyclopedia of Statistical Science*, S. Kotz, N.L. Johnson and C.B. Read (Ed's), Volume 7, 1986, pp. 656 - 658.

On a Simplicial Complex Associated with a Set of Power Products

BERND HERZOG

Department of Mathematics
University of Stockholm
Box 6701, S–11385 Stockholm, Sweden
e-mail: herzogmatematik.su.se.

0 Introduction

The aim of this article is to describe a variant of Hochster-Reisner theory [5] which seems to be useful in connection with Hilbert functions. We give a construction associating a simplicial complex with a set of power products. As an illustration how to use this complex we prove Macaulay's Theorem on the growth of the Hilbert function by estimating the number of linear independent syzygies of a power product ideal in terms of the dimensions of the homology groups of the complex (Proposition 3.6). The central notion here is the notion of 1-component (Definition 3.3). It turns out to be useful also for another problem dealing with Hilbert functions, for Lech's problem. We give a short survey of the results achieved in this context.

Throughout this paper we use the notations and conventions of commutative algebra as in [7]. $\mathbf{N}$ will denote the set non-negative integers.

1 Macaulay's Function

Let $G = \bigoplus_{n=0}^{\infty} G(n)$ be a graded algebra over a field K which is finitely generated in degree 1. In other words, G is the factor ring of a polynomial algebra $K[T_1, \ldots, T_N]$ modulo a homogeneous ideal. The *Hilbert function* of G associates to every non-negative integer d the vector space dimension of the degree d part of G,

$$H_G(d) = dim_K G(d).$$

Partly supported by Swedish Natural Science Research Council.

Macaulay's theorem gives an upper bound for the (d+1)-st value $H_G(d+1)$ of the Hilbert function of G in terms of its d-th value. Before stating Macaulay's result, we define the function occurring in the formulation of the theorem.

Terminology 1.1 The number of elements of a set M will be denoted by $\sharp(M)$. The sequence of indeterminates $T_1, \ldots, T_N$ will be often denoted by T. So we will write $K[T]$ for the polynomials in $T_1, \ldots, T_N$ over the field K and, if S is a set of power products in the variables T, the set obtained multiplying all elements of S with one of the variables will be denoted by

$$T \cdot S = TS := \{T_i s \mid s \in S, i = 1, \ldots, N\}$$

Given two different power products $a = T_1^{a_1} \cdot \ldots \cdot T_N^{a_N}$ and $b = T_1^{b_1} \cdot \ldots \cdot T_N^{b_N}$ we will write $a < b$ and say that a is *(degreewise) lexicographically earlier* or *smaller* than b if either $\sum_{i=1}^{N} a_i < \sum_{i=1}^{N} b_i$ or $\sum_{i=1}^{N} a_i = \sum_{i=1}^{N} b_i$ and the first non-zero member in the sequence

$$a_1 - b_1, a_2 - b_2, \ldots, a_N - b_N$$

is positive. Unless stated otherwise, sets of power products will be equipped with this order.

Given a set S of degree d power products in the indeterminates T the *compression* of S (cf [1]) is defined to be the set $\mathbf{C}(S)$ of the lexicographically first $\sharp(S)$ power products of degree d. A set of power products S will be called *compressed* if it satisfies $\mathbf{C}(S) = S$.

Definition 1.2 Let S be a set of degree d power products in the variables $T_1, \ldots, T_N$ such that the ideal (S) of the polynomial ring $K[T]$ generated by S has prescribed Hilbert function

$$H_{K[T]/(S)}(d) = n$$

in degree d. Then *Macaulay's function* is defined by

$$Q_d(n) := H_{K[T]/(\mathbf{C}(S))}(d+1).$$

In other words, $Q_d(n)$ is the value of the Hilbert function in degree $d+1$ associated with the ideal generated by the $H_{K[T]}(d) - n$ lexicographically first power products of degree d. Equivalently one could say that Macaulay's function is defined by the identities

$$Q_d(H_{K[T]/(S)}(d)) := H_{K[T]/(\mathbf{C}(S))}(d+1)$$

for arbitrary power product ideals (S) of $K[T]$.

It is easy to see that this definition is correct, i.e., does not depend upon the number N of indeterminates T_i (which, however, should be large enough). To see this, adjoin to $K[T]$ one further indeterminate, say T_0, which is lexicographically earlier than all other variables. Then (S) and $(\mathbf{C}(S))$ must be replaced by the ideals containing additionally all power products of degree d which are multiples of T_0. But this means that the graded rings $K[T]/(S)$ and $K[T]/(\mathbf{C}(S))$ don't change in degree d and higher.

2 Macaulay's Theorem

Theorem 2.1 (F.S. Macaulay [6]) *A function $H : \mathbf{N} \to \mathbf{N}$ is a Hilbert function of some graded algebra generated in degree one if and only if $H(d+1) \leq Q_d(H(d))$ for every d.*

In other words, the ideals generated by compressed sets of power products have a Hilbert function with maximal growth, and every function satisfying the growth conditions coming from the growth of the Hilbert functions of compressed power product ideals, is a Hilbert function of some graded algebra.

This theorem has several interesting applications. One which seems to be less well known is Hironaka's theorem which states that in a sequence of germs of permissible blowups,

$$(X_0, x_0) \leftarrow (X_1, x_1) \leftarrow \ldots \leftarrow (X_n, x_n) \leftarrow \ldots$$

the associated sequence of local Hilbert functions stabilizes. Hironaka's theorem is a trivial consequence of Macaulay's theorem and the fact that permissible blowups cannot increase the local Hilbert functions. For details see [2].

The proof of Macaulay's theorem is generally considered to be difficult. All known proofs use heavily combinatorial arguments and do not really help to understand the phenomenon. Since Macaulay's theorem is related to different problems in connection with Hilbert functions, it might be useful to have a less combinatoric and more qualitative proof. Below we give a proof which may be considered as a translation of the proof due to Clements and Lindström [1] into a more geometric language. It is based on the calculation of the first homology group of a simplicial complex. The hard part of Macaulay's theorem may be stated as follows.

Proposition 2.2 *Let S be a set of power products of degree d in the indeterminates $T_1, \ldots, T_N$. Then*

$$T \cdot \mathbf{C}(S) \subset \mathbf{C}(T \cdot S).$$

Proof of Macaulay's theorem using 2.2 Assume H is the Hilbert function of some graded algebra $G = K[T]/I$. Replacing I by the ideal which is generated by the initial terms of the elements from I, we may assume that I is generated by some set $S = \{s_1, \ldots, s_n\}$ of power products. Moreover we may assume that all these generators have degree d. Then

$$H(d+1) = H_{K[T]/(S)}(d+1) = H_{K[T]/(T \cdot S)}(d+1) = H_{K[T]/(\mathbf{C}(T \cdot S))}(d+1).$$

Here (S) denotes the ideal of $K[T]$ generated by the elements of S. Note that the ideals (S) and $(T \cdot S)$ are equal in degree $d+1$ and that the operator $\mathbf{C}$ preserves the number of elements of the set it is applied to.

By Proposition 2.2, the last Hilbert function value above cannot decrease if the ideal $(\mathbf{C}(T \cdot S))$ is replaced by $(T \cdot \mathbf{C}(S))$. So

$$\begin{aligned} H(d+1) &\leq H_{K[T]/(T \cdot \mathbf{C}(S))}(d+1) = H_{K[T]/(\mathbf{C}(S))}(d+1) = \\ &= Q_d(H_{K[T]/(S)}(d)) \\ &= Q_d(H(d)) \end{aligned}$$

This proves, the condition in Macaulay's theorem is necessary.

Conversely, assume that the condition of Macaulay's theorem is satisfied for some function H. For every d, let S_d denote a set of power products of degree d such that $H_{K[T]/(S_d)}(d) = H(d)$. Then

$$\begin{aligned} H_{K[T]/(\mathbf{C}(S_{d+1}))}(d+1) &= H(d+1) \leq Q_d(H(d)) \\ &= H_{K[T]/(\mathbf{C}(S_d))}(d+1) = H_{K[T]/(T \cdot \mathbf{C}(S_d))}(d+1), \end{aligned}$$

hence $(T \cdot \mathbf{C}(S_d)) \subset (\mathbf{C}(S_{d+1}))$ for every d. In other words, the union

$$I = \bigcup_{d=1}^{\infty} (\mathbf{C}(S_d))$$

of the ideals $(\mathbf{C}(S_d)) \subset K[T]$ is an ideal which has Hilbert function H. Q.E.D.

Remark 2.3 The assertion of Proposition 2.2 is trivial for power products in one variable $(N = 1)$, since every set of degree d power products is compressed in this case.

Let us consider the case of two variables $x = T_1, y = T_2$. If S is a set of degree d power products containing n elements, so

$$\mathbf{C}(S) = \{ \; x^d, x^{d-1}y, \ldots, x^{d-n+1}y^{n-1} \; \}$$

hence

$$T \cdot \mathbf{C}(S) = \{ \; x^{d+1}, x^d y, \ldots, x^{d-n+1}y^n \; \}.$$

On the other hand the two n element sets $x \cdot S$ and $y \cdot S$ are different (they have different first elements). So the number of elements of their union $T \cdot S$ is at least $n+1$, and hence so is the number of elements of $\mathbf{C}(T \cdot S)$. Therefore $T \cdot \mathbf{C}(S) \subset \mathbf{C}(T \cdot S)$, and we have proved Proposition 2.2 in the case of two variables.

The proof of the general case will use induction on the number N of variables. In the rest of this paragraph we study the question, what can be deduced from the $N-1$ variable case of Proposition 2.2 for the proof of the N variable case. We will prove an inclusion as in the proposition, but with the compression operator $\mathbf{C}$ replaced by a partial compression operator. The proof of the proposition will be reduced this way to the case of partially compressed sets S. The treatment is essentially the same as in [1] (we use a dual terminology which is more adapted to polynomials).

Terminology 2.4 The set of all degree d power product in these variables will be denoted by T^d. Let S be a set of degree d power products in the variables $T_1, \ldots, T_N$ and e be a non-negative integer. Then we write

$$S_{e_i=e} := \{ T_1^{e_1} \cdot \ldots \cdot T_N^{e_N} \in S \mid e_i = e \}$$

for the set of the power products from S with i-th exponent equal to e. So $T^d_{e_i=e}$ is the set of all degree d power products with i-th exponent equal to e.

Given a subset $S \subset T^d_{e_i=e}$, the set $\mathbf{C}_i(S)$ is defined to be the set of the lexicographically $\sharp(S)$ first elements of $T^d_{e_i=e}$. For an arbitrary subset $S \subset T^d$ we define

$$\mathbf{C}_i(S) := \bigcup_{e=0}^{d} \mathbf{C}_i(S_{e_i=e}).$$

In other words, write S as a disjoint union of sets of type $S_{e_i=e}$ and apply $\mathbf{C}_i$ to each of these subsets. The operator $\mathbf{C}_i$ is called the *i-th (partial) compression operator*, and a set $S \subset T^d$ is called *i-compressed* if $\mathbf{C}_i(S) = S$. It is called *partially compressed* if it is i-compressed for $i = 1, \ldots, N$. The operator $\mathbf{C}_*$ is defined to be the composition of the operators $\mathbf{C}_i$,

$$\mathbf{C}_* := \mathbf{C}_N \cdot \mathbf{C}_{N-1} \ldots \mathbf{C}_2 \cdot \mathbf{C}_1.$$

Let $n(s)$ be the position of s in the lexicographically ordered sequence of power products in T, i.e., $n(s) = k$ if s is the lexicographically k-th power product. Moreover, define $n(S) := \sum_{s \in S} n(s)$ for every finite set S of power products in the given variables T.

Example 2.5 The set consisting of the power products x^6, x^5y, x^5z, x^4y^2, x^4yz, x^4z^2, x^3y^3, x^3y^2z, x^3yz^2, x^3z^3, x^2y^4, x^2y^3z, $x^2y^2z^2$, xy^5, xy^4z, and y^6 is partially compressed, but not compressed (we have used x, y, and z instead of T_1, T_2, and T_3, respectively).

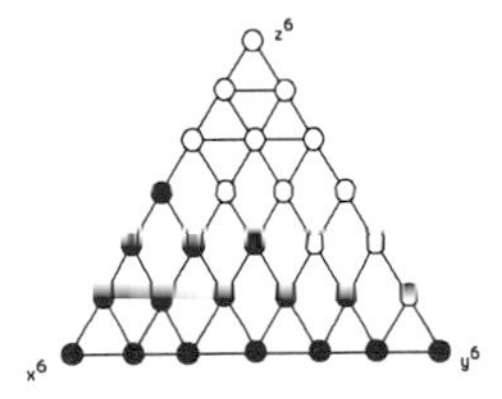

Proposition 2.6 *Let S be a set of degree d power products in the variables $T_1, \ldots, T_N$ (with $N > 1$). Then, if Proposition 2.2 is true for $N - 1$ variables,*

$$T \cdot \mathbf{C}_i(S) \subset \mathbf{C}_i(T \cdot S)$$

Proof. Proof Consider the set obtained from $S_{e_i=e}$ by setting the i-th variable T_i equal to 1. The $N - 1$ variable version of Proposition 2.2 applied to this set gives

$$T \cdot \mathbf{C}_i(S_{e_i=e}) \cap T^d_{e_i=e} \subset \mathbf{C}_i(T \cdot S_{e_i=e}).$$

This is, modulo the power products with i-th exponent equal to $e + 1$, just the inclusion we want to prove. So let us consider the power products of the latter type. One has

$$T \cdot \mathbf{C}_i(S_{e_i=e}) \cap T^d_{e_i=e+1} \subset \mathbf{C}_i(T \cdot S_{e_i=e}),$$

which can be easily seen as follows. To get the left hand side set, one has to multiply the elements of $\mathbf{C}_i(S_{e_i=e})$ with the variable T_i. So the left hand side set consists of the first $\sharp(S_{e_i=e})$ elements of $T^d_{e_i=e+1}$. To obtain the elements with $e_i = e+1$ of the right hand side set, one has to multiply the elements of $S_{e_i=e}$ with the variable T_i and then to apply $\mathbf{C}_i$. As a result one gets again the first $\sharp(S_{e_i=e})$ elements of $T^d_{e_i=e+1}$. This proves the last inclusion.

The elements of $T \cdot \mathbf{C}_i(S_{e_i=e})$ have either i-th exponent e or $e+1$. Therefore the last two inclusions together give

$$T \cdot \mathbf{C}_i(S_{e_i=e}) \subset \mathbf{C}_i(T \cdot S_{e_i=e}).$$

Taking on both sides the union over all possible values of e, one gets the claim. Q.E.D.

Proposition 2.7 *Let S be a set of degree d power products in the variables $T_1, \ldots, T_N$ (with $N > 1$). Then, if Proposition 2.2 is true for $N-1$ variables,*

1. $T \cdot \mathbf{C}_*(S) \subset \mathbf{C}_*(T \cdot S)$
2. $(\mathbf{C}_*)^k(S)$ *is partially compressed for every large k.*

Proof. Proof The first assertion is just an iterated application of Proposition 2.6. Consider the second assertion. By the very definition of the operators $\mathbf{C}_i$ we have $n(\mathbf{C}_i(S)) \leq n(S)$ with equality if and only if S is i-compressed (see 2.4 for the definition of $n(S)$). This implies $n(\mathbf{C}_*(S)) \leq n(S)$ with equality if and only if S is partially compressed. From the latter inequality we see that the sequence $\{n((\mathbf{C}_*)^k(S))\}_{k=1,2,\ldots}$ stabilizes. But this means, $(\mathbf{C}_*)^k(S)$ is partially compressed for large k. Q.E.D.

Remark 2.8 Obviously one has $\mathbf{C}(\mathbf{C}_*(S)) = \mathbf{C}(S)$. This fact, together with Proposition 2.7 reduces the proof of Proposition 2.2 to the case of partially compressed sets.

3 Linear Syzygies and 1-Components

Definition 3.1 Let $S = \{s_1, \ldots, s_n\}$ be a set of power products in the indeterminates $T_1, \ldots, T_N$. The $K[T]$-module of *syzygies* of S will be denoted by $Z(S)$, i.e.,

$$Z(S) := \{(r_1, \ldots, r_n) \in K[T] \mid \sum_{i=1}^{n} r_i s_i = 0\}$$

In view of the multigraded structure of the ideal generated by the elements of S the syzygy module $Z(S)$ is generated by the *pairwise* syzygies of the power products s_i, i.e., by the syzygies

$$r(s_i, s_j) := (0, \ldots, 0, s_j/s_{ij}, 0, \ldots, 0, -s_i/s_{ij}, 0, \ldots, 0)$$

where the only non-zero coordinates are s_j/s_{ij} at position i and $-s_i/s_{ij}$ at position j, and where s_{ij} is the greatest common divisor of s_i and s_j. An element $(r_1, \ldots, r_n)$ of $Z(S)$ will be called *linear* if all the coordinates r_i are homogeneous polynomials of degree one. The linear syzygies of S form a K-linear subspace of $Z(S)$, which will be denoted by

$$LZ(S).$$

Remark 3.2 Our aim is to give a description of a minimal generating set of $LZ(S)$ in terms of data which can be easily obtained from S. The terminology needed for this is based on the notion of 1-component which will be introduced now.

Definition 3.3 As above let $S = \{s_1, \ldots, s_n\}$ be a set of power products in the indeterminates $T_1, \ldots, T_N$. Two elements m and n of S are called *1- related* if there are indeterminates T_i and T_j (i.e., power products of degree one) such that

$$T_i m - T_j n = 0,$$

in other words if they are related by a linear syzygy. The power products m and n are called *1-connected*, if there is a sequence of successive 1-related power products in S beginning with m and ending with n. The set S is called *1-connected* if any two elements of it are 1-connected. Note that this implies that all elements of S have the same degree. A *1-component* of S is a maximal 1-connected subset.

Remark 3.4 The pairs of 1-related elements of S obviously correspond to a generating set of the vector space $LZ(S)$. Usually this will *not* be a minimal set of generators. To describe a minimal one, we introduce the following simplicial complex.

Definition 3.5 Let $S = \{s_1, \ldots, s_n\}$ be as above a set of power products, and denote by $K(S)$ the simplicial complex over S consisting of all simplices $\{s_{i_0}, \ldots, s_{i_k}\}$ with the property that there are indeterminates $x_0, \ldots, x_k \in \{T_1, \ldots, T_N\}$ such that

$$x_0 s_{i_0} = x_1 s_{i_1} = \ldots = x_k s_{i_k}.$$

The oriented chain complex with coefficients in the field K generated by the faces of $K(S)$ will be denoted by $C_*(S, K)$, its homology by $H_*(S, K)$, and the dimension of the i-th homology group over K by

$$h_i(S) := dim_K H_i(S, K).$$

Proposition 3.6 *Let S be a set of power products of equal degree in the indeterminates $T_1, \ldots, T_N$. Then one has the following formula for the dimension of the vector space generated by the linear syzygies of S.*

$$dim\ LZ(S) = h_1(S) - h_0(S) + \sharp(S)$$

Here $\sharp(S)$, as above, is the number of elements in S.

Proof. By definition, the 1-faces of $K(S)$ just correspond to the linear pairwise syzygies of S, i.e., to a generating set of $LZ(S)$. There is a canonical surjective linear map

$$C_1(S,K) \to LZ(S), (s,t) \mapsto r(s,t)$$

defined on the 1-chains of $K(S)$ with coefficients in K mapping each 1-face (s,t) to the corresponding linear syzygy $r(s,t)$. The chains in $C_1(S,K)$ of type

$$(t,u) - (s,u) + (s,t)$$

with (s,t,u) being an oriented 2-face, i.e., such that there are indeterminates x, y, z in $\{T_1, \ldots, T_N\}$ with $xs = yt = zu$, form a generating set for the subspace $B_1(S,K)$ of 1-boundaries. They are mapped to

$$\begin{aligned} r(t,u) - r(s,u) + r(s,t) = & \quad (\ldots,0,\ldots,y,\ldots,-z,\ldots) \\ & - \ (\ldots,x,\ldots,0,\ldots,-z,\ldots) \\ & + \ (\ldots,x,\ldots,-y,\ldots,0,\ldots) \\ = & \quad 0 \end{aligned}$$

The above map induces a mapping

$$C_1(S,K)/B_1(S,K) \to LZ(S), (s,t) \mapsto r(s,t)$$

which is in fact an isomorphism. Therefore,

$$dim\ LZ(S) = dim\ C_1(S,K)/B_1(S,K) = h_1(S) + dim\ C_1(S,K)/Z_1(S,K)$$

where $Z_1(S,K)$ denotes the space of 1-cycles. To calculate the second term on the right, assume for a moment that S is 1-connected (which just means that the complex $K(S)$ is connected) and let $\Gamma(S)$ be a maximal connected subgraph of $K(S)$ which doesn't contain cyclic subgraphs. Then the inclusion of $\Gamma(S)$ into $K(S)$ induces a linear map on the chains over the field K and hence a map

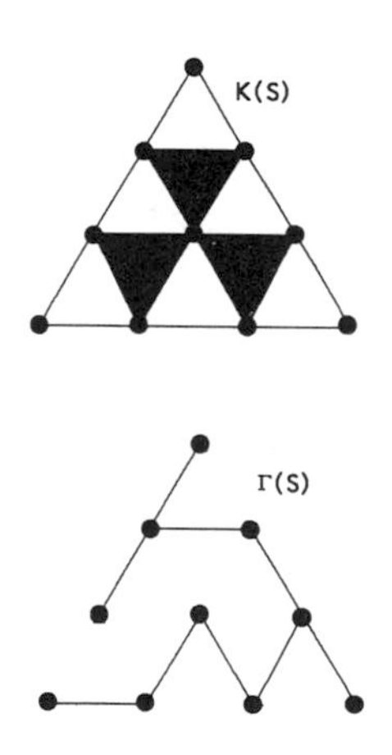

$$C_1(\Gamma(S),K) \to C_1(S,K)/Z_1(S,K)$$

This map must be injective, since otherwise there would be a 1-chain in $\Gamma(S)$ which is a cycle in $K(S)$.

Moreover, the map is surjective. For, given a 1-face $s \in C_1(S,K)$ which is not a face of $\Gamma(S)$, a cycle is produced when s is added to $\Gamma(S)$ (according to the maximality property of $\Gamma(S)$). In other words, there is a chain $c \in C_1(\Gamma(S),K)$ such that $s - c$ is a 1-cycle, i.e., represents the zero element in $C_1(S,K)/Z_1(S,K)$. But this means, s is in the image of the above map. The number of edges of $\Gamma(S)$ is easily seen to be $\sharp(S)-1$ (start with a subgraph Γ consisting of one single edge and, as long as there are vertices in $K(S)$ which are not yet in Γ, add an edge to Γ which connects such a vertex with some vertex already in Γ. This increases both the number of vertices and the number of edges in Γ by one. After a finite number of steps you'll get a maximal connected subgraph without cycles). So we have, in case S is 1-connected,

$$dim\ C_1(S,K)/Z_1(S,K) = dim\ C_1(\Gamma(S),K) = \sharp(S)-1$$

In the general case one has to decompose S into 1-components, say $S = S_1 \cup \ldots \cup S_r$. Then $C_1(S,K)/Z_1(S,K)$ decomposes into a direct sum, and one has

$$\begin{aligned} dim\ C_1(S,K)/Z_1(S,K) &= \sum_{i=1}^{r} dim\ C_1(S_i,K)/Z_1(S_i,K) \\ &= \sum_{i=1}^{r} \sharp(S_i) - 1 \\ &= \sharp(S) - r \\ &= \sharp(S) - h_0(S). \end{aligned}$$

Substituting this into the above formula for $dim\ LZ(S)$, we get the result. Q.E.D.

Examples. 3.7.1 Let $S = \{s_1 \ldots, s_6\}$ with $s_1 = x^3$, $s_2 = x^2y$, $s_3 = y^3$, $s_4 = y^2z$, $s_5 = z^3$,$s_6 = z^2x$. Then S decomposes into the three 1-components $S_1 = \{s_1, s_2\}$,$S_2 = \{s_3, s_4\}$, and $S_3 = \{s_5, s_6\}$. The complex $K(S)$ is the union of three disjoint edges. By the above proposition,

$$dim\ LZ(S) = 6 - 3 + 0 = 3.$$

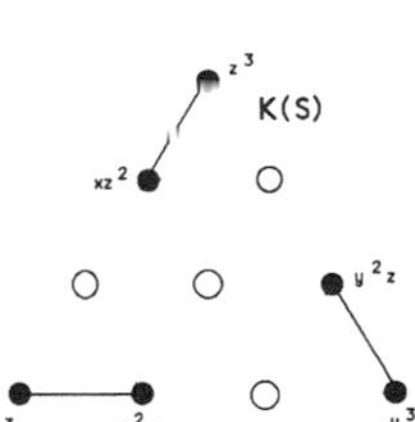

So the linear pairwise syzygies

$$r(s_1,s_2),\ r(s_3,s_4),\ r(s_4,s_6),$$

form a minimal generating set for $LZ(S)$. Note that we consider only the pairwise relations $r(s_i, s_j)$ with $i < j$. Otherwise we would have to use a directed graph whose cycles reflected the fact that for each two indices i, j one of the two relations $r(s_i, s_j)$ and $r(s_j, s_i)$ should be removed.

3.7.2 Let $S = \{s_1 \ldots, s_9\}$ with $s_1 = x^3$, $s_2 = x^2y$, $s_3 = xy^2$, $s_4 = y^3$, $s_5 = y^2z$, $s_6 = yz^2$, $s_7 = z^3$, $s_8 = z^2x$, $s_9 = zx^2$. Then S is 1-connected, $K(S)$ is a polygon with nine edges, so that

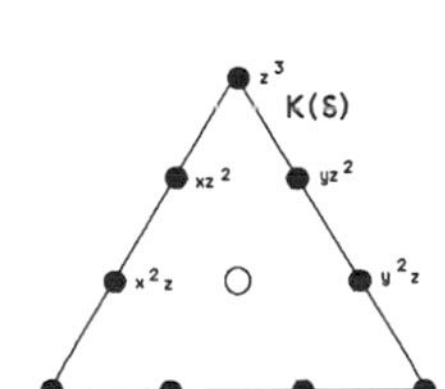

$$dim\ LZ(S) = 1 - 1 + 9 = 9$$

So all the nine pairwise linear syzygies are needed to generate $LZ(S)$.

$$\begin{aligned}
r(s_1,s_2) &= (y,-x,0,0,0,0,0,0,0)\\
r(s_2,s_3) &= (0,y,-x,0,0,0,0,0,0)\\
r(s_3,s_4) &= (0,0,y,-x,0,0,0,0,0)\\
r(s_4,s_5) &= (0,0,0,z,-y,0,0,0,0)\\
r(s_5,s_6) &= (0,0,0,0,z,-y,0,0,0)\\
r(s_6,s_7) &= (0,0,0,0,0,z,-y,0,0)\\
r(s_7,s_8) &= (0,0,0,0,0,0,x,-z,0)\\
r(s_8,s_9) &= (0,0,0,0,0,0,0,x,-z)\\
r(s_1,s_9) &= (z,0,0,0,0,0,0,0,-x)
\end{aligned}$$

3.7.3 Let $S = \{s_1 \ldots, s_{10}\}$ the set of power products of degree 3 in the variables x, y, z. Then $K(S)$ is as in the figure. It is easily seen to be homotopy equivalent to a bucket of 6 circles. As a consequence we have

$$dim\ LZ(S) = 6 - 1 + 10 = 15.$$

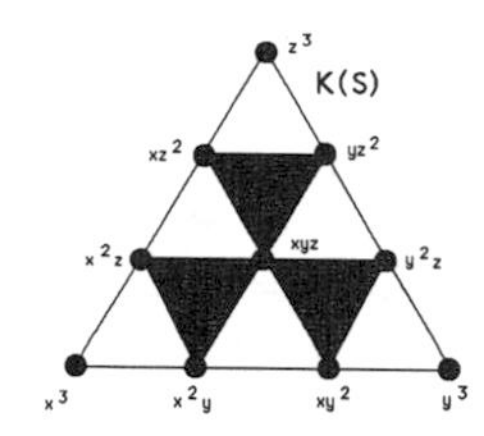

Corollary 3.8 *Let S and S' be two sets of degree d power products in the indeterminates $T_1 \ldots, T_N$ with the same number of elements,*

$$\sharp(S) = \sharp(S').$$

Then the following assertions are equivalent.

1. $H_{K[T]/(S)}(d+1) \leq H_{K[T]/(S')}(d+1)$ *with (S) and (S') denoting the ideal generated by S and S', respectively.*

2. $\sharp(TS) \geq \sharp(TS')$

3. $h_1(S) - h_0(S) \leq h^1(S') - h^0(S')$

In particular, the inclusion $T \cdot \mathbf{C}(S) \subset \mathbf{C}(T \cdot S)$ of Proposition 2.2 is equivalent to the inequality

$$h_1(S) - h_0(S) \leq h^1(\mathbf{C}(S)) - h^0(\mathbf{C}(S)).$$

Proof. The equivalence of the first two assertions follows directly from the definition of the Hilbert function. To see that the first assertion is equivalent to the last, consider the exact sequence

$$0 \to Z(S) \to K[T]^{\sharp(S)} \to K[T] \to K[T]/(S) \to 0.$$

From this sequence and Proposition 3.6 above we deduce,

$$\begin{aligned} H_{K[T]/(S)}(d+1) &= H_{K[T]}(d+1) - \sharp(S) \cdot H_{K[T]}(1) + dim\ LZ(S) \\ &= H_{K[T]}(d+1) + h_1(S) - h_0(S) - \sharp(S)(N-1). \end{aligned}$$

Comparing this formula with the corresponding formula for S' we get the claim.

We have yet to prove the remark at the end of the corollary. Consider the inclusion of Proposition 2.2,

$$T \cdot \mathbf{C}(S) \subset \mathbf{C}(T \cdot S).$$

Both sets involved are compressed. So this inclusion is equivalent to the corresponding inequality between their numbers of elements,

$$\sharp(T \cdot \mathbf{C}(S)) \leq \sharp(\mathbf{C}(T \cdot S))(= \sharp(T \cdot S)).$$

But this is, as we just have proved, equivalent to the inequality

$$h^1(S) - h^0(S) \leq h_1(\mathbf{C}(S)) - h_0(\mathbf{C}(S)).$$

Q.E.D.

We study now which way $h_1(S) - h_0(S)$ may change when a power product is added to S.

Definition 3.9 Let S be a set of power products of degree d in $T_1, \ldots, T_N$ and let a be one further such power product (not in S). Then we say, *a is 1-related in direction i with S* if $\frac{aT_i}{T_j} \in S$ for some $j \neq i$, symbolically, $a \sim^i S$.

Proposition 3.10 *Let $S = \{s_1, \ldots, s_m\}$ be a non-empty set of degree d power products of the indeterminates $T_1, \ldots, T_N$ and let a be one further such power product which is not in S. Then*

$$h_1(S \cup \{a\}) - h_0(S \cup \{a\}) = h_1(S) - h_0(S) + \Delta(S, a) - 1$$

where $\Delta(S, a)$ is the number of different directions, in which a is 1-related with S,

$$\Delta(S, a) = \sharp\{\ i \mid a \sim^i S\}.$$

In particular, if S is the set of all degree d power products preceding a lexicographically and

$$a(1, T_2, \ldots, T_N) = T_{i_1}^{e_1} \ldots T_{i_r}^{e_r}$$

with $1 < i_1 < i_2 < \ldots < i_r$ and all the exponents e_i non-zero, then $\Delta(S, a) = i_r - 1$.

Proof. Assume a is such that

$$a(1, T_2, \ldots, T_N) = T_{i_1}^{e_1} \ldots T_{i_r}^{e_r}$$

with $1 < i_1 < i_2 < \ldots < i_r$ and all the exponents e_i non-zero. The complex $K(S \cup \{a\})$ is obtained from $K(S)$ via adjunction of certain simplexes having a as a vertex. These simplexes are given by power products $s_{k_1}, s_{k_2}, \ldots, s_{k_t} \in S$ such that $T_j a = T_{j_1} s_{k_1} = \ldots = T_{j_t} s_{k_t}$ for some choice of the indices j_ν and j. In other words the simplexes are of type

$$\{a, (T_j/T_{j_1}) \cdot a, (T_j/T_{j_2}) \cdot a, \ldots, (aT_j/T_{j_t}) \cdot a\}.$$

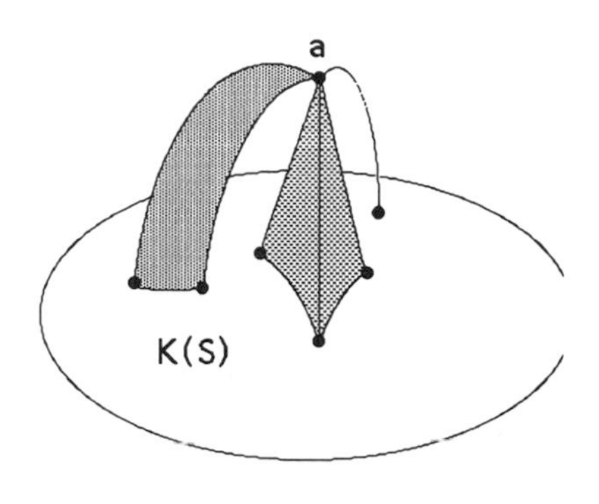

We see from this, for every j such that a is 1-related with S in direction j, we have at least one such simplex. Whenever we have more than one simplex with one and the same j, then these simplexes are common faces of a larger simplex of this type.In other words, the maximal simplexes adjoined to $K(S)$ are in one-one correspondence with the different directions j such that $a \sim^j S$. Moreover, simplexes belonging to different directions j have a as the only common vertex. This means, after having adjoined the first maximal simplex to $K(S)$, any further adjunction causes either $h_1(S)$ to increase by one (if the new simplex is added to the same connected component as some former one) or $h_0(A)$ to decrease by one. So $h_1(S) - h_0(S)$ just increases by $\Delta(S, a) - 1$ as claimed.

Now assume, S is the set of all degree d power products preceding a lexicographically. We may assume that in the above description of the simplexes adjoined to $K(S)$ the variables T_{j_ν} occur in increasing lexicographic order. Since a is lexicographically later than every $aT_j/T_{j_\nu} \in S$, T_j must be earlier than any T_{j_ν},

$$T_j < T_{j_1} < T_{j_2} < \ldots < T_{j_t}.$$

If some variable dividing a can be included into this chain, the simplex is a face of a higher dimensional simplex which is also adjoined to the complex. So the maximal simplexes which are adjoined to $K(S)$ are just those of type

$$\{s, (T_j/T_{i_u}) \cdot s, (T_j/T_{i_{u+1}}) \cdot s, (T_j/T_{i_{u+2}}) \cdot s, \ldots, (sT_j/T_{i_r}) \cdot s\}$$

with $i_{u-1} \leq j < i_u$. The number of possible such simplexes with fixed u (i.e., with fixed dimension $r-u+1$) is $i_u - i_{u-1}$. In case $u = 1$ the condition upon j is $1 \leq j < i_1$. So we get $i_r - 1$ as the total number of maximal simplexes adjoined to $K(S)$, i.e., $\Delta(S, a) = i_r - 1$.

Q.E.D.

Remark 3.11 The proof of Proposition 3.10 above shows even more than stated. For example, it is clear now that $\Delta(S,a) \leq N-1$ if a is lexicographically later than every element of S. Conversely, $\Delta(S,a) \geq N-1$ if $T_N \mid a$ and every degree d power product, which is earlier than a, is in S. These two facts will be crucial in the proof of Proposition 2.2 below.

Proof of Proposition 2.2 Let S be a set of degree d power products in the variables $T_1, \ldots, T_N$ (with $N \geq 3$). We want to prove $T \cdot \mathbf{C}(S) \subset \mathbf{C}(T \cdot S)$. Note that $\mathbf{C}(\mathbf{C}_*(S)) = \mathbf{C}(S)$ and, by 2.7,

$$\mathbf{C}(T \cdot \mathbf{C}_*(S)) \subset \mathbf{C}(\mathbf{C}_*(T \cdot S)) = \mathbf{C}(T \cdot S)).$$

So in proving the claim, the set S may be replaced by $\mathbf{C}_*(S)$. Doing so repeatedly, we may assume that S is partially compressed (see 2.7). By Corollary 3.8, the inclusion to be proved, $T \cdot \mathbf{C}(S) \subset \mathbf{C}(T \cdot S)$, is equivalent to the inequality

$$h_1(\mathbf{C}(S)) - h_0(\mathbf{C}(S)) \geq h^1(S) - h^0(S).$$

We will prove this inequality by induction on $n(S)$ (see 2.4 for the definition of $n(S)$). If S is compressed (in which case $n(S)$ takes its minimal possible value), there is nothing to prove. So assume it is not. To prove the inequality, it is sufficient to find some subset $S' \subset T^d$ with $\sharp(S)$ elements such that

$$h_1(S') - h_0(S') \geq h^1(S) - h^0(S) \quad \text{and} \quad n(S') < n(S).$$

Let $a \in T^d$ be the lexicographically first element which is not in S,

$$a \notin S,$$

and let $b \in S$ the lexicographically last element which is in S. We will show that

$$S' := (S - \{b\}) \cup \{a\}$$

is a set of the kind we need. Since S is not compressed,

$$a < b \quad \text{(lexicographically)}.$$

Therefore $n(a) < n(b)$ hence $n(S') < n(S)$. To prove, $h_1(S') - h_0(S') \geq h^1(S) - h^0(S)$, it will be sufficient to show $T_N \mid a$ (by Remark 3.11). So assume that $a = T_1^{a_1} \cdot \ldots \cdot T_N^{a_N}$ is not a multiple of T_N, i.e., $a_N = 0$. But then the conditions of Lemma 3.12 below are satisfied and, since S is partially compressed, a must be in S. This contradiction to the choice of a, completes the proof. Q.E.D.

Lemma 3.12 *Let $a = T_1^{a_1} \cdot \ldots \cdot T_N^{a_N}$ and $b = T_1^{b_1} \cdot \ldots \cdot T_N^{b_N}$ be degree d power products in the variables $T_1, \ldots, T_N$ such that $a < b$ lexicographically. Suppose $a_N = 0$. Then there is a sequence $s_1, \ldots, s_u$ of degree d power products beginning with a and ending with b such that*

1. $s_i < s_{i+1}$ *lexicographically.*

2. s_i *and* s_{i+1} *have the same* j*-th exponent for some* j*.*

In particular, if b *is an element of some partially compressed set of power products, then so is* a*.*

Proof. We will construct the sequence in question by induction on b_N. If $b_N = 0$, we may choose the sequence to be a, b. So assume $b_N > 0$. Since $a < b$, there is some index $i < N$ with $a_i > b_i$. If i is not the first index with this property or if $a_i > b_i + 1$, define

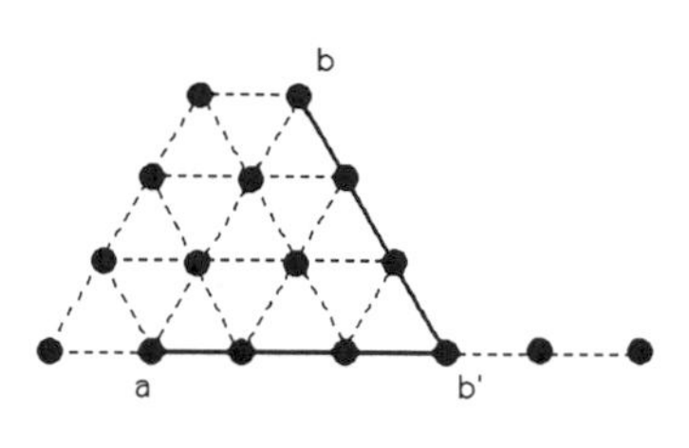

$$b' := \frac{bT_i}{T_N}.$$

Then $a < b' < b$ lexicographically and the induction hypothesis applied to a and b' gives the sequence we are looking for. We have yet to consider the case that $a_i = b_i + 1$ and i is the only index smaller than N such that the i-th exponent of a is greater than that of b, i.e., $a = \frac{bT_i}{T_N}$. But then a and b have the same j-th exponent for some j, so that we may choose the sequence to be a, b. This completes the proof. Q.E.D.

4 On Lech's problem

Another context where the concept of 1-components may be applied is Lech's inequality. Lech's problem is to decide whether for every flat local homomorphism $f : (A, m) \to (B, n)$ of local rings one has for some non-negative integer i the inequality

$$H^i_{(A,m)} \leq H^i_{(B,n)}$$

between the i-th sum transforms of the Hilbert series of A and B, respectively. More explicitly, $H^0_{(A,m)}$ is the formal power series in one indeterminate, say T, with the k-th coefficient equal to the vector space dimension

$$H^0_{(A,m)}(k) = dim_{A/m}\, m^k/m^{k+1}$$

and the i-th sum transform is defined as

$$H^i_{(A,m)} = (1-T)^i H^0_{(A,m)}.$$

The above inequality between the two series means that one has this inequality between any two corresponding coefficients. From [3] one knows a sufficient condition for the above inequality to hold (even with $i = 0$) for all flat local homomorphisms $f : (A, m) \to (B, n)$

with fixed special fiber $B_0 = B/mB$. The condition may be formulated in terms of the normal module of the graded ring $G(B_0) := \bigoplus_{k=0}^{\infty} n^k B_0/n^{k+1}B_0$ (or, which is the same, in terms of Schlessingers module $T^1_{G(B_0)}$). For this write $G(B_0)$ as a factor ring of a polynomial ring modulo some ideal, $G(B_0) = K[T_1, \ldots, T_N]/I$, and equip the normal module

$$N_{G(B_0)} := Hom_{G(B_0)}(I, G(B_0))$$

with the graded structure coming from the gradation of $G(B_0)$. Let $N_{G(B_0)}(k)$ be the subgroup of homogeneous elements of degree k of $N_{G(B_0)}$. Further denote by $N_{G(B_0)}(< k)$ the K-linear subspace of $N_{G(B_0)}$ generated by the homogeneous elements of degree less than k. It is not very hard to see that the vector space $N_{G(B_0)}(< k)$ is for negative integers k independent upon the special representation of $G(B_0)$ as a factor ring of a polynomial ring (see [3] for details). We can now formulate our criterion.

Theorem 4.1 (see [3]) *Let B_0 be a local ring such that $N_{G(B_0)}(< -1) = 0$. Then one has*

$$H^0_{(A,m)} \leq H^0_{(B,n)}$$

for every flat local homomorphism $f : (A, m) \to (B, n)$ of local rings with special fiber isomorphic to B_0.

The condition $N_{G(B_0)}(< -1) = 0$ is easy to check in most cases, when the defining equations of the singularity B_0 are known. For example, one has

Theorem 4.2 (see [3]) *Let $I \subset K[T_1, \ldots, T_N]$ be an ideal generated by a set S of power products of the indeterminates $T_1, \ldots, T_N$. Assume that all elements of S have equal degree and that each 1-component has a greatest common divisor of degree at most one. Then*

$$N_{K[T_1,\ldots,T_N]/I}(< -1) = 0$$

Note that, in view of the lemma below, the restriction to ideals generated by power products is not really essential. In what follows, a *term order* is an irreflexive, anti-symmetric, and transitive linear order "<" on the set of power products in the variables $T_1, \ldots, T_N$ such that $a < b$ implies $ac < ac$. A term order is called *degreewise* if the power products with greater (total) degree are greater with respect to the term order than those with smaller degree. The *initial term* of a polynomial with respect to a given term order is the term with smallest power product among all non-zero terms of the polynomial.

Lemma 4.3 (cf [3],(4.7)) *Let $I \subset K[T_1, \ldots, T_N]$ be a homogeneous ideal and let I_0 be the ideal generated by the initial terms of the elements of I with respect to some degreewise term order. Then there is a descending filtration F on the normal module $N_I = Hom_{K[T]}(I, K[T]/I)$ by K-linear subspaces which is a refinement of the filtration defined by total degree. Moreover, there is an injective degree-preserving K-linear map*

$$gr(N_I) \to N_{I_0}$$

where $gr(N_I)$ is the graded vector space associated with the filtration F. In particular, $N_{I_0}(k) = 0$ for some k implies $N_I(k) = 0$.

Proof. Given a polynomial $f \in K[T]$ let $in(f)$ denote the initial term of f with respect to the given term order. Fix some standard base $f_1, \ldots, f_n$ of I, i.e., homogeneous polynomials generating I and such that the initial terms $in(f_i)$ generate I_0. An element

$$g \in N_I = Hom_{K[T]}(I, K[T]/I)$$

is then defined by an n-tuple of polynomials g_i representing the images $g(f_i)$. We will say in this situation, g is represented by polynomials $g_1, \ldots, g_n$. Let f denote the least common multiple of the initial terms $in(f_i)$. Then the filtration F is defined as follows.

$$g \in N_I \text{ is in } F^k \quad \Longleftrightarrow \quad \begin{cases} g \text{ can be represented by polyno-} \\ \text{mials } g_i \text{ such that } n(\frac{f \cdot in(g_i)}{in(f_i)}) \geq k \\ \text{for every } i \end{cases}$$

Here the function $n()$ is defined in the same way as the function $n()$ of 2.4, but the lexicographic order is replaced by the currently used term order. Moreover, the definition is extended from the set of power products to the monomials by simply ignoring the coefficients, i.e., $n(c \cdot s) = n(s)$ for every power product s and every constant $c \in K$. The *F-order* of an element $g \in N_I$ is defined to be

$$ord_F(g) = sup\{ \ k \in \mathbf{N} \mid g \in F^k \ \}.$$

Note that $ord_F(g) = \infty$ if and only if $g = 0$. An element $g \in N_I$ of finite order k can be represented by polynomials of the type

$$g_i = c_i \cdot s_i + \text{ later terms},$$

where c_i is in K and s_i is the power product with $n(\frac{f \cdot s_i}{in(f_i)}) = k$ (if there is no such s_i let $c_i := 0$ and, likewise, consider s_i to be zero). We may assume that the polynomial g_i is chosen such that $in(g_i)$ is as large as possible with respect to the given term order. Then the monomial $c_i \cdot s_i$ is uniquely determined by g. For, the only possible modification of g_i is one modulo elements of I which have an initial form greater than or equal to s_i. If there are elements in I with initial form s_i, then c_i must be zero, if not $c_i \cdot s_i$ is preserved under any such modification.

Define $\varphi_i(g) := c_i \cdot s_i$. The above arguments show that $\varphi_i : F^k \to K \cdot s_i$ is well defined, and it is obviously linear. Since F^{k+1} is mapped to zero, we have an induced map

$$\bar{\varphi}_i : F^k/F^{k+1} \to K \cdot s_i.$$

Note that a non-zero element $(g \ mod \ F^{k+1})$ is mapped into a non-zero element $\varphi_i(g) = c_i \cdot s_i$ for at least one i (if all coefficients c_i were zero, g would have order $\geq k+1$). Now define

$$\varphi : F^k/F^{k+1} \to N_{I_0}$$

such that $\varphi(g \ mod \ F^{k+1})$ is the element of N_{I_0} represented by the monomials $\varphi_i(g) = c_i \cdot s_i$. Assume for a moment, that there is always an element in N_{I_0} represented by these monomials. Then the map φ is well defined, K-linear, and injective. The latter follows from the fact that, whenever c_i is non-zero for some i, s_i is not in the initial ideal I_0.

Let s be the smallest power product which is a power product of one of the monomials $\frac{f \cdot in(g_i)}{in(f_i)}$. Then, $\varphi_i(g) = c_i \cdot s_i$ is non-zero if and only if the power product of $\frac{f \cdot in(g_i)}{in(f_i)}$ is equal to s.

We have yet to prove that the monomials $\varphi_i(g) = c_i \cdot s_i$ represent an element of N_{I_0}. This means, we have to show that for every n-tuple of monomials $(R_1, \ldots, R_n)$ satisfying $\sum_{i=1}^{n} R_i \cdot in(f_i) = 0$ one has $\sum_{i=1}^{n} R_i \cdot \varphi_i(g) \in I_0$. Suppose we have an n-tuple $(R_1, \ldots, R_n)$ as above. We may assume that the monomials $R_i \cdot in(f_i)$ have all the same power product (at least those which are non-zero). Since the polynomials f_i form a standard base, one can find homogeneous polynomials $r_1, \ldots, r_n$ with $\sum_{i=1}^{n} r_i \cdot f_i = 0$ and $in(r_i) = R_i$ for every i with $R_i \neq 0$. Since the polynomials g_i represent an element of N_I, one has $\sum_{i=1}^{n} r_i \cdot g_i \in I$. Taking the initial terms on both sides we get the required relation,

$$\sum_{i=1}^{n} R_i \cdot \varphi_i(g) \in I_0.$$

To see this, note that, by construction, the products $in(r_i) \cdot in(f_i)$ with $R_i \neq 0$ have all the same power product, and the same is true for the monomials $\frac{f \cdot in(g_i)}{in(f_i)}$ with $\varphi_i(g) \neq 0$. But then this also holds for the non-zero products

$$R_i \cdot \varphi_i(g) = in(r_i) \cdot in(g_i)$$

which are just the minimal ones among the monomials $in(r_i \cdot g_i)$. We have proved the mapping φ is well defined.

We have yet to show that φ is degree preserving. But this simply follows from the fact that it even preserves (by construction) the multigraded structure defined on both modules. Q.E.D.

Using the two last theorems above we constructed in [3] large classes of local rings B_0 such that Lech's inequality holds for all flat local homomorphisms $f : (A, m) \to (B, n)$ with special fiber B_0. Examples of such special fibers are the extremal Cohen-Macaulay singularities (among them singularities defined by the maximal minors of generic matrices and the Cohen-Macaulay singularities of maximal embedding dimension), affine cones over Grassmann varieties, Schubert varieties, flag varieties, determinantal varieties, Veronese varieties (one has always to exclude the complete intersection case).

The concept of 1-components can be generalized in an obvious way to the concept of k-components with k a positive integer (two power products are called k-related if they are related by a syzygy with coordinates of degree k). Using this generalization, T.Richter ([8]) has recently proved the following k-component version of Theorem 4.2.

Theorem 4.4 (see [8]) *Let $I \subset K[T_1, \ldots, T_N]$ be an ideal generated by a set S of power products of the indeterminates $T_1, \ldots, T_N$. Assume that all elements of S have equal degree* *and that for some k each k-component has a greatest common divisor of degree at most k. Then*

$$N_{K[T_1,\ldots,T_N]/I}(< -k) = 0$$

Moreover, T. Richter calculated an upper bound for the homogeneous parts of the normal module in negative degree in terms of the degrees of the greatest common divisors of the k-components.

Theorem 4.5 (see [8]) *Let S be a set of degree d power products in the indeterminates $T_1, \ldots, T_N$ and let $S = S_1 \cup \ldots \cup S_n$ be the decomposition into k-components (with $k \geq 1$). Then*

$$dim_K N_{K[T_1,\ldots,T_N]/(S)}(-(k+1)) \leq \sum_{j=1}^{n} \binom{N + GCD(S_j) - k - 2}{N-1}.$$

This estimation has consequences for Lech's inequality when used together with the following inequality involving Schlessinger's T^1.

Theorem 4.6 (see [4], (8.3)) *Let $f : (A, m) \to (B, n)$ be a flat local homomorphism of local rings which is residually rational. Then*

$$H_A^1 \cdot H_{B/mB}^0 \leq H_B \cdot \prod_{k=2}^{\infty} \left(\frac{1 - T^k}{1 - T} \right)^{n(k)}, \quad n(k) := dim_{B/nB} T_G^1(B_0)(-k).$$

Here $T_G^1(B_0)(-k)$ denotes the degree $-k$ part of Schlessinger's T^1 for the tangent cone of the special fiber B/mB of f.

The following corollary is a more or less immediate consequence of 4.5 and 4.6.

Corollary 4.7 (see [8]) *Let B_0 be a local ring with completion $K[[T_1, \ldots, T_N]]/I$ where I is an ideal such that the associated ideal of initial terms (with respect to some term order) is generated by power products of degree two. Then Lech's inequality holds for every residually rational flat local homomorphism with special fiber B_0.*

Remark 4.8 The sufficient condition of Theorem 4.2 on the degree of the greatest common divisors of the 1-components is far from being necessary. For example, if S is the set of 3.7.1, each 1-component has GCD of degree two. Nevertheless,

$$N_{K[T]/(S)}(< -1) = 0.$$

T. Richter [8] has constructed many such examples (most with interesting symmetries). To study this phenomenon, the following modification of the notion of k-component turns out to be useful.

Definition 4.9 Let S be a set of degree d power products in the variables $T_1, \ldots, T_N$. Two elements $a, b \in S$ are called *weakly related* if there are power products a', b' such that

1. $a \cdot a' = b \cdot b'$ (i.e., a and b are k-related for some k).

2. $a \cdot a'$ and $b \cdot b'$ are not in the ideal of $K[T_1, \ldots, T_N]$ generated by S.

The notion of *weak component* can now be defined in a similar manner as the notion of 1-component (see 3.3).

Theorem 4.10 *Let $I \subset K[T_1, \ldots, T_N]$ be an ideal generated by a set $S = \{s_1, \ldots, s_n\}$ of power products of the indeterminates $T_1, \ldots, T_N$. Assume that*

1. *All elements of S have equal degree d.*

2. *Every 1-component of S has GCD of degree at most two.*

3. *Every weak component of S has GCD of degree at most one.*

Then

$$N_{K[T_1,\ldots,T_N]/I}(< -1) = 0.$$

Proof. Since the 1-components of S have greatest common divisor of degree at most two, the same is true for the 2-components. But then 4.4 implies,

$$N_{K[T_1,\ldots,T_N]/I}(< -2) = 0.$$

All we have to show is that there are no non-zero homogeneous elements of degree -2 in $N_{K[T_1,\ldots,T_N]/I}$. Assume there is some non-zero $g \in N_{K[T_1,\ldots,T_N]/I}(-2)$. Let g be represented by the polynomials $g_1, \ldots, g_N$, i.e., the polynomials g_i are such that $g(s_i) = g_i \ mod \ I$. We may assume that the g_i's are monomials degree $d-2$. Consider two generators s_i and s_j. If they are 1-related, i.e., if one has

$$T_u s_i = T_v s_j$$

for some choice of u and v, then $T_u g_i - T_v g_j \in I$ and, since I is generated in degree d, $T_u g_i - T_v g_j = 0$. Thus,

$$\frac{g_i}{g_j} = \frac{T_v}{T_u} = \frac{s_i}{s_j}.$$

This means, the g_i's, belonging to one and the same 1-component of S, are proportional to the corresponding s_i's. Since their degree is $d-2$, for every 1-component of S there is some constant $c \in K$ and some monomial q of degree two with

$$g_i = c \cdot \frac{s_i}{q}$$

for every i such that s_i is in this component (in particular, c must be zero, if the 1-component has GCD of degree less than two).

Now let s_i and s_j be weakly related. There are monomials a and b such that $as_i = bs_j$ and $as_i, bs_j \notin I$. The former relation implies $ag_i - bg_j \in I$, and by the latter we get, in view of the identities $g_i = cs_i/q$ above, $ag_i, bg_j \notin I$ (except for the case that g_i or g_j is zero). Therefore, $ag_i - bg_j = 0$, hence

$$\frac{g_i}{g_j} = \frac{b}{a} = \frac{s_i}{s_j}.$$

This means, the g_i's, belonging to one and the same weak component of S, are proportional to the corresponding s_i's. So for every weak component there is a constant $c \in K$ and a monomial q of degree two with

$$g_i = c \cdot \frac{s_i}{q}$$

for all i such that s_i is in this component. Since the weak components have GCD of degree one, this is possible only if $c = 0$. So all the polynomials g_i must be zero, i.e., $g = 0$. Q.E.D.

Remark 4.11 A more natural variant of Theorem 4.10 would be one with the second condition replaced by $N_{K[T]/I}(< -2) = 0$. More generally, it would be nice to know what are the conditions a power product ideal $I = (S)$ with $N_{K[T]/I}(< -(k+1)) = 0$ should satisfy so that one even has $N_{K[T]/I}(< -k) = 0$. But this seems to require more delicate invariants than greatest common divisors of components of power product sets.

References

[1] G. F. Clements and B. Lindström: A generalization of a combinatorial theorem of Macaulay, J. Combinatorial Theory 7 (1969) 230–238

[2] B. Herzog: On the Macaulayfication of local rings, J. Algebra 67 (1980) 305–317

[3] B. Herzog: Local singularities such that all deformations are tangentially flat, Trans. Amer. Math. Soc. 324 (1991) 555–601

[4] B. Herzog: Tangential flatness for filtered modules over local rings, Reports of the Mathematical Institute of Stockholm University 4 (1991)

[5] M. Hochster: Topics in the homological theory of modules over commutative rings, CBMS Regional Conf. Ser. in Math. 24 (1975)

[6] F.S. Macaulay: Some properties of enumeration in the theory of modular systems, Proc. London Math. Soc. 26 (1927) 531–555

[7] H. Matsumura: Commutative ring theory, Cambridge Univ. Press 1986

[8] T. Richter: Potenzproduktsingularitäten mit nur tangential flachen Deformationen, Diplomarbeit an der Friedrich-Schiller-Universität Jena 1992

Applications of Massive Computations: The Artin Conjecture

Ian Kiming

IEM, University of Essen
Ellernstr. 29, 45326 Essen
Germany

0. Introduction

This lecture presents an overview of a joint work with J. Basmaji, G. Frey, X. Wang (all at IEM, Univ. Essen) and L. Merel (Paris). Details and numerical results will appear elsewhere.

0.1 The work concerns the simplest in general unproved case of "Langland's philosophy":

Consider equivalence classes of 2-dimensional, irreducible, continuous, odd Galois representations over $\mathbf{Q}$:

$$\rho : \mathrm{Gal}(\overline{\mathbf{Q}}/\mathbf{Q}) \longrightarrow GL_2(\mathbf{C})$$

with Artin conductor $N(\in \mathbf{N})$ and determinant character $\det \rho = \varepsilon$. Here $\mathbf{C}$ is endowed with the discrete topology and $\mathrm{Gal}(\overline{\mathbf{Q}}/\mathbf{Q})$ has the natural topology as a profinite group so that "continuous" means "having finite image". The determinant character $\det \rho$ is the character on $\mathrm{Gal}(\overline{\mathbf{Q}}/\mathbf{Q})$ obtained by composing ρ with the determinant homomorphism

$$\det : GL_2(\mathbf{C}) \longrightarrow \mathbf{C}^{\times}.$$

Then $\varepsilon = \det \rho$ is a character on $\mathrm{Gal}(\overline{\mathbf{Q}}/\mathbf{Q})$ with class field theoretic conductor dividing N, and that ρ is odd means that ε has value -1 at a Frobenius infinity. By class field theory we may identify ε with a Dirichlet character modulo N which we will do in the following. With this identification, ρ is odd if and only if $\varepsilon(-1) = -1$.

It is conjectured that these equivalence classes are in 1-1 correspondence with the normalized newforms $f(z)$ of weight 1 and nebentype ε on the congruence group

$$\Gamma_0(N) = \left\{ \begin{pmatrix} a & b \\ c & d \end{pmatrix} \in SL_2(\mathbf{Z}) | c \equiv o(N) \right\}$$

(see for example [9] for definitions). More explicitly one expects the Fourier coefficients of f at ∞:

$$f(z) = \sum_{n=1}^{\infty} a_n q^n, \quad q = e^{2\pi i z},$$

to coincide with the coefficients of the Artin L-series of ρ:

$$L(s,\rho) = \sum_{n=1}^{\infty} a_n n^{-s}, \quad \mathrm{Re}(s) > 1.$$

The arithmetical interest of this conjecture is that it will, if true, constitute a natural extension of class field theory to a truly non-abelian situation.

A deep theorem of Deligne and Serre (cf. [3]) states that if $f(z) = \sum a_n q^n$ is a normalized newform on $\Gamma_0(N)$ of weight 1 and nebentype ε, then there is a representation ρ of the above type with Artin conductor N, determinant character ε and Artin L-series $L(s,\rho) = \sum a_n n^{-s}$ (for $\mathrm{Re}(s) > 1$). A classical theorem of Hecke (cf. for example [9], chap. 4) now implies that this L-series, enlarged by the usual Γ-factor, has a holomorphic continuation to the whole complex plane.

With this, a theorem of Weil (cf. [16] or [7]) shows that the above conjecture conjunctively for all 2-dimensional, irreducible, continuous, odd representations of Gal $(\overline{\mathbf{Q}}/\mathbf{Q})$ is equivalent to the Artin conjecture for these representations. Recall that the Artin conjecture for the considered representations claims the existence of a holomorphic continuation of the associated (enlarged) Artin L-series. For more information in this connection the reader is referred to [11].

We want to make it clear that from an arithmetical point of view, the interest is not attached to the Artin conjecture itself. The Artin conjecture (in our case) should rather be regarded as a concentrated way of expressing a very deep conjecture of a truly arithmetical (and <u>not</u> analytical) nature.

0.2 Let us now introduce the following notation:

Fix $N \in \mathbf{N}$ and let ε be a Dirichlet character modulo N with $\varepsilon(-1) = -1$. Let $d(N,\varepsilon)$ denote the number of equivalence classes of irreducible, continuous, 2-dimensional representations of Gal$(\overline{\mathbf{Q}}/\mathbf{Q})$ with Artin conductor N and determinant character ε. (This number is obviously finite.) Further let $S_1^+(N,\varepsilon)$ denote the complex vector space generated by the (finite number of) newforms of weight 1 and nebentype ε on $\Gamma_0(N)$. From the discussion in 0.1 it follows that

$$(*) \qquad \dim S_1^+(N,\varepsilon) \leq d(N,\varepsilon),$$

and that equality holds if and only if the conjecture in 0.1 is true for all representations ρ of the above type with Artin conductor N and determinant character ε. Hence this conjecture may be verified for all such representations simultaneously by computing the numbers $\dim S_1^+(N, \varepsilon)$ and $d(N, \varepsilon)$ and showing that they are equal. However, as we shall see, the question of determining these numbers leads to theoretical problems of independent interest. Before we proceed to the consideration of these, we shall briefly review in what cases the conjectured equality in (*) is known to hold.

Together with a representation ρ of the above type, we consider its projectivisation

$$\overline{\rho} : \mathrm{Gal}(\overline{\mathbf{Q}}/\mathbf{Q}) \longrightarrow PGL_2(\mathbf{C})$$

obtained by composing ρ with the canonical homomorphism $GL_2(\mathbf{C}) \longrightarrow PGL_2(\mathbf{C})$. For group theoretical reasons the (finite) image of $\overline{\rho}$ is either isomorphic to a cyclic group, to a dihedral group, to the alternating group A_4, to the symmetric group S_4, or to the alternating group A_5, where however the cyclic case is excluded since ρ was assumed to be irreducible. We shall distinguish these cases by saying that ρ is of dihedral-, A_4-, S_4-, or A_5-type respectively.

Now, if ρ is of dihedral type, then one may by class field theory show that ρ is associated with a newform as in 0.1. By deep results of Langlands and Tunnell (cf. [6] and [14]) this is also the case if ρ is of A_4- or S_4-type. Thus the question of equality in (*) is only of interest if there exists a ρ of A_5-type with Artin conductor N and determinant character ε. Here the question of computational verification of equality in (*) (for some numerical cases) turns up, since the methods in [6] and [14] for associating a newform to a ρ of type A_4 or S_4 do not work, and cannot be made to work, in the case of a ρ of type A_5. We note that there is in the literature such a computational verification only for one single numerical case (cf. [1]; here $N = 800$). The method in [1] can however not be described as an algorithm.

This discussion may now serve as motivation for the following objectives:

(1) To develop an algorithm for verifying equality in (*) (given N and ε). As has already been noted this leads to theoretical problems of independent interest.

(2) To provide more non-trivial examples of equality in (*).

We proceed now to sketch the solution which has been obtained for (1).

1. Computation of $d(N, \varepsilon)$

1.1 We fix $N \in \mathbf{N}$ and ε a character on the idele class group of $\mathbf{Q}$ with conductor dividing N. In this section there is no reason to restrict our attention to the case of an odd ε. Furthermore, we shall fix a 2-dimensional, irreducible, continuous, projective representation

$$\overline{\rho}: \mathrm{Gal}(\overline{\mathbf{Q}}/\mathbf{Q}) \longrightarrow PGL_2(\mathbf{C}).$$

The image of $\overline{\rho}$ is finite, and we let K be the fixed field of the kernel of $\overline{\rho}$. Thus $\overline{\rho}$ may be perceived as a finite Galois extension $K/\mathbf{Q}$ together with an imbedding

$$\mathrm{Gal}(K/\mathbf{Q}) \hookrightarrow PGL_2(\mathbf{C}).$$

In this section we want to ask the following question: What is the number of inequivalent liftings of $\overline{\rho}$, i.e. continuous representations ρ of $\mathrm{Gal}(\overline{\mathbf{Q}}/\mathbf{Q})$ in $GL_2(\mathbf{C})$ such that

$$\begin{array}{ccc} & \mathrm{Gal}(\overline{\mathbf{Q}}/\mathbf{Q}) & \\ & \rho\downarrow \qquad \searrow^{\overline{\rho}} & \\ GL_2(\mathbf{C}) & \longrightarrow & PGL_2(\mathbf{C}) \end{array}$$

commutes, with Artin conductor N and determinant character ε?

1.2 We want to reduce this question to an analogous local one. For this purpose we introduce the following notation: For each prime number p denote by D_p resp. I_p the decomposition resp. inertia group of a place of $\overline{\mathbf{Q}}$ above p; we identify D_p with $\mathrm{Gal}(\overline{\mathbf{Q}}_p/\mathbf{Q}_p)$. Now consider the following theorem of Tate (cf. [11]):

Theorem (Tate): Let θ: $\mathrm{Gal}(\overline{\mathbf{Q}}/\mathbf{Q}) \longrightarrow PGL_n(\mathbf{C})$ be an n-dimensional, continuous, projective representation. Suppose that for each prime number p there is given a continuous representation θ_p: $\mathrm{Gal}(\overline{\mathbf{Q}}_p/\mathbf{Q}_p) \longrightarrow GL_n(\mathbf{C})$ which is a lifting of the restriction $\overline{\theta}|D_p$, i.e.

$$\begin{array}{ccc} & \mathrm{Gal}(\overline{\mathbf{Q}}_p/\mathbf{Q}_p) & \\ & \theta_p\downarrow \qquad \searrow \ \overline{\theta}|D_p & \\ GL_n(\mathbf{C}) & \longrightarrow & PGL_n(\mathbf{C}) \end{array}$$

commutes, and such that the verification $\theta_p|I_p$ is trivial for almost all p.

(Here $GL_n(\mathbf{C}) \longrightarrow PGL_n(\mathbf{C})$ is the canonical projection.) Then there is a unique lifting $\theta : \mathrm{Gal}(\overline{\mathbf{Q}}/\mathbf{Q}) \longrightarrow GL_n(\mathbf{C})$ of $\overline{\theta}$ such that

$$\theta|I_p = \theta_p|I_p \quad \text{for all } p \ .$$

The theorem follows from another theorem of Tate on the vanishing of the Schur multiplier of $\mathrm{Gal}(\overline{\mathbf{Q}}/\mathbf{Q})$ together with the fact that $\mathbf{Q}$ has class number 1 .

This theorem allows us to reduce the question in 1.1 to certain analogous local problems: Let $\overline{\rho}$ be as in 1.1. If p is a prime number, denote by $\overline{\rho}_p$ the restriction of $\overline{\rho}$ to D_p as above. If $\overline{\rho}$ is going to have a lifting with Artin conductor N, then $\overline{\rho}_p$ must be unramified for $p \nmid N$. Assuming this to be the case, each $\overline{\rho}_p$ has a lifting r_p such that r_p is unramified for $p \nmid N$ (cf. [11] and [12]). The above theorem of Tate then gives us a unique lifting ρ of $\overline{\rho}$ such that:

$$\rho|I_p = r_p|I_p \quad \text{for all } p \ .$$

Now, any other lifting of $\overline{\rho}$ has the form $\rho \otimes \chi$, where χ is a character of $\mathrm{Gal}(\overline{\mathbf{Q}}/\mathbf{Q})$. We have:

$$\det(\rho \otimes \chi) = \det(\rho) \cdot \chi^2 \ .$$

Concerning the question of equivalence of these "twists" $\rho \otimes \chi$, one must know for what characters χ the representations ρ and $\rho \otimes \chi$ are equivalent. If χ is non-trivial, this can only happen if $\mathrm{Im}(\overline{\rho})$ is a dihedral group, and this case can be completely analysed, as will become clear from the following, by use of the well-known theorem of Mackey concerning induced representations.

Now, in the above situation the determinant of ρ is given once one knows its restriction to I_p for all p, and this restriction is $\det(r_p)|I_p$. Viewing via local class field theory the character $\det(r_p)$ as a character on $\mathbf{Q}_p^\times$, this restriction is simply the restriction of $\det(r_p)$ to the group of units of $\mathbf{Z}_p$. Furthermore, if χ is a character of $\mathrm{Gal}(\overline{\mathbf{Q}}/\mathbf{Q})$, then we may by global class field theory view χ as an idele class character and consider its restriction χ_p to $\mathbf{Q}_p^\times$ for every p . The Artin conductor of $\rho \otimes \chi$ is the product of the Artin conductors of $r_p \otimes \chi_p$ for all p , and these latter conductors depend only on the restrictions of r_p and χ_p to I_p .

It follows from this discussion that we can answer the question of 1.1, if we can solve the following problem.

Problem: Given a prime number p and a continuous representation:

$$\overline{\varphi} : \mathrm{Gal}(\overline{\mathbf{Q}}/\mathbf{Q}) \longrightarrow PGL_2(\mathbf{C}).$$

determine some lifting φ sufficiently explicitly so that:

(a) the restriction of $\det(\varphi)$ to the inertia group can be determined,

(b) the Artin conductor of $\varphi \otimes \chi$, for χ a character of $\mathrm{Gal}(\overline{\mathbf{Q}}/\mathbf{Q})$, can be computed in "terms" of χ .

1.3 In [5] we have given a complete solution of this problem; in fact a slightly more general problem is considered in [5]. This work is in turn based on results of Buhler and Zink ([1] and [19]; see also [17]). Since both the solution and the methods of obtaining it are rather technical, we shall, due to considerations of space, only give a very brief sketch of the ideas involved together with one (simple) example of the results obtained.

Let $\overline{\varphi}$ be as in the above problem.

Now one has to split up the discussion into two main cases: The image of $\overline{\varphi}$ is either a cyclic group, a dihedral group, or is isomorphic to A_4 or S_4. The "A_5-type" is of course excluded, since any Galois extension of a local field has a solvable Galois group; also, the cyclic case is a priori not excluded since we have not required $\overline{\varphi}$ to be irreducible. The cyclic case is however easily seen to be trivial, so that one has the following two main cases:

(i) $\mathrm{Im}\,\overline{\varphi}$ is a dihedral group,

(ii) $\mathrm{Im}\,\overline{\varphi}$ is isomorphic to A_4 or S_4 .

In the first case one bases oneself on the following description of the liftings of $\overline{\varphi}$ (cf. [11]):

Suppose that the ground field is K. Then $\overline{\varphi}$ provides us with a Galois extension M/K such that $\mathrm{Gal}(M/K)$ is a dihedral group. Except when $[M:K]=4$, which case can be discussed separately, the field M contains a unique quadratic extension L/K, and $\mathrm{Gal}(M/L)$ is cyclic. Now, the restriction of $\overline{\varphi}$ to $\mathrm{Gal}(\overline{\mathbf{Q}}_p/L)$ has the form

$$\overline{\varphi}(g) = \begin{pmatrix} \psi(g) & 0 \\ 0 & 1 \end{pmatrix} \quad \text{modulo } \mathbf{C}^{\times} ,$$

where ψ is a certain character on $\mathrm{Gal}(\overline{\mathbf{Q}}_p/L)$. Suppose that σ is an element of

$\text{Gal}(\overline{\mathbf{Q}}_p/K)$ which modulo $\text{Gal}(\overline{\mathbf{Q}}_p/L)$ generates $\text{Gal}(L/K)$. Now any lifting of $\overline{\varphi}$ is a representation of the form:

$$\varphi = \text{Ind}_{L/K}(\omega)$$

(representation induced by ω), where ω is a character on $\text{Gal}(\overline{\mathbf{Q}}_p/L)$ such that

$$\omega(\sigma g \sigma^{-1}) = \omega(g)\psi(g) \quad \text{for all } g \in \text{Gal}(\overline{\mathbf{Q}}/L) .$$

The conductor of such a φ is:

$$D_{L/K} \cdot N_{L/K}(f(\omega)) ,$$

where $D_{L/K}$ is the discriminant of L/K and $f(\omega)$ is the conductor of ω. Viewing ω via local class field theory as a character on $L^\times$, we may consider its restriction to $K^\times : \omega|K^\times$. Denoting by α the character on $K^\times$ of order 2 corresponding to L/K, the determinant of φ is:

$$\det \varphi = \alpha \cdot (\omega|K^\times) .$$

Using this description one can now by splitting the discussion into numerous cases according to the structure of $K^\times$, of L/K and of ψ, find answers to (a) and (b) above.

In case (ii) (the "primitive" case) one knows that the residue characteristic of the ground field must be 2. Here we restrict the discussion to the case where the ground field is $\mathbf{Q}_2$ (this clearly suffices, as we have seen in 1.2, for the purpose of answering the question in 1.1). One can then find that there are exactly 4 possibilities for $\overline{\varphi}$; There is exactly 1 extension of $\mathbf{Q}_2$ with Galois group isomorphic to A_4, and exactly 3 extensions with Galois group isomorphic to S_4; see [17]. Now, for these cases one has some general theory: In the first place there is the determination of the minimal conductor of a lifting of $\overline{\varphi}$, together with an intrinsic characterisation of those liftings which have minimal conductor; this is due to Buhler and Zink, see [1] and [19] (and also [17]). Secondly we have the answer to (b) above for a φ which has minimal conductor; see [19]. With this and a detailed analysis of each of the 4 cases above, one can find a satisfactory answer to (a) and (b) if one chooses for φ a certain lifting (depending on the case) with minimal conductor. Let us give just one example of the results:

The unique A_4-extension of $\mathbf{Q}_2$ is $K/\mathbf{Q}_2$ with

$$K = \mathbf{Q}_2(\theta, \sqrt{1+2\theta}, \sqrt{1+2\theta^2}, \sqrt{1+2\theta^4}) ,$$

where θ is a primitive 7'th root of unity. This gives us a unique representation:

$$\overline{\varphi} : \mathrm{Gal}(\overline{\mathbf{Q}}_2/\mathbf{Q}_2) \longrightarrow PGL_2(\mathbf{C})$$

of A_4-type, and this $\overline{\varphi}$ has a lifting φ with the following properties:

Let α be the character on $\mathbf{Q}_2^\times$ corresponding to $\mathbf{Q}_2(\sqrt{-1})/\mathbf{Q}_2$. We view a character ψ on $\mathrm{Gal}(\overline{\mathbf{Q}}_2/\mathbf{Q}_2)$ also as a character on $\mathbf{Q}_2^\times$ and write $f(\psi)$ for $\log_2$ of the conductor of ψ. Then the determinant of $\varphi \otimes \psi$ is $\alpha\psi^2$. Furthermore, $\log_2$ of the conductor of $\varphi \otimes \psi$ is 5 if $f(\psi) \leq 2$, and it is $2f(\psi)$ if $f(\psi) \geq 3$.

1.4 We now return to the situation and notation of 1.1. We have seen how to reduce the question asked in 1.1 to a finite number of local problems (one for each prime divisor in N), and we have indicated how one can solve these local problems. This reduces the problem of determining $d(N, \varepsilon)$ to the problem of finding the potential candidates for the field K in 1.1. Now if the representation $\overline{\rho}$ has a lifting with conductor N, then trivially the smallest possible conductor of a lifting of $\overline{\rho}$ is $\leq N$; if one has the results of 1.3 at ones disposal, then one can easily from this derive explicit bounds for the discriminant of $K/\mathbf{Q}$. Thus it is in principle clear that one has an algorithm for determining the possible fields K (given N and ε). The question is how to do this as effectively as possible.

If $\mathrm{Gal}(K/\mathbf{Q})$ is a dihedral group or isomorphic to A_4 or S_4, then the use of class field theory is the most efficient method for determining the candidates K. The use of class field theory is possible since $\mathrm{Gal}(K/\mathbf{Q})$ is in each of these cases a solvable group.

Of course this method does not work if $\mathrm{Gal}(K/\mathbf{Q}) \cong A_5$. In this case one can, given N and ε, find explicit bounds, not only for the discriminant of $K/\mathbf{Q}$, but also for the discriminant of a root field of K by which we mean an extension of degree 5 over $\mathbf{Q}$ contained in K. This gives us the problem to be considered in the next section.

2. Enumeration of A_5-fields

2.1 It is not hard to see that a projective representation $\overline{\rho} : \mathrm{Gal}(\overline{\mathbf{Q}}/\mathbf{Q}) \longrightarrow PGL_2(\mathbf{C})$ with $\mathrm{Gal}(K/\mathbf{Q}) \cong A_5$ where K is the fixed field of the kernel of $\overline{\rho}$, has a lifting $\rho : \mathrm{Gal}(\overline{\mathbf{Q}}/\mathbf{Q}) \longrightarrow GL_2(\mathbf{C})$ with odd determinant character if and only if K is not real.

From this and from the results of section 1 we conclude that we have to confront the following problem: Determine all non-real Galois extensions of $\mathbf{Q}$ with Galois group isomorphic to A_5 and for which the discriminant of a root field is $\leq D^2$ for some fixed number D.

Here the number D has to be chosen so large that such fields actually occur, and secondly so that one of the (2-dimensional) projective representations associated with one of the occurring A_5-field has a lifting with odd determinant character and Artin conductor of a "moderate" size. The reason for the second condition will become apparent in the next section. It turns out (from the table described below) that one must with these requirements choose D to be at least ≈ 500 .

Now, the A_5-fields in the above problem will be obtained as splitting fields of polynomials

$$f(x) = X^5 + a_1X^4 + a_2X^3 + a_3X^2 + a_4X + a_5 \ ,$$

where $a_i \in \mathbf{Z}$, and the problem is to find good bounds for the $|a_i|$ so that we may use computer power to eliminate the uninteresting $f's$. Of course, the classical theory of geometry of numbers according to Minkowski provides us with bounds on the $|a_i|$ in terms of D, but the reader may convince himself that these bounds lead even for $D = 500$ to such a large number of possibilities for $f(x)$ that a computer search on this basis is absolutely impossible. Hence we have to do better than this.

2.2 We took our starting point in the following theorem of Hunter:

Theorem: (cf. [4]) Suppose that $F/\mathbf{Q}$ is an extension of degree 5 with discriminant D. Then there is an algebraic integer θ in F such that $F = \mathbf{Q}(\theta)$, such that

$$|Tr_{F/\mathbf{Q}}(\theta)| \leq 2 \ ,$$

and such that, denoting by $\theta_1 = \theta, \theta_2, \theta_3, \theta_4, \theta_5$ the conjugates of θ ,

$$\Big(\sum_{i=1}^{5} |\theta_i|^2\Big)^4 \leq \frac{8}{5}|D| \ .$$

Using this theorem and the fact that the potential $f's$ in 2.1 must have precisely one real root (since the splitting field should be a non-real extension of $\mathbf{Q}$ with Galois group isomorphic to A_5) it is not hard to derive rather "reasonable" bounds for the $|a_i|$. We shall not here write down these bounds but merely be content to note that they enabled us to solve the problem in 2.1 for the following value of D :

$$D = 2083 \ .$$

The reason for choosing this value, which is a prime number, will be recognized at the end of this section.

With this value of D and the above bounds for the $|a_i|$ one then gets a priori $\sim 7,7 \cdot 10^9$ possibilities for $f(x)$, and one then applies computer power to eliminate the uninteresting $f's$ through the following steps:

Step 1: Require the discriminant of $f(x)$ to be a square modulo the primes $3, 5, 7, 11, 13, 17, 19, 23$ and 29 .

Step 2: Require the discriminant of $f(x)$ to be a square in $\mathbf{Z}$.

Step 3: Require $f(x)$ to be irreducible over $\mathbf{Z}$.

Step 4: Require the discriminant of a root field of the splitting field of $f(x)$ to be $\leq 2083^2$.

Step 5: Require the Galois group of a splitting field of $f(x)$ to be isomorphic to A_5 .

Step 6: Distribute the remaining $f's$ into classes with identical splitting fields.

There is for each step a standard algorithm. Of these, perhaps only the algorithm involved in step 6 is less well known. For this algorithm the reader may consult [18].

The total computing time required to perform the six steps was about 280 hours. The second step was the most time consuming, requiring about 200 hours of computing time. The machine used was an Apollo DN 10000.

The final result of this computation is the complete table of all non-real A_5-extensions of $\mathbf{Q}$ for which the discriminant of a root field is bounded by 2083^2. The table contains 238 fields; we shall not reproduce the table here, but be content to give one example of the use of it.

2.3 Consider $N = 2083$. This is a prime number $\equiv 3$ modulo 4 . Let ε be the Legendre symbol $\varepsilon(\cdot) = \left(\frac{\cdot}{2083}\right)$. It is not hard to deduce that if the representation ρ : $\mathrm{Gal}(\overline{\mathbf{Q}}/\mathbf{Q}) \longrightarrow GL_2(\mathbf{C})$ is of A_5-type with Artin conductor $N = 2083$ and determinant character ε , then the discriminant of a root field of the corresponding A_5-extension of $\mathbf{Q}$ is 2083^2. The aforementioned table shows that there is exactly one such A_5- extension of $\mathbf{Q}$, namely the splitting field of

$$(**) \qquad X^5 + 8X^3 + 7X^2 + 172X + 53 \ .$$

Now, there are exactly two non-equivalent representations $A_5 \hookrightarrow PGL_2(\mathbf{C})$ for which the fixed field of the kernel of $\overline{\rho}$ is the splitting field of (**); furthermore a straightforward analysis shows that each of these has exactly two non-equivalent liftings (to $GL_2(\mathbf{C})$) with Artin concuctor $N = 2083$ and determinant character ε .

It is not a problem to see that there are no representations $\rho : \mathrm{Gal}(\overline{\mathbf{Q}}/\mathbf{Q}) \longrightarrow GL_2(\mathbf{C})$ with Artin conductor 2083 and determinant character ε of type A_4 or S_4 ; on the other hand there are exactly 3 such representations of dihedral type. The last fact is deduced from the fact that $\mathbf{Q}(\sqrt{-2083})$ has class number 7 (cf. [11]). Hence we may conclude that

$$d(2083, \varepsilon) = 2 \cdot 2 + 3 = 7 \ .$$

There are examples of a similar nature for many other values of N , for example

$$4 \cdot 487,\ 4 \cdot 751,\ 4 \cdot 887,\ 4 \cdot 919,\ 2^5 \cdot 73,\ 2^5 \cdot 193 \ .$$

3. Computation of dim $S_1^+(N, \varepsilon)$

3.1 For $k, M \in \mathbf{N}$ and δ a Dirichlet character modulo M with $\delta(-1) = (-1)^k$, we denote by $S_k(M, \delta)$ the complex vector space of cusp forms of weight k and nebentype δ on

$$\Gamma_0(M) = \left\{ \begin{pmatrix} a & b \\ c & d \end{pmatrix} \in SL_2(\mathbf{Z}) | c \equiv o(M) \right\} .$$

These spaces are finite dimensional. The space $S_k(M, \delta)$ always has an <u>integral basis</u> by which we mean a basis $(f_1, \ldots, f_d)$ over $\mathbf{C}$ such that the Fourier coefficients of any f_i are rational integers (cf. [13], Chap. 3).

3.2 Now let us once again fix $N \in \mathbf{N}$ and let ε be a Dirichlet character modulo N with $\varepsilon(-1) = -1$.

In order to compute dim $S_1^+(N, \varepsilon)$ it obviously suffices to have a general algorithm for the determination of dim $S_1(N, \varepsilon)$. The problem here is that a "formula" for this dimension is not available; this is in contrast to the situation for weight ≥ 2 where such formulas exist and are classical (see for example [2]). To compute the dimension of $S_1(N, \varepsilon)$ we shall therefore use the following trick:

For the sake of simplicity, suppose that N has a prime divisor $\equiv 3(4)$; let p be such a prime divisor and let η be the Legendre symbol $\eta(\cdot) = \left(\frac{\cdot}{p}\right)$. Consider the Eisenstein series

$$E = \frac{1}{2}h + \sum_{n=1}^{\infty}\Big(\sum_{d|n}\eta(d)\Big)q^n \ , \quad q = e^{2\pi i z} \ ,$$

where h is the class number of $\mathbf{Q}(\sqrt{-p})$. Then E is a modular form of weight 1 and nebentype η on $\Gamma_0(p)$. The form $2E$ has integers as Fourier coefficients at ∞ and does not vanish at ∞.

On the other hand we consider the series

$$\theta_2(z) = \sum_{m\equiv 1(2)} q^{m^2/8} \ , \quad q = e^{2\pi i z} \ .$$

Then θ_2^8 is a modular form of weight 4 and trivial nebentype on $\Gamma_0(2)$; furthermore, θ_2^8 does not vanish in the upper half plane, and among the cusps it vanishes only at ∞ , where it has integers as Fourier coefficients (cf. [10], chap. 1). Defining N' to be N if N is even, and $N' = 2N$ if N is odd, it is now clear that the map

$$f \longmapsto (2Ef, \theta_2^8 f)$$

maps $S_1(N,\varepsilon)$ isomorphically onto the subspace of

$$S_2(N,\varepsilon\eta) \times S_5(N',\varepsilon)$$

consisting of pairs (h_1, h_2) with

$$(+) \qquad h_1\theta_2^8 = h_2 \cdot 2E \ .$$

Here we have viewed η as a Dirichlet character modulo N and ε also as a Dirichlet character modulo N' . Now, for $h_1 \in S_2(N,\varepsilon\eta)$ and $h_2 \in S_5(N',\varepsilon\eta)$ the forms $h_1\theta_2^8$ and $h_2 \cdot 2E$ are elements of $S_6(N',\varepsilon\eta)$. Denote for $M \in \mathbf{N}$ by $m(M)$ the index of $\Gamma_0(M)$ in $SL_2(\mathbf{Z})$; one has

$$m(M) = M \cdot \prod_{p|M}\Big(1 + \frac{1}{p}\Big)$$

(cf. [13], chap. 1). For a non-zero element of $S_k(M,\delta)$ its order of 0 at ∞ is less than $km(N)/12$ (see [10], chap. 1). Hence condition (+) simply means that the first $\frac{1}{2}m(N')$ Fourier coefficients of $h_1\theta_2^8 - h_2 \cdot 2E$ are all 0 . If now d and e are the dimensions of the spaces $S_2(N,\varepsilon\eta)$ and $S_5(N',\varepsilon)$ respectively, we then see that if we have an algorithm for finding (i.e. computing Fourier coefficients of) an integral basis for each of these spaces, then the problem of computing the dimension of $S_1(N,\varepsilon)$ has been reduced to the computation of the rank of a certain $d+e$ by $\frac{1}{2}m(N')$ matrix with coefficients in $\mathbf{Z}$.

If N does not have a prime divisor $\equiv 3(4)$, one can replace the level N by $8N$ in the above, but then the discussion becomes somewhat more complicated.

3.3 We then turn to the problem of determining an integral basis for a space $S_k(M,\delta)$ where $k \geq 2$.

Here again we have to be very brief and give only a rough sketch of the ideas involved.

Let $R = \mathbf{Z}[\frac{1}{6}]$. We consider a certain $R[\Gamma_0(M)]$-module $L_{k,\delta}$. As R-module $L_{k,\delta}$ is the R-module of homogeneous polynomials of degree $k-2$ in two variables over R :

$$L_{k,\delta} = \left\{ \sum_{i=0}^{k-2} a_i x^i y^{k-2-i} | a_i \in R \right\} .$$

The action of $\Gamma_0(N)$ on $L_{k,\delta}$ is given by:

$$\begin{pmatrix} a & b \\ c & d \end{pmatrix} x^i y^{k-2-i} = \delta(d)(ax+cy)^i(bx+dy)^{k-2-i} .$$

Now, the Eichler-Shimura isomorphism gives us an exact sequence:

$$0 \to S_k(M,\delta) \oplus \overline{S_k(M,\delta)} \to H^1(\Gamma_0(M), L_{k,\delta} \otimes \mathbf{C}) \xrightarrow{r} L \to 0 .$$

Here, $\overline{S_k(M,\delta)}$ is the space obtained from $S_k(M,\delta)$ by complex conjugating Fourier coefficients at ∞ , and L is a certain space which will not be described here.

This gives an embedding of $S_k(M,\delta)$ into the group of fixed points of $\begin{pmatrix} -1 & 0 \\ 0 & 1 \end{pmatrix}$ in $H^1(\Gamma_0(M), L_{k,\delta} \otimes \mathbf{C})$. Now one constructs a concrete model W of the latter group. The Hecke operators T_m can be defined on the above cohomology group in a way which is compatible with the embedding of $S_k(M,\delta)$ in W .

In the model W one can find a basis $\{b_i\}$ such that $T_m b_i \in \bigoplus_i \mathbf{Z} b_i$ for all i and m . It is now shown (L. Merel) that if $w \in W$ is a $\mathbf{Z}$-linear combination of the b_i's and if we define the integers $a_m^{(i)}$ by:

$$T_m w = \sum_i a_m^{(i)} b_i \ ,$$

then for all i, $a_0^{(i)} + \sum_m a_m^{(i)} q^m$ defines a modular form on $\Gamma_0(M)$ with weight k and nebentype δ for some constant $a_0^{(i)}$. We have here ignored some technicalities concerning the definition of T_m if m is not prime to M . Furthermore, it is shown that these forms for "generic" w provide us with an integral basis for the space of modular forms on $\Gamma_0(M)$ with weight k and nebentype δ .

The constant $a_0^{(i)}$ in the above cannot be computed directly, but by explication of the homomorphism r above, one can easily isolate the cusp forms, and one thus obtains an integral basis for $S_k(M, \delta)$.

This algorithm has been completely implemented (by X. Wang) in Essen and is working very effectively.

Following the strategy outlined in the above, we have obtained 7 new non-trivial verifications of the Artin conjecture (corresponding to the 7 values of N mentioned in 2.3). We shall report on this elsewhere.

Acknowledgements

This work was supported by the Deutsche Forschungsgemeinschaft.

References

[1] **J. P. Buhler:** Icosahedral Galois Representations.
Lecture Notes in Mathematics 654, Springer-Verlag 1978.

[2] **H. Cohen, J. Oesterlé:** Dimensions des espaces de formes modulaires.
In: J.-P. Serre, D.B. Zagier: Modular Functions of One Variable VI.
Lecture Notes in Mathematics 627, Springer-Verlag 1977.

[3] **P. Deligne, J.-P. Serre:** Formes modulaires de poids 1.
Ann.Sci.Ec.Norm.Sup. 7 (1974), pp. 507-530.

[4] **J. Hunter:** The minimum discriminants of quintic fields.
Proc. Glasgow Math. Assoc. 3 (1957), pp. 57-67.

[5] **I. Kiming:** On the liftings of 2-dimensional, projective Galois representations over **Q**. Preprint, 1993.

[6] **R.P. Langlands:** Base change for GL(2). Annals of Mathematics Studies 96, Princeton University Press 1980.

[7] **W. Li:** On converse theorems for GL(2) and GL(1). Amer. J. Math. 103, No. 5 (1981), pp. 851-885.

[8] **L. Merel:** An elementary theorem about Hecke operators and periods of modular forms. Preprint, 1992.

[9] **T. Miyake:** Modular Forms. Springer-Verlag, 1989.

[10] **H. Petersson:** Modulfunktionen und quadratische Formen. Ergebnisse der Mathematik und ihrer Grenzgebiete 100. Springer-Verlag, 1982.

[11] **J.-P. Serre:** Modular forms of weight one and Galois representations. In: A. Fröhlich: Algebraic Number Fields. Academic Press 1977.

[12] **J.-P. Serre:** Cohomologie Galoisienne. Lecture Notes in Mathematics 5, Springer-Verlag, 1986.

[13] **G. Shimura:** Introduction to the arithmetical theory of automorphic functions. 1970.

[14] **J. Tunnell:** Artin's conjecture for representations of octahedral type. Bull.Amer.Math.Soc., Vol. 5, No. 2 (1981), pp. 173-175.

[15] **X. Wang:** The Hecke algebra on the cohomology of $\Gamma_0(p_0)$. Nagoya Math. J. 121 (1991), pp. 97-125.

[16] **A. Weil:** Über die Bestimmung Dirichletscher Reihen durch Funktionalgleichungen. Math. Ann. 168 (1967), pp. 149-156

[17] **A. Weil:** Exercises dyadiques. Invent. Math. 27 (1974), pp. 1-22.

[18] **Weinberger, Rothschild:** Factoring polynomials over algebraic number fields. ACM Transactions on Mathematical Software, Vol. 2, No. 4 (1976), pp. 335-350.

[19] **E.-W. Zink:** Ergänzungen zu Weils Exercises dyadiques. Math. Nachr. 92 (1979), pp. 163-183.

Next Generation Computer Algebra Systems AXIOM and the Scratchpad Concept: Applications to Research in Algebra

Larry Lambe

Department of Mathematics
and Computer Science
Kent State University
Kent, OH 44242, USA

Abstract

One way in which mathematicians deal with infinite amounts of data is symbolic representation. A simple example is the quadratic equation $x = \frac{-b \pm \sqrt{b^2-4ac}}{2a}$, a formula which uses symbolic representation to describe the solutions to an infinite class of equations. Most computer algebra systems can deal with polynomials with symbolic coefficients, but what if symbolic exponents are called for (e.g., $1+t^i$)? What if symbolic limits on summations are also called for (e.g., $1+t+\ldots+t^i=\sum_j t^j$)? The "Scratchpad Concept" is a theoretical ideal which allows the implementation of objects at this level of abstraction and beyond in a mathematically consistent way. The AXIOM computer algebra system is an implementation of a major part of the Scratchpad Concept. AXIOM (formerly called Scratchpad) is a language with extensible parameterized types and generic operators which is based on the notions of domains and categories **[Lambe1]**, **[Jenks-Sutor]**. By examining some aspects of the AXIOM system, the Scratchpad Concept will be illustrated. It will be shown how some complex problems in homological algebra were solved through the use of this system.

§1 Introduction

New paradigms are evolving in computer science. There is a thrust towards typed languages in which object oriented methodologies may be conveniently implemented. In mathematical programming, such concepts were proposed in the 1970's by Richard Jenks **[Jenks]** through the theoretical design of MODLISP. Not much later, at IBM Research, Yorktown Heights, NY, Jenks embarked on the implementation of a system which was intended to allow the mathematical programmer a very high level language in which he could, for example, implement only one algorithm for several types of rings if indeed that algorithm was valid for such rings. For example, since the Euclidian algorithm works over any Euclidian domain, one should not have to implement it once for the integers, again for the Gaussian integers, and yet another time for polynomials in one variable over a field. The theoretical basis for this system was MODLISP and the system was called Scratchpad. Over Scratchpad's more than 20 year history at Yorktown Heights it underwent many design changes and had many contributors (see **[Jenks-Sutor]**). It has allowed a style of mathematical programming that represents the way many mathematicians actually think

Supported by NSF Grant No. CCR-9207241 and the Department of Mathematics, Stockholm University. The author also wishes to express his gratutude to Professor P-C. Ramusino for his hospitality at the Universities of Trent and Milan where a part of this report was prepared.

about the inter-relationships between the sub-areas within mathematics. This style of mathematical programming will be called the *Scratchpad Concept.* A formal definition will not be given. Instead, examples of this style will be highlighted.

In the early 1990's, the Scratchpad Research Project evolved into a system called AXIOM (a trade mark of NAG, Ltd.) and is now commercially available through the offices of NAG. The system consists of an interpreter, a compiler, an extensive library, and a sophisticated graphical user interface. The Scratchpad Concept will be first illustrated abstractly and then through the use of AXIOM. Examples are taken from the speaker's research in homological algebra.

Some well-known mathematical paradigms are followed, but with some very new twists. For example, one might have a need to deal with formulae given recursively in a situation where closed formulae would yield an immediate solution. One method, employed since the beginning of mathematics, is to simply try some symbolic manipulation aided by the calculation of several cases to get a handle on formulae which might then be proven correct by an induction. In our cases, the symbolic manipulation as well as the calculation of special cases was done using AXIOM, but the resulting formulae are almost as complex as the original forms. However, much progress has been made through this process since the resulting closed forms allow the construction of extremely efficient routines to deal with further calculation and manipulation in AXIOM (or other systems) involving the original structures. It is believed that this very high level of "code generation", so convenient in a system like AXIOM, will be one of the most important contributions to modern computational mathematics of the new generation computer algebra systems.

1.1 Procedures With Mathematical Objects as Parameters

Consider matrix row reduction. It can be accomplished by using the three basic operations

$ero1(i,j)$: Interchange rows i and j

$ero2(r,i,j)$: Add r times row i to row j

$ero3(r,i)$: Multiply row j by r

Algorithms for matrix row reduction using these operations can easily be implemented in Pascal, for example. One might have

```
MTX = matrix[1..maxRow, 1..maxCol] of Field;
Field = real;
Int = integer;
```

The procedure call might look like

```
rowReducedEchelon(var m : MTX; n: Int; m : Int);
```

But clearly the *ero*'s can be performed over a number of rings and, indeed, sometimes one encounters problems that call for matrix row reduction over one of these other rings. For example, what if $Ring = GF(3,2)$, the field with 9 elements? Furthermore, if elements are not in a field, but in a principal ideal domain, then what can one do? It would be nice to be able to write

rowReducedEchelon(M : Matrix(n,m,R), R : Ring, n, m : Int).

Not only is is more convenient for the user, it is also more convenient for the implementer if only one procedure needs to be written for all admissible types.

Here is a segment that gives natural expression of the algorithm:

```
Ring : Category
R : Ring
n, m : Int
M : Matrix(n,m,R)

rowReducedEchelon(M,R,n,m) ==
  if R is a Field then
      Block1
      Block2
         .
         .
  else if R is a EuclidianDomain then
        Block1'
        Block2' ...
```

1.2 Polynomials of All Sorts

For another example, suppose that R is some ring and $X = \{1, t_1, t_2, \ldots\}$ is a set of indeterminates. We can form the free R-module on X. It has X as a basis and so consists of all linear combinations $\sum a_i t_i$ where $a_i \in R$ and the sum is finite. Let's denote this mathematical object by

$$FreeModule(R, X)$$

We could define an operation on the elements of X and extend it bilinearly to obtain an algebra over R. For example, define $1t_i = t_i$, $t_i 1 = t_i$, and $t_i t_j = t_{i+j}$. Clearly this algebra is isomorphic to the polynomial ring in one variable t. Call the result $Poly(R, t)$.

Notice that $Poly(R, t)$ is an extension of $FreeModule(R, X)$. It has all the operations from $FreeModule(R, X)$ and an extra one, viz., a multiplication. Notice also that all of this makes sense for any binary operation on X. A set M with an operation

$$* : M \times M \to M$$

that is associative and has an identity element 1 is called a monoid. For any monoid, there is an operation on $FreeModule(R, M)$ which is obtained by extending the multiplication of M bilinearly. The resulting algebra over R is called the monoid ring. Let's denote that algebra by

$$MonoidRing(R, M).$$

Going back to our polynomial example, notice that except for cosmetics, $Poly(R, t)$ is just $MonoidRing(R, NonNegativeInteger)$ where the second parameter denotes the monoid of non-negative integers under addition of integers and identity 0.

If an element of $FreeModule(R, M)$ is represented by a list of terms (with addition implied) then a correspondence of the form

$$r_1 m_1 + \ldots + r_k m_k \leftrightarrow [[r_1, m_2], \ldots [r_k, m_k]]$$

has been set up between formal linear combinations and lists of two element lists where the first member of the two element sublist is the coefficient r_i from R and the second element is the basis element m_i from M. Another abstract correspondence can be given by simply choosing a dummy variable t and matching the list of lists above to linear combinations of expressions of the form t^{m_i}. Thus one can also use the representation

$$[[r_1, m_2], \ldots [r_k, m_k]] \leftrightarrow r_1 t^{m_1} + \ldots + r_k t^{m_k}.$$

With this representation, and when the monoid M is the non-negative integers, one clearly gets ordinary polynomial expressions from $MonoidRing(R, M)$. When the monoid M is the full group of integers, one gets "polynomials" with non-negative *or* negative exponents, i.e., Laurent polynomials. To get polynomials in several indeterminates, several approaches can be taken. Given the discussion of polynomials in one variable, it is easily seen that polynomials in two variables are (except for cosmetics) given by

$$MonoidRing(POL, NonNegativeInteger)$$

where $POL = MonoidRing(R, NonNegativeInteger)$. In fact, one can simply iterate this procedure to obtain polynomials in n variables for any $n \geq 1$.

For another representation, let NNI stand for the non-negative integers with its additive monoid operation. The cartesian product, NNI^n, of n copies of NNI, can be given an induced monoid operation in the usual way by adding coordinate-wise. Denote NNI^n by

$$DirectProduct(n, NNI).$$

The polynomial ring in n indeterminates (except for cosmetics) is then given also by

$$MonoidRing(R, DirectProduct(n, NNI)).$$

We will come back to the cosmetic issues later.

1.3 Skew Symmetric Polynomials

The exterior (or Grassman) algebra is another useful construction. The exterior algebra can be constructed from the free R module with basis in bijective correspondence with the set of all subsets of $\{1, 2, \ldots, n\}$. Think of a basis element as $e_{i_1} \wedge \ldots \wedge e_{i_k}$.

So there is an algebra structure on the module $FreeModule(R, P(X))$ where $P(X)$ is the set of all subsets of X. Thus here is another example of an algebra constructed from $FreeModule(R, Y)$ (for $Y = P(X)$) by ***adding on operations***. The new algebra constructor will be called $AntiSymm(R, X)$. This algebra is commonly denoted by $E[e_1, \ldots, e_n]$ and this notation will be used in what follows.

The organization of the construction of the ring of antisymmetric polynomials might be accomplished by a scheme like

AntiSymm(R,X) : Algebra(R) ==
FreeModule(R,P(X)) add
.
.

Think of an element of $AntiSymm(R, X)$ as having "type" $AntiSymm(R, X)$. Think of $AntiSymm(R, X)$ as having "type" $Algebra(R)$. Note that, in this sense, the "category" of algebras is parameterized by rings R.

§2. The AXIOM Language

The AXIOM language allows "categories" and "domains" to be constructed in a natural way.

2.1 Categories, Domains, and Packages

To understand what categories are in AXIOM, think of groups. All groups have a binary operation $*$, a unary operation inv, and a constant 1

$$* : G \times G \longrightarrow G$$
$$inv : G \longrightarrow G$$
$$1 \in G$$

which satisfy certain axioms. Roughly speaking, in the AXIOM language, the specification of the range and domain of a function as given for $*$ and inv above is called a "signature". Also speaking roughly, a category in AXIOM is a collection of signatures for potential domains which are to be of that category. Every domain of a given category should implement each of the signatures in the category, but might contain additional signatures. An example of how one might give the category of groups in AXIOM is given below. (See **[Jenks-Sutor]** for precise definitions of category, domain, and package.)

In the following example, the first line says that the *Group*() constructor is a category (so a list of signatures is expected) and that this category is built upon the category *Set*, i.e., it will have all the *Set* signatures along with those that are going to be specified in the lines that follow.

The second line is a "−−" comment. It is a comment line that is private to the source code and is intended for the implementer.

The third line gives the signature for a group operation. The $ in source code denotes the domain or category under construction.

The fourth line is a "++" comment (also called "documentation"). It is a line that is public and can be used by various data base and browser facilities when a user asks for information about a given domain or category.

The operation "/" will be given by a "default definition" (see below)

The line following the internal comment "attributes" sets a system-wide attribute for all operations which are implemented as group operations. Attributes are qualities that an object might have which can be tested in future source code which builds upon previous source code. Recall the tests "If R is a Field then" and "else if R is a EuclidianDomain" from (1.1).

Finally, there is a "default definition" given by the last two lines. Any domain which implements the *Group*() signatures will automatically have an implementation of "division" It should be remarked that this discussion has been for the purposes of illustration and while the category *Groups*() actually exists in AXIOM, it has a somewhat different (but similar in spirit) implementation.

```
Group() : Category == Set with
       -- operations
       * : ($, $) -> $
       ++  x * y returns the product

       inv : $ -> $
       ++ inv(x) returns the inverse

       1 : () -> $
       ++ 1 is the identity element

       / : ($, $) -> $
       ++ x/y  = x * inv(y)
       ++ it is defined in all groups by a default definition

       --  attributes
       associative(*)

     add
       x:$ / y:$ == x * inv(y)
```

Given the AXIOM category Group, one does not necessarily have any group! To obtain a group, one must

i. implement the group signatures or

ii. call some "group constructor".

The same holds for all constructors. To explain, one might write

$$Q := Fraction(Integer)$$
$$G := Heisenberg(Q)$$

to obtain the group of 3×3 upper triangular matrices with ones along the diagonal and rational number entries. This is an example of how one calls a group constructor (viz., $Heisenberg(R)$) which has already been implemented in the system. In passing, also note that the constructor $Fraction(R)$ where R is an integral domain already exists in AXIOM. It returns a domain which represents the field of fractions of R.

Given a domain such as G above, one works with it as follows. To declare some elements to be of type G write

$$(x, y, z) : G$$

To make these variables the standard group generators write

$$x := generator(1)$$
$$y := generator(2)$$
$$z := generator(3)$$

(this assumes that the signature $generator : NNI \to G$ with the obvious usage exists in the domain G). To create some words in G, for example, write

$$x * inv(y) * x.$$

There is also the notion of a "package" in AXIOM. A package is exactly like a domain in the sense that certain signatures are implemented but it only exports operations that work on objects supplied by other domains (it has no signatures with "$" in them). The purpose of a package is to group together a collection of procedures which may be used for various specialized tasks. For example, one might have a package to implement isomorphisms (coercions) of several types between one domain and another.

2.2 Cosmetic Issues

There is a very rich domain in AXIOM called $OutputForm$. Its purpose is to provide the implementor with convenient facilities for displaying an object in a desired form. For example, the free abelian group on n generators $\{t_1, \ldots, t_n\}$ written multiplicatively might be implemented by defining operations on lists of integers $[i_1, \ldots, i_n]$. In any output, it is desirable to represent this list of integers as $t_1^{i_1} \ldots t_n^{i_n}$. AXIOM provides for this through the $OutputForm$ domain in connection with a *coercion mechanism*.

Every domain must either inherit or implement a signature of the form

$$coerce : \$ \to OutputForm.$$

The purpose of this is to let the AXIOM system know how to output an expression from the given domain. For example, in the free abelian group case above, an implementor might require parameters in the free abelian group constructor which the user is to supply and which will be used to determine the generators of the group. The constructor and its parameters might look like $FreeAbelian(var, dim)$ where var is of type $Symbol$ and dim is of type NNI. A call to this constructor might look like

$$fab4 := FreeAbelian(t, 4).$$

In order to accomplish the desired output, the implementor could write something like the following lines.

$$macro\ O ==> OutputForm$$
$$listGens := [sub(t :: O, k :: O)\ for\ k\ in\ 1..dim]$$

.

.

$$coerce(x : \$) : O ==$$
$$x = 1 => 1 :: O$$
$$reduce(*, [listGens.k ** x.k :: O\ for\ k\ in\ 1 \ldots dim \mid\ not\ zero?\ x.k]).$$

The first line is a macro. It allows the substitution of O for the longer $OutputForm$ (macros are also possible in the interpreter).

The second line takes the user input symbol t and coerces it to O and also coerces (coercions can be forced by the "::" construct) the NNI, k to O and then forms a list of t subscripted by k as k ranges through the values 1 to dim. In other words, it simply forms the list of "output forms" $[t_1, \ldots, t_n]$.

The next few lines indicate that there may be some other bits of code which the implementor may have written and the next line starts the implementation of the coercion routine for the domain which is under construction.

The next line states that if the expression is equal to the identity element of the group, then the expression "1" should be output.

The next line applies the multiplication operator $*$ *from OutputForm* to the list $[t_1^{x.1}, \ldots, t_n^{x.n}]$ which is itself formed using the operation of exponentiation from O.

The multiplication operator from O takes two output forms and returns an output form with just enough space between them to look reasonable for representing the product of two elements. Similarly, the exponential of an output form by another creates an output form which will display in the usual fashion.

As was mentioned earlier, the domain $OutputForm$ is quite rich. It contains operations to put boxes around objects, place spaces horizontally or vertically, overline or underline, and so on. In addition, the AXIOM system provides for TEX output.

§3. Applications to Research in Algebra

A general method which can be used to attack problems in several areas of mathematics is called "perturbation theory". The basic idea is to realize a given problem as a "perturbation" of a problem with a known solution and then perturb the known solution to obtain a solution to the original problem.

Homological perturbation theory can be applied to obtain small resolutions over several classes of algebras. In this context, one has an algebra A' which has the same underlying R-module structure as an algebra A. One also has a module M' over A' with the same underlying R-module structure as a module M over A (examples will be given later). The objective is to "perturb" a free resolution of M over A to a free resolution of M' over A'. Thus, one is given

$$(A \otimes X, d) \longrightarrow M \longrightarrow 0$$

and one seeks

$$(A' \otimes X', d') \longrightarrow M' \longrightarrow 0$$

obtained from the original resolution by some process. A formal solution to this problem exists [**Lambe2**], [**Lambe3**], [**Barnes-Lambe**]. In order to discuss these concepts further, some notation will be necessary.

3.1 Homological Perturbation Theory

Let (X, d) denote a differential graded module X over some commutative ring R with unit. Thus $X = \{X_n\}_{n \geq 0}$ is a sequence of R-modules and there are R-module maps

$$\ldots X_{n+1} \xrightarrow{d_{n+1}} X_n \xrightarrow{d_n} X_{n-1} \ldots$$

with

$$d_n d_{n+1} = 0.$$

The differential module (X, d) will often be simply denoted by X. When several modules are involved and the differentials need to be distinguished, they will be written as d_X, etc.

Given two differential modules M and X, M is said to be a *strong deformation retraction* (SDR) of X when there exist maps

$$M \xrightarrow{\nabla} X, \quad X \xrightarrow{f} M$$

and

$$X \xrightarrow{\phi} X$$

such that the following identities hold.

$$\nabla f = 1_M, \quad f\nabla = 1_X - (d\phi + \phi d),$$

$$f\phi = 0, \quad \phi\nabla = 0, \quad \phi\phi = 0.$$

A fundamental theorem on transferring differentials from one object to another is the "basic perturbation lemma" (see, for example, [**Shih**], [**Brown**], [**Gugenheim**], [**Lambe-Stasheff**], [**Hübschmann**], [**Barnes-Lambe**]).

3.1.1. Theorem (Basic Perturbation Lemma): *Given SDR–Data*

$$((X, d) \underset{f}{\overset{\nabla}{\rightleftarrows}} (Y, d), \phi)$$

and a new differential D on Y, let $t = D - d$ and more generally

$$t_{n+1} = (t\phi)^n t, \quad n \geq 0.$$

For each n, define (on X)

$$\partial_n = d + f(t_1 + t_2 + \dots t_{n-1})\nabla$$
$$\nabla_n = \nabla + \phi(t_1 + t_2 + \dots t_{n-1})\nabla$$

and (on Y)

$$f_n = f + f(t_1 + t_2 + \dots + t_{n-1})\phi$$
$$\phi_n = \phi + \phi(t_1 + t_2 + \dots t_{n-1})\phi.$$

If the limit exists, new SDR–data

$$((X, \partial_\infty) \underset{f_\infty}{\overset{\nabla_\infty}{\rightleftarrows}} (Y, d+t), \phi_\infty)).$$

is obtained.

Remark: There are situations in which the limit does not exist, but there are many classes of examples for which it does. One criterion that is often used is the existence of a filtration for which $t\phi$ can be shown to be nilpotent in any finite degree. There are however other criteria ([**Gugenheim**], [**Barnes-Lambe**]).

To apply this algorithm to the case of "perturbing" a given resolution as mentioned above, the following method can be attempted.

(3.1.2) Find an SDR of X into the bar construction (see below) of A

$$(A \otimes X \underset{f}{\overset{\nabla}{\rightleftarrows}} B(A), \phi)$$

(3.1.3) Use the additive isomorphism $B(A) \cong B(A')$ to "transfer" the differential of $B(A')$ down to $A \otimes X'$ using the formulae in (3.1.1).

This method has been discussed in some detail in [**Lambe2**] and [**Lambe3**].

3.2. Application to Group Cohomology

Consider the following class of groups G_q and H from [**Lambe2**]. For a fixed integer q, let

$$G_q = \left\{ \begin{pmatrix} 1 & x & q^{-1}z \\ 0 & 1 & y \\ 0 & 0 & 1 \end{pmatrix} \mid x, y, x \in Z \right\}$$

and give G_q the operation of matrix multiplication. This is an example of a polynomial group law on $\mathbb{Z} \times \mathbb{Z} \times \mathbb{Z} = \mathbb{Z}^3$.

Let

$$H = \mathcal{G} \times \mathbb{Z}$$

where $\mathbb{Z}$ is the additive group of integers and $\mathcal{G} = \mathbb{Z} + i\mathbb{Z}$ is the additive group of Gaussian integers. The multiplication is given as follows. Let $g_1 = x_1 + x_2 i$, $g_2 = y_1 + y_2 i$.

$$(g_1, x_3)(g_2, y_3) = (g_1 + i^{x_3} g_2, x_3 + y_3).$$

Note that

$$i^{x_3} g_2 = y_1 s_1 - \frac{x_3 y_2 \pi}{2} s_2 + (y_2 s_1 + \frac{x_3 y_1 \pi}{2} s_2) i$$

where

$$s_1 = \sum \frac{(-1)^k (x_3^2 \pi^2)^k}{4^k (2k)!} \quad s_2 = \sum \frac{(-1)^k (x_3^2 \pi^2)^k}{4^k (2k+1)!}.$$

Thus, here is an example of a convergent power series group law on $\mathbb{Z}^3$.

$$(x_1, x_2, x_3)(y_1, y_2, y_3) =$$

$$(x_1 + y_1, x_2 + y_2, x_3 + y_3) + (\frac{-x_3 y_2 \pi}{2} + (\frac{x_3 y_1 \pi}{2}) i, 0) + O(\geq 3).$$

More generally, think of groups K with underlying set $\mathbb{Z}^n$ with group law either a polynomial or a convergent power series In any case, think of the group ring $\mathbb{Z}(K)$ as a perturbation of the ring of finite Laurent polynomials

$$A = Z[t_n^{-1}, \ldots, t_1^{-1}, t_1, \ldots, t_n]$$

which is itself the group ring of the free abelian group $(\mathbb{Z}^n, +, 0)$. In the case of the family G_q above, since there is clearly an isomorphism with the family of operations ρ_q on $\mathbb{Z}^3$ given by

$$(x_1, x_2, x_3)(y_1, y_2, y_3) = (x_1 + y_1, x_2 + y_2, x_3 + y_3 + q x_1 y_2),$$

one sees that

$$\lim_{q \to 0} G_q \cong \mathbb{Z}^3$$

and $\mathbb{Z}(G)$ is actually a deformation of the ring of Laurent polynomials [**Lambe3**].

The homology of a group G is the homology of the complex

$$\mathbb{Z} \otimes_{\mathbb{Z}(G)} B(\mathbb{Z}(G))$$

where it is recalled that the bar construction resolution $B(A)$ of an algebra A is a free A-module and resolution $B(A) \longrightarrow \mathbb{Z} \to 0$. It is given by

$$B(A) = A \otimes \bar{B}(A), \quad \bar{B}(A) = \sum_{n=0}^{\infty} \bar{B}_n(A)$$

$$\bar{B}_n(A) = \otimes^n \bar{A}$$

where it is assumed that A is augmented with unit

$$0 \to R \xrightarrow{\sigma} A \xrightarrow{\epsilon} R \to 0$$

and $\bar{A} = A/\sigma(R)$. The differential in $B(A)$ is given by

$$\partial[b_1 | \ldots | b_n] = [b_2 | \ldots | b_n] + \sum \pm [b_1 | \ldots | b_i b_{i+1} | \ldots | b_n]$$
$$\pm [b_1 | \ldots | b_{n-1}].$$

See [**MacLane**] for details about the bar construction.

To compute the necessary formulae in the case of the perturbations of the abelian group law above, one requires the following objects

i.) The abelian group $\mathbb{Z}^n$.

ii.) The group $G = (\mathbb{Z}^n, \rho)$, $\rho(x,y) = x + y + \wp(x,y)$.

iii.) The group ring A of $\mathbb{Z}^n$ (Laurent polynomials).

iv.) The group ring of G, a perturbation of the Laurent polynomials.

v.) The bar constructions $B(A)$ and $B(\mathbb{Z}(G))$.

vi.) An SDR $(A \otimes E[u_1, \ldots, u_n] \underset{f}{\overset{\nabla}{\rightleftarrows}} B(A), \phi)$ (see [**Lambe2**], [**Lambe3**] and the references given there), and

vii.) The additive isomorphism $B(\mathbb{Z}(G)) \cong B(A)$.

These objects and the calculations necessary for (3.1.1) have been implemented and carried out in AXIOM. To describe what it looks like, suppose that there are domain constructors

FreeAb(n), (the free abelian group of rank n),

CEMTKRes(n), (Cartan-Eilenberg-MacLane-Tate-Koszul Resolution),

BarCons(R, G), (the bar construction of $R(G)$),

SdrFreeAb(n), (the strong deformation retraction mentioned above),

and

LagRes$(n, xVar, yVar, p)$, (a package, LinearAffineGroupResolution), which implements the additive isomorphism needed in (3.1.3).

The domain constructor *FreeAb*(n) represents the free abelian group on n generators $t_1, \ldots, t_n$ written multiplicatively. It has signatures like

$$generator : NNI -> \$$$

where $generator(i) = t_i$, etc. Let

$$A = Z[t_n^{-1}, \ldots, t_1^{-1}, t_1, \ldots, t_n]$$

be the group ring of the free abelian group on n generators. The group ring constructor exists in AXIOM, but so does the monoid ring and, for our purposes, the monoid ring suffices. Thus, take

$$A = MonoidRing(Integer, FreeAb(n)).$$

The domain constructor *CEMTKRes*(n) represents the resolution

$$(A \otimes E[u_1, \ldots, u_n], d, \varphi) \xrightarrow{\epsilon} \mathbb{Z} \to 0$$

where the differential d is given by

$$d(p) = 0, \quad d(u_i) = t_i - 1$$

for $p \in A$ (d is extended as a derivation of the whole algebra).

The augmentation ϵ is given by

$$\epsilon(t^i) = 1, \quad d(t^i u_I) = 0$$

where $u_I = u_{i_1} \ldots u_{i_k}$ for an index set $I = \{i_1, \ldots, i_k\}$ with $k \geq 1$. An explicit contracting homotopy φ exists, and in fact a general formula for it was derived through the use of AXIOM (see [**Lambe3**] and below). This was done through a sort of "bootstrap" process – recursive formulae were known, and by calculating enough examples a general formula was found. The recursive formulae are rather complex and unwieldy so that hand calculation is overly tedious. In fact, as can be seen from examples below, the closed formulae for φ are also rather complex and tedious to deal with by hand. One may still use a computer to deal with them but, if one has closed formulae (as opposed to recursive formulae), one can write routines that are tremendously more efficient than recursive forms. This is an important issue and we will surely see more and more of this sort of bootstrap process in many areas of application in the future.

The domain $CEMTKRes(n)$ has the signatures

$$\begin{aligned} &generator : NNI \rightarrow \$ \\ &diff : \$ \rightarrow \$ \\ &phi : \$ \rightarrow \$ \\ &terms : \$ \rightarrow List(\$) \\ &homogeneous? : \$ \rightarrow Boolean \end{aligned}$$

among others. They have the obvious meanings. The function $diff$ gives the differential d above, $terms$ gives a list of the terms in a given expression, and $homogeneous?$ checks to see if a given expression is of one degree (and not a linear combination of terms of mixed degrees).

The domain constructor $BarCons(R, G)$ has the expected signatures, including the standard contracting homotopy s.

The package constructor $SdrFreeAb(n)$ implements the SDR

$$(A \otimes E[u_1, \ldots, u_n] \underset{f}{\overset{\nabla}{\rightleftarrows}} B(A), \phi)$$

mentioned above. It has signatures

$$\begin{aligned} &inclusion : CEMTKRes(n) \rightarrow BarCons(Integer, FreeAb(n)) \\ &projection : BarCons(Integer, FreeAb(n)) \rightarrow CEMTKRes(n) \\ &homotopy : BarCons(Integer, FreeAb(n)) \rightarrow BarCons(Integer, FreeAb(n)) \end{aligned}$$

among others.

The package constructor $LagRes(dim, xVar, yVar, p)$ implements several auxiliary signatures necessary to carry out the steps of the perturbation method mentioned above. It has

$$barG2barZ : barG \rightarrow barZ$$
$$barZ2barG : barZ \rightarrow barG$$
$$basis : () \rightarrow List(CEMTKRes(n))$$
$$t : barZ \rightarrow barZ$$
$$thpi : barZ \rightarrow barZ$$

among other signatures, where

$$barZ = BarCons(Integer, FreeAb(n)),$$
$$barG = BarCons(Integer, G)$$

and G is the group whose polynomial group law is given by the polynomial p in arguments $xVar$ and $yVar$. The function $barG2barZ$ implements the additive isomorphism mentioned above in (3.2) vii.), and $barZ2barG$ is its inverse. The function $basis$ gives the list of basis (over A) elements for the exterior algebra

$$\{1, u_1, \ldots, u_n, u_1u_2, \ldots, u_1u_2 \ldots u_n\}$$

(2^n elements). It takes no arguments. The function t implements the difference of the abelian group law differerential and the "perturbed" group law differential needed for (3.1.1) and the function $thpi$ is the composite $t\phi$ mentioned in (3.1.1).

It should be pointed out that these domains and packages as well as some of the categories and domains on which they were built were not a part of the basic AXIOM system distributed by NAG. The author needed to implement all the necessary objects, but it was possible to do this in a very convenient manner following the ideas already outlined as well as those outlined in [**Lambe1**]. The complete source code is available from the author.

Assuming that we have all the above, here is an actual AXIOM input file which does the calculation for the group G of upper triangular 3×3 matrices in (3.2) at the value $q = 1$.

```
-- alternate names for some domains used
L := List
P := Polynomial
I := Integer

-- Set up for the Heisenberg Group
dim := 3
fab := FreeAb dim
cem := CEMTKRes dim
bar := BarCons(I, fab)
```

```
sdr := SdrFreeAb dim

xVar: L Symbol := [x1, x2, x3]
yVar: L Symbol := [y1, y2, y3]

-- Give the group law
p : L P I := [x1 + y1, x2 + y2, x3 + y3 + x1 * y2]

respkg    := LagRes(dim, xVar, yVar, p, inv)
extBasis  := basis()$respkg
image     := [inclusion(l)$sdr  for l in extBasis]
timage    := [t(l)$respkg for l in image]

extRank := 2**dim - 1

-- We now need to apply tphi iteratively
-- until each term vanishes, i.e., until
-- (tphi)**n(timage) = 0

lim: L bar := [ [ ] for i in 1..extRank]

for i in 1..extRank repeat
   tmp   := timage.i
   tmp   := tphi(tmp)$respkg
   lim.i := cons(tmp, lim.i)
   while not(tmp = 0) repeat
      tmp   := tphi(tmp)$respkg
      lim.i := cons(tmp,lim.i)

-- Now compute the projections
ftimage := [projection(l)$sdr for l in 1..extRank]

partials: L L cem := [ [ ] for i in 1..extRank]

for i in 1..extRank repeat
   partials.i := [proj(k)$sdr for k in lim.i]

-- Now the differentials
dcem   := [diff l for l in extBasis]
dinfty := [ dcem.i + ftimage.i + reduce(+, partials.i)
                     for i in 1..extRank]
```

A similar calculation can be set up and run for the power series group law for H above. The results of these calculations are quite orderly. For the power series group law we have obtained the resolution

$$\begin{aligned}
&(Z(G) \otimes E[u_1, u_2, , u_3], d) \\
d(u_i) &= t_i - 1 \\
d(u_1 u_2) &= (t_1 - 1)u_2 - (t_2 - 1)u_1 \\
d(u_1 u_3) &= (t_2 - 1)u_3 + u_2 + t_3 u_1 \\
d(u_2 u_3) &= -(t_1^{-1} - 1)u_3 - t_3 u_2 - t_1^{-1} u_1 \\
d(u_1 u_2 u_3) &= u_2 u_3 - (t_1^{-1} - 1)u_1 u_3 - (t_1 - t_3)u_1 u_2.
\end{aligned}$$

Clearly, the homology (or cohomology), for example, of the group H with $\mathbb{Z}$ coefficients can be easliy read from this.

§4. Abstractly Symbolic Domains of Computation

Given the power and convenience of a system like AXIOM, one is tempted to wonder if the type of calculation done above for an individual group can be achieved for a parameterized class of groups like the class G_q in (3.2) which is parameterized by q. At least a part of the issue in doing that involves some difficulties in manipulating the SDR data from (3.2) vi.) in a way that allows one to keep q "symbolic". To understand this, consider the case $n = 1$ of the SDR in (3.2) vi.).

$$(A \otimes E[u] \underset{f}{\overset{\nabla}{\rightleftarrows}} B(A), \phi)$$

where

$$A = \mathbb{Z}[t^{-1}, t].$$

The maps f and ϕ crucial to the perturbative method outlined above are completely determined by a contracting homotopy φ on $A \otimes E[u]$. In fact, this is true in much more generality (see [**Lambe3**] for details). The equation that determines a contracting homotopy is

$$d\phi + \phi d = 1.$$

In this one-dimensional case, it is not hard to solve this equation for φ given d and the augmentation already mentioned (for $CEMTKRes(n)$ in (3.2)). In fact, a solution is

$$\varphi(t^i) = \frac{t^i - 1}{t - 1} u, \quad \varphi(t^i u) = 0.$$

Now ∇, f, and ϕ are determined by the formulae

$$\nabla(u_I) = s\nabla(du_I), \quad f(\bar{b}) = sf(\partial \bar{b}), \quad \phi(\bar{b}) = -s(\nabla f(\bar{b}) + \phi\partial(\bar{b}))$$

defined on "reduced elements", i.e., elements of $E[u_1, \ldots, u_n]$ and $\bar{B}(A)$ and extended A-linearly over all their domains. In the one-dimensional case, it is easy to see that a formula for f is therefore

$$\begin{aligned}
f([t^i]) &= \frac{t^i - 1}{t - 1} u \\
f([t^{i_1} | \ldots | t^{i_k}]) &= 0, \quad \text{if } k > 1.
\end{aligned}$$

Now the higher-dimensional cases of this are also needed. For example, the 3-dimensional case is needed for the class G_q, In dealing with the formulae in (3.1.1) some functions need to be iterated over quantities involving the coefficient $\frac{t^i-1}{t-1}$. Of course, to evaluate these iterations, one can expand

$$\frac{t^i - 1}{t - 1} = \sum_{j=0}^{i-1} t^j,$$

but then one has to wonder about representing not only the symbolic exponent, but now also the symbolic limit on the summation! This is a serious issue, particularly because in the iterations necessary one does indeed encounter applications of φ to coefficients like $\frac{t^i-1}{t-1}$. One solution that has been quite successful is to encode the "symbolic summation" as

$$t^{\{i\}} = \frac{t^i - 1}{t - 1}$$

for i any integer. Thus,

$$f(t^i) = t^{\{i\}_1} u$$

and *by definition* we take

$$f(t^{\{i\}_1}) = t^{\{i\}_2} u$$

and so on to recursively define $t^{\{i\}_n}$. For the purposes of the perturbation calculations necessary in this class of problems, there is an "algebra of exponents" for such "symbolically exponentiated" quantities and, given the flexible extensibility of the AXIOM system, it has been possible to develop an actual domain of computation for such symbolic calculations. With these "symbolic domains", it has been possible to calculate closed formulae for the perturbation steps needed for the resolution calculations mentioned above. For example, a class of resolutions parameterized by q for the class of groups G_q parameterized by q may be derived. These 3-dimensional results are summarized below. See [**Lambe2**] and [**Lambe3**] for more complete results and details.

The SDR

$$(A \otimes E[u_1, u_2, u_3] \underset{f}{\overset{\nabla}{\rightleftarrows}} B(A), \phi)$$

for the group ring

$$A = Z[t_3^{-1}, t_2^{-1}, t_1^{-1}, t_1, t_2, t_3]$$

is given as follows. The differential d is

$$d(p) = 0$$
$$d(u_i) = t_i - 1.$$

The contracting homotopy φ on $A \otimes E[u_1, u_2, u_3]$ is given by

$$\begin{aligned}
\varphi(t^{i_1}t^{i_2}t^{i_3}) &= t^{\{i_1\}}u_1 + t_1^{i_1}t_2^{\{i_2\}}u_2 + t_1^{i_1}t_2^{i_2}t_3^{\{i_3\}}u_3 \\
\varphi(t^{i_1}t^{i_2}t^{i_3}u_1) &= t_1^{i_1}t_2^{\{i_2\}}u_2u_1 + t_1^{i_1}t_2^{i_2}t_3^{\{i_3\}}u_3u_1 \\
\varphi(t^{i_1}t^{i_2}t^{i_3}u_2) &= t_1^{i_1}t_2^{i_2}t_3^{\{i_3\}}u_3u_2 \\
\varphi(t^{i_1}t^{i_2}t^{i_3}u_3) &= 0 \\
\varphi(t^{i_1}t^{i_2}t^{i_3}u_1u_2) &= t_1^{i_1}t_2^{i_2}t_3^{\{i_3\}}u_1u_2u_3 \\
\varphi(t^{i_1}t^{i_2}t^{i_3}u_1u_3) &= 0 \\
\varphi(t^{i_1}t^{i_2}t^{i_3}u_2u_3) &= 0 \\
\varphi(t^{i_1}t^{i_2}t^{i_3}u_1u_2u_3) &= 0.
\end{aligned}$$

The inclusion, (related to the well-known "shuffle product"), is

$$\begin{aligned}
\nabla(u_i) &= [t_i] \\
\nabla(u_iu_j) &= [t_i|t_j] - [t_j|t_i] \\
\nabla(u_1u_2u_3) &= [t_1|t_2|t_3] - [t_1|t_3|t_2] - [t_2|t_1|t_3] \\
&\quad + [t_2|t_3|t_1] + [t_3|t_1|t_2] - [t_3|t_2|t_1].
\end{aligned}$$

The projection in one degree is

$$f([t^{i_1}t^{i_2}t^{i_3}]) = t_1^{i_1}t_2^{i_2}t_3^{\{i_3\}}u_3 + t_1^{i_1}t_2^{\{i_2\}}u_2 + t_1^{\{i_1\}}u_1.$$

The other degrees for the projection as well as a formula for the homotopy will be omitted here since they are given in the references already mentioned. In this dimension it is not overly difficult, but still tedious, to compute these formulae by hand. This case and higher dimensions were successfully (and conveniently) computed entirely in AXIOM.

Consider the class of groups

$$G_q = \left\{ \begin{pmatrix} 1 & x & q^{-1}z \\ 0 & 1 & y \\ 0 & 0 & 1 \end{pmatrix} \mid x, y, z \in \mathbb{Z} \right\}.$$

once again. A resolution of $\mathbb{Z}$ over $\mathbb{Z}(G_q)$ is given by the complex (computed entirely in AXIOM)

$$(\mathbb{Z}(G_q) \otimes E[u_1, u_2, u_3], d)$$

where

$$\begin{aligned}
d(u_i) &= t_i - 1 \\
d(u_1u_2) &= (-t_1t_2t_3^{\{q\}})u_3 + (t_1 - 1)u_2 - (t_2 - 1)u_1 \\
d(u_1u_3) &= (t_1 - 1)u_3 - (t_3 - 1)u_1 \\
d(u_2u_3) &= t_2 - 1)u_3 - (t_3 - 1)u_2 \\
d(u_1u_2u_3) &= (t_1 - 1)u_2u_3 - (t_2 - 1)u_1u_3 + (t_3 - 1)u_1u_2.
\end{aligned}$$

From this it is easy, for example, to see that the second homology group of G_q with coefficients in the integers is

$$H^2(G_q;\mathbb{Z}) \cong \mathbb{Z} \oplus \mathbb{Z} \oplus \mathbb{Z}/q$$

for any q.

This calculation was achieved through the use of several symbolic domains of computation. First, there is *SymbSdrFreeAb* with signatures

$$pow : (\$, ExpressionInteger) \to \$$$
$$sExp : (ExpressionInteger, NonNegativeInteger) \to \$$$

where, given any symbolic group element $g = t_1^{i_1} \dots t_n^{i_n}$ and any expression e (with integer coefficients), the operation $pow(g,e)$ returns $t_1^{i_1 e} \dots t_n^{i_n e}$ and given an expression e, and a non-negative integer i such that $1 \leq i \leq n$, $sExp(e,i) = t_i^{\{e\}}$ (iterated brackets $t_i^{\{e\}_k}$ are handled by simply iterating brackets – they are counted up on output and a sensible output form is chosen). Of course, this domain also has all the usual operations as well. Next, there is *SymbCEMTKRes*. This domain has all the same signatures as its counterpart *CEMTKRes*, except things have been rearranged internally so that if one calls the contracting homotopy *phi* on a "symbolic group ring element" t_j^i, then the correct *symbolic expression* involving $t_j^{\{i\}}$ is returned. There are also *SymbBarCons*, and *SymbSdrFreeAb*. Finally, there is the symbolic domain constructor

$$SymbGroup(dim, groupLaw)$$

where

$$dim : NonNegativeInteger$$
$$groupLaw : (LEI, LEI) \to LEI$$

and $LEI := List\ Expression\ Integer$. It contains signatures

$$symbDiff : SBAR \to SBAR$$
$$t : SBAR \to SBAR$$

among others, where $SBAR$ is the symbolic bar construction domain for this (symbolically represented) group. The coefficient ring for $SBAR$ is *ExpressionInteger*. As before, t is the function which gives the difference between the bar construction differential for this group and the (symbolic) free abelian group.

With these domains in place, the above output for the class G_q is quite easily computed using essentially the same input as in (3.2). One needs to change the first few lines to

```
-- Set up for general groups of rank 3
--
dim := 3
```

```
fab := SymbFreeAb dim
cem := SymbCEMTKRes dim
bar := SymbBarCons(Expression Integer, fab)
sdr := SymbSdrFreeAb dim
```

Finally, it should be pointed out that the perturbation method does not just give rise to resolutions that can be used to compute *Ext* and *Tor* abstractly, it actually gives a strong comparison with the bar construction, i.e., this comparison is actually a part of what gets computed. In the case of the types of group rings presented here (polynomial and convergent power series group laws), it actually *computes* an SDR of the new small resolution in the bar construction. From this, one may conveniently read off bar construction cycles that represent homology classes of the group. In cases where the group is actually a $K(\pi,1)$-manifold, as is the case for the classes mentioned here, this calculation then actually leads to an explicit cell structure for the manifold (by realizing the simplicial version of the bar construction). We will not go into details here, but will simply end with an example that arose in some conversations with Ronnie Brown and Graham Ellis at the University of Wales, Bangor, UK recently. It concerns the group $F_{3,2}$ which is the free group F_3 on three generators modulo the third term of its lower central series (thus $F_{3,2}$ is the free-nilpotent 3-generated 2-step nilpotent group).

First the complex (just in degrees 3 and 4) for computing the homology is given. It was derived from a small resolution of the integers over the group ring (obtained by homological perturbation theory). In fact, the complex is so small and sparse, that one can compute the homology easily by sight. Here is the complex:

$$\bar{d}(u_1\ u_2\ u_3) = u_5\ u_6 + u_4\ u_6 + u_4\ u_5 + u_3\ u_4 - u_2\ u_5 + u_1\ u_6$$
$$\bar{d}(u_1\ u_2\ u_5) = -u_4\ u_5$$
$$\bar{d}(u_1\ u_2\ u_6) = -u_4\ u_6$$
$$\bar{d}(u_1\ u_3\ u_4) = +u_4\ u_5$$
$$\bar{d}(u_1\ u_3\ u_6) = -u_5\ u_6$$
$$\bar{d}(u_2\ u_3\ u_4) = +u_4\ u_6$$
$$\bar{d}(u_2\ u_3\ u_5) = +u_5\ u_6$$

$$\bar{d}(u_1\ u_2\ u_3\ u_4) = +u_4\ u_5\ u_6 + u_2\ u_4\ u_5 - u_1\ u_4\ u_6$$
$$\bar{d}(u_1\ u_2\ u_3\ u_5) = -u_4\ u_5\ u_6 + u_3\ u_4\ u_5 - u_1\ u_5\ u_6$$
$$\bar{d}(u_1\ u_2\ u_3\ u_6) = +u_4\ u_5\ u_6 + u_3\ u_4\ u_6 - u_2\ u_5\ u_6$$
$$\bar{d}(u_1\ u_2\ u_5\ u_6) = -u_4\ u_5\ u_6$$
$$\bar{d}(u_1\ u_3\ u_4\ u_6) = +u_4\ u_5\ u_6$$
$$\bar{d}(u_2\ u_3\ u_4\ u_5) = -u_4\ u_5\ u_6$$

Clearly, the homology in dimension three has generators

$$u_1u_2u_4, u_1u_2u_5 + u_1u_3u_4, u_1u_2u_6 + u_2u_3u_4, u_1u_3u_5,$$

$$u_1u_3u_6 + u_2u_3u_5, u_1u_4u_5, u_1u_4u_6, u_1u_5u_6, u_2u_3u_6, u_2u_4u_6, u_2u_5u_6, u_3u_5u_6$$

Thus the rank of H^3 is 12. Cycle representatives in the bar construction for all the generators of all the homology have been computed in AXIOM using homological perturbation. They are obtained by computing the image of the map ∇_∞ from (3.1.1). Only the image of the cycle $u_1u_2u_5 + u_1u_3u_4$ will be given here. Complete results for several other classes of groups and algebras have also been obtained.

$$\nabla_\infty(u_1u_2u_5 + u_1u_3u_4) =$$

$$[{t_1}^1|{t_2}^1|{t_5}^1] - [{t_1}^1|{t_5}^1|{t_2}^1] - [{t_2}^1|{t_1}^1|{t_5}^1] + [{t_2}^1|{t_5}^1|{t_1}^1] + [{t_5}^1|{t_1}^1|{t_2}^1] - [{t_5}^1|{t_2}^1|{t_1}^1]$$

$$+$$

$$-[{t_1}^1\ {t_2}^1|{t_4}^1|{t_5}^1] + [{t_1}^1\ {t_2}^1|{t_5}^1|{t_4}^1] - [{t_5}^1|{t_1}^1\ {t_2}^1|{t_4}^1]$$

$$+$$

$$[{t_1}^1|{t_3}^1|{t_4}^1] - [{t_1}^1|{t_4}^1|{t_3}^1] - [{t_3}^1|{t_1}^1|{t_4}^1] + [{t_3}^1|{t_4}^1|{t_1}^1] + [{t_4}^1|{t_1}^1|{t_3}^1] - [{t_4}^1|{t_3}^1|{t_1}^1]$$

$$+$$

$$[{t_1}^1\ {t_3}^1|{t_4}^1|{t_5}^1] - [{t_1}^1\ {t_3}^1|{t_5}^1|{t_4}^1] - [{t_4}^1|{t_1}^1\ {t_3}^1|{t_5}^1]$$

Current address: DIMACS, Rutgers University
CoRE Bldg., Frelinghuysen Rd.
Piscataway, NJ 08855-1179
llambe@cesl.rutgers.edu

References

[Barnes-Lambe] Barnes, D., and Lambe, L., *Fixed point approach to homological perturbation theory*, Proc. Amer. Math. Soc., 112(1991), 881-892.

[Brown] Brown, R., *The twisted Eilenberg-Zilber theorem*, Celebrazioni Archimedee del secolo XX, Simposio di topologia 34-37(1967).

[Gugenheim] Gugenheim, V.K.A.M., *On the chain complex of a fibration*, IL. J. Math. 3(1972), 398-414.

[Hübschmann] Hübschmann, J., *The homotopy type of $F\Psi^q$, the complex and symplectic cases*, Cont. Math. 55(1986), 487-518.

[Jenks] Jenks, R., *MODLISP: An Introduction*, Springer Lecture Notes in Comp. Sci., vol. 72, 466-480.

[Jenks-Sutor] Jenks, R., and Sutor, R., **AXIOM**, Springer Verlag, NY, 1992.

[Lambe1] Lambe, L., *Scratchpad II as a tool for mathematical research*, Jon Barwise's column in Notices of the Amer. Math. Soc., February, 1989.

[Lambe2] Lambe, L., *Homological perturbation theory, Hochschild homology and formal groups*, Proc. Conference on Deformation Theory and Quantization with Applications to Physics Amherst, MA June 1990, Cont. Math., AMS (to appear).

[Lambe3], Lambe, L., *Resolutions that split off of the bar construction*, J. Pure & Appl. Alg., (to appear).

[Lambe-Stasheff] Lambe, L., and Stasheff, J., *Applications of perturbation theory to iterated fibrations*, Manuscripta Math., 58(1987), 363-376.

[MacLane] MacLane, S., **Homology**, Die Grundlehren der Math. Wissenschaften, Band 114, Springer Verlag, NY, 1967.

[Shih] Shih, W., *Homology des espaces fibrés*, Inst. des Hautes Études Sci., vol. 13 (1962), pp. 93-176.

A Note on the Nonlinear Rayleigh Quotient

Peter Lindqvist

Department of Physics and Mathematics
Norwegian Institute of Technology
N-7034 Trondheim, NORWAY

Abstract

The "power concavity" of the first eigenfunction to the equation $div(|\nabla u|^{p-2}\nabla u) + \lambda |u|^{p-2}u = 0$ is studied in a ball. An introduction to this non-linear eigenvalue problem is given. A conjecture is made, in particular, for the equation $\Delta u + \lambda u = 0$.

1 Introduction

If u is a solution to the eigenvalue problem

$$div(|\nabla u|^{p-2}\nabla u) + \lambda |u|^{p-2}u = 0 \tag{1.1}$$

in the domain Ω in $\boldsymbol{R}^n$, then, at least formally,

$$v(x,t) = u(x)|a\cos(\sqrt{\lambda}t) + b\sin(\sqrt{\lambda}t)|^{\frac{2-p}{p-1}}(a\cos(\sqrt{\lambda}t) + b\sin(\sqrt{\lambda}t))$$

satisfies the non-linear wave equation

$$\frac{\partial^2 |v|^{p-2}v}{\partial t^2} = div(|\nabla v|^{p-2}\nabla v)$$

and

$$v(x,t) = u(x)\, e^{-\lambda t/(p-1)}$$

is a solution to the non-linear evolution equation

$$\frac{\partial |v|^{p-2}v}{\partial t} = div(|\nabla v|^{p-2}\nabla v).$$

Acknowledgement: At the 21st Nordic Congress of Mathematicians 1992 the address was Institute of Mathematics, Helsinki University of Technology, SF-02150 Esbo, Finland

Unfortunately, such solutions cannot be superposed, except in the linear case $p = 2$, when the above equations reduce to $\Delta u + \lambda u = 0$, $v_{tt} = \Delta v$, and $v_t = \Delta v$.

The so-called p-harmonic operator $div(|\nabla u|^{p-2}\nabla u)$ appears in many contexts in physics: non-Newtonian fluids (dilatant fluids have $p > 2$, pseudoplastics have $1 < p < 2$), reaction-diffusion problems, non-linear elasticity (torsional creep), and glaceology ($p = \frac{4}{3}$), just to mention a few applications.

The first eigenvalue $\lambda_p = \lambda_p(\Omega)$ of the p-harmonic operator $div(|\nabla u|^{p-2}\nabla u)$ is here defined as the least number λ for which the equation (1.1) has a non-trivial solution u with zero boundary data in a given bounded domain Ω in the n-dimensional Euclidean space. The first eigenvalue is the minimum of the Rayleigh quotient

$$\lambda_p = \min_u \frac{\int_\Omega |\nabla u|^p \, dx}{\int_\Omega |u|^p \, dx}. \tag{1.2}$$

Here $1 < p < \infty$, and in the linear case $p = 2$ one obtains the principal frequency, cf. [PS]. We shall often use the term *principal frequency* for the non-linear cases as well. The differential equation (1.1) is interpreted in the weak sense and all functions in the Sobolev space $W_0^{1,p}(\Omega)$ are admissible in (1.2). See Definition 2.1.

The solutions to Eqn (1.1) share many properties with solutions to more general quasi-linear eigenvaluc problems. But here we would like to emphasize the following specific features:

I *Uniqueness.* The first eigenfunctions are essentially unique in any bounded domain: they are merely constant multiples of each other. Moreover, they have no zeros in the domain and they are the only eigenfunctions not changing signs.

II *Stability.* For any bounded regular domain $\lim_{s\to p} \lambda_s = \lambda_p$ [L1, Theorem 6.1].

III *Concavity.* For any bounded convex domain $\log u_p$ is concave, u_p denoting a positive eigenfunction [S, Theorem 1].

IV *"Isoperimetric" property.* Among all domains with the same volume (area) the ball (the disc) has the smallest λ_p.

The uniqueness (I) for *arbitrary* bounded domains was apparently first proved in [L2]. The radial case has been studied by F. de Thélin in [Th2] and a good reference for C^2-domains is [S, Theorem A.1]. The logarithmic concavity mentioned in (III) is due to S. Sakaguchi [S], when $p \neq 2$, and the linear case is credited to H. Brascamp & E. Lieb. Property (IV) follows by spherical symmetrization (Schwarz symmetrization), cf. [Ka, p. 90]. For $p = 2$ this is the celebrated conjecture of Lord Rayleigh, proved by E. Krahn and G. Faber.

Because of property (IV), the ball is certainly important. Indeed, the radially symmetric case has been studied in [Th1], [Th2], [KL], and [B]. The purpose of this note is to *study* *a concavity question for the first eigenfunction u_p in a ball.* By Sakaguchi's result $\log |u_p|$ is concave for any convex domain. On the other hand $|u_p|$ itself is never concave, the one-dimensional case being an exception [Ko, Remark 2.7]. However, for a ball much more than logarithmic concavity holds, e.g., $\sqrt[n]{|u_p|}$ is concave (Lemma 3.8). We shall construct an exponent $\alpha = \alpha(n,p) > 1/n$ such that $|u_p|^\alpha$ is concave (Theorem 3.9). In the linear case $\Delta u + \lambda u = 0$ numerical calculations with explicitly known solutions show that our construction yields nearly the right power concavity exponent (Remark 3.12).

All this suggests together with some other evidence that among all (convex) domains the ball might have the greatest power concavity exponent. The greater the exponent α is for which u^α, $u > 0$, is concave, the better the concavity is, cf. [Ke, Property 2, p. 689]. The case $\alpha = 0$ should here mean that $\log u$ is concave. In conclusion, to each convex domain there corresponds a largest concavity exponent α, depending only on n, p, and the domain itself: $|u_p|^\alpha$ is concave. Moreover, $0 \le \alpha < 1$ (in the one dimensional case $\alpha = 1$). We cannot resist making a conjecture.

1.3 Conjecture *Among all convex domains the ball has the best power concavity exponent.*

We do not have a proof, not even in the case $\Delta u + \lambda u = 0$, but it seems likely that suitable symmetrization procedures related to level sets are relevant.

Let us finally mention that the best constant in the Fiedrichs-Poincaré inequality

$$\int_\Omega |\phi|^p \, dx \le C \int_\Omega |\nabla \phi|^p \, dx$$

is the reciprocal of the principal frequency: $C = 1/\lambda_p$. Here $\phi \in C_0^\infty(\Omega)$, Ω being a bounded domain. See [GT, Eqn (7.44), p. 164].

2 Preliminaries

In defining the eigenvalues λ for the p-harmonic operator in a given bounded domain $\Omega \subset \boldsymbol{R}^n$ we shall interpret Eqn (1.1) in the weak sense:

2.1. Definition *We say that λ is an eigenvalue, if there exists a continuous function $u \in W_0^{1,p}(\Omega)$, $u \not\equiv 0$, such that*

$$\int_\Omega |\nabla u|^{p-2} \nabla u \cdot \nabla \eta \, dx = \lambda \int_\Omega |u|^{p-2} u \eta \, dx \tag{2.2}$$

whenever $\eta \in C_0^\infty(\Omega)$. The function u is called an eigenfunction.

The Sobolev space $W_0^{1,p}(\Omega)$ is the completion of $C_0^\infty(\Omega)$ with respect to the norm

$$||\eta|| = \left\{ \int_\Omega (|\eta|^p + |\nabla \eta|^p) \, dx \right\}^{1/p}.$$

See [Z]. The continuity of u is a redundant requirement in the definition: the weak solutions of (2.2) can be made continuous after a redefinition in a set of measure zero. This is standard elliptic regularity theory. One even has $u \in C^{1,\alpha}_{loc}(\Omega)$ for some $\alpha > 0$, cf. [DB] and [To], but this Hölder continuity of the *gradient* is a deep result.

The eigenvalues are positive and the least of them, say λ_p, is obtained as the minimum of the Rayleigh quotient

$$\lambda_p = \inf_v \frac{||\nabla v||^p_{L^p(\Omega)}}{||v||^p_{L^p(\Omega)}} = \inf_v \frac{\int_\Omega |\nabla v|^p\,dx}{\int_\Omega |v|^p\,dx} \tag{2.3}$$

the infimum being taken among all $v \in W^{1,p}_0(\Omega)$, $v \not\equiv 0$. Alternatively, one can further restrict the class of admissible functions to $C^\infty_0(\Omega)$. This minimization problem is equivalent to Eqn (2.2) with $\lambda = \lambda_p$.

To any bounded domain Ω there is a first eigenfunction $u_p > 0$ corresponding to the least eigenvalue. The existence is standard Calculus of Variations, cf. [Th2], and the strict positivity follows from the Harnack inequality [Tr, Theorem 1.1] applied to the non-negative minimizing function $|u_p|$.

In very irregular domains the boundary values are not attained in the classical sense, when $p \le n$, but one always has $u_p \in W^{1,p}_0(\Omega)$.

Note that, if $\Omega_1 \subset \Omega_2$, then we have $\lambda_p(\Omega_1) \ge \lambda_p(\Omega_2)$ for the principal frequencies. This can be read off from the Rayleigh quotient.

3 Power concavity in the radial case

If the domain is a ball, the eigenfunctions are radial by [Th2], [KL] and [B] and so the symmetry reduces the problem to an ordinary differential equation.

In the unit ball $|x| < 1$ the differential equation for the radially symmetric eigenfunction $u = u(r)$ is

$$\frac{d}{dr}\left(r^{n-1}|\dot{u}|^{p-2}\dot{u}\right) + \lambda_p r^{n-1}|u|^{p-2}u = 0 \tag{3.1}$$

where $0 \le r \le 1$, $\dot{u}(0) = 0$, and $u(1) = 0$. Observe that $u = u(r)$ is smooth, when $0 < r < 1$. Multiplying the solution u by a suitable constant we may require the normalization $\dot{u}(1) = -1$. From now on u (or occasionally u_p) will always denote the *normalized first eigenfunction.* For $p = 2$ we have the well-known Bessel equation $\ddot{u} + (n-1)\dot{u}/r + \lambda_2 u = 0$. The solution $u_2(r)$ is a normalizing constant times

$$(r\sqrt{\lambda_2})^{(n-2)/2} J_{(n-2)/2}(r\sqrt{\lambda_2}) \tag{3.2}$$

and $\sqrt{\lambda_2}$ is the first positive zero of the Bessel function $J_{(n-2)/2}$.

Let us start by noting two useful identities:

$$r^{n-1}|\dot{u}|^{p-2}\dot{u}u = \int_0^r |\dot{u}|^p t^{n-1}\,dt - \lambda_p \int_0^r u^p t^{n-1}\,dt \tag{3.3}$$

and

$$(p-1)r^n|\dot{u}|^p = -\lambda_p r^n u^p + \lambda_p n \int_0^r u^p t^{n-1}\,dt + (p-n)\int_0^r |\dot{u}|^p t^{n-1}\,dt \tag{3.4}$$

for the normalized $u = u_p$. The first formula follows immediately by integration by parts of the differential equation (3.1) multiplied by u. To obtain the second one, we eliminate the term $r^n|\dot{u}|^{p-2}\dot{u}\ddot{u}$ from the expressions

$$\begin{aligned}
\frac{d}{dr}(r^n|\dot{u}|^p) &= \dot{u}r\frac{d}{dr}(r^{n-1}|\dot{u}|^{p-2}\dot{u}) + r^{n-1}|\dot{u}|^p + r^n|\dot{u}|^{p-2}\dot{u}\ddot{u},\\
\frac{d}{dr}(r^n|\dot{u}|^p) &= pr^n|\dot{u}|^{p-2}\dot{u}\ddot{u} + nr^{n-1}|\dot{u}|^p
\end{aligned}$$

and then we integrate the resulting identity

$$\begin{aligned}
(p-1)\frac{d}{dr}(r^n|\dot{u}|^p) &= p\dot{u}r\frac{d}{dr}(r^{n-1}|\dot{u}|^{p-2}\dot{u}) + (p-n)r^{n-1}|\dot{u}|^p\\
&= -\lambda_p pr^{n-1}u^{p-1}\dot{u} + (p-n)r^{n-1}|\dot{u}|^p\\
&= -\lambda_p r^n\frac{d}{dr}u^p + (p-n)r^{n-1}|\dot{u}|^p.
\end{aligned}$$

Here the differential equation was used. (Needless to say, almost any calculations would sooner or later lead to the same result, and the derivation above is included just because it is short and direct.)

Let us illustrate the meaning of these formulae by first proving Sakaguchi's theorem in the radial special case. Our proof is elementary.

3.5 Proposition *If* u_p *is the first (positive) eigenfunction in a ball, then* $\log u_p$ *is concave.*

Proof: Inserting the expressions

$$v = \log u, \quad \dot{v} = \frac{\dot{u}}{u}, \quad \ddot{v} = \frac{\ddot{u}}{u} - \frac{\dot{u}^2}{u^2}$$

in the equation

$$|\dot{u}|^{p-2}\left\{(p-1)\ddot{u} + \frac{n-1}{r}\dot{u}\right\} + \lambda_p|u|^{p-2}u = 0$$

we arrive at

$$-(p-1)r^n u^2|\dot{u}|^{p-2}\ddot{v} = (p-1)r^n|\dot{u}|^p + (n-1)r^{n-1}|\dot{u}|^{p-2}\dot{u}u + \lambda_p r^n u^p.$$

Using (3.3) and (3.4) we obtain the simple identity

$$-r^n|\dot{u}|^{p-2}u^2\ddot{v} = \int_0^r \left[|\dot{u}|^p + \frac{\lambda_p}{p-1}u^p\right] t^{n-1}\,dt. \tag{3.6}$$

(This is a kind of energy integral.) Clearly $\ddot{v} \leq 0$ and the inequality is strict, when $0 < r < 1$. This proves the desired concavity for v.

Returning to the question of power concavity, we find that similar calculations for the function

$$\phi = u^\alpha \qquad (0 < \alpha < 1)$$

yield

$$\begin{aligned} -(p-1)r^n|\dot{u}|^{p-2}u^2\ddot{\phi} &= [p-1+\alpha(n-p)]\int_0^r |\dot{u}|^p t^{n-1}\,dt \\ &+ \lambda_p(1-n\alpha)\int_0^r u^p t^{n-1}\,dt + \alpha\lambda_p r^n u^p. \end{aligned} \tag{3.7}$$

The factor $p-1+\alpha(n-p)$ of the first integral is always positive. In establishing that $\ddot{\phi} \leq 0$, the factor $1-n\alpha$ in front of the second integral causes some difficulties, if $\alpha > 1/n$. At this stage we had better read off a miscellaneous result.

3.8 Lemma *The function $\sqrt[n]{u_p}$ is concave for a ball in $\boldsymbol{R}^n$.*

The value $1/n$ of the power concavity exponent can be considerably improved. (However, the right order is $0(\frac{1}{n})$).

3.9 Theorem *Suppose that u_p is the positive eigenfunction for a ball. Then there is an exponent $\alpha = \alpha(n,p)$ in the interval $]1/n, 1[$ such that u_p^α is concave. The exponent*

$$\alpha = \frac{pq\varepsilon - p + 1}{pq\varepsilon - p + n} \qquad \left(\frac{1}{p} + \frac{1}{q} = 1\right) \tag{3.10}$$

will do where ε is the greater positive root of the equation

$$\varepsilon^p - \frac{np}{n-1}\varepsilon + \frac{n(p-1)^2}{p(n-1)} = 0. \tag{3.11}$$

3.12 Remarks 1° Clearly $\frac{n}{n-1} < \varepsilon^{p-1} < \frac{np}{n-1}$. The function $f(\varepsilon) = \varepsilon^p - np\varepsilon/(n-1) + n(p-1)^2/p(n-1)$ decreases strictly in $[0, (n/(n-1))^{1/(p-1)}]$ from the value $n(p-1)^2/p(n-1)$ to the negative minimum

$$\frac{n(p-1)}{n-1}\left\{1 - \frac{1}{p} - \left(1 + \frac{1}{n-1}\right)^{1/(p-1)}\right\}.$$

For $\varepsilon \geq (n/(n-1))^{1/(p-1)}$ the function increases strictly. Thus Eqn (3.11) has two positive roots.

2° The obtained theoretical power concavity exponent α in (3.10) is not a mere artifact of the method. It is surprisingly accurate in the linear case $p = 2$. For example, in the two dimensional case the solution is $u(r) = J_0(r\sqrt{\lambda})$, $\sqrt{\lambda} = 2,4048255577....$ The best power concavity exponent for this Bessel function is $0,9457984....$ Our recipe gives $0,9330127....$ The difference is only $0,0127857....$ On the other hand Lemma 3.8 yields the comparatively poor result $0,5$.

Proof of Theorem 3.9. Proceeding from (3.7) we have to determine those values of α for which the right hand member is positive. Since

$$\int_0^r |\dot{u}|^p t^{n-1}\,dt = \lambda_p \int_0^r u^p t^{n-1}\,dt - \lambda_p u(r) \int_0^r u^{p-1} t^{n-1}\,dt$$

according to the differential equation, we can state the problem in a form not containing the derivative $\dot{u}$. The function $\phi = u^\alpha$ is concave precisely for those values of α that satisfy the inequality

$$p(1-\alpha)\int_0^r u^p t^{n-1}\,dt + \alpha r^n u^p(r) > [p-1+\alpha(n-p)]u(r)\int_0^r u^{p-1}t^{n-1}\,dt \tag{3.13}$$

when $0 < r < 1$. Fortunately, neither λ_p nor $\dot{u}$ do occur explicitly in (3.13). It is clear that the inequality cannot hold for $\alpha = 1$.

To improve α from $1/n$ it is essential to use the information hidden in the small term $\alpha r^n u^p(r)$ as $r \approx 1$. (The crucial value is $r \approx 0,8852$ for the case $p = n = 2$ mentioned in Remark 3.12.) To this end, use Young's inequality

$$ab \leq \varepsilon^{p-1}\frac{a^p}{p} + \varepsilon^{-1}\frac{b^q}{q} \qquad \left(\frac{1}{p} + \frac{1}{q} = 1\right)$$

to obtain $u(r)t^{n-1}u(t)^{p-1} < \varepsilon^{p-1}u(r)^p t^{n-1}/p + u(t)^p t^{n-1}/q\varepsilon$. By integration with respect to t we obtain

$$u(r)\int_0^r u^{p-1}t^{n-1}\,dt \leq \frac{\varepsilon^{p-1}u(r)^p r^n}{pn} + \frac{1}{q\varepsilon}\int_0^r u(t)^p t^{n-1}\,dt \tag{3.14}$$

where ε is any positive number.

The right thing to do is to choose the auxiliary parameter ε so that (3.13) and (3.14) are balanced against each other. This means that we must have

$$\begin{cases} \varepsilon^{p-1}[p-1+\alpha(n-p)] \leq pn\alpha, \\ p-1+\alpha(n-p) \leq pq\varepsilon(1-\alpha). \end{cases}$$

Checking and ruling out the cases $pn < \varepsilon^{p-1}(n-p)$ and $pq\varepsilon < p-n$, we are left with the double inequality

$$\frac{\varepsilon^{p-1}(p-1)}{pn - \varepsilon^{p-1}(n-p)} \leq \alpha \leq \frac{pq\varepsilon - p + 1}{pq\varepsilon - p + n}. \tag{3.15}$$

The optimal choice of $\varepsilon > 0$ is easily seen to be prescribed by Eqn (3.11). By Remark 3.12 the numerators in (3.15) are positive for this ε. This concludes our proof. □

References

[B] T. BHATTACHARYA, "Some results concerning the eigenvalue problem for the p-Laplacian", Annales Academiae Scientiarum Fennicae, Series A.I. Mathematica, **14** (1990), pp. 325-343.

[DB] E. DIBENEDETTO, "$C^{1+\alpha}$ local regularity of weak solutions of degenerate elliptic equations", Nonlinear Analysis, Theory, Methods & Applications **7** (1983), pp. 827-859.

[GT] D. GILBARG & N. TRUDINGER, "Elliptic Partial Differential Equations of Second Order", 2nd Edition, Springer-Verlag, Berlin 1983.

[Ka] B. KAWOHL, "Rearrangements and Convexity of Level Sets in PDE" (Lecture Notes in Mathematics 1150), Springer-Verlag, Heidelberg 1985.

[KL] B. KAWOHL & M. LONGINETTI, "On radial symmetry and uniqueness of positive solutions of a degenerate elliptic eigenvalue problem", Journal of Applied Mathematics and Physics (ZAMP) **68** (1988), T459-T460.

[Ke] A.U. KENNINGTON, "Power concavity and boundary value problems", Indiana University Mathematics Journal **34** (1985), pp. 687-704.

[Ko] N.J.KOREVAAR, "Convex solutions to nonlinear elliptic and parabolic boundary value problems", Indiana University Mathematics Journal **32** (1983), pp. 603-614.

[L1] P. LINDQVIST, "On non-linear Rayleigh quotients", To appear in Potential Analysis.

[L2] P. LINDQVIST, "On the equation $\operatorname{div}(|\nabla u|^{p-2}\nabla u) + \lambda|u|^{p-2}u = 0$", Proceedings of the American Mathematical Society **109** (1990), pp. 157-164.

[PS] G. PÓLYA & G. SZEGÖ, "Isoperimetric Inequalites in Mathematical Physics", Princeton University Press, Princeton 1951.

[RN] F. RIESZ & B. SZ.-NAGY, "Vorlesungen über Funktionalanalysis", Deutscher Verlag der Wissenschaften, Berlin 1956.

[S] S. SAKAGUCHI, "Concavity properties of solutions to some degenerate quasilinear elliptic Dirichlet problems", Annali della Scuola Normale Superiore de Pisa, Serie IV (Classe di Scienze), **14** (1987), pp. 403-421.

[Th1] F. de THÉLIN, "Quelques résultats d'existence et de non-existence pour une E.D.P. Elliptique non linéaire", C.R. Acad. Sc. Paris **299** (1984), Série I, pp. 911-914.

[Th2] F. de THÉLIN, "Sur l'espace propre associé à la première valeur propre du pseudo-laplacien", C.R. Acad. Sc. Paris **303** (1986), Série I, pp. 355-358.

[To] P. TOLKSDORF, "Regularity for a more general class of quasi-linear elliptic equations", Journal of Differential Equations **51** (1984), pp. 126-150.

[Tr] N.TRUDINGER, "On Harnack type inequalities and their applications to quasilinear elliptic equations", Communications on Pure and Applied Mathematics **20** (1967), pp. 721-747.

[Z] W.P. ZIEMER, "Weakly Differentiable Functions: Sobolev Spaces and Functions of Bounded Variation", Springer-Verlag, New York 1989.

Global Stability of a Model for Competing Predators

Torsten Lindström

Department of Applied Mathematics
University of Luleå
S-97187 Luleå, Sweden

Abstract

In this paper we state and prove a new theorem for global stability of a model for competing predators. We compare our theorem to other theorems in the literature in several examples at the end of the paper.

1 Introduction

S. B. Hsu, S. P. Hubell and P. Waltman [13] introduced and made a careful examination of the following model for competing predators

$$\begin{aligned} \dot{s} &= rs(1 - s/K) - x_0\frac{c_0 s}{s + a_0} - x_1\frac{c_1 s}{s + a_1} \\ \dot{x}_0 &= \left(m_0\frac{c_0 s}{s + a_0} - d_0\right) x_0 \\ \dot{x}_1 &= \left(m_1\frac{c_1 s}{s + a_1} - d_1\right) x_1. \end{aligned} \tag{1}$$

Here s is the prey, x_0 and x_1 are predators competing for the same prey. The parameter r is the intrinsic growth rate of the prey, K is the carrying capacity for the prey, c_0 and c_1 are the search rates, a_0 and a_1 the search rates multiplied by the handling times [20], m_0 and m_1 the conversion rates and d_0, d_1 the death-rates of the predators x_0 and x_1, respectively. Most of the basic properties of the system (1), including a detailed classification of the stationary points of the system, were examined in [13]. Concerning this model, these results support the ecological principle of competitive exclusion [11], in the sense that the two predators cannot possess equilibrium coexistence except for exceptional parameter values.

However, as already noted in [13], parameter values corresponding to nonequilibrium coexistence exist, so the ecological principle of competitive exclusion is violated here. In some works this is shown by the singular perturbation argument (see e. g. [21] and [24]). J. P. Keener used multiparameter bifurcation analysis to prove oscillatory coexistence for a certain parameter range [14]. R. McGehee and R. A. Armstrong [19] showed by an elegant construction in a slightly more general case than (1) that coexistence is possible. On the other hand, it should be noted that the range of parameter values giving rise to coexistence is quite narrow and that the possibility for coexistence decreases with the number of predators competing for the same prey [4].

The small range of parameters giving rise to coexistence is related to the non-slanted predator isocline in the related two-dimensional model

$$\begin{aligned} \dot{s} &= rs(1-s/K) - x\frac{cs}{s+a} \\ \dot{x} &= \left(m\frac{cs}{s+a} - d\right)x. \end{aligned} \tag{2}$$

This "non-slanted predator isocline property" has been criticized by biologists because, if the carrying capacity of the prey decreases and the system is persistent, then the equilibrium number of prey remains the same and the equilibrium number of predators decreases. One would expect that the equilibrium number of prey also decreases, see e. g. [9].

Arditi and Ginzburg suggested in [1] and [2] the use of a ratio-dependent functional response instead of a prey-dependent functional response. This approach has been supported in [9] with the reservation that slanted predator-isoclines passing through the origin do not fit with (the author's) intuition. However, this approach has also been seriously criticized, see e. g. [5] and [23].

Another way to get slanted predator-isoclines is to use the logistic growth equation with a carrying capacity directly related to prey density in the predator equation, see e. g. [10] and [22]. What essentially is lost here, is the control over how prey biomass is transferred into predator biomass. However, these models have interesting mathematical properties, which make it easier to control the motion of the limit cycles, see e.g. [17]. In this case the predator isocline also passes through the origin.

A possible real mechanism which may give rise to slanted isoclines in nature is the possibility for the generalist predators to use the specialist predators as nutrient, too. This will give rise to a slanted predator isocline, see example 1.1. This type of slanted predator isocline does not pass through the origin. In nature generalist predators are usually mixed among the specialist predators. For mathematical results applicable to this case, see e.g. [7] and [8].

Example 1.1 Put

$$\psi_0(s) = m_0\frac{c_0 s}{s+a_0} - d_0.$$

If the generalist predators are able to feed on the specialist predator species x_0, we get, when the functional response is given by ϵx_0^2:

$$\dot{x}_0 = x_0\psi_0(s) - \epsilon p x_0^2.$$

If we solve the predator isocline from this equation we get $x_0 = \psi_0(s)/p\epsilon$.

The presence of generalist predators has been connected with stabilization of limit cycles and more complicated dynamical behavior in predator-prey systems (see e. g. [6], [10], [17]). If the generalist predator class uses only a negligible part of the specialist predators as nutrient we may expect that, because interior equilibria exist only for exceptional parameter values, that one of the specialist predators becomes extinct before stabilization.

It is, however, difficult to verify extinction and global stability for some of the parameter values in the model (1). Some partial answers to this question were given in [12], [13] and [16]. In this paper we shall take a look at some of these problems.

2 The model

We consider the following model for competing predators

$$\begin{aligned} \dot{s} &= h(s) - x_0 f_0(s) - x_1 f_1(s) \\ \dot{x}_0 &= x_0\psi_0(s) \\ \dot{x}_1 &= x_1\psi_1(s). \end{aligned} \tag{3}$$

Here s denotes the prey density, x_0 and x_1 are the densities of the specialist predators feeding on the same prey s.

We shall analyze the model (3) qualitatively under the following general conditions:

(A–I) All the functions h, f_0, f_1, ψ_0 and ψ_1 are continuously differentiable of any required order (at least $C^1[0,\infty[$).

(A–II) There exists a constant K, $K > 0$, such that h satisfies $h(s) > 0$ if $0 < s < K$ and $h(s) < 0$ if $s < 0$ or $s > K$

(A–III) The functions f_0 and f_1 are increasing and have unique zeros at $s = 0$.

(A–IV) The functions ψ_0 and ψ_1 are increasing and there exist λ_i such that ψ_i satisfy $\psi_i(s) < 0$ if $0 < s < \lambda_i$ and $\psi_i(s) > 0$ if $s > \lambda_i$ for $i = 0, 1$. Moreover, we assume that $\lambda_0 < \lambda_1$.

(A–V) We have

$$\int_{\lambda_0}^{0} \frac{\psi_0(s)}{f_0(s)} ds = \infty \text{ and } \int_{\lambda_0}^{\infty} \frac{\psi_0(s)}{f_0(s)} ds = \infty.$$

The following theorem is essential.

Theorem 2.1 *Assume* (A-I)-(A-V). *If* $s(0) > 0$, $x_0(0) > 0$ *and* $x_1(0) > 0$ *then the solutions of the system* (3) *remain positive and bounded.*

Proof The uniqueness theorem implies positivity of solutions. If $\lambda_0 > K$, we have $\dot{s} < 0$ when $s < K$ and $\dot{x_0} < 0$, $\dot{x_1} < 0$, when $s < \lambda_0$, so the theorem holds when $\lambda_0 > K$.

Assume $\lambda_0 < K$. We note that all solutions will enter the infinite parallelepiped $0 < s < s_{\max}$, $s_{\max} > K$. Define

$$F_0(s) = \frac{h(s)}{f_0(s)} \text{ and } F_1(s) = \frac{h(s)}{f_1(s)}. \tag{4}$$

We introduce the Lyapunov functions

$$\begin{aligned} V_1(s, x_0 + x_1) &= \textstyle\int_{\lambda_0}^{s} \frac{ds'}{f_0(s')} + \int_{\lambda_0}^{s} \frac{ds'}{f_1(s')} + \int_{F_0(\lambda_0)}^{x_0+x_1} \frac{d(x_0+x_1)'}{(x_0+x_1)'}, \quad \lambda_0 < s < s_{\max} \\ V_2(s, x_0 + x_1) &= x_0 + x_1, \quad 0 < s < \lambda_0. \end{aligned}$$

First note that $V_1(s, x_0 + x_1) \to \infty$ and $V_2(s, x_0 + x_1) \to \infty$ when $x_0 + x_1 \to \infty$. A calculation of the total time derivative of the first function shows that

$$\begin{aligned} \dot{V}_1(s, x_0 + x_1) &= \frac{1}{f_0(s)}(h(s) - x_0 f_0(s) - x_1 f_1(s)) + \\ &\quad \frac{1}{f_1(s)}(h(s) - x_0 f_0(s) - x_1 f_1(s)) + \\ &\quad \frac{1}{x_0 + x_1}(x_0 \psi_0(s) + x_1 \psi_1(s)) = \\ &\quad F_0(s) - x_0 - x_1 \frac{f_1(s)}{f_0(s)} + F_1(s) - x_0 \frac{f_0(s)}{f_1(s)} - x_1 + \\ &\quad \frac{x_0}{x_0 + x_1}\psi_0(s) + \frac{x_1}{x_0 + x_1}\psi_1(s) \leq \\ &\quad F_0(s) - x_0 + F_1(s) - x_1 + \psi_0(s) + \psi_1(s). \end{aligned}$$

The last expression is less than zero when $\lambda_0 < s < s_{\max}$ if

$$x_0 + x_1 > \sup_{\lambda_0 < s < s_{\max}} (F_0(s) + F_1(s) + \psi_0(s) + \psi_1(s)) = q$$

Put

$$V_0 = \int_{\lambda_0}^{s_{\max}} \frac{ds'}{f_0(s')} + \int_{\lambda_0}^{s_{\max}} \frac{ds'}{f_1(s')} + \int_{F_0(\lambda_0)}^{q} \frac{d(x_0 + x_1)'}{(x_0 + x_1)'}.$$

Because the integral $\int \frac{d(x_0+x_1)}{x_0+x_1}$ diverges we may choose

$$q_{\max} = \left\{ x_0 + x_1 \middle| \int_{F_0(\lambda_0)}^{x_0+x_1} \frac{d(x_0 + x_1)'}{(x_0 + x_1)'} = V_0 \right\}.$$

By the implicit function theorem and the positivity of $f_0(s)$, $f_1(s)$, x_0 and x_1 the level surface $V_1(s, x_0 + x_1) = V_0$ defines a function, $s = v(x_0 + x_1)$. A calculation of the total time derivative of the second Lyapunov function, $V_2(s, x_0 + x_1)$, shows that it is negative when $0 < s < \lambda_0$. Now all trajectories will enter and remain in the region

$$\begin{aligned}\mathcal{R} &= \{(s, x_0, x_1) | 0 < x_0 + x_1 < q, 0 < s < s_{\max}\} \cup \\ &\quad \{(s, x_0, x_1) | q < x_0 + x_1 < q_{\max}, 0 < s < v(x_0 + x_1)\}\end{aligned}$$

□

We remark that the most usual specific forms of the functions included in the model (3) are $h(s) = rs(1 - s/K)$, $f_i(s) = \frac{c_i s}{s + a_i}$, $\psi_i(s) = m_i f_i(s) - d_i$, $i = 1, 2$.

3 Results on existence and stability of equilibria

We present some results concerning the existence and stability of the equilibria of the system (3). If (A-I)-(A-V) are satisfied, which they are in most cases of biological relevance, we can deduce that the only possible equilibria are $(0,0,0)$, $(K,0,0)$, $(\lambda_0, F_0(\lambda_0), 0)$, $(\lambda_1, 0, F_1(\lambda_1))$, i.e. no interior equilibria exist. We remark that λ_0, λ_1 were defined by (A-IV) and that F_0, F_1 were defined by (4). Concerning the existence and stability of these equilibria we have:

Lemma 3.1 *(Stability property of the origin). The origin is a saddle point if* (A-I)-(A-V) *hold.*

Proof A calculation of the eigenvalues of the Jacobian of the system (3) at the origin shows that they are $h'(0) > 0$, $\psi_0(0) < 0$ and $\psi_1(0) < 0$. □

Lemma 3.2 *(Stability property of the point $(K, 0, 0)$). Let* (A-I)-(A-V) *hold. The equilibrium $(K, 0, 0)$ is a saddle point if $\lambda_0 < K$ and a stable equilibrium, if $\lambda_0 > K$.*

Proof The Jacobian of the system (3), calculated at $(K, 0, 0)$, is given by

$$J(K,0,0) = \begin{pmatrix} h'(K) & -f_0(K) & -f_1(K) \\ 0 & \psi_0(K) & 0 \\ 0 & 0 & \psi_1(K) \end{pmatrix}$$

Now $h'(K) < 0$, $\psi_0(K) > 0$, when $K > \lambda_0$ and $\psi_1(K) > 0$, when $K > \lambda_1$. □

Lemma 3.3 *Assume* (A-I)-(A-V). *The equilibrium $(\lambda_0, F_0(\lambda_0), 0)$ exists if $\lambda_0 < K$. The equilibrium $(\lambda_1, 0, F_1(\lambda_1))$ exists if $\lambda_1 < K$. If $F_0'(\lambda_0) < (>) 0$ then some neighborhood of the point $(\lambda_0, F_0(\lambda_0), 0)$ in the (s, x_0)-coordinate plane belongs to the (un)stable manifold of the equilibrium $(\lambda_0, F_0(\lambda_0), 0)$. If $F_1'(\lambda_1) < (>) 0$ then some neighborhood of the*

$(\lambda_1, 0, F_1(\lambda_1))$ in the (s, x_1)-coordinate plane belongs to the (un)stable manifold of the equilibrium $(\lambda_1, 0, F_1(\lambda_1))$. The equilibrium $(\lambda_0, F_0(\lambda_0), 0)$ has at least a one-dimensional stable manifold in the interior of $\mathbf{R}^3_+$. The equilibrium $(\lambda_1, 0, F_1(\lambda_1))$ has at least a one-dimensional unstable manifold in the interior of $\mathbf{R}^3_+$.

Proof The Jacobian matrix of the system (3) calculated in the point $(\lambda_0, F_0(\lambda_0), 0)$ is given by

$$J(\lambda_0, F_0(\lambda_0), 0) = \begin{pmatrix} f_0(\lambda_0)F_0'(\lambda_0) & -f_0(\lambda_0) & -f_1(\lambda_0) \\ F_0(\lambda_0)\psi_0'(\lambda_0) & 0 & 0 \\ 0 & 0 & \psi_1(\lambda_0) \end{pmatrix}.$$

The eigenvalues of this matrix is given by the characteristic equation

$$\begin{vmatrix} f_0(\lambda_0)F_0'(\lambda_0) - \mu & -f_0(\lambda_0) \\ F_0(\lambda_0)\psi_0'(\lambda_0) & -\mu \end{vmatrix} (\psi_1(\lambda_0) - \mu) = 0. \tag{5}$$

From the first factor (the determinant) we get that some neighborhood of the equilibrium $(\lambda_0, F_0(\lambda_0), 0)$ in the (s, x_0)-plane belongs to the (un)stable manifold of the point $(\lambda_0, F_0(\lambda_0), 0)$ if $F_0'(\lambda_0) < (>) 0$. The factor $\psi_1(\lambda_0) - \mu$ gives that the equilibrium point $(\lambda_0, F_0(\lambda_0), 0)$ has at least a one dimensional stable manifold in the interior of $\mathbf{R}^3_+$. We get the conclusions of the equilibrium $(\lambda_1, 0, F_1(\lambda_1))$ in a similar way. □

Remark 3.4 The determinant in the expression (5) leads to the classical Rosenzweig-MacArthur graphical criterion for the two-dimensional system

$$\begin{aligned} \dot{s} &= f_0(s)(F_0(s) - x_0) \\ \dot{x}_0 &= x_0\psi_0(s) \end{aligned}$$

see e.g. [7] and [25].

4 Extinction and global stability

We state one known theorem before our main theorem.

Theorem 4.1 *Let* (A-I)-(A-V) *hold. If the condition*

$$\beta\psi_1(s) - \alpha\psi_0(s) < 0 \text{ for some } \alpha > 0 \text{ and some } \beta > 0 \text{ when } 0 < s < K, \tag{6}$$

is satisfied, then the specialist predator x_1 becomes extinct, except for initial conditions in the (s, x_1)-coordinate-plane.

Proof The equation for the coordinate $\eta = \frac{x_1^\beta}{x_0^\alpha + x_1^\beta}$ is given by $\dot{\eta} = \eta(1-\eta)(\beta\psi_1(s) - \alpha\psi_0(s))$. When the conditions in the theorem are satisfied, we have that $\dot{\eta} < 0$ when $0 < \eta < 1$. Together with boundedness of solutions (theorem 2.1), this gives the theorem. □

Remark 4.2 The above theorem contains exactly the same conclusion as the main theorem in [16], but the proof presented here is shorter.

Example 4.3 We show that theorem 3.6 in [13] is included in theorem 4.1. Introduce $b_0 = \frac{c_0 m_0}{d_0}$, $b_1 = \frac{c_1 m_1}{d_1}$ and consider the model (1). We show that $0 < \lambda_0 < \lambda_1 < K$, $a_0 < a_1$, $b_0 < b_1$ and

$$K < \frac{b_0 a_1 - b_1 a_0}{b_1 - b_0},$$

imply

$$\frac{\psi_0(s)}{d_0} - \frac{\psi_1(s)}{d_1} > 0$$

for $0 < s < K$.

If we start from $0 < \lambda_0 < \lambda_1$, we get that $b_0 - 1 > 0$ and $b_1 - 1 > 0$. After this we note that $\lambda_0 < \lambda_1$ implies $a_0(b_1 - 1) < a_1(b_0 - 1)$ and we get $b_0 a_1 - a_0 b_1 > a_1 - a_0 > 0$. Using the facts that $0 < K < \frac{b_0 a_1 - b_1 a_0}{b_1 - b_0}$ and $b_1 - b_0 > 0$ we note that

$$(b_0 - b_1)s + (b_0 a_1 - b_1 a_0)$$

is greater than zero in the interval $0 < s < K$. But this implies

$$\frac{\psi_0(s)}{d_0} - \frac{\psi_1(s)}{d_1} > 0$$

in the interval $0 < s < K$. That is, if the system (1) satisfies the assumptions of theorem 3.6 in [13], then the system (1) will satisfy the assumptions of theorem 4.1 if we choose $\alpha = \frac{1}{d_0}$ and $\beta = \frac{1}{d_1}$.

The above theorem implies that it is possible to reduce the system to a two-dimensional predator-prey model of the form

$$\begin{aligned} \dot{s} &= f_0(s)(F_0(s) - x_0) \\ \dot{x}_0 &= x_0\psi_0(s), \end{aligned} \tag{7}$$

provided that the conditions given in the theorem hold, but not that the cycles disappear. However, there are well known results concerning global stability for Gause-type predator systems like (7). See e.g. [3], [15], [18] and [26]. Loosely speaking, systems like (7) possess global stability when F_0 decreases enough. A trivial result which is contained in the results of the above mentioned works states that if F_0 is decreasing, then the system (7) is globally asymptotically stable. So this is not a real shortcoming with theorem 4.1. However, as the following example will show, the condition (6) will appear to be a too strong condition for global stability in some cases. We remark that these cases do not belong to the group of special cases completely examined in [27].

Example 4.4 We consider a system where $h(s) = s(1 - s/2.5) - ps^2$, $f_0(s) = \frac{s}{s+1}$, $f_1(s) = \frac{\frac{13}{10}s}{s+\frac{3}{2}}$, $\psi_0(s) = \frac{1}{5}f_0(s) - \frac{1}{20}$, $\psi_1(s) = \frac{1}{5}f_1(s) - \frac{1}{20}$. The parameter p represents generalist predators, and we note that if we increase the generalist predator density, then the carrying capacity of the system will decrease. In the absence of generalist predators the system has an interior limit cycle, see figure 1. A straightforward application of

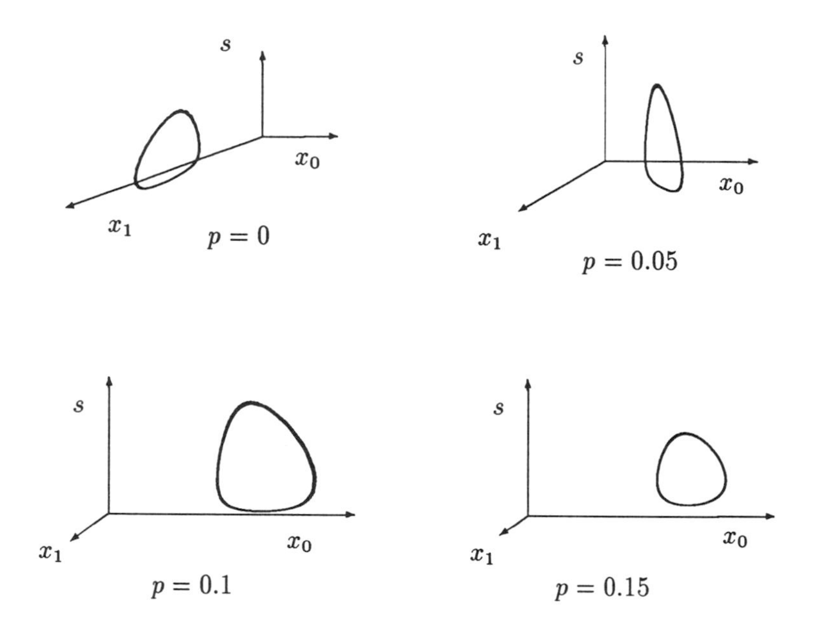

Figure 1: Variation of the limit cycles of the system (3) for different parameter values of p, $p = 0$ (a), $p = 0.05$ (b), $p = 0.1$ (c) and $p = 0.15$ (d).

theorem 4.1 tells us that when $p > \frac{11}{10}$ then $x_1 \to 0$. In this case it will be of no help to make a better choice of the numbers α and β than the already given ones, $\alpha = \beta = 1$. We remark here that theorem 3.6 in [13] was included in theorem 4.1 according to example 4.3 and that theorem 3.4 in [13] does not apply here. However, numerical solution of the system tells us that a much lower generalist predator population will lead to extinction of the specialist predator x_1.

Example 5.2 will show that the following theorem proves global stability in example 4.4 when $\frac{3}{5} < p < \frac{11}{10}$. We remark that $\lambda_1 > K$ implies $x_1 \to 0$ before we state the theorem.

Theorem 4.5 *Assume* (A-I)-(A-V), $\lambda_1 < K$ *and that* F_0 *is a decreasing function. If the condition*

$$\alpha\psi_1(s)f_0(s) - \beta\psi_0(s)f_1(s) < 0 \text{ for some } \alpha > 0 \text{ and some } \beta > 0 \tag{8}$$

is satisfied for $0 < s < K$ *then the equilibrium* $(\lambda_0, F_0(\lambda_0), 0)$ *is globally asymptotically stable, except for initial conditions in the* (s, x_1)*-coordinate plane.*

Remark 4.6 Forthcoming examples (section 5), will show that $\lambda_0 < \lambda_1$ implies condition (8) in several specific examples of biological relevance. This is, for example the case in the model (1). We remark that we can always assume $\lambda_0 < \lambda_1$, compare condition (A-IV).

Proof Introduce the Lyapunov function

$$W(s, x_0, x_1) = \beta\left(\int_{\lambda_0}^{s} \frac{\psi_0(s')}{f_0(s')} ds' + \int_{F_0(s)}^{x_0} \frac{x_0 - F_0(\lambda_0)}{x_0} dx_0'\right) + \alpha \int_0^{x_1} dx_1'.$$

This Lyapunov function is continuously differentiable and positive definite. Moreover we have that $W(s, x_0, x_1) \to \infty$ when $s \to \infty$, $s \to 0$ (A-V), $x_0 \to \infty$, $x_0 \to 0$ or $x_1 \to \infty$. The total time derivative of this Lyapunov function multiplied by $f_0(s) > 0$ (A-III) is given by

$$\begin{aligned} f_0(s)\dot{W} &= f_0(s)\beta\left(\frac{\psi_0(s)}{f_0(s)}(h(s) - x_0 f_0(s) - x_1 f_1(s))\right) + \\ & \quad f_0(s)\beta\left(\frac{x_0 - F_0(\lambda_0)}{x_0} x_0\psi_0(s)\right) + \alpha x_1\psi_1(s)f_0(s) \\ & \quad f_0(s)\beta\psi_0(s)(F_0(s) - x_0) - \beta\psi_0(s)f_1(s)x_1 + \\ & \quad f_0(s)\beta\psi_0(s)(x_0 - F_0(\lambda_0)) + \alpha x_1\psi_1(s)f_0(s) = \\ & \quad \beta f_0(s)\psi_0(s)(F_0(s) - F_0(\lambda_0)) + x_1(\alpha\psi_1(s)f_0(s) - \beta\psi_0(s)f_1(s)). \end{aligned}$$

This quantity is negative, except on the lines $s = \lambda_0$, $x_1 = 0$ and $s = 0$, $x_1 = 0$ where it approaches the value zero, if the conditions of thc theorem are satisfied. □

5 Examples

We will apply theorem 4.5 to the specific models,

$$\begin{aligned} \dot{s} &= rs(1 - s/K) - x_0 c_0 s - x_1 c_1 s \\ \dot{x}_0 &= m_0(s - \lambda_0)x_0 \\ \dot{x}_1 &= m_1(s - \lambda_1)x_1 \end{aligned} \tag{9}$$

and

$$\begin{aligned}\dot{s} &= h(s) - x_0\frac{c_0 s^n}{s^n + a_0} - x_1\frac{c_1 s^n}{s^n + a_1} \\ \dot{x}_0 &= \left(m_0\frac{c_0 s^n}{s^n + a_0} - d_0\right)x_0 \\ \dot{x}_1 &= \left(m_1\frac{c_1 s^n}{s^n + a_1} - d_1\right)x_1,\end{aligned} \tag{10}$$

in this section.

It was proved in [12] that the model (9) is globally asymptotically stable for all parameter values. In the example 5.1 we will use theorem 4.5 to provide a different proof of this fact. We will also show that $\lambda_0 < \lambda_1$ implies condition (8) for the models (9)-(10) and verify that theorem 4.5 proves global stability of the system in example 4.4 in the case $\frac{3}{5} < p < \frac{11}{10}$.

Example 5.1 We use theorem 4.5 to show global stability for the model (9). Put $f_0(s) = c_0 s$, $f_1(s) = c_1 s$, $\psi_0(s) = m_0(s - \lambda_0)$, $\psi_1(s) = m_1(s - \lambda_1)$, $h(s) = rs(1 - s/K)$. Let $\lambda_0 < \lambda_1 < K$. We note first that $F_0(s) = \frac{h(s)}{f_0(s)} = \frac{r}{c_0}(1 - s/K)$ is decreasing. We also have that $\lambda_0 < \lambda_1$ implies condition (8) because

$$\begin{aligned}\alpha\psi_1(s)f_0(s) - \beta\psi_0(s)f_1(s) &= \\ \alpha m_1(s - \lambda_1)c_0 s - \beta m_0(s - \lambda_0)c_1 s &= \\ s\left((\alpha m_1 c_0 - \beta m_0 c_1)s + (\beta m_0 \lambda_0 c_1 - \alpha m_1 \lambda_1 c_0)\right).\end{aligned}$$

Now put $\alpha = m_0\lambda_0 c_1$ and $\beta = m_1\lambda_1 c_0$. We get

$$\begin{aligned}\alpha\psi_1(s)f_0(s) - \beta\psi_0(s)f_1(s) &= \\ s^2\left(m_0\lambda_0 c_1 m_1 c_0 - m_1\lambda_1 c_0 m_0 c_1\right) &= \\ s^2 m_0 m_1 c_0 c_1(\lambda_0 - \lambda_1) &< 0.\end{aligned}$$

Example 5.2 We treat the model (10) here. We show first that $\lambda_0 < \lambda_1$ implies (8). We have

$$\begin{aligned}\alpha\psi_1(s)f_0(s) - \beta\psi_0(s)f_1(s) &= \\ \alpha\left(\frac{m_1 c_1 s^n}{s^n + a_1} - d_1\right)\frac{c_0 s^n}{s^n + a_0} - \beta\left(\frac{m_0 c_0 s^n}{s^n + a_0} - d_0\right)\frac{c_1 s^n}{s^n + a_1} &= \\ s^n\frac{1}{s^n + a_0}\frac{1}{s^n + a_1}\left(\alpha(m_1 c_1 - d_1)c_0 s^n - \alpha d_1 c_0 a_1 - \beta(m_0 c_0 - d_0)c_1 s^n - \beta d_0 c_1 a_0\right)\end{aligned}$$

Now put $\alpha = d_0 c_1 a_0$ and $\beta = d_1 c_0 a_1$. We get

$$\alpha\psi_1(s)f_0(s) - \beta\psi_0(s)f_1(s) =$$

$$s^{2n}\frac{1}{s^n+a_0}\frac{1}{s^n+a_1}(d_0c_1a_0m_1c_1c_0 - d_0c_1a_0d_1c_0 - d_1c_0a_1m_0c_0c_1 + d_1c_0a_1d_0c_1) =$$

$$d_0d_1\frac{c_0s^n}{s^n+a_0}\frac{c_1s^n}{s^n+a_1}\left(\frac{c_1m_1}{d_1}a_0 - a_0 - \frac{c_0m_0}{d_0}a_1 + a_1\right) =$$

$$d_0d_1\left(\frac{c_0m_0}{d_0}-1\right)\left(\frac{c_1m_1}{d_1}-1\right)\frac{c_0s^n}{s^n+a_0}\frac{c_1s^n}{s^n+a_1}\left(\frac{a_0}{\frac{c_0m_0}{d_0}-1} - \frac{a_1}{\frac{c_1m_1}{d_1}-1}\right) =$$

$$d_0d_1\left(\frac{c_0m_0}{d_0}-1\right)\left(\frac{c_1m_1}{d_1}-1\right)\frac{c_0s^n}{s^n+a_0}\frac{c_1s^n}{s^n+a_1}(\lambda_0^n - \lambda_1^n) < 0$$

Hence $\lambda_0 < \lambda_1$ implies condition (8).

Theorem 4.5 can now be applied if F_0 is decreasing. If $n = 1$ and $h(s) = rs(1 - s/K)$, theorem 4.5 applies when $a_0 > K$, so the system in example 4.4 is globally asymptotically stable when $p > \frac{3}{5}$.

Remark 5.3 Global stability of the model (10) with $n = 1$ and $h(s) = r(1 - s/K)$ was proved in [12]. We have not considered this model here because (A-II) is not satisfied, so the origin is not an equilibrium. However, the proof of theorem 4.5 can be extended to cover this case. Analysis analogous to example 5.2 will prove global stability here, since F_0 is decreasing.

6 Summary

We have stated and proved a new theorem for global stability of the following model for competing predators,

$$\begin{aligned} \dot{s} &= h(s) - x_0f_0(s) - x_1f_1(s) \\ \dot{x}_0 &= x_0\psi_0(s) \\ \dot{x}_1 &= x_1\psi_1(s), \end{aligned}$$

in this paper. Our theorem was motivated by example 4.4 and we showed in example 5.2 that our theorem removes a large part of the problems in example 4.4 and that it contains new results. Furthermore the examples 5.1-5.2 and remark 5.3 show that theorem 4.5 applies to a wide range of examples of biological relevance and that the conditions of the theorem are easy to check.

Acknowledgements This research was done at Luleå University of Technology and during my stay at the International Institute for Applied Systems Analysis under the direction of Prof. Mats Gyllenberg, to whom the author owes a debt of gratitude for his supervision. I want to thank Prof. Alexander Osipov for some valuable discussions connected to this work. This work was supported in part by, The Bank of Sweden Tercentenary Foundation, the Royal Swedish Academy of Sciences, the Swedish Council for Planning and Coordination of Research and the Wallenberg Foundation. Generous support in form of computer time from SDCN is gratefully acknowledged.

References

[1] R. Arditi, L. R. Ginzburg. Coupling in predator-prey dynamics: Ratio dependence. *J. Theoret. Biol.*, 139:311–326, 1989.

[2] R. Arditi, L. R. Ginzburg, H. R. Akçakaya. Variation in plankton densities among lakes: a case for ratio-dependent models. *Amer. Natur.*, 138:1287–1296, 1991.

[3] Cheng K.-S., S.-B. Hsu, S.-S. Lin. Some results on global stability of a predator-prey system. *J. Math. Biol.*, 12:115–126, 1981.

[4] J. Coste. Dynamical regression in many species ecosystems: The case of many predators competing for several prey. *SIAM J. Appl. Math.*, 45:555–564, 1985.

[5] S. Diehl, P. A. Lundberg, H. Gardfell, L. Oksanen, L. Persson. Daphnia-phytoplankton interactions in lakes: Is there a need for ratio-dependent consumer-resource models. Amer. Natur., in Press, 1992.

[6] S. Erlinge, G. Göransson, L. Hansson, G. Högstedt, O. Liberg, I. N. Nilsson, T. Nilsson, T. von Schantz and M. Sylvén. Predation as a regulating factor on small rodent populations in southern Sweden. *OIKOS*, 40:36–52, 1983.

[7] M. Farkas, H. I. Freedman. Stability conditions for two-predators one prey systems. *Acta Appl. Math.*, 14(1-2):3–10, 1989.

[8] M. Farkas, H. I. Freedman. The stable coexistence of competing species on a renewable resource. *J. Math. Anal. Appl.*, 138:461–472, 1989.

[9] I. Hanski. The functional response of predators: Worries about scale. *Trends in Ecology and Evolution*, 6(5):141–142, 1991.

[10] I. Hanski, L. Hansson, H. Henttonen. Specialist predators, generalist predators and the microtine rodent cycle. *J. Animal Ecology*, 60, 1991.

[11] G. Hardin. The competitive exclusion principle. *Science*, 131:1292–1298, 1960.

[12] S. B. Hsu. Limiting behavior for competing species. *SIAM J. Appl. Math.*, 34(4):760–763, June 1978.

[13] S. B. Hsu, S. P. Hubell, P. Waltman. Competing predators. *SIAM J. Appl. Math.*, 35(4):617–625, December 1978.

[14] J. P. Keener. Oscillatory coexistence in the chemostat. *SIAM J. Appl. Math.*, 43(5):1005–1018, October 1983.

[15] Y. Kuang. Global Stability of Gause-Type Predator-Prey Systems. *J. Math. Biol.*, 28:463-474, 1990.

[16] **С. Н. Кустаров, А. В. Осипов. Достаточные условия вымирания вида для одного класса систем типа "два хищника - одна жертва"** (Conditions for extinction of a species in a dynamical system of two predators-one prey type). **Деп. в ВИНИТИ, 4931-В86**, 1986.

[17] T. Lindström. Qualitative analysis of a predator-prey system with limit cycles. Accepted for publication in J. Math. Biol., 1992.

[18] Liou L.-P., Cheng K.-S. Global stability of a predator-prey system. *J. Math. Biol.*, 26:65–71, 1988.

[19] R. McGehee, R. A. Armstrong. Some mathematical problems concerning the ecological principle of competitive exclusion. *J. Differential Equations*, 23:30–52, 1977.

[20] J. Metz, O. Diekmann. *The Dynamics of Physiologically Structured Populations.* Springer-Verlag, Berlin, Heidelberg, 1986.

[21] S. Muratori, S. Rinaldi. Remarks on competitive coexistence. *SIAM J. Appl. Math.*, 49(5):1462–1472, October 1989.

[22] J. D. Murray. *Mathematical Biology.* Springer-Verlag, New York, 1989.

[23] L. Oksanen, J. Moen, P. A. Lundberg. The time scale problem in exploiter-victim models: Does the solution lie in ratio-dependent exploitation? *Amer. Natur.*, 151, 1993.

[24] **А. В. Осипов,** G. Söderbacka, T. Eirola. **О существовании внутреннего периодического решения для одной динамической системы типа "два хищника - одна жертва"** (On the existence of positive periodic solutions in a dynamical system of two predators-one prey type). **Деп. в ВИНИТИ, 4305-В86**, 1986.

[25] M. L. Rosenzweig, R. H. MacArthur. Graphical representation and stability conditions of predator-prey interactions. *The American Naturalist*, XCVII:209–223, July-August 1963.

[26] **В. В. Виноградов, А. В. Осипов. О предельных циклах одной двумерной системы.** (Uniqueness of a limit cycle in a two-dimensional system). **Деп. в ВИНИТИ, 1160-В87**, 1987.

[27] D. R. Wilken. Some remarks on the competing predators problem. *SIAM J. Appl. Math.*, 42:895–902, 1982.

Inverse Scattering Problem for a Quantum Mechanical Two-Body System

Anders Melin

Lund Institute of Technology
Box 118
S-221 01 Lund, Sweden

Abstract

In this paper we give a brief survey of some recent progress in inverse scattering theory for some classes of Schrödinger operators describing interaction between two particles. In particular, we construct a family of operators which intertwine the Schrödinger operator with the corresponding operator obtained when the potential is removed. The scattering data are easily described in terms of the intertwining operators, and we discuss some methods that are used to reconstruct the potential from these data. Complete proofs will be given in a forthcoming paper.

1. The scattering matrix and the intertwining operators A_θ.

We consider in $\mathbf{R}^n$, when $n \geq 3$, a real valued function $v(x)$ and the corresponding Schrödinger operator $H_V = H_0 + V$, where V denotes the operator which is multiplication by $v(x)$ and $H_0 = -\Delta$. For the sake of simplicity we assume that $v^{(\alpha)} \in L^1(\mathbf{R}^n)$ for every α. This implies in particular that $v(x)$ is bounded, and H_V is self-adjoint with domain $\mathcal{H}_{(2)}$, the space of functions with derivatives of order ≤ 2 in L^2.

The time dependent Schrödinger equation has the form

$$i\partial_t u = H_V u, \; u(x,0) = u_0(x), \tag{1.1}$$

where $u_0 \in L^2$. We may view this as the model for the motion of particles in a field $-\nabla v$ caused by some particle, which is kept fixed at the origin. The solution $u(\cdot, t) = e^{-itH_V} u_0$ to (1.1) is a continuous function of t with values in L^2. We observe that if $V = 0$, then

$$u(tx,t) = (2\pi)^{-n} \int e^{it(\langle x,\xi\rangle - \xi^2)} \widehat{u}_0(\xi)\, d\xi,$$

where

$$\widehat{u}_0(\xi) = \mathcal{F}u_0(\xi) = \int e^{-i\langle x,\xi\rangle} u_0(x)\, dx$$

is the Fourier transform of u_0. An application of the method of stationary phase when $\widehat{u} \in C_0^\infty(\mathbf{R}^n \setminus 0)$ gives

$$u(x,t) = (\pi|t|)^{-n/2} e^{-in\pi \operatorname{sgn} t/4} e^{ix^2/4t} \widehat{u}_0(x/2t) + r_t(x), \tag{1.2}$$

where

$$|t|^{-n/2}\|r_t\|_{L^1} + |t|^{n/2}\|r_t\|_{L^\infty} + \|r_t\|_{L^2} = O(t^{-1}) \quad \text{as } |t| \to \infty.$$

Thus the solution to (1.1) when $V = 0$ may be described as a wave packet which travels to infinity in directions determined by the spectrum of the initial state u_0.

In the case when V is present, then any solution to (1.1), such that u_0 is orthogonal to the space spanned by the eigenvectors of H_V, will for large $|t|$ have the same asymptotic behaviour as a solution to the free equation. In fact, it is easy to prove that the wave operators

$$W_\pm = \lim_{t\to\pm\infty} e^{itH_V} e^{-itH_0}$$

exist in the strong sense. They are isometric with the same range. If

$$u_0 = W_+ u_+ = W_- u_-,$$

where $u_\pm \in L^2$, then by the definition of $W_\pm$ it is true that

$$u(\cdot, t) \sim e^{-itH_0} u_-, \quad t \to -\infty,$$

while

$$u(\cdot, t) \sim e^{-itH_0} u_+, \quad t \to +\infty.$$

The scattering operator S which sends u_- to u_+ is unitary. It measures the change of momentum caused by the potential field on particles, which have passed near the origin at some time but for large negative and positive time move as free particles near infinity.

It is important to observe that the wave operators have the following intertwining property:

$$H_V W_\pm = W_\pm H_0.$$

We are going to construct a family $\{A_\theta;\ \theta \in S^{n-1}\}$ of operators having this intertwining property and also satisfying some other conditions, so that in particular A_θ is close to the identity when acting on wave packets travelling to infinity in the direction of $-\theta$. (Such operators were constructed already by Faddeev [5], and we have tried to develop further his ideas. In particular, we want to obtain existence results for the A_θ in as general situations as possible and investigate in more detail its relations to various scattering data.)

In order to describe the properties of the A_θ more precisely it is necessary to introduce some terminology. First, we let $\mathcal{M}$ be the space of operators U on $L^1(\mathbf{R}^n) + L^\infty(\mathbf{R}^n)$ with the following properties:

(i) Any repeated commutator of U with constant coefficient vector fields is continuous on $L^p(\mathbf{R}^n)$ when $1 \le p \le \infty$. The distribution kernel $U(x, y)$ of such an operator is then a measure. In fact, $U(x, y + x)$ and $U(x - y, x)$ are smooth functions of x with values in the space $M(\mathbf{R}^n)$ of bounded measures on $\mathbf{R}^n$.

(ii) If K is compact, then the distribution $U'(x,y)$ of any repeated commutator of U with constant coefficient vector fields satisfies the condittion

$$\int_K (|U'(x,y)| + |U'(y,x)|)\, dy \to 0 \quad \text{as } |x| \to \infty.$$

(The integrals make sense when interpreted as integrals of measures.)

A typical operator in $\mathcal{M}$ is the identity mapping with distribution kernel $U_0(x,y) = \delta(x-y)$, and more generally, any operator which is convolution by a function in L^1 belongs to $\mathcal{M}$.

Some subspaces of $\mathcal{M}$ will also be needed. $\mathcal{M}_\theta$, where $\theta \in S^{n-1}$, is the set of all $U \in \mathcal{M}$ so that with U' as above

$$\begin{aligned} &\int |U'(x,y)|\, dy \to 0 \quad \text{as } |x| \to \infty,\ x/|x| \to \theta \\ &\int |U'(x,y)|\, dx \to 0 \quad \text{as } |y| \to \infty,\ y/|y| \to -\theta. \end{aligned} \tag{1.3}$$

Finally $\mathcal{V}$ is the space of U such that $(\partial_x + \partial_y)^\alpha U \in M(\mathbf{R}^n \times \mathbf{R}^n)$ for every α, where $M(\mathbf{R}^n \times \mathbf{R}^n)$ is the Banach space of bounded measures on $\mathbf{R}^n \times \mathbf{R}^n$. It is easy to see that $\mathcal{M}$, $\mathcal{M}_\theta$ and $\mathcal{V}$ are Fréchet spaces, $\mathcal{V} \subset \mathcal{M}_\theta \subset \mathcal{M}$, and $\mathcal{E}' \cap \mathcal{V}$ is dense in $\mathcal{V}$, if $\mathcal{E}'$ is the space of compactly supported distributions.

In what follows we always identify an operator U, which is continuous from $C_0^\infty(\mathbf{R}^n)$ to $\mathcal{D}'(\mathbf{R}^n)$, with its distribution kernel $U(x,y)$. Note that, if $1 \le p \le \infty$ and $W^{p,k} = W^{p,k}(\mathbf{R}^n)$ is the space of all u such that $u^{(\alpha)} \in L^p$ when $|\alpha| \le k$, then every $U \in \mathcal{M}$ is continuous on $W^{p,k}$, and $(\partial_{x_j} + \partial_{y_j})U(x,y)$ is the distribution kernel of the commutator $[\partial_j, U] = \partial_j U - U\partial_j$, where $\partial_j = \partial_{x_j}$. Thus $(\partial_x + \partial_y)^\alpha U(x,y)$ is the distribution kernel of a repeated commutator, and we have Leibniz' rule.

$$(\partial_x + \partial_y)^\alpha (U_1 \circ U_2)/\alpha! = \sum_{\beta+\gamma=\alpha} ((\partial_x + \partial_y)^\beta U_1) \circ ((\partial_x + \partial_y)^\gamma U_2)/\beta!\gamma!, \quad U_1, U_2 \in \mathcal{M}.$$

These observations show that $\mathcal{M}$ is closed under composition, and that the mapping $(U_1, U_2) \mapsto U_1 \circ U_2$ is continuous. The corresponding statements hold for $\mathcal{V}$, which is a two-sided ideal in $\mathcal{M}$. We observe also that, if V is the potential operator considered above, then $V(x,y) = v(x)\delta(x-y)$ and the condition that $v^{(\alpha)} \in L^1$ for every α means precisely that $V \in \mathcal{V}$. The condition that our potential operator is local means, of course, that $x = y$ in $\operatorname{supp}(V)$.

Theorem 1. *Assume that $V \in \mathcal{V}$ is local and that $\theta \in S^{n-1}$. Then there exists a unique $A_\theta \in \mathcal{D}'(\mathbf{R}^n \times \mathbf{R}^n)$ with the following properties*

(i) If λ is large, then $e^{-\lambda\langle y-x,\theta\rangle} A_\theta(x,y) - \delta(x-y) \in \mathcal{M}_\theta$.

(ii) $\langle y-x,\theta\rangle \geq 0$ *in* $\text{supp}\,(A_\theta)$.
(iii) A_θ *has the intertwining property, i.e.*

$$H_V A_\theta = A_\theta H_0.$$

Moreover, if V *is small in the sense that*

$$p_n(v) = p_n(V) = \sum_{|\alpha|\leq n} \|v^{(\alpha)}\|_1$$

is smaller than some constant, which depends on the dimension only, then the conclusion holds with $\lambda = 0$. *Also,* A_θ *depends continuously on* θ.

Let $W^{p,k}_{\lambda,\theta}$ be the space of all u such that $e^{\lambda\langle x,\theta\rangle}u(x) \in W^{p,k}$. Then the theorem asserts that A_θ is a continuous linear operator on $W^{p,k}_{\theta,\lambda}$ when $\lambda \geq c_n(v)$. Here $0 \leq c_n(v)$ depends on $p_n(v)$, and $c_n(v) = 0$ when $p_n(v)$ is sufficiently small.

The uniqueness assertion of the theorem follows from some uniqueness results for distributions U satisfying the ultra-hyperbolic equation $(\Delta_x - \Delta_y)U(x,y) = 0$ together with the condition that $\langle y-x,\theta\rangle \geq 0$ in $\text{supp}\,(u)$ and some growth conditions at infinity. This uniqueness result implies also that $A^*_{-\theta}$ is the inverse of A_θ. Hence the A_θ are bijective on $W^{p,k}_{\lambda,\theta}$ when λ is large. Note that the existence of bound state will force λ to be positive in general for this statement to be true.

The proof of the existence part of the theorem is rather complicated. The details will be presented in a forthcoming paper, and we refer to Melin [8] for the odd-dimensional case. One solves by iteration a singular integral equation

$$A_\theta(x,y) = U_0(x,y) + E_\theta * (V \circ A_\theta)(x,y),$$

where $U_0(x,y) = \delta(x-y)$ as before, and E_θ is a temperate fundamental solution of the ultra-hyperbolic operator, that is

$$(\Delta_x - \Delta_y)E_\theta(x,y) = \delta(x,y).$$

It turns out that there exists a unique such fundamental solution, which satisfies the additional conditions

$$\langle y-x,\theta\rangle \geq 0 \ \text{ in } \ \text{supp}\,(E_\theta),$$
$$E_\theta(x+t\theta, y+t\theta) \to 0 \ \text{ as } \ |t| \to \infty.$$

An explicit formula for E_θ may be given as follows. Set

$$\langle \Phi, L_{\omega,r}\rangle = \int_{\mathbf{R}^n} \lambda_n^{(n-2)}(x\cdot\omega)\Phi(x - r\omega/2, x + r\omega/2)\,dx, \quad \Phi \in C_0^\infty(\mathbf{R}^n \times \mathbf{R}^n)$$

when $r > 0$, $\omega \in S^{n-1}$, and

$$\lambda_n(s) = 2^{-1}\big((s+i0)^{-1} + (-1)^n(s-i0)^{-1}\big).$$

Then $L_{\omega,r}$ belongs to the space $\mathcal{S}'(\mathbf{R}^n \times \mathbf{R}^n)$ of temperate distributions, and $y-x \in \overline{\mathbf{R}_+\omega}$ in its support. The integral

$$L_\omega = \int_0^\infty L_{\omega,r}\,dr$$

converges in $\mathcal{S}'(\mathbf{R}^n \times \mathbf{R}^n)$ and depends continuously on ω. We have

$$E_\theta = (2\pi i)^{-n} \int_{\langle\omega,\theta\rangle\geq 0} L_\omega\,d\omega.$$

In the literature, it is more common that one studies some Fourier transformed version of E_θ. (See [5], [6], [11]). In order to see the relation between the two approaches we define $E(p;x,y)$ for $p \in \mathbf{C}^n \setminus \mathbf{R}^n$ as the inverse Fourier transform of the locally integrable function

$$\mathbf{R}^n \times \mathbf{R}^n \ni (\xi,\eta) \mapsto \big((\eta-p)^2 - (\xi+p)^2\big)^{-1}.$$

Then

$$E(p;x,y) = e^{i\langle y-x,p\rangle} E_\theta(x,y), \qquad \theta = \operatorname{Im} p/|\operatorname{Im} p|.$$

Hence, $E_\theta(x,y)$ is the limit in $\mathcal{D}'$ as $\varepsilon \to +0$ of $E(i\varepsilon\theta;x,y)$, and the Fourier transform of E_θ is a principal value distribution

$$\frac{1}{\eta^2 - \xi^2 \pm i0}$$

where the choice of + or − depends on the sign of $\langle \xi-\eta,\theta\rangle$.

In addition to the properties mentioned above the A_θ will have another important property, which will be needed in the considerations below. Namely, $A_\theta(x,y) - \delta(x-y)$ is a locally integrable function, and if $\lambda \geq c_n(v)$, then one can find a locally integrable function $U(x,y)$, which is independent of θ, so that

$$e^{-\lambda\langle y-x,\theta\rangle}|A_\theta(x,y) - \delta(x-y)| \leq U(x,y), \qquad \theta \in S^{n-1}.$$

Moreover, the corresponding operator U is continuous on L^p when $1 \leq p \leq \infty$. (We say that a family A_θ with this property is dominated.) From now on we extend A_θ to a function homogeneous of degree 0 in $\theta \in \mathbf{R}^n \setminus 0$.

In order to relate the conditions on A_θ to the dynamical properties of the solutions to (1.1) we assume for a while that v is small, so that all assertions above are true when

$\lambda = 0$. Let us consider a function $u_0 \in \mathcal{S}(\mathbf{R}^n)$ with its Fourier transform supported in a small neighbourhood of some point $\theta \in \mathbf{R}^n \setminus 0$. The estimates for r_t in (1.2) and the condition that $A_{-\theta} \in \mathcal{M}_{-\theta}$, lead us to expect that

$$R(\cdot, t) \equiv A_{-\theta} e^{-itH_0} u_0 - e^{-itH_0} u_0$$

is small for large t in the sense that $t^{-n/2} \|R(\cdot,t)\|_1$ is small then. This condition implies that the norm in L^2 of $R(\cdot, t)$ is small, for $t^{n/2} \|R(\cdot,t)\|_\infty$ is bounded from above. Since e^{itH_V} is unitary, and

$$e^{itH_V} R(\cdot, t) = A_{-\theta} u_0 - e^{itH_V} e^{-itH_0} u_0,$$

by the intertwining property, our arguments indicate that $A_{-\theta} u_0$ is for large t a good approximation to $e^{itH_V} e^{-itH_0} u_0$. This is actually the case. In order to state a more precise result one considers, when $\widehat{u_0} \in C_0^\infty(\mathbf{R}^n \times \mathbf{R}^n)$, the functions

$$f_t(x) = \chi(t^{-1/2} x) u_0(x) = (2\pi)^{-n} \int \chi_{t,\eta} \widehat{u}_0(\eta)\, d\eta, \quad \chi_{t,\eta}(x) = e^{i\langle x, \eta \rangle} \chi(t^{-1/2} x).$$

where $\widehat{\chi} \in C_0^\infty(\mathbf{R}^n)$, $\chi(0) = 1$. Then $f_t \to u_0$ in L^2 when $t \to \infty$, and if $\phi = e^{-iH_0} \chi$, then

$$e^{-itH_0} \chi_{t,\eta}(x) = e^{-it|\eta|^2 + \langle x, \eta \rangle} \phi(t^{-1/2} x - 2t^{1/2} \eta).$$

The function to the right is for large positive t concentrated near the half-ray determined by η, and using that the family $\{A_\theta\}$ is dominated one can prove that

$$e^{-itH_0} f_t - (2\pi)^{-n} \int \widehat{f}(\eta) A_{-\eta} (e^{-itH_0} \chi_{t,\eta})\, d\eta \to 0 \quad \text{in } L^2 \quad \text{as } t \to \infty.$$

Multiplying the left-hand side by e^{itH_V} and using the intertwining property, we find that

$$W_+ u_0 = (2\pi)^{-n} \int \widehat{u}_0(\eta) A_{-\eta} e^{\langle \cdot, \eta \rangle}\, d\eta.$$

These consideration and similar arguments applied to W_- lead to the following result.

Theorem 2. *Let $V \in \mathcal{V}$ be local and small and $\phi_\pm(x, \xi)$ be the distribution kernel of $W_+ \mathcal{F}^*$, where $\mathcal{F}$ is the Fourier transformation. Then*

$$\phi_\pm(x, \xi) = \int A_{\mp\xi}(x, y) e^{i\langle y, \xi \rangle}\, dy. \tag{1.4}$$

The scattering operator is unitary and commutes with H_0. In order to introduce the scattering matrix we consider the mapping $C_0(\mathbf{R}^n) \ni u \mapsto \widetilde{u}(\lambda)$, where $\widetilde{u}(\lambda)$ is a function on S^{n-1} defined for $\lambda > 0$ by

$$\widetilde{u}(\omega; \lambda) = (2\pi)^{-n/2} \lambda^{(n-1)/2} \widehat{u}(\lambda\omega).$$

The mapping $u \mapsto \widetilde{u}$ extends to a unitary operator from $L^2(\mathbf{R}^n)$ to $L^2(\mathbf{R}_+; S^{n-1})$. We say that an operator T on $L^2(\mathbf{R}^n)$ is local with respect to energy, if there is a family $\widetilde{T}(\lambda)$ of operators on $L^2(S^{n-1})$ such that $\widetilde{f}(\lambda) = \widetilde{T}(\lambda)\widetilde{u}(\lambda)$ for almost ever $\lambda > 0$ when $u \in L^2$ and $f = Tu$. It is well known that the scattering operator $S = W_+^* W_-$ is local with respect to energy; the operator $\widetilde{S}(\lambda)$ is called the scattering matrix. For the convenience of the reader we give a proof in the case when v is small as above.

Recall that

$$I - S = I - W_+^* W_- = W_+^*(W_+ - W_-),$$

since W_+ is isometric. Now

$$\begin{aligned} W_+^*(W_+ - W_-) &= W_+^* \lim_{N\to\infty} \int_{-N}^{N} \frac{d}{dt} e^{itH_V} e^{-itH_0}\, dt \\ &= \lim_{N\to\infty} i \int_{-N}^{N} W_+^* e^{itH_V} V e^{-itH_0}\, dt \\ &= \lim_{N\to\infty} i \int_{-N}^{N} e^{itH_0} W_+^* V e^{-itH_0}\, dt, \end{aligned}$$

where the limits exist in the strong sense. In fact, since $v \in L^2$ and $n \geq 3$, it is obvious from (1.2) that $\|V e^{-itH_0} u_0\|_2$ is integrable for a dense set of functions $u_0 \in L^2$. This proves also that we may replace the limit in the right-hand side above by the limit as $\varepsilon \downarrow 0$ of

$$i \int_{-\infty}^{\infty} e^{-\varepsilon|t|} e^{itH_0} W_+^* V e^{-itH_0}\, dt.$$

Hence the distribution kernel of $\mathcal{F}(I - S)\mathcal{F}^*$ is given by

$$\mathcal{F}(I - S)\mathcal{F}^*(\xi, \eta) = \lim_{\varepsilon\to 0} i \int_{-\infty}^{\infty} e^{-\varepsilon|t|} e^{it\xi^2} (\mathcal{F} W_+^* V \mathcal{F}^*)(\xi, \eta) e^{-it\eta^2}\, dt. \tag{1.5}$$

It follows from Theorem 2 and some simple computations, that the distribution kernel of $\mathcal{F} W_+^* V \mathcal{F}^*$ is given for $\eta \neq 0$ by the formula

$$(\mathcal{F} W_+^* V \mathcal{F}^*)(\xi, \eta) = \widehat{W_\xi}(-\eta, \xi), \tag{1.6}$$

if we set

$$W_\xi(x, y) = (V \circ A_{-\xi})(x, y) = v(x) A_{-\xi}(x, y).$$

We observe that the right-hand side of (1.6) is continuous when $\xi \neq 0$, and in that set

$$\lim_{\varepsilon \to 0} \int_{-\infty}^{\infty} e^{-\varepsilon|t|} e^{it(\xi^2-\eta^2)}\, dt = 2\pi\delta(\xi^2 - \eta^2) = \frac{\pi}{|\xi|}\delta(|\xi| - |\eta|),$$

where the left-hand side converges weakly as a measure. We conclude that the distribution kernel of $\mathcal{F}(S - I)\mathcal{F}^*$ equals $-(\pi i/|\xi|)\delta(|\xi| - |\eta|)\widehat{W}_\xi(-\eta, \xi)$ outside the origin in $\mathbf{R}^n \times \mathbf{R}^n$. It follows that S is local and also that

$$\widetilde{S}(\phi, \phi'; \lambda) = \delta(\phi - \phi') - (i\pi)(2\pi)^{-n}\lambda^{n-2}\widehat{W}_\phi(-\lambda\phi', \lambda\phi), \quad \lambda > 0,\ \phi,\ \phi' \in S^{n-1}. \tag{1.7}$$

Thus the information given by the scattering matrix is the restriction of $\widehat{W}_\eta(\xi, \eta)$ to the set where $|\xi| = |\eta|$. Note that the scattering data depends on $2n - 1$ real parameters, while the potential depends on n parameters only. Therefore, the inverse scattering problem, i.e. the problem to determine v from S is overdetermined.

In the considerations above we assumed that V was small. To handle the case when V is large it is convenient to extend the previous construction of intertwining operators also to non-local potential operators V, and one may then also allow V to depend on θ.

Now let $V \in \mathcal{V}$ be any local potential. Then, by using the more general construction of intertwining operators in several steps, one may construct a family of operators B_θ with the same properties as the A_θ in Theorem 1, but with $\lambda = 0$ in (i), so that

$$B_\theta^{-1} H_V B_\theta = H_0 + K_\theta,$$

where the K_θ are of finite rank and depend continuously on θ. More precisely,

$$K_\theta(x, y) = \sum_1^N f_{j,\theta}(x) g_{j,\theta}(y),$$

where the $f_{j,\theta}$ and $g_{j,\theta}$ are continuous functions of θ with values in $W^{1,\infty} = \cap_{k \geq 0} W^{1,k}$. Moreover $\langle y - x, \theta \rangle \geq 0$ in the support of each term in the right-hand side above.

The problem of constructing an intertwining operator is now reduced to the corresponding problem for $H_0 + K_\theta$, and this will only involve elementary operations in some L^1-algebras of functions supported in half-spaces. In particular, one can construct a certain determinant $\delta - q_\theta$ associated to the sequence $(f_{j,\theta} \otimes g_{j,\theta})_{j=1}^N$. The function q_θ is supported in the set where $\langle x, \theta \rangle \leq 0$, and it is a continuous function of θ with values in $W^{1,\infty}$. Moreover, if A_θ is the intertwing operator in Theorem 1 and $[q_\theta]$ denotes the convolution operator with distribution kernel $q_\theta(x - y)$, then $A_\theta \circ (I - [q_\theta]) \in \mathcal{M}$.

We observe that

$$\mathcal{A}_\theta = A_\theta(I - [q_\theta]) \tag{1.8}$$

is also an intertwining operator, and since $A^*_{-\theta}$ is the inverse of A_θ, it follows that

$$\mathcal{A}^*_{-\theta}\mathcal{A}_\theta = (I - [q_{-\theta}])^* A^*_{-\theta} A_\theta (I - [q_\theta]) = (I - [q_{-\theta}])^*(I - [q_\theta]).$$

In particular, if $\delta - q_\theta$ is for each θ an invertible element in the Banach algebra $\mathcal{B}_\theta$ of L^1-functions supported in the set where $\langle x, \theta\rangle \leq 0$ (with the unit adjoined), then the $\mathcal{A}_\theta$ are invertible intertwining operators in the class $\mathcal{M}$. When H_V has a bound state, then $\mathcal{A}_\theta$ can not be invertible in $\mathcal{M}$. Hence, at least for some θ it must be true that $\delta - q_\theta$ can not be invertible in $\mathcal{B}_\theta$. In fact, $\delta - q_\theta$ is not invertible for any θ then, for it follows from our constructions that the function

$$\mathbf{C} \ni z \mapsto 1 - \widehat{q_\theta}(z\theta), \tag{1.9}$$

which is analytic in the upper complex half-plane and continuous in its closure, has a zero for every z such that z^2 is an eigenvalue for H_V. (On the other hand, it is true that $1 - \widehat{q_\theta}(s\theta) \neq 0$ when $0 \neq s \in \mathbf{R}$.)

Using the ideas sketched above, one obtains a formula for the scattering matrix also when V is not small. Let us set

$$\mathcal{W}_\xi(x, y) = V \circ \mathcal{A}_{-\xi}(x, y) = v(x)\mathcal{A}_{-\xi}(x, y).$$

Then we may replace (1.7) by the equation

$$\widetilde{S}(\phi, \phi'; \lambda) = \delta(\phi - \phi') + c_n(\lambda)\widehat{\mathcal{W}_\phi}(-\lambda\phi', \lambda\phi)/(1 - \widehat{q}_{-\phi}(-\lambda\phi)), \tag{1.10}$$

where

$$c_n(\lambda) = -i\pi(2\pi)^{-n}\lambda^{n-2}. \tag{1.11}$$

2. Some preliminary results in inverse scattering.

In this section we discuss some results in inverse scattering that are rather immediate from the analysis of the intertwining operators in the preceding section.

Versions of the following observations can be found in classical papers of Berezanskii [3] and Faddeev [4].

Theorem 3. *Assume that $V \in \mathcal{V}$ is local and real-valued. Then V is uniquely determined from its scattering matrix.*

PROOF: We let $a(\phi, \phi'; \lambda)$ be the integral kernel of $(\widetilde{S}(\lambda) - I)/c_n(\lambda)$. By (1.10) this is a locally integrable function given by

$$a(\phi, \phi'; \lambda) = \widehat{\mathcal{W}}_\phi(-\lambda\phi', \lambda\phi)/(1 - \widehat{q}_{-\phi}(-\lambda\phi)). \tag{2.1}$$

Let $(\lambda_\nu)_1^\infty$ be a sequence of positive numbers so that $\lambda_\nu \to \infty$ as $\nu \to \infty$. Since

$$\mathbf{R}^n = \cap_{N>0}\left(\cup_{\nu>N}(\lambda_\nu S^{n-1} + \lambda_\nu S^{n-1})\right),$$

an arbitrary vector ξ in $\mathbf{R}^n$ may always be written as a limit

$$\xi = \lim_{\nu \to \infty} \lambda_\nu(\phi_\nu - \phi'_\nu), \tag{2.2}$$

where ϕ_ν, ϕ'_ν are sequences of unit vectors. We claim now that

$$a(\phi_\nu, \phi'_\nu; \lambda_\nu) \to \widehat{v}(\xi) \quad \text{as } \nu \to \infty. \tag{2.3}$$

To see that this is the case, we first observe that, since q_θ is a continuous function of θ with values in $W^{1,\infty}$, it is true that

$$(\mathcal{F}q_{-\phi_\nu})(-\lambda_\nu\phi_\nu) \to 0 \quad \text{as } \nu \to \infty.$$

Thus the limit in (2.3) will not be changed, if the last factor in the right-hand side of (2.1) is removed.

Next we write

$$\mathcal{W}_\theta(x, y) = v(x)\delta(x - y) + R_\theta(x, y),$$

where $R_\theta(x, y) = v(x)(\mathcal{A}_\theta(x, y) - \delta(x - y))$ is a bounded measure on $\mathbf{R}^n \times \mathbf{R}^n$, which depends continuously on θ. Since $A_\theta(x, y) - \delta(x - y)$ is locally integrable, it is true that R_θ is a continuous function of θ with values in $L^1(\mathbf{R}^n \times \mathbf{R}^n)$. It follows therefore from the Riemann-Lebesgue lemma that

$$(\mathcal{F}R_{\phi_\nu})(-\lambda_\nu\phi'_\nu, \lambda_\nu\phi_\nu) \to 0 \quad \text{as } \nu \to \infty.$$

Therefore, in the left-hand side of (2.3) one may now replace $a(\phi_\nu, \phi'_\nu; \lambda_\nu)$ by

$$b(\phi_\nu, \phi'_\nu; \lambda_\nu) = \widehat{V}(-\lambda_\nu \phi'_\nu, \lambda_\nu \phi_\nu) = \widehat{v}(\lambda_\nu(\phi_\nu - \phi'_\nu)).$$

This proves (2.3), and therefore that v is uniquely determined from the scattering operator. In fact, we have proved that v can be reconstructed as soon as one knows $\widetilde{S}(\lambda_\nu)$ for some sequence λ_ν which tends to infinity.

The method described above is not applicable in practice, since the determination of the Fourier transform of v at a point ξ would require much data from high energy. In inverse scattering on the real line (see [7], [9]) it is well-known that the potential can be recovered from the scattering data (which in that case in addition to the scattering matrix also consists of some finite-dimensional data related to the bound states) by means of an integral equation, the Gelfand-Levitan equation. Similar equations are also important in n-dimensional scattering, and they are often referred to as the Newton-Marchenko equations. (See Newton [11].) Related to this is another method which we describe below.

Let us assume first that V is small. Then it follows that $\|S(\lambda) - I\|$ is small. Let $\theta \in S^{n-1}$ and $\mathcal{N}_\theta$ be the set of operators on the form $I + K$, where K is an operator on $L^2(S^{n-1})$ with a continuous integral kernel $K(\phi, \phi')$ supported in the set $\pm\langle \phi' - \phi, \theta \rangle \geq 0$. Then one may write $S(\lambda)$ in a unique way as a product

$$S(\lambda) = N_\theta^+(\lambda) N_\theta^-(\lambda), \tag{2.4}$$

where $N_\theta^\pm(\lambda) \in \mathcal{N}_{\pm\theta}$. In fact, we have

$$S = W_+^* W_- = W_+^*(A_\theta A_{-\theta}^*) W_- = (W_+^* A_\theta)(W_-^* A_{-\theta})^* = T_\theta^+ (T_\theta^-)^*,$$

where $T_\theta^\pm = W_\pm^* A_{\pm\theta}$ commute with H_0 and are local with respect to energy. If the operators $N_\theta^\pm(\lambda)$ correspond to $T_\theta^\pm$ in the same way as $\widetilde{S}(\lambda)$ corresponds to S, then one can prove that $N_\theta^\pm(\lambda) \in \mathcal{N}_\theta$. Therefore, if one knows the factorization of the scattering matrix into upper and lower triangular matrices with respect to the ordering of S^{n-1} induced by θ, then one knows also

$$A_\theta^* A_\theta = (W_+ A_\theta)^* (W_+ A_\theta) = (T_\theta^+)^* T_\theta^+. \tag{2.5}$$

The potential is easily recovered from the left-hand side of (2.5). In fact, by analysing the singularities of the distribution kernel of $A_\theta^* A_\theta$ along the diagonal in $\mathbf{R}^n \times \mathbf{R}^n$ one can easily prove that $v(x)$ can be computed in a simple way from $A_\theta^* A_\theta$. (The corresponding equation in the one-dimensional case is the Gelfand -Levitan equation.)

In the discussion above we assumed that the potential was small. In the general case there are obstructions to factorization of the scattering matrix into upper and lower triangular factors. Such obstructions are due to zeros of $\widehat{q}_\theta(\xi)$ for certain values of (θ, ξ) which are called exceptional points. (See [6], [10], [11].) It can be shown that there are exceptional points as soon as the point spectrum of H_V is not empty.

3. A $\overline{\partial}$-equation for generalized scattering data.

In this section we shall briefly comment on a method, the so called $\overline{\partial}$-method, which has been important for the reconstruction problem and which at the same time gives (at least partially) a solution to the characterization problem. The last problem consists of finding necessary and sufficient conditions for a family $S(\lambda)$ of operators on $L^2(S^{n-1})$ to be a scattering matrix corresponding to some class of potentials. The $\overline{\partial}$-method was introduced by Ablowitz–Nachman and then developed further by Henkin–Novikov. For systems of ordinary differential equations it had been considered earlier by Beals–Coifmann. (See [1], [2], [6] and also the monograph [11] of Newton for more references.) We also want to mention here that parallel to this, considerations of other functional equations, such as for example the so-called Newton-Marchenko equations, already mentioned above, have been of great importance in solving inverse problems for the Schrödinger operator. The pioneering contributions here were given by Faddeev and Newton. We refer to Newton [11] for further references, and point out that the results have been generalized further by Weder [12].

It turns out that the generalized scattering matrix, which is the main ingredient in the $\overline{\partial}$-approach is naturally described in terms of the intertwining operators A_θ. Our main observation is, that there is a simple expression for $\partial_\theta A_\theta$ in terms of V and A_θ. This gives us a new functional equation, and we are going to sketch how this identity leads to the functional equation used in the $\overline{\partial}$-approach. We hope to discuss in a forthcoming paper in more detail the physical interpretations of these equations and their consequences for inverse scattering.

We have seen in the previous section that $W_+^* A_\theta$ corresponds to an upper triangular factor in a factorization of the scattering matrix. This was true at least when V is small. In the case when V was large one had to modify this statement by replacing A_θ by the operator $\mathcal{A}_\theta$ in (1.8). In what follows we shall assume for simplicity that V is small, since similar modifications will also allow us to pass to general V in the statements made below.

In the following we shall consider the distributions

$$E_{\theta,j} = \partial_{\theta_j} E_\theta, \quad \theta \in \mathbf{R}^n \setminus 0, \ 1 \leq j \leq n \tag{3.1}$$

where E_θ is considered as a function homogeneous of degree 0 in θ. It is easily seen that

$$E_{\theta,j} = (2\pi i)^{-n} \int_{\langle \omega, \theta \rangle = 0} \omega_j L_\omega \, d\omega / |\theta|.$$

This implies that $E_{\theta,j} * U$ is defined for every $U \in \mathcal{V}$, in particular when $U = (V \circ A_\theta)$. The same remark applies of course to

$$E_\theta = (2\pi i)^{-n} \int_{\langle \omega, \theta \rangle \geq 0} L_\omega \, d\omega,$$

and we recall from Section 1 that A_θ solves the equation

$$A_\theta = \delta(x-y) + E_\theta * (V \circ A_\theta). \tag{3.2}$$

The new identity we want to discuss is the following one:

Theorem 4. *Assume that $V = v(x)\delta(x-y) \in \mathcal{V}$ is local and small. Then*

$$\partial_{\theta_j} A_\theta = A_\theta \circ (E_{\theta,j} * (V \circ A_\theta)). \tag{3.3}$$

We remark that this identity is far from being trivial, and it does not follow directly by differentiating (3.2). In its proof one has to combine the fact that $A_\theta - \delta(x-y) \in \mathcal{M}_\theta$ with some uniqueness results for solutions to the ultra-hyperbolic equation.

Let v be as in Theorem 4. In order to discuss how the fundamental equation in the $\overline{\partial}$-approach follows from (3.3), we introduce

$$A_p(x,y) = e^{i\langle y-x,p\rangle} A_\theta(x,y), \quad \theta = \operatorname{Im} p,\ p \in \mathbf{C}^n \setminus \mathbf{R}^n.$$

Let us consider the mapping

$$\mathbf{R}^n \times \overline{\mathbf{R}_+} \times S^{n-1} \ni (\xi, \lambda, \theta) \mapsto \xi + i\lambda\theta \in \mathbf{C}^n,$$

which is surjective and a diffeomorphism when restricted to the set where $\lambda > 0$. This allows us to consider $\mathbf{C}^n \setminus \mathbf{R}^n$ as the interior of a smooth manifold $\widetilde{\mathbf{C}^n}$ with boundary $\partial\widetilde{\mathbf{C}^n}$ equal to $\mathbf{R}^n \times S^{n-1}$. In the interior of $\widetilde{\mathbf{C}^n}$ we have the differential operators

$$\overline{\partial}_{p_j} = \frac{1}{2}\left(\frac{\partial}{\partial \operatorname{Re} p_j} + i\frac{\partial}{\partial \operatorname{Im} p_j}\right),$$

and it follows from the considerations in Section 1, that A_p may be considered as a continuous function of $p \in \widetilde{\mathbf{C}^n}$ with values in $\mathcal{M}$. The equations (3.3) are obviously equivalent to

$$\overline{\partial}_{p_j} A_p = A_p \circ (E_{p,j} * (V \circ A_p)), \quad p \in \mathbf{C}^n \setminus \mathbf{R}^n \tag{3.4}$$

if we set

$$E_{p,j}(x,y) = \frac{i}{2} e^{i\langle y-x,p\rangle} E_{\operatorname{Im} p,j}(x,y).$$

It follows from the definition of $E(p;x,y)$ in Section 1 that $E_{p,j} = \partial_{p_j} E(p;\cdot,\cdot)$, and the formula for its Fourier transform there gives $\widehat{E}_{p,j}(\xi,\eta) = \mu_{p,j}(\xi,\eta)$, where the right-hand side is a measure given by

$$\mu_{p,j}(\zeta,\eta) = -2\pi(\xi_j + \eta_j)\delta((\eta-p)^2 - (\xi+p)^2).$$

If we set $X = \eta - \xi$, $Y = \eta + \xi$, then the right-hand side is interpreted as

$$-\pi Y_j \delta(X \cdot Y - 2Y \cdot \operatorname{Re} p)\delta(Y \cdot \operatorname{Im} p).$$

Note that the right-hand side is well-defined where $Y \neq 0$, for then $X \cdot Y - 2Y \cdot \operatorname{Re} p$ and $Y \cdot \operatorname{Im} p$ have linearly independent differentials. Since the right-hand side is homogeneous of degree -1 in Y, it extends uniquely to a measure in $\mathbf{R}^n \times \mathbf{R}^n$. One can prove also that $\mu_{p,j}(\xi, \eta)$ may be considered as a continuous measure-valued function of (η, p). The value of this measure when $\eta = 0$ will then be denoted $\nu_{p,j}$.

We introduce now the *generalized scattering data*

$$G(\xi, p) = (\mathcal{F}(V \circ A_p))(\xi, 0) = \int\int e^{-i\langle x, \xi \rangle} v(x) A_p(x, y)\, dx dy, \quad \xi \in \mathbf{R}^n,\ p \in \widetilde{\mathbf{C}^n}.$$

Here the integral is the integral of a bounded measure. We observe that $G(\xi, p)$ is continuous on $\mathbf{R}^n \times \widetilde{\mathbf{C}^n}$, and it is easy to prove that it is a rapidly decreasing function of ξ, uniformly with respect to p. If $a(\phi, \phi'; \lambda)$ is the function in (2.1) then

$$a(\phi, \phi'; \lambda) = \lim_{\varepsilon \to +0} G(\lambda(\phi - \phi')), -(\lambda + i\varepsilon)\phi).$$

Recall that the function to the left carries all information about the scattering matrix. It follows from the Riemann-Lebesgue lemma and some arguments used in the previous section that

$$G(\xi, p) \to \widehat{v}(\xi), \quad \text{as } |p| \to \infty.$$

Thus v can be computed as soon as we know $G(\cdot, p^{(\nu)})$ for some sequence $p^{(\nu)}$ tending to infinity in $\widetilde{\mathbf{C}^n}$. In particular one may choose the sequence so that the complex kinetic energy $p^{(\nu)} \cdot p^{(\nu)}$ stays bounded.

We now state the following important result.(Cf. [6], Theorem 1.4.)

Theorem 5. *Let $V = v(x)\delta(x - y)$ be as in Theorem 3 and assume that v is small. Then*

$$\overline{\partial}_{p_j} G(\xi, p) = (2\pi)^{-n} \int G(\xi - \eta, p + \eta) G(\eta, p) d\nu_{p,j}(\eta), \quad p \in \mathbf{C}^n \setminus \mathbf{R}^n \tag{3.5}$$

PROOF: We multiply the equation (3.4) from the left by V and introduce

$$B_p = V \circ A_p,$$

which is now a continuous function of $p \in \widetilde{\mathbf{C}^n}$ with values in $\mathcal{V}$. It follows that

$$\overline{\partial}_{p_j} B_p = B_p \circ (E_{p,j} * B_p). \tag{3.6}$$

One can prove now that the Fourier transform of $B_p \circ (E_{p,j} * B_p)$ equals

$$(\xi, \eta) \mapsto (2\pi)^{-n} \int \widehat{B}_p(\xi, -\eta')\widehat{B}_p(\eta', \eta)\mu_{p,j}(\eta', \eta)\, d\eta'.$$

From the definition of G follows that

$$\widehat{B_p}(\xi, \eta) = G(\xi + \eta, p - \eta).$$

Hence

$$\overline{\partial}_{p_j} G(\xi + \eta, p - \eta) = (2\pi)^{-n} \int G(\xi - \eta', p + \eta')G(\eta + \eta', p - \eta)\mu_{p,j}(\eta', \eta)\, d\eta',$$

and the theorem follows if one lets η tend to 0 in this formula.

REFERENCES.

1. Ablowitz, M. and Nachman A., A multidimensional inverse scattering method, *Studies in Appl. Math.*, 71 (1984), 243-250.
2. Beals, R. and Coifman, R., Linear spectral problems, nonlinear equations and the $\overline{\partial}$-method, *Inverse problems*, 5 (1989), 87-130.
3. Berezanskii, Yu.M., The uniqueness theorem in the inverse problem of spectral analysis for the Schrödinger equation, *Trudy Moskov. Mat. Obshch.*, 7 (1958), 3-62.
4. Faddeev, L.D., Uniqueness of the solution of the inverse scattering problem, *Vestnik Leningrad. Univ*, 7 (1956), 126-130.
5. Faddeev, L.D., Inverse problem of quantum scattering theory, II, *J. Sov. Math.*, 5 (1976), 334-396.
6. Henkin, G.M. and Novikov, R.G., The $\overline{\partial}$-equation in multi-dimensional inverse scattering, *Uspeckhi Mat. Nauk*, 42(1987), 93-152 (also in *Russian Math. Surveys*, 42 (1987), 109-180.
7. Marchenko, V.A., *Sturm-Liouville operators and applications in operator theory: Advances and applications*, Vol 22, Birkhäuser Verlag, 1986.
8. Melin, A., Intertwining methods in multi-dimensional scattering theory I, in *University of Lund and Lund Institute of Technology preprint series*, No 13, 1987, p. 1-81.
9. Melin, A., Operator methods for inverse scattering on the real line, *Comm. in Partial Diff. Eqs.*, 10 (1985), 677-766.
10. Melin, A., Inverse problems and microlocal analysis, in *Journées 'Équations aux dérivées partielles'*, *Saint-Jean-de Monts*, 1991.
11. Newton R.G, *Inverse Schrödinger scattering in three dimensions*, Springer-Verlag, 1989.
12. Weder, R., Multidimensional inverse scattering: the reconstruction problem, *Inverse problems*, 6 (1990), 267-298.

Projective Algebraic Geometry in Positive Characteristic

Ragni Piene

Matematisk institutt
University of Oslo
P.B.1053 Blindern
0316 Oslo, Norway

Abstract

The geometry of projective algebraic varieties defined over a field of positive characteristic may be quite different from the usual characteristic 0 geometry. Here we have a look at some of the phenomena that may occur — in particular we consider the geometry of plane curves; the theory of tangent and osculating spaces and the Gauss map for arbitrary varieties; the special behavior of the Fermat hypersurfaces. We also discuss a geometric approach to the number theoretic problem of finding a bound for the number of integer (or rational) solutions to a polynomial equation.

This paper, which is based on the talk given at the 21st Nordic Congress of Mathematicians, is meant as a brief introduction to the subject. The only new material included is a corrected proof of Proposition 11 in the paper "On the inseparability of the Gauss map" by Kleiman and the author [**15**, p.120] — as Kaji has pointed out, there was an error in the previous proof.

1. Introduction

Let k be an algebraically closed field, denote by $\mathbf{P}_k^n$ projective n-space over k, and suppose $X \subset \mathbf{P}_k^n$ is a variety. In the classical case, i.e., if the characteristic of k is 0, there is a well known geometric theory of X, including the tangent spaces and osculating spaces, the Gauss map, the conormal variety of X, the dual variety, the points of hyperosculation (or generalized Weierstrass points), and so forth. Moreover, one can give enumerative formulas — like the generalized Plücker formulas that relate the characters of X to those of its dual variety, or in some cases formulas counting the number of hyperosculating points. One of the classical results is that *biduality*, or *reciprocity*, holds, i.e., the double dual variety of X is equal to X. In characteristic 0 this is equivalent to *reflexivity*, i.e., the conormal variety of a variety X is equal to the conormal variety of its dual variety.

In the case that the characteristic of k is positive, however, the "geometry" of X may be quite different. Some of the phenomena that are new to positive characteristic are: X may have concurrent tangent spaces, all tangents to X may be multitangents, the Gauss map may be inseparable, the gap sequence may be non-classical.

We first consider the case of plane curves, where these phenomena are fairly well understood.

It was shown by Lluis [**18**] that the only nonsingular curves with concurrent tangents are conics in characteristic 2. This was generalized in [**15**], where we showed that the only nonsingular hypersurfaces with concurrent tangent spaces are odd-dimensional quadrics in characteristic 2.

The Gauss map of a given variety is the map that sends a nonsingular point to the tangent space at that point, considered as a point in the Grassmann variety. To say that the Gauss map is purely inseparable, is to say that a general tangent space is tangent at only one point, i.e., that most tangent spaces are not "multitangents". For curves there is a sufficient condition for the Gauss map to be purely inseparable ([**12**], [**15**]). In higher dimensions, not much is known. It was shown in [**15**] that complete intersection surfaces have purely inseparable Gauss maps — however, Kaji pointed out an error in the proof of Proposition 11 of [**15**], which is used to prove the above mentioned result. A corrected proof of this proposition is presented below. A recent result by Kaji [**13**] gives another sufficient condition for arbitrary dimensional varieties to have purely inseparable Gauss map, namely that their cotangent bundle be generically ample.

Even in characteristic 0, it is well known that the geometry of Fermat hypersurfaces is a bit special. Here we consider Fermat hypersurfaces of degree $q+1$, where $q = p^e$ and p is the characteristic of k, i.e., hypersurfaces projectively equivalent to the one given by $\Sigma T_i^{q+1} = 0$. For example, Beauville [**2**] has given several characterizations of such Fermat hypersurfaces, related to hyperplane sections, polar divisors, and dual varieties. In [**15**] it was conjectured — and proved in the case of surfaces — that these characterizations are equivalent to the condition that the dual variety is nonsingular.

For curves in positive characteristic, Laksov [**16**] defined, and gave a formula for the number of hyperosculating points. Xu [**26**] gave a formula for the number of hyperosculating points in the case that X is a Fermat surface as above, and he obtained a bound for this number in the case of higher dimensional Fermat hypersurfaces [**27**].

Finally we shall give an example of some work of the "Brazilian school" of how the geometry of curves in positive characteristic can be used to deduce bounds for the number of rational points of a curve defined over a finite field.

2. Geometry of plane curves

Let k be an algebraically closed field and let $\mathbf{P}_k^2$ denote the projective plane over k. The zero set of a homogeneous polynomial $F(x, y, z)$ defines a plane curve $X \subset \mathbf{P}_k^2$, of degree equal to the degree of F. Suppose F has no multiple factors. Then the *dual map* of X is the rational map

$$\gamma : X \to (\mathbf{P}_k^2)^*,$$

which sends a nonsingular point $x \in X$ to the tangent line to X at x, considered as a point in the dual projective plane. In other words, γ is the map given by the partial derivatives of F: $\gamma = (F_x, F_y, F_z)$. The (closure of the) image X^* of γ is called the *dual curve* of X.

If the base field k has characteristic 0, it is well known that the map γ is birational onto the dual curve X^*. The birational inverse is the dual map of X^* and $(X^*)^* = X$ holds (note that the dual space of $(\mathbf{P}_k^2)^*$ is equal to $\mathbf{P}_k^2$). In particular the curve X and the dual curve X^* have the same geometric genus. It also follows that a general tangent line to the curve X is tangent at only one point. Moreover, the inflection points of X are the (nonsingular) points of ramification of the map γ and are finite in number, or, most tangent lines have order of contact 2 with the curve at the point of tangency. In fact, the inflection points are the nonsingular points of the intersection of X with its *Hessian curve*, the curve defined by the determinant of the Hessian matrix of F.

Let d denote the degree of X and d^* the degree of X^* — also called the *class* of X. Then d^* is equal to the number of tangent lines to X that pass through a given (general) point of $\mathbf{P}_k^2$, or to the number of lines through a general point that are tangent to X.

The enumerative characters of X and X^* are related by the *Plücker formulas*. One of them is the following:

$$d^* - d(d-1) \quad \sum_{P \in \mathrm{Sing} X} e_P, \qquad (I)$$

where e_P denotes the multiplicity of the Jacobian ideal in the local ring of X at a point P. Moreover, if g denotes the geometric genus of X and X^*, and if X has only δ nodes and κ ordinary cusps as singularities, then

$$d^* = d(d-1) - 2\delta - 3\kappa = 2d + 2g - 2 - \kappa,$$

and there are corresponding dual formulas

$$d = d^*(d^* - 1) - 2\tau - 3\iota = 2d^* + 2g - 2 - \iota,$$

where τ is the number of bitangents and ι the number of inflection points (flexes).

Suppose now that the field k has characteristic $p > 0$. Then the "geometry" of X may be quite different:

(1) X may have concurrent tangent lines (i.e., all tangents to X at nonsingular points pass through a common point).
(2) X may have infinitely many multitangents (i.e., almost all tangents are tangent at several points of X).
(3) All points of X may be inflection points (i.e., every tangent has order of contact > 2 with X).

In fact, it turns out that the dual map γ is not necessarily birational — it may have inseparable degree > 1, or both inseparable and separable degrees > 1 — and X and X^* need not have the same geometric genus. This phenomenon was first studied seriously by Wallace [**25**] (see also [**14**] and references therein). Curves with concurrent tangents are often called *strange* curves — they have recently been studied in [**1**] and [**9**].

The Plücker formula (I) still holds, provided the class d^* is replaced by $\deg\gamma \cdot d^*$.

Example 1. Let X be the curve given by the equation

$$F(x,y,z) = xy^{p-1} - z^p = 0.$$

Then the partial derivatives of F are $F_x = y^{p-1}$, $F_y = -xy^{p-2}$, and $F_z = 0$. The tangent line at the point $x = (a,b,c)$ is the line $bx - ay = 0$ — hence all tangents pass through the point $(0,0,1)$. Therefore, the dual curve X^* is the line in $(\mathbf{P}^2)^*$ dual to the point $(0,0,1)$, and the dual map is just the projection of X from the point $(0,0,1)$.

In this case, we have $d^* = 1$, $\deg\gamma = p$, and $e_P = p(p-2)$, which checks with the Plücker formula (here $P = (1,0,0)$ is the only singular point of X, if $p > 2$).

Example 2. ([**14**]) Let X be the curve given by the equation

$$F(x,y,z) = xy^{q-1}z^{q^2-q} + x^q z^{q^2-q} + y^{q^2} = 0,$$

where $q = p^e$, with $p^e > 2$. One sees that all tangents pass through $(0,0,1)$, but one also checks that every tangent line is tangent at $q-1$ distinct points, and has order of contact q at each of these points. In this case the dual map has separable degree $q-1$ and inseparable degree q. Note that the singular points of X are a high order cusp at $(1,0,0)$ and an ordinary multiple point at $(0,0,1)$.

We have $\deg\gamma = q(q-1)$. Since the multiplicity of the Jacobian ideal at an ordinary q-tuple point is $q(q-1)$, we conclude from the Plücker formula that the multiplicity of the Jacobian ideal at the cusp $(1,0,0)$ is equal to $q(q-1)^2(q+2)$.

The curves of Examples 1 and 2 have singular points. Indeed, it was shown by Lluis ([**18**], see also Samuel [**21**]) that the only *nonsingular* curves with concurrent tangents are conics in characteristic 2. A similar result holds for hypersurfaces (of arbitrary dimension):

Theorem. (Kleiman–Piene [**15**, Th. 7, p.118]) *The only nonsingular hypersurfaces $X \subset \mathbf{P}^n$, of degree $d \geq 2$, with concurrent tangent hyperplanes, are odd-dimensional quadrics in characteristic* 2.

The idea of the proof is the following: if X has concurrent tangent hyperplanes, then the dual variety X^* is a hyperplane in $\mathbf{P}^{n*}$, in particular it is nonsingular. Now if the dual variety is nonsingular, then one can show that it has the same top Chern class as X — since the dual variety has degree 1, this is only possible if X has degree 1 or 2.

Let us consider the case $p = 2$, i.e., $X \subset \mathbf{P}^2_k$ is a plane curve over an algebraically closed field of characteristic 2 ([**4**], [**23**], [**3**]). In this case the dual map is never separable (the Hessian determinant is 0). If X is a conic, X has concurrent tangents. If X is a nonsingular cubic, then so is X^*, and $X^{**} \neq X$ unless X is projectively equivalent to a Fermat cubic, given by $x^3 + y^3 + z^3 = 0$. This last condition is equivalent to the j-invariant of X being 0, or to the Hasse invariant of X being 0. Moreover if X is not projectively equivalent to a Fermat curve, then $X \cap X^{**}$ is equal to the set of inflection points, so that in this case the double dual curve X^{**} plays the role of the Hessian curve in the characteristic 0 case. If X is projectively equivalent to a Fermat curve, then $X^{**} = X$ holds, and X "behaves" like a conic in characteristic 0. If X is a nodal cubic, then X^{**} is the line through the three inflection points of X!

The basic problem in the theory of *enumerative geometry* of plane curves is the following: find the number $N_{\alpha,\beta}$ of curves in a given family that go through α given points and are tangent to β given lines, where $\alpha + \beta$ is equal to the dimension of the family. The numbers $N_{\alpha,\beta}$ are called the *characteristic numbers* of the family. Even in characteristic 0 these numbers are known only for families of curves of degree $d \leq 4$ — of these, only the characteristic numbers for families of curves of degree 2 and 3, originally determined by Chasles, Zeuthen, and Maillard, have been rigorously verified according to modern standards.

For families of curves in positive characteristic p, where p is small with respect to the degree d of the curves, not much is known. The case of conics in characteristic 2 was done by Vainsencher [**23**]. The case of cubics in characteristic 2 was treated by Berg [**3**], who obtained the following characteristic numbers for families of cubics with j-invariant 0 (i.e., the Fermat cubics):

$$
\begin{aligned}
N_{8,0} = N_{0,8} &= 1 \\
N_{7,1} = N_{1,7} &= 2 \\
N_{6,2} = N_{2,6} &= 4 \\
N_{5,3} = N_{3,5} &= 8 \\
N_{4,4} &= 10
\end{aligned}
$$

The symmetry of the numbers reflects the fact that the dual of a cubic curve with j-invariant 0 is again a cubic curve with j-invariant 0.

3. Gauss maps

Let $X \subset \mathbf{P}^n$ be a variety of dimension r. The *Gauss map* of X is the rational map

$$\gamma : X \to \mathrm{Grass}(r, n)$$

sending a nonsingular point x of X to the (projective) tangent space to X at x, considered as a point in the Grassmann variety of r-dimensional linear subspaces of $\mathbf{P}^n$. If X is a hypersurface, i.e., if $r = n - 1$, then the Gauss map is the same as the dual map. Set $\mathcal{L} = \mathcal{O}_{\mathbf{P}^n}(1)|_X$, let $\mathcal{P}^1_X(\mathcal{L})$ denote the 1st order sheaf of principal parts of $\mathcal{L}$, and let

$$a^1 : \mathcal{O}_X^{n+1} \to \mathcal{P}^1_X(\mathcal{L})$$

denote the Taylor map, taking a coordinate function to its 1st jet [**20**]. On the open subvariety where X is nonsingular, a^1 determines an $(r+1)$-dimensional quotient, and the Gauss map is the map defined by this quotient. In characteristic 0 it is known that the Gauss map is birational, and that it is finite if X is nonsingular. In positive characteristic the Gauss map need not be birational — it may have inseparable degree strictly greater than 1, and possibly also separable degree strictly greater than 1.

For curves we have the following result:

Theorem. (Kaji [**12**, Th.0.2], Kleiman–Piene [**15**, Th.8, p.118]) *Let $X \subset \mathbf{P}^n$ be a (reduced and irreducible) curve of geometric genus g. Let $X' \to X$ denote the normalization map and κ the degree of the ramification divisor of this map. If $2g-2 > \kappa$ holds, then the Gauss map of X is purely inseparable.*

It is reasonable to believe, as was conjectured in [**15**, Conj.1, p.107], that many nonsingular projective varieties should have purely inseparable Gauss map. The two known results are the following.

Theorem. (Kleiman–Piene [**15**, Th.13, p.124]) *If X is a nonsingular complete intersection surface of degree ≥ 2, then its Gauss map is purely inseparable.*

Theorem. (Kaji [**13**]) *Every nonsingular projective variety with generically ample cotangent bundle has purely inseparable Gauss map.*

Kaji observed that there is an error in the proof of Proposition 11 of [**15**, p.120]. There we claimed that for the extension of function fields $K(X)/K(X^\vee)$ defined by the Gauss map, if X' denotes the normalization of $X^\vee$ in the purely inseparable closure of $K(X^\vee)$ in $K(X)$, then the induced map $X \to X'$ is separable. This claim only holds for function fields of dimension 1. Kaji shows that for a finite extension L/K of fields, if K_s and K_i denote the separable and purely inseparable closures of K in L, then L is not necessarily separable over K_i though L is purely inseparable over K_s. Indeed, he gives an example of 2-dimensional function fields where this statement fails, namely: $K = k(x,y)$, $L = K(z)$, where z is a root of the equation $Z^{2p} + xZ^p + y = 0$, and k is a field of characteristic $p > 2$. (Note that this field extension does not come from a Gauss map.)

Here is the statement of the above mentioned proposition, and a new proof:

Proposition. [**15**, Prop.11, p.120] *Let X be a smooth projective surface of degree $d \geq 2$, with very ample invertible sheaf $\mathcal{L}$. Set $h := c_1(\mathcal{L})$. Assume that $c_1(\mathcal{P}^1_X(\mathcal{L}))$ is numerically equivalent to a rational multiple of h and that $c_1(X) \leq 4h$ and $c_2(X) > 4d$. Then the Gauss map of X is purely inseparable.*

Proof. Let $X^\vee$ denote the image of the Gauss map, and X'' the normalization of $X^\vee$ in the separable closure of its function field in the function field of X. Then the Gauss map factors through a map $\epsilon : X \to X''$ which is finite, purely inseparable, and of degree i; moreover, ϵ is flat off the finite number of singular points of X''.

As in [**15**, Setup 1, p.112], let $Y = \mathbf{P}(\mathcal{P}^1_X(\mathcal{L}))$ denote the (abstract) tangent developable of X, so that Y is the pullback to X of the incidence variety, under the Gauss map. Let Y'' and $Y^\vee$ denote the pullbacks of the incidence variety to X'' and $X^\vee$, and $\eta : Y \to Y''$ the induced map. The tangent developable $Y \to X$ has an obvious section — let $S \subset Y$ denote its image. Let D' denote the image of S in $Y^\vee$ and D the pullback of D'

to Y; hence, D is equal to the pullback under η of D'', where D'' is the pullback of D' to Y''. Therefore,

$$\eta_*[D] = i[D'']. \tag{$*$}$$

Moreover, D'' is reduced, because $X'' \to X^\vee$, and hence $Y'' \to Y^\vee$, is separable. Hence, if S'' denotes the image of S in Y'', then the cycle $[D'']$ contains $[S'']$ with multiplicity 1. Hence $[D]$ contains $\eta^*[S'']$ with multiplicity 1; that is, if, say, $\eta^*[S''] = j[S]$, then $[D]$ contains $[S]$ with multiplicity j. Finally, $j = i$; indeed, $(*)$ implies that $\eta_* j[S] = i[S'']$, and $\eta_*[S] = [S'']$ because $S \to S''$ is birational (in fact, an isomorphism). Thus we have shown that $[D]$ contains $[S]$ with multiplicity i.

Let $\tilde{\pi} : \tilde{Y} \to Y$ be the blowup along S, and let $\tilde{S}$ be the strict transform of D. It now follows that the direct image $\tilde{\pi}_*[\tilde{S}]$ is equal to the difference $[D] - i[S]$. By [**15**, Lemma 2, p.114], [D] is numerically equivalent to $is[S]$, where s denotes the separable degree of $X \to X^\vee$. Hence $\tilde{\pi}_*[\tilde{S}]$ is numerically equivalent to $is[S] - i[S] = i(s-1)[S]$. The rest of the proof [**15**, pp.120–121] now goes through as given, after replacing each occurrence of $(s-1)$ by $i(s-1)$.

4. Fermat varieties

Suppose $X \subset \mathbf{P}^2_k$ is a Fermat curve given by the equation

$$x^{q+1} + y^{q+1} + z^{q+1} = 0,$$

where $q = p^e$ is a power of the characteristic p of k. Then the Gauss map is just the iterated Frobenius: $\gamma(x, y, z) = (x^q, y^q, z^q)$, hence it has separable degree 1 and inseparable degree q. In this case, the dual curve X^* is a Fermat curve of the same type, and $X^{**} = X$ holds.

It has been shown that the only nonsingular plane curves of degree $d \geq 4$ such that the dual curve is also nonsingular are the ones projectively equivalent to Fermat curves of degree $q + 1$ as above ([**19**], [**10**], [**11**], [**6**]). It was conjectured in [**15**, Conj 2, p.107] that the same holds for hypersurfaces in any dimension. Beauville gave several charactersizations of hypersurfaces projectively equivalent to Fermat ones; using this, and the pure inseparability of the Gauss map, it was shown in [**15**, Th.14, p.124] that the conjecture holds for surfaces.

Given a variety $X \subset \mathbf{P}^n$ of dimension r, set $\mathcal{L} = \mathcal{O}_{\mathbf{P}^n}(1)|_X$, let $\mathcal{P}^m_X(\mathcal{L})$ denote the sheaf of principal parts of order m of $\mathcal{L}$, and let

$$a^m : \mathcal{O}_X^{n+1} \to \mathcal{P}^m_X(\mathcal{L})$$

denote the Taylor map, taking a coordinate function to its mth jet [**20**]. The rank of each a^m is constant, equal to $s(m)+1$, say, on some open subvariety of X. This gives a sequence

$$0 = s(0) < r = s(1) \leq s(2) \leq \ldots \leq s(\bar{m}) = n,$$

where $\bar{m}$ is the smallest integer m such that a^m has generic rank $n+1$. Discarding those $s(j)$ such that $s(j) = s(j-1)$, we obtain a strictly increasing sequence

$$0 = s(0) < r = s(1) < s(b_2) < \ldots < s(b_t) = n.$$

The sequence

$$b_0 = 0 < b_1 = 1 < b_2 < ... < b_t$$

is called the *gap sequence* of $X \subset \mathbf{P}^n$. It was introduced for curves by Laksov [**16**, p.49] and used for hypersurfaces by Xu ([**26**], [**27**]). For varieties in characteristic 0, not contained in a linear subspace, the gap sequence is always $0 < 1 < 2 < ...$ — such a sequence is called *classical.*

A *point of hyperosculation*, also called a *generalized Weierstrass point*, of X, is a point where the rank of a^m is smaller than $s(m)$, for some m.

Laksov [**16**] gave generalized Plücker formulas for curves in $\mathbf{P}^n$ in positive characteristic, in particular he found formulas for the (weighted) number of hyperosculating points. For example, if $X \subset \mathbf{P}^2_k$ is a Fermat curve of degree $q+1$, $q = p^e$, then the gap sequence is $0 < 1 < q$, and the total number of points of hyperosculation is

$$(q+1)(q^2 - q + 1).$$

There is no reason to expect a variety $X \subset \mathbf{P}^n$ of dimension ≥ 2 to have only finitely many points of hyperosculation, but if it does, one would like a formula for their number as in the case of curves. For example, a *general* surface in $\mathbf{P}^3$ has *no* points of hyperosculation, since no plane intersects such a surface in a point of multiplicity ≥ 3. On the other hand, one can give a formula for the number of hyperosculating points of a *ruled surface, assuming* that it only has a finite number.

Suppose next that $X \subset \mathbf{P}^n$ is a Fermat hypersurface of degree $q+1$, with $q = p^e > 2$ equal to a power of the characteristic $p > 0$. Then almost every tangent hyperplane to X has order of contact q at the point of tangency. The gap sequence is $0 < 1 < q$, and the points of hyperosculation of X are the points such that the tangent hyperplane to X at that point has order of contact strictly greater than q (hence equal to $q+1$). For $n = 3$, Xu [**26**, Th.2.2, Th.3.3] showed that a general surface of degree $d = tq+1$, where $q = p^e$, and with gap sequence $0 < 1 < q$, has finitely many points of hyperosculation, their (weighted) number being equal to

$$((q^2 + q + 1)d^2 - 4q(q+1)d + 6q^2)d.$$

If X is a Fermat surface, then the number of hyperosculating points is finite and equal to

$$(q+1)(q^4 - q^3 + 2q^2 - q + 1).$$

In the case of arbitrary dimensional hypersurfaces, Xu gave a bound for the number of hyperosculating points [**27**], provided the number is finite.

5. An application to number theory

The geometry of curves in positive characteristic can be used to answer number theoretical questions (see e.g. [**22**], [**7**], [**24**], [**17**], [**5**], [**8**]) — as an example, let us look at the following situation [**24**].

Let X be a nonsingular curve defined over a finite field $\mathbf{F}_q$, $q = p^e$, and let

$$N := \#X(\mathbf{F}_q)$$

denote the number of rational points of X. Then Weil proved the Riemann Hypothesis for curves over finite fields, namely

$$|N - q - 1| \leq 2g\sqrt{q}.$$

One can show this by considering an embedding $X \subset \mathbf{P}_k^n$, where k is an algebraic closure of $\mathbf{F}_q$. Let $F : X \to X$ denote the iterated Frobenius

$$F(x_0, ..., x_n) = (x_0^q, ..., x_n^q).$$

Then the rational points of X are just the fixed points of F, and their number can be computed using l-adic cohomology and the Lefschetz Trace Formula.

One can also give a proof along the following lines ([**22**], [**24**]): There exists an embedding $X \subset \mathbf{P}_k^n$ such that X has a classical gap sequence [**24**, 3. Prop., p.123]; moreover, we may assume X is nonspecial and linearly normal, so that $d = g + n$ holds, where d is the degree and g is the genus of X. The fixed points of F can be bounded by the number of points $x \in X$ such that $F(x)$ is contained in an osculating hyperplane to X at x. This last number can be computed as follows. Set $\mathcal{L} = \mathcal{O}_{\mathbf{P}^n}(1)|_X$, and consider the Taylor map

$$a^{n-1} : \mathcal{O}_X^{n+1} \to \mathcal{P}_X^{n-1}(\mathcal{L}).$$

Since the gap sequence is classical, the map a^{n-1} is generically surjective. Its kernel $\mathcal{K}$ is locally free (since X is a nonsingular curve) with rank 1. We have $F^*\mathcal{L} \simeq \mathcal{L}^{\otimes q}$, and the composition

$$\mathcal{K} \to \mathcal{O}_X^{n+1} \to F^*\mathcal{L}$$

gives a section

$$\mathcal{O}_X \to \mathcal{K}^{-1} \otimes \mathcal{L}^{\otimes q}.$$

This section is not identically zero, and obviously its zeros are the points considered above, each ocurring with multiplicity $\geq n$. Therefore, we get

$$\begin{aligned} N &\leq \frac{1}{n}\deg\mathcal{K}^{-1} \otimes \mathcal{L}^{\otimes q} \\ &= \frac{1}{n}(q(g+n) + n(g+n) + (n-1)(g-1)) \\ &= (n-1)(g-1) + \frac{1}{n}(q+n)(n+g). \end{aligned}$$

We may assume q is a square and take $n = \sqrt{q}$; this gives

$$N \leq 1 + n^2 + 2gn = 1 + q + 2g\sqrt{q}.$$

References

[1] V. Bayer, A. Hefez, *Strange curves,* Comm. Alg. **19** (1991), 3041–3059.

[2] A. Beauville, *Sur les hypersurfaces dont les sections hyperplanes sont à module constant,* in "The Grothendieck Festschrift, Vol.I," P.Cartier, L. Illusie, N. Katz, G. Laumon, Y. Manin, K. Ribet (eds.) Prog. Math. **86**, Birkhäuser, 1990, pp. 121–133.

[3] A. H. Berg, "Enumerativ geometri for plane kubiske kurver i karakteristikk 2," Hovedfagsoppgave, University of Oslo, 1991.

[4] P. Boughon, J. Nathan, P. Samuel, *Courbes planes en caractéristique* 2, Bull. Soc. Math. France **83** (1955), 275–278.

[5] A. Garcia, J. Voloch, *Fermat curves over finite fields,* J. of Number Theory **30** (1988), 345–356.

[6] A. Hefez, *Nonreflexive curves,* Compositio Math. **69** (1989), 3–35.

[7] A. Hefez, N. Kakuta, "New bounds for Fermat curves over finite fields", Preprint.

[9] A. Hefez, I. Vainsencher, *Varieties of strange plane curves,* Comm. Alg. **19** (1991), 333–345.

[8] A. Hefez, J. Voloch, *Frobenius non-classical curves,* Arch. Math. **54** (1990), 263–273.

[10] M. Homma, *Funny plane curves in characteristic* $p > 0$, Comm. Alg. **15** (1987), 1469–1501.

[11] M. Homma, *A souped-up version of Pardini's theorem and its application to funny curves,* Compositio Math. **71** (1989), 295–302.

[12] H. Kaji, *On the Gauss maps of space curves in characteristic p, II,* Compositio Math. **78** (1991), 261–269.

[13] H. Kaji, "On the pure inseparability of the Gauss map," Preprint 1992.

[14] S. Kleiman, *Multiple tangents of smooth plane curves (after Kaji),* in "Algebraic Geometry: Sundance 1988," B. Harbourne, R. Speiser (eds.), Contemp. Math. **116**, Amer. Math. Soc., 1991, pp. 71–84.

[15] S. Kleiman, R. Piene, *On the inseparability of the Gauss map,* in "Proc. of the 1989 Zeuthen Symp.," S. Kleiman, A. Thorup (eds.), Contemp. Math. **123**, Amer. Math. Soc., 1991, pp. 107–129.

[16] D. Laksov, *Weierstrass points on curves,* Astérisque **87–88** (1981), 221–247.

[17] D. Levcovitz, *Bounds for the number of fixed points of automorphisms of curves,* Proc. London Math. Soc. **62** (1991), 133–150.

[18] E. Lluis, *Variedades algebraicas con ciertas condiciones en sus tangentes,* Bol. Soc. Mat. Mex. **7** (1962), 47–56.

[19] R. Pardini, *Some remarks on plane curves over fields of finite characteristic,* Compositio Math. **60** (1986), 3–17.

[20] R. Piene, *Numerical characters of a curve in projective n-space,* in "Real and complex singularities," P. Holm (ed.), Proc. Conf., Oslo 1976, Sitjhoff & Noorhoof, 1977, pp. 475–496.

[21] P. Samuel "Lectures on old and new results on algebraic curves," Tata Institute of Fundamental Research, Bombay, 1966.

[22] K.-O. Stöhr, J. Voloch, *Weierstrass points and curves over finite fields,* Proc. London Math. Soc. **52** (1986), 1–19.

[23] I. Vainsencher, *Conics in characteristic 2,* Compositio Math. **36** (1978), 101–112.

[24] I. Vainsencher, *On Stöhr–Voloch's proof of Weil's theorem,* manuscripta math. **64** (1989), 121–126.

[25] A. Wallace, *Tangency and duality over arbitrary fields,* Proc. London Math. Soc. **(3) 6** (1956), 321–342.

[26] Xu Mingwei, *The hyperosculating points of surfaces in* $\mathbf{P}^3$, Compositio Math. **70** (1989), 27–49.

[27] Xu Mingwei, *The hyperosculating spaces of hypersurfaces,* Math. Z. **211** (1992), 575–591.

Intrinsic Geometry of Convex Ideal Polyhedra in Hyperbolic 3-Space

Igor Rivin

Institute for Advanced Study, School of Mathematics, Princeton, NJ 08540

Abstract

I describe a simple relationship between triangulations in $\mathbb{C}$ and ideal polyhedra in $\mathbb{H}^3$. I produce a complete intrinsic characterization of convex polyhedra in hyperbolic 3-space $\mathbb{H}^3$ with all vertices on the sphere at infinity. I also show that such polyhedra are uniquely determined by their intrinsic metric.

1 Introduction

In this paper I study the intrinsic geometry of convex polyhedra in three-dimensional hyperbolic space $\mathbb{H}^3$, with all vertices on the sphere at infinity S^2_∞. Such a polyhedron P is homeomorphic to the sphere $\mathbf{S}^2$ with a number of punctures (corresponding to the vertices of P). It is not hard to see that P is a complete hyperbolic surface of finite area.

In this paper I prove the following converse:

Theorem 1.1 Characterization of ideal polyhedra *Let M_N be a complete hyperbolic surface of finite area, homeomorphic to the N times punctured sphere. Then M_N can be isometrically embedded in H^3 as a convex polyhedron P_N with all vertices on the sphere at infinity.*

Furthermore,

Theorem 1.2 Uniqueness of realization *The polyhedron P_N promised by Theorem 1.1 is unique, up to congruence.*

In fact, Theorem 1.2 is a special case of the following more general result:

Theorem 1.3 Uniqueness of generalized polyhedra *A generalized polyhedron P_N in H^3 is determined by its intrinsic metric, up to congruence.*

A *generalized polyhedron* is a polyhedron some of whose vertices are inside $\mathbb{H}^3$, some are on the ideal boundary of $\mathbb{H}^3$ and some are beyond the ideal boundary. This is best visualized in the projective model of $\mathbb{H}^3$ as a (Euclidean) polyhedron, all of whose edges intersect the unit ball $\mathbb{H}^3$. The derivation of Theorem 1.3 from Theorem 4.4 is essentially the same as that of Theorem 1.2.

The proof of Theorem 1.1 uses the Invariance of Domain Principle of A. D. Aleksandrov. This is as follows:

Given a map $f : A \to B$ between topological spaces A and B, then f is onto, provided the following criteria are satisfied:

1. The image of f is non-empty.
2. f is continuous.
3. f maps open sets in A to open sets in B.
4. f maps closed sets in A to closed sets in B.
5. B is connected.

Outline of the proof of Theorem 1.1. Let M_μ be a surface homeomorphic to the N-times punctured 2-sphere, with a *marking* μ, that is, a labelling of the punctures. Let $\mathcal{P}^N$ be the space of convex ideal polyhedra in $\mathbb{H}^3$, parametrized by the positions of their vertices on the sphere at infinity of $\mathbb{H}^3$ (interpreted as the Riemann sphere $\overline{\mathbb{C}}$). Three of the vertices are fixed at 0, 1 and ∞, which eliminates the action of the isometry group of $\mathbb{H}^3$. $\mathcal{P}^N$ is easily seen to be a $2N-6$ dimensional manifold. $\mathcal{P}^N$ plays the role of A in the invariance of domain principle. We will abuse notation and view a polyhedron P both as a geometric object and as a polyhedral isometric embedding of M into $\mathbb{H}^3$. P will inherit the labelling of vertices from μ.

Let $\mathcal{T}^N$ be the set of complete, finite volume hyperbolic structures on M_μ.

This set is parametrized by *shears* along the edges of a geodesic triangulation. This parametrization is explained in detail in section 2. It will also be shown (Theorem 2.10) that $\mathcal{T}^N$ is a $2N-6$ dimensional contractible manifold. Although many of the results of section 2 are known to Teichmüller theorists, they are so elementary in this particular setting that it was impossible to resist including a full exposition.

$\mathcal{T}^N$ will play the role of B in the invariance of domain principle.

The role of the map f will be played by the map $\mathfrak{g} : \mathcal{P}^N \to \mathcal{T}^N$. $\mathfrak{g}(P)$ is P viewed as an abstract Riemannian manifold. The continuity of $\mathfrak{g}$ with respect to the chosen coordinate systems on $\mathcal{P}^N$ and $\mathcal{T}^N$ is the content of Theorem 3.7.

Since $\mathcal{P}^N$ and $\mathcal{T}^N$ are manifolds of the same dimension and $\mathfrak{g}$ is continuous, Theorem 1.2 shows that $\mathfrak{g}$ is an open map. That $\mathfrak{g}$ is closed is the content of Theorem 3.9. □

The theory developed in Section 3 is of independent interest. In particular, it leads to a very simple derivation of a set of conditions satisfied by dihedral angles of an ideal polyhedron (Theorem 3.12).

2 Hyperbolic geometry of the N-times punctured sphere

2.1 Geometry of triangles

Let ABC be an ideal triangle in $\mathbb{H}^2$. Pick a point p on the geodesic AB. How far is p from A? This question turns out to make sense in $\mathbb{H}^2$:

Consider the unique horocycle h_A centered at A and passing through p. h_A will intersect AC in a point q. Define $\mathfrak{D}_{ABC}(p)$ to be the distance along h_A between p and q. $\mathfrak{D}_{ABC}$ has the following important property:

Lemma 2.1 *Let p_1 and p_2 be two points on AB. The hyperbolic distance between p_1 and p_2 is equal to $|\log(\mathfrak{D}_{ABC}(p_1)/\mathfrak{D}_{ABC}(p_2))|$.*

Proof. Use the upper half-space model of $\mathbb{H}^2$. Recall that the hyperbolic metric is related to the Euclidean metric on the upper half-space by $ds_h = |dz|/\text{Im}z$. This means, in particular, that the hyperbolic distance between $z_1 = x + iy_1$ and $z_2 = x + iy_2$ is $|\log(y_1/y_2)|$.

By a hyperbolic isometry A can be sent to ∞, B to 0 and C to 1. Horocycles centered on A are then simply horizontal lines, and if $p = x + iy$, then $\mathfrak{D}(p) = 1/y$, since the metric on the horocycle around infinity through p is then simply the standard metric on the real line, rescaled by $1/y$. The assertion of the Lemma now follows. □

Lemma 2.1 gives a way to quantify the ways in which two ideal triangles can be joined together along a side to form an ideal quadrilateral. Intuitively, ideal triangles ABC and ADC can slide with respect to each other along the common side AC. Pick a point p on AC. If $\mathfrak{D}_{ACB}(p) = \mathfrak{D}_{ACD}(p)$ ($\mathfrak{D}$ is taken with respect to the vertex A), then we say that ABC and ADC are joined without a shear (and it is easy to see that reflection in AC will send B to D and vice versa). Otherwise, ABC and ADC are joined with shear $\log(\mathfrak{D}_{ACB}(p)/\mathfrak{D}_{ACD}(p))$. It is clear that shear doesn't depend on which of the vertices A or C is taken as the center of the horocycles. Henceforth the shear between triangles t_1 and t_2 will be denoted by $\mathfrak{s}(t_1, t_2)$. By abuse of notation $\mathfrak{s}(ABCD) = \mathfrak{s}(ABC, ADC)$.

All of the above is quite easily seen in the upper half-space model of $\mathbb{H}^2$ – see Figure 1. Figure 1 also demonstrates Lemma 2.2.

Lemma 2.2 *If the shear is α, and the triangle ABC is positioned so that $A = \infty$, $B = 1$, $C = 0$, then* $\log|D| = \alpha$.

This also shows that the following fundamental lemma:

Lemma 2.3 *$\mathfrak{s}(ABCD)$ is equal to the log of the absolute value of the* **cross ratio** $[C, B, D, A]$.

Recall that cross-ratio of z_1, z_2, z_3, z_4 is defined to be

$$[z_1, z_2, z_3, z_4] = \frac{(z_1 - z_3)(z_2 - z_4)}{(z_1 - z_2)(z_3 - z_4)}$$

Proof. Both $\mathfrak{s}(ABCD)$ and $[C, B, D, A]$ are invariant under the Möbius group, and since any three points can be transformed to 0, 1, and ∞ by a Möbius transformation, the Lemma follows from the obvious special case where $A = \infty$, $B = 1$, and $C = 0$. □

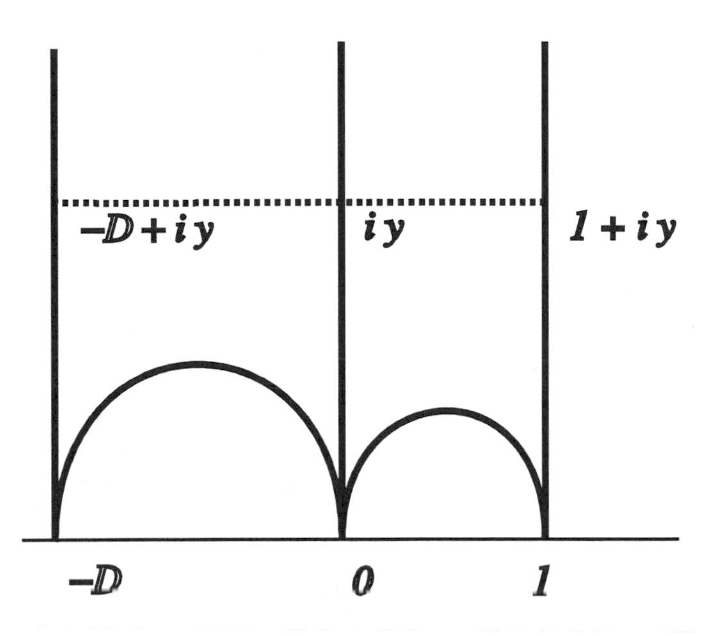

Figure 1 Shear in the upper half-space model of $\mathbb{H}^2$

2.2 Geometry of triangulations

Recall that M_μ is a surface homeomorphic to the 2-sphere with N punctures, together with its marking.

Let T be a triangulation of M_μ. Then the above discussion serves to parametrize all the complete hyperbolic structures on (M, μ) where the faces of T are ideal triangles — to each edge of T we associate the shear of the two abutting faces of T. This information specifies the geometry completely. On the other hand, it is not quite true that any assignment of real numbers to edges of T corresponds to a complete hyperbolic structure on M_μ with those numbers as shears – it is necessary and sufficient that the sum of shears around any cusp add up to zero (pick a horocycle h centered at a vertex v of T. In order for a hyperbolic structure to be complete, h must close up). Thus if there are V vertices and E edges of T, the set of hyperbolic structures on M_μ such that T is an ideal triangulation is naturally parametrized as R^{E-V}. Note that an Euler's formula computation yields $E - V = 2V - 6$, so the dimension of the space of hyperbolic structures depends only on the number of cusps.

The following lemma shows that the reliance on the particular triangulation T in the above discussion is not critical – any other topological ideal triangulation will do as well:

Lemma 2.4 straightening *Any topological triangulation $\tilde{T}$ with all vertices at cusps of M_μ can be straightened to a geodesic triangulation.*

Proof. All that is necessary to show is that if v_1, v_2, v_3 and v_4 are cusps and there are non-intersecting curves γ_1 connecting v_1 and v_2 and γ_2 connecting v_3 and v_4, then the corresponding geodesics also don't intersect. That this last statement is true can be observed by examining the geometry of universal cover of S. The postulated curves γ_1 and γ_2 exist if and only if the lifts of v_1 and v_2 do not separate the lifts of v_3 and v_4 on the circle at infinity of H^2. In that case, however, it is seen that the corresponding geodesic segments do not intersect either. □

Thus, every topological triangulation of M_μ with vertices at the cusps (*topological ideal triangulation*) corresponds to a coordinate system on the space of hyperbolic metrics on S with cusps at the prescribed vertices.

Definition 2.5 *The space of complete hyperbolic structures on M_μ shall be denoted by $\mathcal{T}^N$. A* shear coordinate system *corresponding to a triangulation T on $\mathcal{T}^N$ is the map $C_T : \mathcal{T}^N \longrightarrow \mathbb{R}^{2N-6}$ associating to a particular metric its shears along the straightened edges of T.*

Theorem 2.6 *For any two triangulations T_1 and T_2, the map $c_{T_1T_2} = C_{T_2} \circ C_{T_1}^{-1}$ is a continuous function from $\mathbb{R}^{2N-6}$ to itself.*

To prove Theorem 2.6 it will first be necessary to understand *triangulation graph* of the sphere with N vertices.

Definition 2.7 *Let T be a triangulation, and ABC and ADC be two of the triangles of T sharing the edge AC. Then the* Whitehead move *w_{ABCD} transforms T into a triangulation T', where the triangles ABC and ADC are replaced by triangles BAD and BCD (in other words the diagonal of the quadrilateral $ABCD$ is "flipped").*

Definition 2.8 *The triangulation graph $\mathbf{T}_N$ is a graph whose vertices are isotopy classes of triangulations of S^2 on N vertices, and there is an edge joining nodes corresponding to T_1 and T_2 if and only if there exists a Whitehead move transforming T_1 and T_2.*

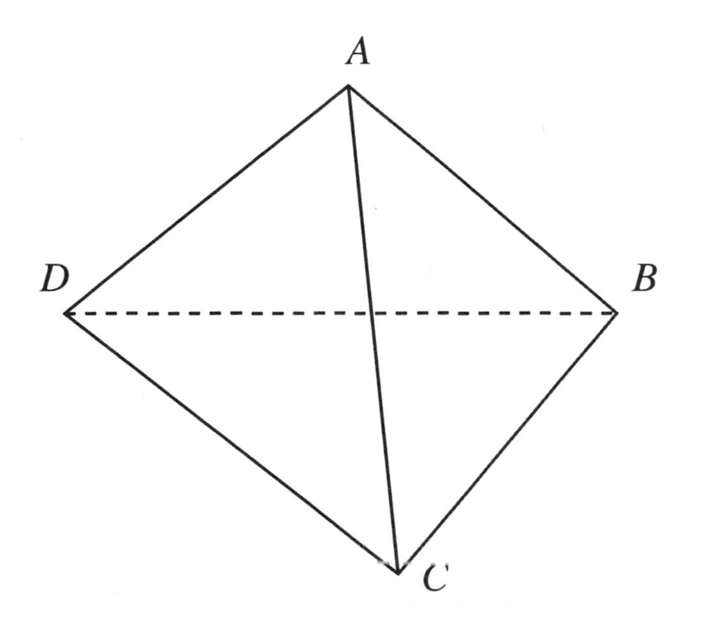

Figure 2 A Whitehead move.

Theorem 2.9 *The graph* $\mathbf{T}_N$ *is connected.*

Proof. Pick two distinguished vertices v_1 and v_2. For any starting triangulation T there is a sequence of Whitehead moves transforming T into a triangulation T_{v_1} where v_1 is connected to every other vertex. Now consider the complement in T_{v_1} of v_1 and all the edges incident to it. This will be a triangulation of an $(N-1)$-gon, all of whose vertices are those of the $(N-1)$-gon. By a similar argument, this can be transformed by a sequence of Whitehead moves into a triangulation where every vertex is connected to v_2. Thus, it is seen that by a sequence of Whitehead moves, every triangulation can be transformed to a particular triangulation (where both v_1 and v_2 have valence $N-1$), and thus $\mathbf{T}_N$ is connected. □

Proof of Theorem 2.6. It is enough to note that the cross ratio of any permutation $z_{\sigma(1)}, z_{\sigma(2)}, z_{\sigma(3)}, z_{\sigma(4)}$ is a rational function of the cross ratio of z_1, z_2, z_3, z_4 (this is most easily seen when $z_1 = 0, z_2 = 1, z_3 = \infty$). By Lemma 2.3 flipping the diagonal of $ABCD$ corresponds to permuting the arguments of the cross ratio. Since a permutation of the arguments corresponds to a fractional linear transformation of the cross-ratio itself, Theorem 2.6 follows.

□

We summarize the results of this section for convenience:

Theorem 2.10 *$\mathcal{T}^N$ is a contractible $2N-6$ dimensional manifold, as evidenced by coordinate systems coming from ideal triangulations of S^2_N. Any two such coordinate systems are analytically equivalent.*

3 Geometry of ideal tessellations

First, let us review briefly the geometry of the upper half-space model of $\mathbb{H}^3$. We will think of the ideal boundary S^2_∞ of $\mathbb{H}^3$ as the Riemann sphere $\overline{\mathbb{C}}$. Hyperbolic planes are represented by hemispheres whose equatorial circles are in S^2_∞. In the present context we think of straight lines as circles passing through ∞. The corresponding hemispheres are vertical planes rising above the lines.

Let P be a convex polyhedron with all vertices on the ideal boundary of $\mathbb{H}^3$. P is the intersection of the half-spaces defined by its faces. By an isometry of $\mathbb{H}^3$ and relabelling we can transform P so that the face f_1 of P lies in the plane rising above the real axis in S^2_∞, and the vertices v_1, v_2 and v_3 are 0, 1 and ∞ respectively. Furthermore, without loss of generality we assume that P lies above the half-plane $\mathrm{Im}(z) \geq 0$.

The rest of the faces of P are then oriented in such a way that the interior of the corresponding hemispheres lie *outside* of P.

P defines a Euclidean tessellation of $\overline{\mathbb{C}}$ in the natural way: P casts a shadow on the ideal boundary of $\mathbb{H}^3$ under the orthogonal projection. The edges of P are then mapped to straight-line segments, and the faces of P to convex polygons. Denote the resulting tessellation of $\mathbb{C}$ by T_P. This tessellation has the following properties:

Condition 1. Every face F of T_P is inscribed in the circle C_F.

Condition 2. No vertices of T_P are contained in the interior of C_F.

Condition 3. T_P is contained in the upper half-plane of $\mathbb{C}$.

In the sequel we will assume for simplicity that T_P is a triangulation (unless otherwise indicated). Any more general tessellation can be subdivided until it is a triangulation. First we note the following:

Lemma 3.1 *Condition 2 of is equivalent to the following:*

Condition 2′. *for any two abutting triangles ABC and ADC of T_P, D is not in the interior of C_{ABC}.*

Proof. Consider P. Lemma 3.1 is equivalent to the observation that the polyhedron P is convex (Condition 2) if and only if all of its edges are convexly bent (Condition 2′). □

Note. A simple direct Euclidean proof of Lemma 3.1 is also possible. This will be left as an exercise for the reader.

Corollary 3.2 *Given an arbitrary triangulation T' on the same vertex set as T_P, T' can be transformed into T_P by a finite sequence of Whitehead moves of the following kind: whenever ABC and ADC are abutting triangles of T' such that D lies inside C_{ABC}, we change ABC and ADC into ABD and CBD.*

Proof. A Whitehead move of the described type corresponds to filling in a missing tetrahedron $ABCD$ of a polyhedron lying above T'. Every time a move as above happens, the edge AC is buried, never to be seen again. Since the number of possible edges is finite, the result follows. □

The following facts from elementary Euclidean geometry will be needed in the sequel.

Lemma 3.3 *Let C be a circle with center O and let ABC be a triangle inscribed in C. Then the $\angle ACB = \frac{1}{2}\angle AOB$ if C and O are on the same side of AB and $\angle ACB = \pi - \frac{1}{2}\angle AOB$ otherwise.*

Lemma 3.4 *Let $ABCD$ be a quadrilateral.*

D *is* outside C_{ABC} *if* $\angle D + \angle B < \pi$.

D *is* inside C_{ABC} *if* $\angle D + \angle B > \pi$.

D *is* on C_{ABC} *if* $\angle D + \angle B = \pi$.

The following important fact follows from Lemma 3.3:

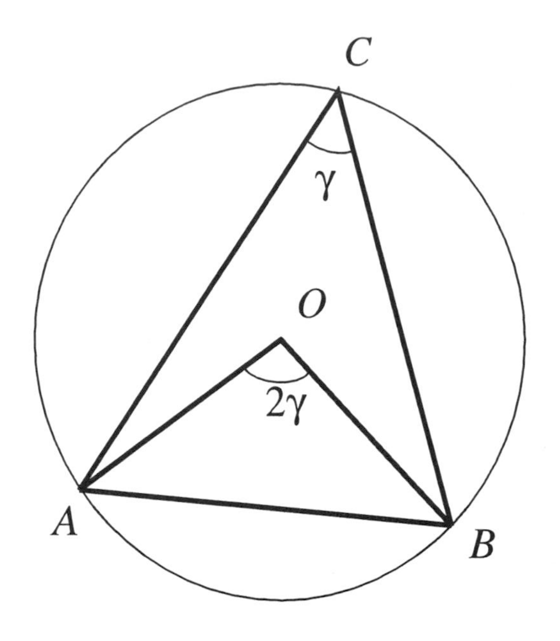

Figure 3

Theorem 3.5 *The dihedral angle between the faces ABC and ADC of P is equal to the sum of angles $\angle ABC$ and $\angle ADC$.*

Proof. First, observe that the angle between the the faces ABC and ADC is equal to the angle between the circles C_{ABC} and C_{ADC}. The rest of the proof is contained in Figure 4. □

The sum of the angles $\angle ABC$ and $\angle ADC$ is easily seen to equal the argument of the cross-ratio

$$\mathfrak{c}(A, B, C, D) = [B, C, A, D] = \frac{(B - A)(C - D)}{(B - C)(A - D)}$$

If A, B, C, and D are transformed by a hyperbolic isometry in such a way that $A = \infty$, $B = 1$, $C = 0$, and $D = z$, then $\mathfrak{c}(A, B, C, D) = z$, then Theorem 3.6 below follows from the discussion of section 2.

Theorem 3.6 *With notation as above* $\mathfrak{s}(ABC, ADC) = \log|c(A, B, C, D)|$.

Proof. This is just a "bent" three-dimensional version of Lemma 2.3.

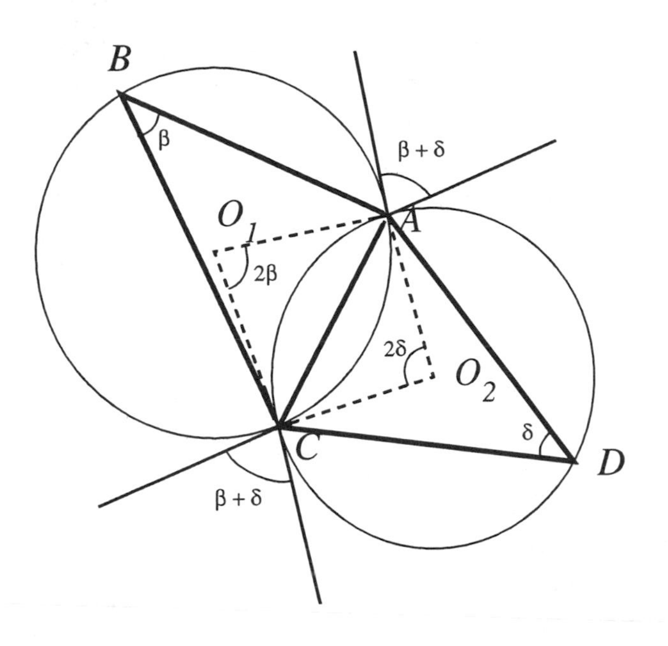

Figure 4 Dihedral angle

The above observations allow us to prove Theorems 3.7 and 3.9, which are two of the steps of the proof of Theorem 1.1.

Theorem 3.7 *The map $\mathfrak{g}$ is continuous.*

Proof. The simpler case is one where T_P is a genuine triangulation. Then, it is clear that a small perturbation of the vertices of T_P doesn't change the combinatorics of T_P, and so continuity follows from Theorem 3.6 and the continuity of the cross-ratio. Things are very slightly more complicated when T_P has non-triangular faces. Then, T_P is combinatorially unstable: a small perturbation in the vertices changes the combinatorial structure, but only in the following simple fashion:

Lemma 3.8 *For a sufficiently small ϵ, a perturbation T_{P^ϵ} of T_P is combinatorially equivalent to T_P with some diagonals added to the non-triangular faces.*

Proof of Lemma. There exists an $\epsilon > 0$, such that if a point D is closer than ϵ to C_{ABC} then A, B, C, and D are co-circular, for any triangle ABC and vertex D of T_P. □

Every way of adding diagonals to T_P until we get a triangulation corresponds to a different coordinate system of T_N (where N is the number of vertices of P), and Lemma 3.8 shows that every sufficiently small perturbation of T_P is close to T_P in at least one of the coordinate systems. Since the transition maps between the various coordinate systems are continuous (Theorem 2.6), Theorem 3.7 follows. □

Theorem 3.9 *The image of $\mathfrak{g}$ is closed.*

Proof. Let $\mathfrak{H}_1, \ldots, \mathfrak{H}_k, \ldots$ be a sequence of metrics on S^2_N converging to a metric $\mathfrak{H}$. Let P_i be such that $\mathfrak{g}(P_i) = \mathfrak{H}_i$. We will show that there exists a P_∞ such that $\mathfrak{g}(P_\infty) = \mathfrak{H}$. First choose a subsequence $P'_1, \ldots, P'_i, \ldots$ such that all of the P'_i have the same combinatorics. This is possible since the number of possible combinatorial structures is finite. As before, the vertices and faces of P'_i are labelled in such a way that $v_1(P'_i) = 0$, $v_2(P'_i) = 1$, $v_3(P'_i) = \infty$, and $f_1(P'_i)$ lies above the real axis. By compactness of the sphere $\overline{\mathbb{C}}$, there exists a limiting tessellation $T_{P'}$. If $T_{P'}$ is non-degenerate (that is, no two vertices of a triangle have coalesced into one), then by Theorem 3.7 it follows that $\mathfrak{g}(P') = \mathfrak{H}$.

We will show that $T_{P'}$ is always non-degenerate. If this is not the case, let t_i be a collapsing face of $T_{P'}$ which is abutting a non-collapsing face t_j. Such a pair of faces must exist, since at least one face (f_1) is not collapsing. By relabelling and a hyperbolic isometry, send t_j to the triangle $0, 1, \infty$, and t_i to $0, \infty, z$. Since $\mathfrak{s}(t_j, t_i) = \log|z|$, and $\mathfrak{H}$ is a non-degenerate metric it follows that z stays away from 0 and ∞, and so t_i is not collapsing after all. □

Remark 3.10 Theorem 3.5 and subsequent discussion is easily seen to lead to the following pleasing "hyperbolic" interpretation of planar triangulations. Consider a (not necessarily convex) polygon Q in the plane, such that the interior of Q is triangulated in such a way that edges of Q are edges of the triangulation $T(Q)$. Then $T(Q)$ is the projection of an ideal polyhedron $\tilde{Q}$ onto the plane $\mathbb{C}$ at infinity of H^3, such that:

1. $\tilde{Q}$ has vertices at the vertices of $T(Q)$, plus one vertex at the point ∞ of $\bar{\mathbb{C}}$.

2. Q is similar to the link of the vertex v_∞ of $\tilde{Q}$ at ∞.

3. $\tilde{Q}$ is star-shaped with respect to v_∞.

4. If $v_1, \ldots, v_k$ are the vertices of Q, then the dihedral angle of $\tilde{Q}$ corresponding to the edge $v_k v_\infty$ is the Euclidean angle of Q at v_k.

5. The dihedral angle of $\tilde{Q}$ corresponding to the boundary edge $v_i v_{i+1}$ is equal to the euclidean angle at the third vertex w of the (unique) triangle wv_iv_{i+1} of $T(Q)$ containing the edge v_iv_{i+1}.

6. The dihedral angle of $\tilde{Q}$ corresponding to a non-boundary edge AB of $T(Q)$ is equal to the sum of the angles at C and D of the two triangles ACB and ADB abutting along the edge AB

The triangulation $T(Q)$ could also be taken to be *immersed*, in which case all of the above statements still hold, with the obvious changes in interpretation.

Definition 3.11 *A set of edges* $C = \{e_1, \ldots, e_k\}$ *in a graph* G *is called* a cutset, *if the removal of those edges disconnects* G. *A cutset* C *is called* minimal, *if no subset of* C *is a cutset.*

The simplest example of a cutset is the set of edges incident to a single vertex of G.

The correspondence above can be used to prove the following result:

Theorem 3.12 *Let* $e_1, \ldots, e_k$ *be a minimal cutset of the 1-skeleton* $\tilde{Q}$ *Then the sum* Σ *of dihedral angles at* $e_1, \ldots, e_k$ *is strictly smaller than* $(k-2)\pi$ *if* $e_1, \ldots, e_k$ *are not all incident to one vertex. If* $e_1, \ldots, e_k$ *are all incident to one vertex then* Σ *is exactly* $(k-2)\pi$.

Proof. (also see figure 5) It is easy to see that any minimal cutset as above is actually the set of internal edges of a triangulation $T(\mathcal{A})$ of an annulus $\mathcal{A}$ (possibly with one boundary component collapsed to a point v , if all of the e_i are incident to v.) From now on all references will be to quantities in $T(Q)$ Let the inner and outer boundary components of $\mathcal{A}$ be $\mathcal{A}_1$ and $\mathcal{A}_2$, respectively. The edges of $T(\mathcal{A})$ naturally fall into three categories – outer boundary edges, inner boundary edges and internal edges. Similarly, divide the angles of the triangles of $T(\mathcal{A})$ into the three sets – A (angles opposite outer boundary), B (angles opposite inner boundary) and Γ (angles opposite inner edges). Obviously,

$$\sum A + \sum B + \sum \Gamma = k\pi \tag{1}$$

(where k is the number of triangles and the cardinality of the cutset). Furthermore, by Theorem 3.5,

$$\sum \Gamma = \Sigma. \tag{2}$$

Now, note that the sum of the angles incident (not opposite) to $\mathcal{A}_2$ is $(\operatorname{card} \mathcal{A}_2 - 2)\pi$, and further note that this sum is equal to $\sum B + (\operatorname{card} \mathcal{A})_2\pi - \sum A$. That is true since if α is *opposite* to $\mathcal{A}_2$, then the other angles of the triangle containing α are *incident* to $\mathcal{A}_2$. Now, since

$$(\operatorname{card} \mathcal{A}_2 - 2)\pi = \sum B + \pi \operatorname{card} \mathcal{A}_2 - \sum A, \tag{3}$$

it follows that $\sum A - \sum B = 2\pi$. Since $\sum B$ is greater than zero precisely when the inner boundary of $\mathcal{A}$ is non-degenerate, it follows that $\sum A + \sum B > 2\pi$ whenever the inner boundary of $\mathcal{A}$ is non-degenerate and $\sum A + \sum B = 2\pi$ otherwise. The statement of the theorem then follows from equations 1, 2, and 3. □

The above theorem is slightly stronger than Theorem 1 of [3], which is stated only for convex polyhedra. It turns out that the conditions of Theorem 3.12 together with the convexity conditions (dihedral angles are between 0 and π) completely characterize the sets of dihedral angles of convex polyhedra. Proof of sufficiency is given in an upcoming paper of the author.

4 Ideal polyhedra are determined by their metric

The purpose of this section is to prove Theorem 1.2. First, a couple of definitions:

Definition 4.1 (Generalized polyhedra and polygons) A *generalized* convex hyperbolic polyhedron is represented in the projective model of $\mathbf{H}^3$ by a Euclidean convex polyhedron which may have some vertices on or outside the sphere at infinity (called "infinite" and "hyperinfinite" vertices respectively). However, each edge must contain some points inside hyperbolic

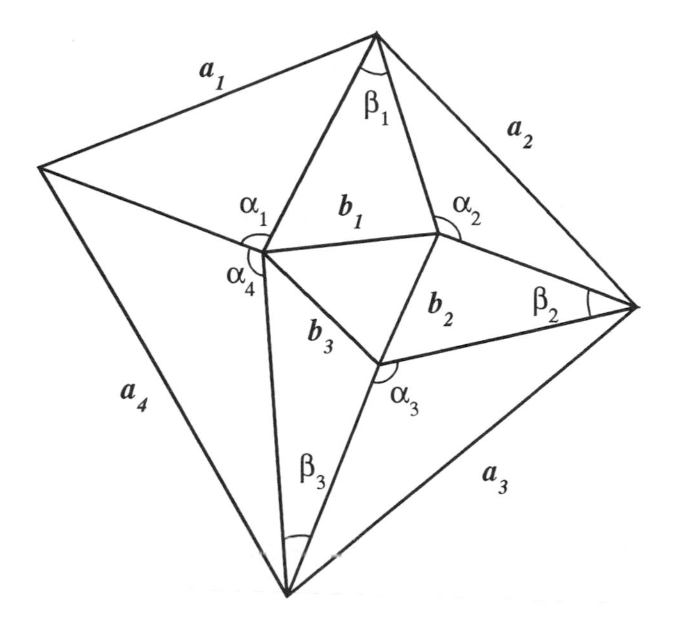

Figure 5 Cutset sum

space. We will usually only be concerned with the part of a generalized polyhedron lying within $\mathbb{H}^3$.

Generalized convex polygons in $\mathbb{H}^2$ are defined similarly.

Definition 4.2 (Links of vertices) A generalized hyperbolic polyhedron has vertices of three types: finite, hyperinfinite and infinite vertices.

The "link" of a finite vertex of a polyhedron is the spherical polygon obtained by intersecting a small sphere centered at the vertex with the polyhedron, and rescaling so the sphere has radius 1. So the edge lengths in the link are precisely the face angles at the vertex.

For each hyperinfinite vertex there is a unique hyperbolic plane orthogonal to the faces meeting at the vertex. The intersection of this plane with these faces is a hyperbolic polygon which we will call the "link" of the vertex. The edge lengths in the link are precisely the lengths of common perpendiculars to adjacent sides meeting at the hyperinfinite vertex.

For each infinite vertex there a 1-parameter family of horospheres centered at the vertex. Each small horosphere intersects the polyhedron in a Euclidean polygon, which we will call the "link" of the vertex. In this case the link is only well defined up to Euclidean similarities.

Remark. The link of an infinite or hyperinfinite vertex then determines the corresponding *end* of the polyhedron up to congruence.

Remark 4.3 The link of an ideal vertex of P is a Euclidean convex polygon. Theorem 3.6 shows that if all vertices of P are ideal, then the logarithm of the ratio of two adjacent sides of the link of a vertex v is equal to the shear between the two corresponding faces.

The following result is obtained in [2] (see [4], Theorem 4.10):

Theorem 4.4 *A generalized convex polyhedron P in hyperbolic 3-space is determined up to congruence by the type of its vertices and the edge lengths of the links of its vertices.*

Note 4.5 The edges of P are not required to be non-degenerate, so some of the dihedral angles may be π.

This theorem means that two combinatorially equivalent polyhedra P_1 and P_2 such that the corresponding sides of corresponding links of P_1 and P_2 are equal are congruent.

Proof of Theorem 1.2.

Let M be a complete finite-volume hyperbolic surface homeomorphic to S^2_N. Let P_1 and P_2 be two different embeddings of M into $\mathbb{H}^3$ as convex polyhedra. If $P_1(M)$ and $P_2(M)$ are combinatorially equivalent, Theorem 4.4 implies that $P_1(M)$ and $P_2(M)$ are congruent.

Assume that $P_1(M)$ and $P_2(M)$ are not combinatorially equivalent. Then P_1 and P_2 induce two different cell decompositions Q_1 and Q_2 of M, where the edges of Q_i are preimages of corresponding edges of $P_i(M)$. Produce a new cell decomposition Q of M by *superimposing* Q_1 and Q_2. The vertex set of Q is the union of V (the cusps of M) with the set V' of intersections of edges of Q_1 with those of Q_2. The image of Q under P_i will be $P_i(M)$ with some extra edges and vertices drawn on it. Then we can treat $P_1(M)$ and $P_2(M)$ as being of the same type (that of Q), and Theorem 4.4 may be applied. Thus $P_1(M)$ and $P_2(M)$ are congruent. □

5 Directions for further research

To the author, the most painful shortcoming of the results presented in this paper is the lack of any constructive method of producing an embedding of a hyperbolic N-punctured sphere $\mathfrak{S}$ into $\mathbb{H}^3$ as a convex ideal polyhedron. In particular, the tesselation of $\mathfrak{S}$ induced by such an embedding is clearly canonical (in view of Theorem 1.2), and yet there seems no known method of producing it.

The simple Theorem 3.5 turns out to be very useful. A number of consequences are given in the author's paper [1]). An efficient algorithm for producing a convex ideal polyhedron with prescribed dihedral angles is contained in an upcoming joint paper of the author and Warren D. Smith.

Acknowledgements. The author would like to thank Craig Hodgson and Warren D. Smith for valuable comments on earlier drafts of this paper.

Bibliography

[1] Igor Rivin. Euclidean structures on simplicial surfaces and hyperbolic volume. *Annals of Mathematics*. to appear.

[2] Igor Rivin. *On geometry of convex polyhedra in hyperbolic 3-space*. PhD thesis, Princeton, June 1986.

[3] Igor Rivin. On geometry of convex ideal polyhedra in Hyperbolic 3-space. *Topology*, 32(1), January 1993.

[4] Igor Rivin and C. D.Hodgson. A characterization of compact convex polyhedra in hyperbolic 3-space. *Inventiones Mathematicae*, 111(1), January 1993.

A Remark on a Theorem Concerning Besov Spaces on Subsets of R^n

Mikael Stenlund
Department of Mathematics
University of Umeå
S-901 87 Umeå, Sweden

Abstract
A counterexample to Theorem 4.1 in [2] is found for a certain A_d-set. The theorem was intended to give a definition of Besov spaces on A_d-sets.

0. Introduction

The classical Besov spaces are defined on $\mathbb{R}^n$. On d-sets F Besov spaces $B_l^{p,q}(F)$, $l > 0$, $1 \leq p, q \leq \infty$, can be defined by means of function families [1, p.123]. In Definition 4 and 6, below, we define A_d, the class of A_d-sets and the function space $(int)\widetilde{B}^l_{p,q}(F)$. The formulation of [2, Th.4.1] is as follows. Let $F \in A_d$. Then $(\text{int})\widetilde{B}^l_{p,q}(F) \hookrightarrow B_l^{p,q}(F) \hookrightarrow (\text{int})\widetilde{B}^l_{p,q}(F)$, $l > 0$, $1 \leq p, q \leq \infty$. In Section 1 it is shown that the imbedding $B_l^{p,q}(F) \hookrightarrow (\text{int})\widetilde{B}^l_{p,q}(F)$ is false for a certain A_d-set.

For further discussion we need some useful definitions. Let a closed cube with centre x and side-length δ be denoted by $Q(x, \delta)$.

Definition 1.[1, p.28] Let F be a non-empty closed subset of $\mathbb{R}^n$ and d a real number satisfying $0 < d \leq n$. A positive Borel measure μ with support F, supp$\mu = F$, is called a *d-measure on F* if there are two positive constants C_1 and C_2 which depend only on n and F such that $C_1\delta^d \leq \mu(Q(x,\delta)) \leq C_2\delta^d$, for all $x \in F$, $0 < \delta \leq 1$.

Definition 2. A closed non-empty subset F of $\mathbb{R}^n$ is a *d-set,* $0 < d \leq n$, if there exists a d-measure on F.

Definition 3. A closed non-empty set $F \subset \mathbb{R}^n$ *preserves Markov's inequality* if, for all positive integers k, all polynomials $P \in \mathcal{P}_k$, and all cubes $Q = Q(x, \delta)$, where $x \in F$ and $0 < \delta \leq 1$, we have the Markov inequality on F

$$(0\text{–}1) \qquad \max_{y \in F \cap Q(x,\delta)} |\nabla P(y)| \leq C_3 \delta^{-1} \max_{y \in F \cap Q(x,\delta)} |P(y)|,$$

where the constant C_3 depends only on F, n, and k.

This definition may also be formulated in terms of balls [1, p.34-35, Def.2] as is clear from [1, p.35, Prop.2, Remark 1].

Remark 1: If we let F be equal to $\mathbb{R}^n$ in Definition 3, then the inequality (0–1) is a special case of the classical Markov inequality. For details see [1, p.34].

In the following α, β, and γ are multi-indices. Let $[l]$ denote the integer part of l. Also let k be the unique non-negative integer satisfying $k < l \leq k+1$. A function family in $B_l^{p,q}(F)$ can be denoted by $\{f^{(\alpha)}\}_{|\alpha|\leq k}$, $f^{(\alpha)} : F \to \mathbb{R}$, where, for $|\alpha| \neq 0$, $f^{(\alpha)}$ is the formal derivative of $f^{(0)}$.

The following notation and terminology will be used to formulate the concept of an A_d-set (see Definition 4 below). Let F be a closed subset of $\mathbb{R}^n$, $x \in F$, $x^{(m)} \in F$, $m = 1, ..., \rho$, and let $\rho = \binom{n+[l]}{[l]}$ be the number of distinct monomials in n variables of degree at most $[l]$.

Let $X^{(m)}$ and X denote the column vectors with elements which are all monomials of the form $1^{k_0} \cdot (x_1^{(m)})^{k_1} \cdot ... \cdot (x_n^{(m)})^{k_n}$, $k_i \geq 0$, $\sum_{i=0}^{n} k_i = [l]$, and $1^{k_0} \cdot (x_1)^{k_1} \cdot ... \cdot (x_n)^{k_n}$, $k_i \geq 0$, $\sum_{i=0}^{n} k_i = [l]$, respectively. The elements in $X^{(m)}$ and X are arranged in lexicographical order in the columns according to the numbers $k_0, k_1, ..., k_n$. The matrix with the jth column vector equal to $X^{(j)}$, $j = 1, \ldots, \rho$, is denoted by $[X^{(1)}, X^{(2)}, ..., X^{(\rho)}]$.

In terms of the quantities ([2]) $\Delta^\rho := det[X^{(1)}, ..., X^{(\rho)}]$,

$$\Delta^{\rho,m} := det[X^{(1)}, ..., X^{(m-1)}, X, X^{(m+1)}, ..., X^{(\rho)}], \quad m = 1, 2, ..., \rho,$$

and if $\Delta^\rho \neq 0$, we define the Lagrange interpolation polynomial $P^\rho f$ of a function f at the interpolation points $x^{(1)}, ..., x^{(\rho)}$ by

$$(P^\rho f)(x) := \sum_{m=1}^{\rho} f(x^{(m)}) \frac{\Delta^{\rho,m}}{\Delta^\rho}.$$

Let the set N_r denote the class of all r-element ordered subsets of $\{1, 2, ..., \rho\}$, $1 \leq r \leq \rho$. Let $T := \{\tau = (r, I, J) : I, J \in N_r,\ r = 1, ..., \rho\}$. For every $\tau = (r, I, J) \in T$ the rth order minor of the matrix $[X^{(1)}, ..., X^{(\rho)}]$, denoted by Δ^τ, is obtained by intersecting the rows with indices $i_1, ..., i_r \in I$ and the columns with indices $j_1, ..., j_r \in J$. The expression $\Delta^{\tau,m}$ denotes the minor which differ from Δ^τ only by replacement of the mth column, $m = 1, ..., r$, of Δ^τ by the corresponding column with respect to the coordinates $x_1, ..., x_n$ of the point x.

Let $Q(\xi, h)$ be a closed cube in $\mathbb{R}^n$ with edges of length h parallel to the coordinate axes and centre ξ. For $\kappa > 1$ and $\tau = (r, I, J) \in T$ we define

$$F_\kappa(\xi, h)^\tau := \Big\{ (x^{(1)}, ..., x^{(\rho)}) : x^{(1)}, ..., x^{(\rho)} \in F \cap Q(\xi, h),$$

$$\max_{\substack{m=1,...,r \\ |\alpha|\geq 0}} \max_{x \in F\cap Q(\xi,h)} h^{|\alpha|} |D^\alpha \Delta^{\tau,m}| \leq \kappa |\Delta^\tau|,$$

$$|\Delta^\tau| > 0, \quad \operatorname{rank}[X, X^{(1)}, ..., X^{(\rho)}] = r, \ x \in F \cap Q(\xi, h) \Big\}.$$

In general this set may be empty.

Definition 4.[2, p.228] The set F belongs to the class of *A_d-sets*, $F \in A_d$, if F is a d-set, μ is a d-measure on F, and there exist positive constants κ, c, and h_0 such that

$$\sum_{\tau \in T} \int_{F_\kappa(\xi,h)^\tau} d\mu(x^{(1)})...d\mu(x^{(\rho)}) \geq c h^{\rho d}$$

for d–a.e., $\xi \in F$, and all h such that $0 < h \leq h_0$.

Theorem 6.2 of [2] guarantees the existence of A$_d$-sets. Let $\tau = (r, I, J) \in T$, $J = \{j_1, ..., j_r\}$, and $\Delta^\tau \neq 0$. The Lagrange type operator is defined in terms of the function $f^{(0)}$ according to

$$P^\tau f^{(0)} := P^\tau_{[l]}(x, f^{(0)}; x^{(1)}, ..., x^{(\rho)}) := \sum_{m=1}^{r} f^{(0)}(x^{(j_m)}) \frac{\Delta^{\tau,m}}{\Delta^\tau},$$

where $f^{(0)}$ is the first function in the family $\{f^{(\alpha)}\}_{|\alpha| \leq k}$. Let $P^\tau_{[l]} f^{(0)} := P^\tau f^{(0)}$. In the following definitions an interpolation is denoted by (int).

Definition 5.[c.f.2, Def.2.2] Let $l > 0$ and let k be the unique integer satisfying $0 \leq k < l \leq k+1$. Let $f = \{f^{(\alpha)}\}_{|\alpha| \leq k}$. Let

$$(\mathrm{int})\omega_{[l+1]}(f;\delta)_{p,F} := \sup_{0<h\leq\delta} \sum_{|\alpha| \leq k} h^{|\alpha|} \sum_{\tau \in T} h^{-(\rho+1)d/p} \Bigg(\int_F d\mu(\xi) \int_{F \cap Q(\xi,h)} d\mu(x) \int_{F_\kappa(\xi,h)^\tau} |f^{(\alpha)} - D^\alpha P^\tau_{[l]} f^{(0)}|^p d\mu(x^{(1)})...d\mu(x^{(\rho)}) \Bigg)^{1/p}.$$

Let

$$\|f^{(\alpha)}\|_{p,F} := \left(\int_F |f^{(\alpha)}(x)|^p d\mu(x) \right)^{1/p}.$$

Definition 6. Let μ be a d-measure with support a d–set F. For $0 \leq k < l \leq k+1$, $1 \leq p, q \leq \infty$, and $h_0 > 0$ let $(int)\widetilde{B}^l_{p,q}(F)$ be the Besov space of function families $f = \{f^{(\alpha)}\}_{|\alpha| \leq k}$ with norm

$$\|f; (\mathrm{int})\widetilde{B}^l_{p,q}(F)\| := \sum_{|\alpha| \leq k} \|f^{(\alpha)}\|_{p,F} + \left(\int_0^{h_0} \left(\frac{(\mathrm{int})\omega_{[l+1]}(f;t)_{p,F}}{t^l} \right)^q \frac{dt}{t} \right)^{1/q}.$$

1. A counterexample

The following is Theorem 4.1 in [2].

Theorem 1. *Let* $F \in A_d$. *Then*
$(int)\widetilde{B}^l_{p,q}(F) \hookrightarrow B^{p,q}_l(F) \hookrightarrow (int)\widetilde{B}^l_{p,q}(F), \quad l > 0,\ 1 \le p, q \le \infty.$

If Theorem 1 was true, the function spaces $(\mathrm{int})\widetilde{B}^l_{p,q}(F)$ and $B^{p,q}_l(F)$ would have equivalent norms. The purpose of this section is to show that in fact these norms are not equivalent. First it is necessary to give some more notations from [2];

$$\left[\{f^{(\alpha)} - D^\alpha P^\tau f^{(0)}\}_{\tau\in T}\right]_{p,F,h} := \sum_{\tau\in T} h^{-(\rho+1)d/p} \left(\int_F d\mu(\xi) \int_{F\cap Q(\xi,h)} d\mu(x) \int_{F_\kappa(\xi,h)^\tau} |f^{(\alpha)} - D^\alpha P^\tau_{[l]} f^{(0)}|^p d\mu(x^{(1)})...d\mu(x^{(\rho)}) \right)^{1/p}.$$

Let $F_\kappa(y,h)^{0,\tau} := F \cap Q(y,h) \times F_\kappa(y,h)^\tau$, $F(y,h) := F \cap Q(y,h)$, and

$$||\psi||_{p,|x-y|\le h} := \left(\iint_{|x-y|\le h} |\psi(x,y)|^p \, d\mu(x) d\mu(y) \right)^{1/p}.$$

The Taylor remainders are defined by:

$$R^k_\alpha(x,y) := f^{(\alpha)}(x) - \sum_{|\alpha+\beta|\le k} f^{(\alpha+\beta)}(y) \frac{(x-y)^\beta}{\beta!}, \quad |\alpha| \le k. \tag{1–1}$$

For $A, B > 0$ the notation $A \asymp B$ will signify the existence of constants $c_1, c_2 > 0$, depending only on n, k, and F, such that $c_1 A \le B \le c_2 A$.

For non-integer l the proof of Theorem 4.1 in [2] depends upon Lemma 4.1 in [2]. This lemma is stated below, together with the essential part of the proof. (Certain notational changes have been made.)

"**Lemma 4.1.**[2] *Let* $F \in A_d$, *let* $k := [l]$ *for non-integer* l, *and let* $1 \le p, q \le \infty$. *Then*

$$\sum_{|\alpha|\le k} h^{|\alpha|} \left[\{f^{(\alpha)} - D^\alpha P^\tau f^{(0)}\}_{\tau\in T}\right]_{p,F,h} \asymp \sum_{|\alpha|\le k} h^{|\alpha|-d/p} ||R^k_\alpha(x,y)||_{p,|x-y|\le h},$$

where $R^k_\alpha(x,y)$ *is defined by (1–1).*"

"Proof". Suppose that $(x, x^{(1)}, ..., x^{(\rho)}) \in F_\kappa(y,h)^{0,\tau}$, $y \in F$ and $\tau \in T$. Let us define a polynomial q_k of degree at most k in the variable x as follows:

$$q_k(x,y) := \sum_{|\gamma| \leq k} f^{(\gamma)}(y) \frac{(x-y)^\gamma}{\gamma!}.$$

For $R_\alpha(\{f^{(\gamma)}\}_{|\gamma|\leq k}; x, y) := R_\alpha^k(x,y)$ we have that

$$f^{(\alpha)}(x) - D^\alpha P^\tau(x, f^{(0)}; x^{(1)}, ..., x^{(\rho)}) =$$

$$= \left(f^{(\alpha)}(x) - D_x^\alpha q_k(x,y)\right) - D_x^\alpha P^\tau(x, f^{(0)} - q_k; x^{(1)}, ..., x^{(\rho)}) \tag{1-2}$$

$$= R_\alpha(x,y) - D_x^\alpha P^\tau(x, R_0; x^{(1)}, ..., x^{(\rho)}).$$

This was a part of the proof of Lemma 4.1. The crucial formula is (1–2) above and it is true if we in the lemma replace A_d-sets by sets preserving Markov's inequality (see [3, Paper B]). Below we will exemplify that $D^\alpha P^\tau f^{(0)}$ in (1–2) can be zero. But first it is necessary to give Definition 7 and Theorem 2. Let $\Pi = \Pi^m$ be a hyperplane in $\mathbb{R}^n$ with dimension m, $1 \leq m \leq n$. Let $P_{r\Pi}$ be a perpendicular projection of vectors in $\mathbb{R}^n$ into Π.

Definition 7.[2, Def.6.1] The closed non–empty set F, lying in a plane $\Pi = \Pi^m$ of dimension m, $1 \leq m \leq n$, *has the Markov inequality on* Π if the inequality

$$\max_{F\cap B} |P_{r\Pi}\text{grad}P| \leq ch^{-1} \max_{F\cap B} |P|$$

where $c = c(k, n, F) \geq 0$, holds for any kth degree polynomial P, all positive integers k and all closed balls $B(\xi, h)$, $\xi \in \Pi \cap F$, $0 < h \leq 1$.

Theorem 2.[2,Th.6.2] *If F is a d-set with the Markov inequality on Π^m, $1 \leq m \leq n$, $F \in A_d$*

The set F which gives the counterexample will be the first coordinate axis in $\mathbb{R}^2$. The set F is then a d-set with $d = 1$. Let D_u be the operator which gives the directional derivative in the direction of the vector u. Let P be a polynomial. Assume that $P_{r\Pi}\text{grad}P \neq \bar{0}$ and let $u := (P_{r\Pi}\text{grad}P)/|P_{r\Pi}\text{grad}P|$, then we have

$$\max_{F\cap B} |P_{r\Pi}\text{grad}P| = \max_{F\cap B} |D_u P|.$$

But $P(x)\big|_F$ can be seen as a polynomial on $\mathbb{R}$. By using Markov's inequality on $\mathbb{R}$ (see Def.3 Remark 1) one can see that the conditions in Definition 7 also are true for F. Then, according to Theorem 2, F is an A_d-set.

Let l be the real number in Theorem 3 and let $1 < l < 2$. According to Lemma 4.1. we have $k = 1 = [l]$. Notice that $\rho = \binom{n+k}{k} = \binom{2+1}{1} = 3$. The definition of $[X^{(1)}, X^{(2)}, ..., X^{(\rho)}]$ given in the first part of this section gives

$$[X^{(1)}, X^{(2)}, X^{(3)}] = \begin{pmatrix} 1 & 1 & 1 \\ x_1^{(1)} & x_1^{(2)} & x_1^{(3)} \\ 0 & 0 & 0 \end{pmatrix}, \quad x^{(1)}, x^{(2)}, x^{(3)} \in F.$$

Let $\tau = (r, I, J) \in T$. Notice that $|\Delta^\tau| > 0$ is one condition in the definition of $F_\kappa(\xi, h)^\tau$. If $r = 3$, then $\Delta^\tau = 0$ because in this case Δ^τ is the determinant of $[X^{(1)}, X^{(2)}, X^{(3)}]$. Furthermore, in the definition of $F_\kappa(\xi, h)^\tau$ we have that $\operatorname{rank}[X, X^{(1)}, X^{(2)}, X^{(3)}] = r$ for all $x \in F \cap Q(\xi, h)$. The number r cannot be equal to 1 because if $r = 1$, then $x_1 = x_1^{(1)} = x_1^{(2)} = x_1^{(3)}$ for all $x = (x_1, x_2) \in F \cap Q(\xi, h)$, $\xi \in F$, and this cannot be true if F is the first coordinate axis in $\mathbb{R}^2$.

Consequently, $F_\kappa(\xi, h)^\tau$ is empty for every $\tau = (r, I, J)$, where r is different from 2. The only elements τ in T which we need to consider are therefore $\tau = (2, \{1,2\}, \{1,2\})$, $(2, \{1,2\}, \{2,3\})$ and $(2, \{1,2\}, \{1,3\})$. Let us define $T' := \{ (2, \{1,2\}, \{1,2\}), (2, \{1,2\}, \{2,3\}), (2, \{1,2\}, \{1,3\}) \}$. We will need this definition below.

Let us assume that $\tau = (2, \{1,2\}, \{1,2\})$ is a τ, such that $F_\kappa(\xi, h)^\tau$ is non-empty. Then we have

$$\Delta^\tau = x_1^{(2)} - x_1^{(1)} \neq 0,$$

$$\Delta^{\tau,1} = x_1^{(2)} - x_1, \qquad \Delta^{\tau,2} = x_1 - x_1^{(1)} \quad \text{where } (x_1, x_2) = x.$$

With this τ we have $J = \{j_1, j_2\} = \{1, 2\}$ and

$$P^\tau f^{(0)} = \sum_{m=1}^{2} f^{(0)}(x^{(j_m)}) \frac{\Delta^{\tau,m}}{\Delta^\tau} = f^{(0)}(x^{(1)}) \frac{x_1^{(2)} - x_1}{x_1^{(2)} - x_1^{(1)}} + f^{(0)}(x^{(2)}) \frac{x_1 - x_1^{(1)}}{x_1^{(2)} - x_1^{(1)}}.$$

Notice that above we have a Lagrange interpolation formula for two points on the real line. We can also see that the formula for $P^\tau f^{(0)}$ does not depend on the second coordinate x_2. With the other possible elements τ in T, and if we assume that $F_\kappa(\xi, h)^\tau$ is non-empty, we obtain formulas similar to that above and not depending on the second coordinate.

However, the first equality in (1–2) cannot be true in our example, because $\alpha = (0, 1)$, $(|\alpha| \leq k)$ gives

$$D_x^\alpha P^\tau(x, f^{(0)}, x^{(1)}, ..., x^{(3)}) = 0.$$

This is also the case when we replace $f^{(0)}$ with $f^{(0)} - q_k$ in the formula above. Consequently, the first equality in (1–2) takes the form

$$f^{(0,1)}(x) = f^{(0,1)}(x) - D_x^{(0,1)} q_k(x, y),$$

and this cannot be true. In our counterexample in this section we will not use this observation.

Why is the theorem false? In Theorem 3 we have the imbedding $B_l^{p,q}(F) \hookrightarrow (\text{int})\widetilde{B}^l_{p,q}(F)$. For the set F defined above we will show that this imbedding does not hold.

If we consider the first part of the proof of Lemma 4.1, it is necessary to assume that $\{f^{(\alpha)}\}_{|\alpha|\le k} \in B_l^{p,q}(F)$. The definition of $(\text{int})\omega_{[l+1]}(f;t)_{p,F}$, where $f = \{f^{(\alpha)}\}_{|\alpha|\le k}$, now gives

$$(\text{int})\omega_{[l+1]}(f;t)_{p,F} = \sup_{0<h\le t} \sum_{|\alpha|\le k} h^{|\alpha|} \sum_{\tau\in T} h^{-(\rho+1)d/p} \left(\int_F d\mu(\xi) \int_{F\cap Q(\xi,h)} d\mu(x) \int_{F_\kappa(\xi,h)^\tau} \left| f^{(\alpha)} - D^\alpha P^\tau_{[l]} f^{(0)} \right|^p d\mu(x^{(1)}) \dots d\mu(x^{(\rho)}) \right)^{1/p} .$$

This formula will be estimated below, but first we need to do some preparatory calculations. By using the inequality $\left(\sum a_i\right)^{1/p} \le \sum a_i^{1/p}$, for $a_i \ge 0$, and $p \ge 1$, we have

$$\sum_{\tau\in T} \left(\int_F d\mu(\xi) \int_{F\cap Q(\xi,h)} d\mu(x) \int_{F_\kappa(\xi,h)^\tau} |f^{(0,1)} \quad D^{(0,1)} P^\tau_{[l]} f^{(0)}|^p \, d\mu(x^{(1)}), \dots , d\mu(x^{(3)}) \right)^{1/p} \ge$$

$$\ge \left(\int_F d\mu(\xi) \int_{F\cap Q(\xi,h)} d\mu(x) \sum_{\tau\in T} \int_{F_\kappa(\xi,h)^\tau} |f^{(0,1)} - D^{(0,1)} P^\tau_{[l]} f^{(0)}|^p \, d\mu(x^{(1)}), \dots , d\mu(x^{(3)}) \right)^{1/p} .$$

Let T' be the set of those elements in T such that T' does not contain elements τ, where $F_\kappa(\xi,h)^\tau$ is empty; see the discussion above. If we in addition use that $D^{(0,1)} P^\tau f^{(0)} = 0$ for $\tau \in T'$, then the last expression is equal to

$$\left(\int_F d\mu(\xi) \int_{F\cap Q(\xi,h)} d\mu(x) \sum_{\tau\in T'} \int_{F_\kappa(\xi,h)^\tau} |f^{(0,1)}|^p \, d\mu(x^{(1)}), \dots, d\mu(x^{(3)}) \right)^{1/p} .$$

Notice that the set T' excludes only such elements τ in T for which $F_\kappa(\xi,h)^\tau$ is empty. The definition of an A_d-set and that we are summing over all τ in T' give that the formula above is at least

$$\left(\int_F d\mu(\xi) \int_{F\cap Q(\xi,h)} ch^{\rho d} |f^{(0,1)}|^p \, d\mu(x) \right)^{1/p} .$$

If we want to estimate $(int)\omega_{[l+1]}(f;t)_{p,F}$ from below, we can take $\alpha = (0,1)$ instead of summing over all α such that $|\alpha| \le 1$. If we use this observation together with our estimate above, we have

$$(\mathrm{int})\omega_{[l+1]}(f;t)_{p,F} \ge \sup_{0<h\le t} h \cdot h^{-(\rho+1)d/p} .$$

$$\cdot c \left(h^{\rho d} \int_F d\mu(\xi) \int_{F\cap Q(\xi,h)} \left| f^{(0,1)}(x) \right|^p d\mu(x) \right)^{1/p} .$$

Let $d = 1$ and let μ be the Lebesgue measure on $\mathbb{R} = F$. Then we can replace $d\mu(x)d\mu(\xi)$ in the integral by $dxd\xi$. Since the first integration is on the set $F \cap Q(\xi,h)$ we can write

$$\int_{F\cap Q(\xi,h)} |f^{(0,1)}(x)|^p dx = \int_F |f^{(0,1)}(x)|^p \mathcal{X}_{Q(\xi,h)}(x) \, dx ,$$

where $\mathcal{X}_{Q(\xi,h)}(x)$ is the characteristic function on the set $Q(\xi,h)$.

The function $g(x,\xi) := |f^{(0,1)}(x)|^p \mathcal{X}_{Q(\xi,h)}(x)$ is a non-negative measurable function in $\mathbb{R}^2$, because $|f^{(0,1)}(x)|^p$ and $\mathcal{X}_S(\xi,x) := \mathcal{X}_{Q(\xi,h)}(x)$, where the set S is defined by $S := \{(\xi,x) : (\xi,x) \in \mathbb{R}^2, \ -\frac{h}{2}+x \le \xi \le \frac{h}{2}+x\}$, are measurable functions in $\mathbb{R}^2$. By using the theorem of Tonelli we can now change the order of integration;

$$\iint_{\mathbb{R}^2} \left| f^{(0,1)}(x) \right|^p \mathcal{X}_S(\xi,x) \, d\xi dx = h \int_{\mathbb{R}} \left| f^{(0,1)}(x) \right|^p dx.$$

If $\|f^{(0,1)}\|_{p,\mu} \ge c > 0$ and $d = 1$, we get

$$(\mathrm{int})\omega_{[l+1]}(f;t)_{p,F} \ge \sup_{0<h\le t} c \cdot h = ct.$$

But, then we have

$$\left(\int_0^{h_0} \left(\frac{(\text{int})\omega_{[l+1]}(f;t)_{p,F}}{t^l} \right)^q \frac{dt}{t} \right)^{1/q} \geq \left(\int_0^{h_0} \frac{c^q}{t^{(l-1)q}} \cdot \frac{dt}{t} \right)^{1/q} = \left(\int_0^{h_0} \frac{c^q}{t^{(l-1)q+1}} dt \right)^{1/q} .$$

In our example we have $1 < l < 2$ so $(l-1)q + 1 > 1$ and the integral is divergent. Then the norm $\|f; (\text{int})\widetilde{B}^l_{p,q}(F)\|$ must be infinite. Thus Theorem 4.1 in [2] is false, because the imbedding $B^{p,q}_l(F) \hookrightarrow (\text{int})\widetilde{B}^l_{p,q}(F)$ does not hold for all A_d–sets.

We have discussed above Theorem 4.1 in [2] when l is not an integer. In [3] it is also proved that Theorem 4.1 must be false also in the integer case, i.e. when l is an integer. The idea of the counterexample in the integer case ($l = 2$) is similar to the idea in the non-integer case.

2. Concluding remarks

The imbeddings in Theorem 3 are true if we replace A_d-sets by d-sets preserving Markov's inequality (see [3]). In this case Theorem 7 in [3, Paper A] has similarities with Theorem 1 (see [3, Paper A]). But Theorem 7 in [3, Paper A] is more adapted for calculations because the non-smoooth spline functions which are used there approximates the Besov space functions in $L^p(\mu)$ norm. In [3, Paper B] it is discussed for which sets the imbeddings in Theorem 1 will be true.

References

[1] A.Jonsson and H.Wallin, *Function Spaces on Subsets of* $\mathbb{R}^n$, Math. Reports, Vol. 2, Part 1, Harwood Academic Publishers, 1984.

[2] G.A.Mamedov, *A constructive description of traces of spaces of differentiable functions of several variables*, Trudy Mat. Inst. Steklov **173** (1986); Proc. of the Steklov Inst. of Math. (1987), 221–251, AMS.

[3] M.Stenlund, *Besov spaces on subsets of* $\mathbb{R}^n$ *and polynomial interpolation*, Dep. of Math., Univ. of Umeå, Ph.D. Thesis No.3, 1992.

Wavelet Transforms with the Franklin System and Application in Image Processing

Jan-Olov Strömberg

Institute of Mathematical an Physical Sciences
University of Tromsø
N-9037 Tromsø, Norway

Abstract

An orthogonal wavelet system $\{\psi_{jk}\}_{j,k\in\mathbf{Z}}$ is an orthonormal basis generated by a single function ψ by the formula

$$\psi_{jk} = 2^{-j/2}\psi(2^{-j}x - k).$$

The Franklin wavelet system is generated by a piecewise linear function g. The numerical algorithms for the Franklin wavelet system are very fast and numerically very stable. I will present some of these algorithms. I will also give examples showing how this system can be used in image processing.

0 Introduction

The classical Franklin system from the 1920's (see ([5]) is an orthonormal system of continuous piecewise linear functions on the unit interval $I = [0,1]$. It Is an orthonormal basis in in $L^2(\mathbf{I})$. In ([8]) I made a modification of the Franklin system to $\mathbf{R}$ and $\mathbf{R}^n$. By the invariance proportion of $\mathbf{R}$, I got an orthonormal basis of continuous, piecewise linear functions in $L^2(\mathbf{R})$ generated by a single function by means of integer translations and dilations by integer powers of 2. More precisely, the orthonormal basis $\{\psi_{jk}\}_{j,k\in\mathbf{Z}}$ is generated by a continuous piecewise linear function ψ by the formula

$$\psi_{jk}(x) = 2^{-j/2}\psi(2^{-j}x - k).$$

On $\mathbf{R}^n$ a similar orthonormal basis is generated by a single function by means of integer translations and a combined dilation-rotations.

When wavelet theory started to develop around 1985, and one got especially interested in functions ψ generating an orthonormal basis as in the formula above. Such a function is called an orthonormal wavelet. Many orthonormal wavelets have been constructed in the last few years. Most famous are the compactly supported, orthonormal wavelets constructed by I. Daubechies ([3]).

The orthonormal wavelet systems have become a very useful tool in many applications for instance in image processing an signal analysis.

In this paper I will give a description of the Franklin wavelet system (and some variations of it), and then show how numerical calculations can be done with it. I will also illustrate how it can be applied to image processing.

1 Construction of the Franklin wavelet system

1.1 The Franklin wavelet on the real line.

Let for $\langle \cdot, \cdot \rangle$ denote the inner product on $L^2(\mathbf{R}^n)$ defined by

$$\langle f, g \rangle = \int_{\mathbf{R}^n} f(\mathbf{x}) g(\mathbf{x}) d\mathbf{x}.$$

Let b be the basic triangular function defined by

$$b(x) = (1 - |x|)_+.$$

and let b_k and b_{jk} be translation resp. translation and dilation of the function b defined by

$$b_k(x) = b(x-k) \text{ and } b_{jk}(x) = 2^{-j/2} b(2^{-j}x - k).$$

Let $\mathbf{V}_j$ be the closure of the space of continuous piecewise linear functions spanned by $\{b_{jk}\}_{k \in \mathbf{Z}}$. Let

$$r = \sqrt{3} - 2,$$

and define

$$\rho^+(x) = \sum_{k=0}^{\infty} r^k b_k(x),$$

and let

$$\rho^-(x) = \rho^+(-x).$$

The constant r is chosen such that $\langle \rho^+, b_k \rangle = 0$ when $k \notin \{-1, 0\}$ and $\langle \rho^-, b_k \rangle = 0$ when $k \notin \{0, 1\}$. We set

$$\rho_k^{\pm}(x) = \rho^{\pm}(x-k) \text{ and } \rho_{jk}^{\pm}(x) = 2^{-j/2} \rho^{\pm}(2^{-j}x - k).$$

It follow that each of the sets of functions $\{\rho_k^+\}_{k \in \mathbf{Z}}$ and $\{\rho_k^-\}_{k \in \mathbf{Z}}$ is an orthogonal basis in $\mathbf{V}_0$ and that for each fixed integer j each of the sets of functions $\{\rho_{jk}^+\}_{k \in \mathbf{Z}}$ and $\{\rho_{jk}^-\}_{k \in \mathbf{Z}}$ is an orthogonal basis in $\mathbf{V}_j$. Note that $\rho_j^{\pm} k$ satisfies the the recursion formula

$$\rho_{jk}^{\pm} = b_{jk} + r \rho_{j,k \pm 1}^{\pm}. \tag{1}$$

Set

$$\begin{aligned} \psi &= c_1\rho^+_{0,1} + c_2\rho^-_{-1,0} + c_3\rho^-_{-1,1}, \\ \phi &= d_0\rho^-, \end{aligned}$$

and for all integers j and k let

$$\begin{aligned} \psi_{jk}(x) &= 2^{-j/2}\psi(2^{-j}x - k), \\ \phi_{jk}(x) &= 2^{-j/2}\phi(2^{-j}x - k). \end{aligned}$$

By looking at each separate term of the sum in the definition of ψ above, we see that $\langle \psi, b_k \rangle = 0$ for all $k \notin \{0, 1\}$. Also we have $\langle \psi, \rho^-_{k,-1} \rangle = 0$ for all $k < 0$. We want to have ψ orthogonal to the space $\mathbf{V}_0$ and to have ψ and ϕ normalized to norm one. In order to obtain that the constants $c_i, i = 1, 2, 3$ and d_0 are chosen such such that ψ and ϕ satisfy the system of equations

$$\left\{ \begin{array}{l} \langle \psi, b_0 \rangle = 0 \\ \langle \psi, b_1 \rangle = 0 \\ \langle \psi, \psi \rangle = 1 \\ \langle \phi, \phi \rangle = 1 \end{array} \right. . \tag{2}$$

The system (2) can be written out explicitly as system of equations in c_1, c_2 and c_3 and d_0. The solution is $c_1 = \sqrt{2} - \sqrt{6}$, $c_2 = 2r = 2(\sqrt{3} - 2)$, $c_3 = 2$ (factor modulo 1); $d_1 = 3 - \sqrt{3}$ (factor modulo 1).

Since $\rho^+_{0,k}$ is in $\mathbf{V}_0$ for all integers k and $\rho^-_{-1,k}$ is orthogonal to ψ for all $k < 0$ we conclude that $\psi_{0,k}$ is orthogonal to ψ for all $k < 0$. Since $\rho^{\pm}_{jk}$ is in $\mathbf{V}_0$ when $j > 0$ we also conclude that ψ_{jk} is orthogonal to ψ when $j > 0$. By translation and dilation arguments we see that $\{\psi_{jk}\}_{j,k \in \mathbf{Z}}$ is an orthonormal set of functions.

Note that ψ is the (unique modulo 1 factor) continuous piecewise linear function with nodes at the points $\{\frac{k}{2} : k \in \mathbf{Z}, k < 0\} \cup \{k : k \in \mathbf{Z}, k \geq 0\} \cup \{\frac{1}{2}\}$ which is orthogonal to the space of continuous piecewise linear functions with nodes in the set $\{\frac{k}{2} : k \in \mathbf{Z}, k < 0\}$ $\cup \{k : k \in \mathbf{Z}, k \geq 0\}$ and with L^2-norm one. That was the way the function ψ originally was defined in ([8]).

We will show that for fixed j, the space $\mathbf{V}_{j-1}$ is the closure of the span of functions $\{\psi_{jk}\}_k$ and the functions in $\mathbf{V}_j$. We have $\psi_{j,k} - r^2\psi_{j,k-1}$ is equal to $(1 - r)(b_{j-1,2k+1} - rb_{j-1,2k-1})$ (modulo functions in $\mathbf{V}_j$). The sum $\sum_{k=-\infty}^{k_0} r^{k_0 - k}(b_{j-1,2k+1} - rb_{j-1,2k-1})$ converges to $b_{j-1,2k_0+1}$. We also have $b_{j-1,2k}$ is equal to $\frac{1}{2}(b_{j-1,2k-1} + b_{j-1,2k+1})$ (modulo functions in $\mathbf{V}_j$). Since $\mathbf{V}_{j-1}$ is the closure of the span of the functions $\{b_{jk}\}_{k \in \mathbf{Z}}$ we get the desired conclusion.

To summarize, the set of functions $\{\psi_{jk}\}_{j,k \in \mathbf{Z}}$ is an orthonormal basis in $L^2(\mathbf{R})$ which we call the Franklin wavelet system. For each integer j the set of functions $\{\phi_{jk}\}_{k \in \mathbf{Z}}$ is an orthonormal basis in the space $\mathbf{V}_j$. Furthermore, if $\mathbf{W}_j$ is the orthogonal complement to the the subspace $\mathbf{V}_j$ in $\mathbf{V}_{j-1}$ then $L^2(\mathbf{R})$ can be split into a direct sum of orthogonal spaces

$$L^2(\mathbf{R}) = \bigoplus_{j \in Z} \mathbf{W}_j.$$

The spaces $\mathbf{W}_j$ (and also the spaces $\mathbf{V}_j$) satisfy the invariance properties

$$f(x) \in \mathbf{W}_j \Longleftrightarrow f(x - 2^j) \in \mathbf{W}_j,$$
$$f(x) \in \mathbf{W}_j \Longleftrightarrow f(2x) \in \mathbf{W}_{j-1}.$$

For each fixed integer j the set of functions $\{\psi_{jk}\}_{k\in\mathbf{Z}}$ is a complete orthonormal basis in $\mathbf{W}_{j-1}$.

1.2 The Franklin wavelet system on $\mathbf{R}^n$

We use the notation $\mathbf{x} = (x_1, \ldots, x_n)$ and define the function $\mathbf{\Psi}$ by

$$\mathbf{\Psi}(\mathbf{x}) = \psi(x_1)\phi(x_2)\cdots\phi(x_n)$$

and the matrix $\mathbf{A}$ by

$$\mathbf{A} = \begin{pmatrix} 0 & 1 & 0 & \ldots & \ldots & 0 \\ 0 & 0 & 1 & 0 & \ldots & 0 \\ \vdots & \vdots & \ddots & \ddots & \ddots & \vdots \\ 0 & \ldots & \ldots & 0 & 1 & 0 \\ 0 & \ldots & \ldots & \ldots & 0 & 1 \\ 2 & 0 & \ldots & \ldots & \ldots & 0 \end{pmatrix}.$$

Let for each $j \in \mathbf{Z}$ and each $\mathbf{k} \in \mathbf{Z}^n$

$$\mathbf{\Psi}_{j\mathbf{k}} = 2^{-j/2}\mathbf{\Psi}(\mathbf{A}^{-j}x - \mathbf{k}).$$

Then $\{\mathbf{\Psi}_{j\mathbf{k}}\}_{j\in\mathbf{Z},\mathbf{k}\in\mathbf{Z}^n}$ is a complete orthonormal system in $L^2(\mathbf{R}^n)$ which we call the Franklin wavelet system in dimension n.

Remark The extension was originally ([8]) done by using tensor products of the functions ψ and ϕ (not all of factors equal to ϕ). Thus $2^n - 1$ different tensor products were used. This has become the standard way of extending wavelets to higher dimension. In ([9]) we did the extension to higher dimension by using one single tensor product as described above by combining the changing of scale by permutation of the coordinates. Note that the total number of coefficients (on a bounded set of data) will be the same if we use 2^n tensor products or one single tensor product. It has several advantages to do the extension to higher dimension in this way. (See the remark at the end of the paper for more comments on this.)

1.3 The Franklin system on a bounded interval

The classical Franklin system on a bounded interval I is very similar to the Franklin wavelet system on the real line defined as above. However, the scale invariance and dilation invariance properties are only true in an asymptotical way far away from the endpoints of the bounded interval. We assume that $I = [0, 2^N]$ for a positive integer N

The inner product will now be defined by

$$\langle f, g\rangle = \int_{\mathbf{I}} f(\mathbf{x})g(\mathbf{x})d\mathbf{x}.$$

We will use the notations b,$\rho^{\pm}$,ϕ and ψ together with the indices j and k for for dilation and translation as before in the real line case. However, the formulas that were stated in that case are no longer true. The constants r, will now change from place to place , i.e. they depend on j and k. More precisely recursion formula (1) of ρ_{pj}^{+} is replaced by

$$\rho_{jk}^{\pm} = b_{jk} + r_{jk}^{\pm}\rho_{j,k\pm 1}^{\pm}, \tag{3}$$

starting with $\rho_{j,2^{N-j}}^{+} = b_{j,2^{N-j}}$ resp. $\rho_{j,0}^{-} = b_{j,0}$.

Furthermore we set

$$\begin{aligned} \psi_{jk} &= c_1\rho_{j,k+1}^{+} + c_2\rho_{j-1,2k}^{-} + c_3\rho_{j-1,2k+1}^{-}, & (4)\\ \phi_{jk} &= d_0\rho_{jk}^{+}, & (5) \end{aligned}$$

where the constants $c_i, i = 1,2,3$ and d_0 depends on j and k. They are chosen such that ψ_{jk} and ϕ_{jk} satisfy the following system of equations

$$\left\{ \begin{array}{c} \langle\psi_{jk}, b_{jk}\rangle = 0 \\ \langle\psi_{jk}, b_{j,k+1}\rangle = 0 \\ \langle\psi_{jk}, \psi_{jk}\rangle = 1 \\ \langle\phi_{jk}, \phi_{jk}\rangle = 1 \end{array} \right. . \tag{6}$$

The values of constants c_1,c_2,c_3 and d_1 can be determined from these equations once we know the values of the constants r_{jk}^{+}, $r_{j-1,2k}^{-}$ and $r_{j,2k+1}^{-}$. The constants $r_{jk}^{\pm}$ satisfy the recursion formula

$$r_{jk}^{\pm} = -\frac{1}{4 + r_{j,k\pm 1}^{\pm}}, \tag{7}$$

starting with $r_{j,2^{N-j}-1}^{+} = -\frac{1}{2}$ and $r_{j,1}^{-} = -\frac{1}{2}$.

The recursion formulas (1) and (3 for $\phi_{jk}^{\pm}$ and (7) for the constants $r_{jk}^{\pm}$ are direct consequences of the Gram-Schmidt orthogonalization procedure.

Notice that mapping $r \mapsto -r/(4+r)$ has two fixed points of which $r = -(2-\sqrt{3})$ is a stable fixed point .

The number $r_{jk}^{\pm}$ will approach $r = -(2-\sqrt{3})$ in this recursion. In fact we have

$$\left. \begin{array}{r} |r_{j,-k+2^{N-j}}^{+} - r| \\ |r_{j,k}^{-} - r| \end{array} \right\} \leq cr^{2k}. \tag{8}$$

This means in practice that the Franklin functions $\psi_{jk}(x)$ (defined on a bounded interval $[0, 2^N]$) can be very well approximated by the Franklin wavelet function $\psi_{jk}(x)$ (defined on $\mathbf{R}$) when the point x is far away from the endpoints of the interval . The error term is bounded by $C2^{-j/2}r^{2d(x)}$ where $d(x) = \min(2^{-j}x, 2^{-j}(2^N - x))$, when $x \in [0, 2^N]$.

1.4 Other versions of the Franklin system

The generating Franklin wavelet function ψ defined above, which is supported on the whole interval, may be replaced by a one-sided version $\tilde{\psi}$, which is supported on the interval $(-\infty, 2]$. The function $\tilde{\psi}$ a similarly defined as ψ with $\rho_{0,1}^{+}$ replaced by $\rho_{\bar{0},0}^{-}$ and the constants c_j adjusted, i.e.

$$\tilde{\psi} = \tilde{c}_1\rho_{\bar{1},0}^{-} + \tilde{c}_2\rho_{\bar{1}}^{-} + \tilde{c}_3\rho_{\bar{2}}^{-},$$

where the constant $\tilde{c}_1$,$\tilde{c}_2$ and $\tilde{c}_3$ are chosen such that the equations (2) are satisfied with $\tilde{\psi}$ replacing ψ. Let $\tilde{\psi}_{jk}(x) = 2^{-j/2}\tilde{\psi}(2^{-j}x - k)$. Then $\{\tilde{\psi}_{jk}\}_{j,k\in\mathbf{Z}}$ is a complete orthonormal basis in $L^2(\mathbf{R})$ which we call the **one-sided Franklin wavelet system.** For each integer j the set of functions $\{\phi_{j+1,k}\}_{k\in\mathbf{Z}} \cup \{\tilde{\psi}_{jk}\}_{j,k\in\mathbf{Z}}$ is a complete orthonormal basis in $\mathbf{V}_j$. Thus the one-sided Franklin wavelet system and the Franklin wavelet system have the same subspaces $\mathbf{V}_j$ in what is called their multiscale analysis.

It is also possible to define a **one-sided Franklin system on a bounded interval.**

Another possibility is to define a **weighted Franklin system** where the Lebesgue measure is replaced by a more general positive measure μ on $\mathbf{R}$ This means that the inner products are defined by the measure μ. As a consequence the constants in the formulas changes from place to place and we do not have asymptotical behavior as we have in the case with the Lebesgue measure on a bounded interval on $\mathbf{R}$.

We can find similar formulas for the **weighted Franklin system in higher dimension** only in the cases when the measure is a product of one dimensional positive measures.

We can also define **non-dyadic Franklin systems on R** where dyadic intervals on $\mathbf{R}$ are replaced by intervals of arbitrary length.

2 Numerical experiments

Recently, I have done numerical calculations with the Franklin wavelet system $\{\psi_{jk}\}$, establishing identities similar to

$$f(x) = \sum_{jk}\langle\psi_{jk}, f\rangle\psi_{jk}(x), \tag{9}$$

i.e.

1. Given the function f, find its Franklin coefficients $\{\langle\psi_{jk}, f\rangle\}$.

2. Given the coefficients $\{c_{jk}\}$ find the sum $\sum_{jk} c_{jk}\psi_{jk}(x)$.

In practice, it is only possible to consider a finite number of coefficients. Let us assume that f is a continuous, piecewise linear function with nodes only at integer points and that it is supported in the interval $[0, 2^N]$. Using only a finite number of scales we can write

$$f(x) = \sum_{j=1}^{N} \sum_k \langle \psi_{jk}, f \rangle \psi_{jk} + \sum_k \langle \phi_{Nk}, f \rangle \phi_{Nk}(x). \tag{10}$$

In the case of the Franklin system on the bounded interval $[0, 2^N]$ the sum above has $2^N + 1$ coefficients and the function f is completely determined by its values int the $2^N + 1$ integer points $0, \dots, 2^N$.

In the case of the Franklin wavelet system on $\mathbf{R}$, f is determined by its values at the $2^N - 1$ integer points $1, \dots, 2^N - 1$ but there are still an infinite number of coefficients on each scale (j) in the sum above. However, we may use the geometric decay of the functions ψ_{jk} and ϕ_{jk} outside the interval $[k2^j, (k+1)2^j]$ and that f is supported in the interval $[0, N]$. By use of the summation formula of a geometric series ($\sum_{k=0}^{\infty} t^k = 1/(1-t)$ when $|t| < 1$) we obtain the formula

$$\begin{aligned} f(x) &= \langle \phi_{N,0}, f \rangle \phi_{N,0}(x) + \frac{1}{1-r^2} \langle \phi_{N,1}, f \rangle \phi_{N,1}(x) \\ &+ \sum_{j=1}^{N} \left(\frac{1}{1-r^2} \langle \psi_{j,-1}, f \rangle \psi_{j,-1}(x) + \left(\sum_{k=0}^{2^{N-j}-1} \langle \psi_{jk}, f \rangle \psi_{jk}(x) \right) \right. \\ &\left. + \frac{1}{1-r^4} \langle \psi_{j,2^{N-j}}, f \rangle \psi_{j,2^{N-j}}(x) \right), \end{aligned} \tag{11}$$

holding for all $x \in [0, 2^N]$. Thus we can recover the values of f from $2^N + 2N + 2$ coefficients.

3 Description of the algorithm

3.1 On the real line

We will describe briefly the algorithm for thc Franklin wavelet system, i.e. in the translation and dilation invariant case. We assume that f is a function in $\mathbf{V}_0$ with support in the interval $[0, 2^N]$. Let $r = \sqrt{3} - 2$.

First by simple integration we get

$$\langle f, b_{0,k} \rangle = \frac{1}{6}(f(k-1) + 4f(k) + f(k+1)).$$

The changing of scale is done by the formula (our scaling filter)

$$\langle f, b_{j,k} \rangle = \frac{1}{2\sqrt{2}}(\langle f, b_{j-1,2k-1} \rangle + 2\langle f, b_{j-1,2k} \rangle + \langle f, b_{j-1,2k+1} \rangle).$$

The key ingredients in algorithm are the following simple recursion formulas

$$\begin{aligned} \langle f, \rho_{j,k}^{+} \rangle &= \langle f, b_{j,k} \rangle + r \langle f, \rho_{j,k+1}^{+} \rangle, \\ \langle f, \rho_{j,k}^{-} \rangle &= \langle f, b_{j,k} \rangle + r \langle f, \rho_{j,k-1}^{-} \rangle. \end{aligned}$$

Because of these recursions, the algorithm goes extremely fast in spite of the fact that the functions ψ_{jk} are not compactly supported. Finally we get the coefficients as linear combinations

$$\begin{aligned} \langle f, \psi_{j,k} \rangle &= c_1 \langle f, \rho^+_{j,k} \rangle + c_2 \langle f, \rho^-_{j-1,2k} \rangle + c_3 \langle f, \rho^-_{j-1,2k+1} \rangle, \\ \langle f, \phi_{jk} \rangle &= d_0 \langle f, \rho^-_{jk} \rangle. \end{aligned}$$

The last formula is used in higher dimensions.

The algorithm to restore the function from the coefficients by the summation formula also uses recursion as a basic ingredient. Given any sequence of real numbers $\{c_k\}_{k=m}^{\infty}$ then

$$\sum_{k=m}^{\infty} c_k \rho^+_{jk}(x) = \sum_{k=m}^{\infty} \tilde{c}_k b_{jk}(x),$$

where the numbers $\tilde{c}_k$ are given by the recursion formula

$$\begin{aligned} \tilde{c}_k &= c_k && \text{for } k = m, \\ \tilde{c}_k &= c_k + r\tilde{c}_{k-1} && \text{for } k > m. \end{aligned}$$

For sums of ρ^-_{jk} there is a similar recursion formula going from right to left. The remaining details are straightforward and left to the readers.

3.2 In higher dimensions

The algorithm is easily extended to higher dimension. In dimension 2 the function f is given by its values along a $(2^N + 1) \times (2^N + 1)$ matrix corresponding to points with integer coordinates in a square in the plane. We assume that the function is linear in each variable, i.e. of the form $ax + by + cxy + d$ on any unit squares between four such points.

We apply the one dimensional procedures for computing $\{\langle f, \psi_{jk} \rangle\}_k$ and $\{\langle f, \phi_{jk} \rangle\}_k$ along the row vectors resp. column vectors (or along column vector resp. row vectors) alternating the order as the scales (j) is increased. In the procedure of computing $\{\langle f, \psi_{jk} \rangle\}_k$ is built in the subprocedure (our scaling filter) of filtering the data for the next scale.

3.3 On the bounded interval

In this case we have to calculate the constants $r\pm_{jk}$ and the constants c_1, c_2, c_3 and d_0, which also depend on j and k, using (7) and (6). Because of (8) it is enough to do this for the L values of k nearest to each endpoint in order to have an accuracy of $Cr^{2L} < C13.9^{-L}$. In practice I used $L = 14$ giving an accuracy of 10^{-16} which was the order of magnitude of the double precision of the computer. The extra time for computing these coefficients is not noticeable when the data set is large.

3.4 The weighted case

The algorithms for the weighted and the non-dyadic Franklin system will be as above but the constants in the formulas as r, c_1, c_2, c_3, d_0 must be calculated for all (j, k) as we have no asymptotic behavior far away from the endpoints.

4 Application to Image processing

We have applied the Franklin wavelet transform to images. We have used the version bounded on a bounded square with $(2^N + 1) \times (2^N + 1)$ integer points. With $N = 9$ this corresponds to an image given by 513×513 pixels. At each pixel the function f was given a value between 0 and 255 according to the grey scale. From this function f its $263,169 = 513 \times 513$ wavelet coefficients were calculated. (This takes about 0.6 second on a HP 9000 750.) The coefficient can be manipulated in some way. Using the inverse wavelet transform we will obtain a new function $\tilde{f}$ from the manipulated coefficients.

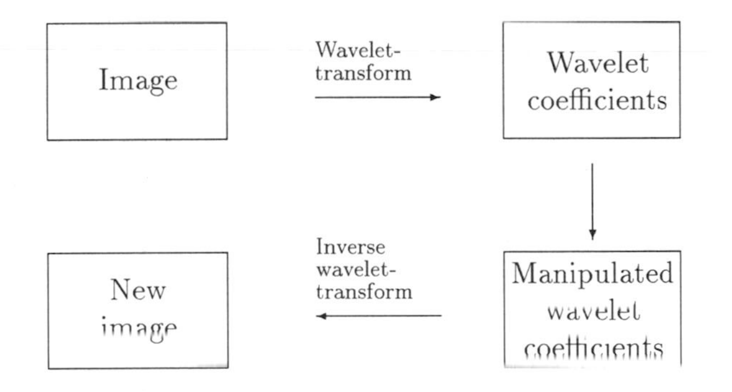

When the image is represented by the wavelet coefficients we can easily split the image information into the orthogonal spaces $\mathbf{W}_j$ and thus sort the information in different scales. For example, by emphasizing small scales we can **detect edges** or **get more contrasts** in the image. Two important properties of the wavelet functions is

1. They are well localized. They live essentially on a square with side length and position corresponding to the indices (j, k). Outside this square thee are rapidly decreasing.

2. Their inner products with the functions $1, x, y$ and xy are zero.

Because of this the wavelet by the inner product $\langle f, \Psi_{j\mathbf{k}} \rangle$ will read information on one scale while background, large scale information is sorted out and smaller scale details are smoothed out by the integration.

In an image their is usually a lot of redundance. An image is build up of objects or details in different sizes. Most information about larger objects is carried by coefficients of larger scale functions while information about small details is carried essentially by smaller scale functions. On smooth parts of the image the coefficients of smaller scale functions will be very small. Since at most positions in the image there usually are neither edges nor small details , a large number of the wavelet coefficients will be small. This makes it possible to do **image compression** by throwing away information form the image and still keeping an image that looks very similar to the original to the human eye. This can be done by truncation of the coefficient matrix of the image. Let C_0 be a positive constant and and let α be a number between 1 and $\sqrt{2}$. We may define the truncated coefficients $c_{j\mathbf{k}}$ by

$$c_{j\mathbf{k}} = \begin{cases} \langle f, \Psi_{j\mathbf{k}} \rangle & \text{when } |\langle f, \Psi_{j\mathbf{k}} \rangle| > C_0 \alpha^j \\ 0 & \text{otherwise} \end{cases} .$$

For suitable choices of C_0 and α their will be only a small fraction of non-zero coefficients $c_{j\mathbf{k}}$ from which we can retrieve a function $\tilde{f}$ approximating f such that the new image look very similar to the original one. The number α is related to which norm we want to use when estimating the error between the original and the compressed image. When $\alpha = 1$ we get the best L^2-norm approximation (comparing all sums with the same number of coefficients). For other values of α the L^2-norm is generalized to Besov space norms. For $\alpha = \sqrt{2}$ the corresponding Besov space norm is in some way " close" to the supremum norm. In my experiments using $\alpha = 1$ in the truncation seemed to be too sensitive for details. With $\alpha = \sqrt{2}$ the truncation often creates an impression of the wrong shade of colors. The value $\alpha = 1.22$ used below was just chosen as it seemed to be a good compromise.

The sparse matrix with the remaining non-zero coefficients may be compressed into a small file. I have not spent much effort to do this coding in the most effective way. I have only used standard compression procedures for this (such as compress in unix). For a more detailed discussion about this method of using wavelet coefficients for image compression and how the coefficients should be coded in an effective way we refer to ([4]).

The original (fig. 1), 513×513 image, is from a painting by Barton. It is fully represented by $263,169$ coefficients. In the following four pictures (figs. 2 – 5) we have truncated the coefficients with as above with $\alpha = 1.22$ in all cases and with the constant $C_0 = 15, 25, 50$ and 100.

Figure 1: *Painting by Barton. Original: 263,169 coefficients*

Figure 2: *Image reconstructed from 26,773 coefficients (*10.2%*).*

Figure 3: *Image reconstructed from 11,697 coefficients (*4.45%*).*

Figure 4: *Image reconstructed from 2,516 coefficients (*0.96%*).*

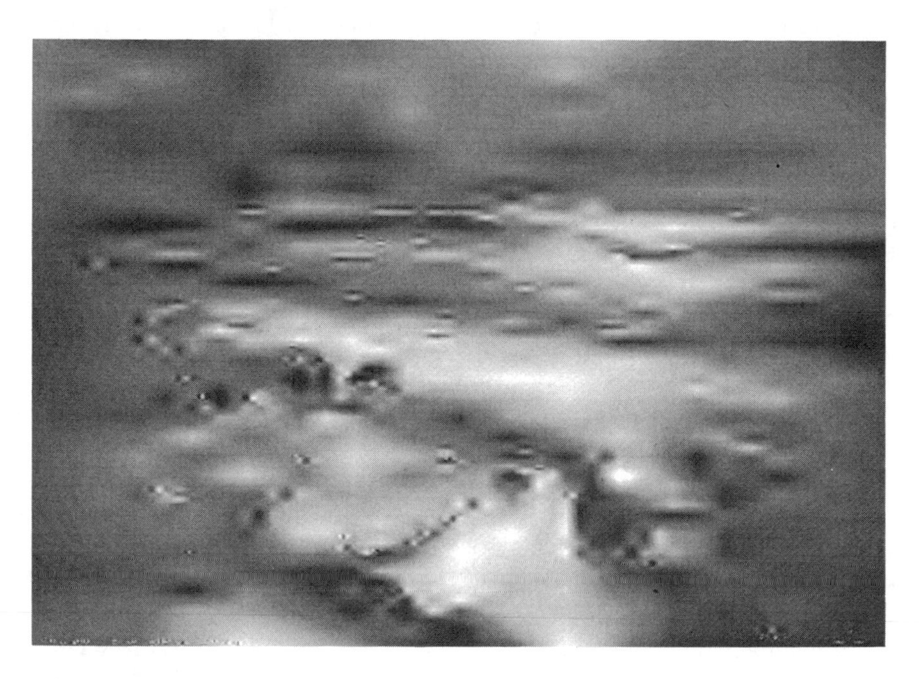

Figure 5: *Image reconstructed from 318 coefficients* (0.12%).

5 Comments

The numerical algorithm for the Franklin wavelet system is very fast. The complexity is linear. In order to calculate the wavelet coefficients it uses in average 6 multiplications and 10 additions per point of input data In in one dimension and in average 8 multiplications and 13 additions per point of input data in dimension two. The calculation of the coefficients of an 513×513 matrix image took about 0.6 seconds on a HP 9000 750 or about $450,000$ coefficients per second (using scalar operations only).

In a special case with a weight corresponding to the point measure and with the one-sided Franklin system on $\mathbf{R}$ the algorithm uses only in average 4 multiplications and 6 additions per data point in order to calculate the wavelet coefficients. In higher dimensions it uses less than in average 5 multiplications and 7 additions per point of input data. The algorithms for inverse wavelet transformations have similar complexity.

This is very fast compared to most other wavelet systems even if the Fast Wavelet Algorithm is used. Only the wavelets with the shortest filters as Daubechies 4 coefficients and 6 coefficient filters could compete with about the same speed. The Haar system has of course a faster algorithm with only 1 multiplication (normalization) per point of input data.

The algorithms described in this paper are very stable. The error term is bounded by $C\epsilon \log_2 M$. Where ϵ is the floating point error of the computer and M is the numbers of points of input data. This is much better than that which is obtained by other wavelet systems using QMF (Quadratic Mirror Filtering) methods. There the error usually is of the order of magnitude ϵM^β for some $\beta > 0$.

The algorithms can easily be vectorized. The calculation of the $263,169$ wavelet coefficients of an 513×513 image takes about six million operations. Theoretically, this can be reduced to about one hundred vector operations.

The way to extend wavelets from one dimension to higher dimensions using a single function as described in section 1 and in ([9]) can be used in general with all other one dimensional wavelet systems. (ψ is the wavelet function and ϕ os the scaling function. This is a way I would strongly recommend. Especially when using QMF methods the algorithm extends extremely simply to higher dimension. One simply do as the "h and g-filtering" on the n-dimensional data tensor in one direction and before we do the filtering in the next scale we change the direction of filtering. After n times of filtering the tensor we are back to the first direction again and we just go on with next scale. From this we see that the complexity of the algorithm will be quite independent of the dimension. Using the extension with $2^n - 1$ tensor products the complexity is proportional with the dimension and one needs extra memory to store a lot of intermediate results. With the single function extension method it is possible to work in the same memory as used by the input data, using only a small memory stack (with size about the sum of lengths of the "h- and g-filters).

The example above indicates that an image is well represented by the wavelet coefficients of the Franklin system and that this representation may be used for data compression on images. Compared to the Haar system (piecewise constant functions), which typically causes boxes to appear the image, the functions in the Franklin system seem to be smooth enough not to give such effects in the image that are easy to detect by the human eye. As a pure mathematician I do not have enough experience in image analysis to be able to compare this method with other image compression methods, and I have not spent much effort to code the wavelet coefficients in the most effective way. It seems, however, that there is a good potential for doing good data compression on images using the Franklin system which is more than 60 years old.

References

[1] Ciesielsky, Z. "Properties of the orthonormal Franklin system, I," *Studia Mathematica,* 23 (1963), 141 - 157.

[2] Ciesielsky, Z. "Properties of the orthonormal Franklin system, II," *Studia Mathematica,* 27 (1966), 284 - 323.

[3] I.Daubechies, "Orthonormal bases of compactly supported wavelets." *Comm. in Pure an Applied Math.,* 41 (1988), 909-996.

[4] DeVore, R.A, Jawerth. B, and Lucier, B.J. "Image Compression Through Wavelet Transform Coding," *IEEE Transactions on Information Theory,* 38 (1992) no. 2, 719 -746.

[5] Franklin, Ph. "A set of continuous orthogonal functions," *Math. Ann. 100 (1928), 522 - 529*

[6] Mallat, S. "Multiresolution approximation and wavelet orthonormal bases of $L^2(R)$," *Trans. Amer. Math. Soc.* 315 (1989), 69-88.

[7] Meyer, Y. "Ondelettes, fonctions splines et analyses graduées," *Cahiers Mathématique de la DéciSion,* no. 8703, Ceremade.

[8] J.-O. Strömberg. "A modified Franklin system on and higher order spline systems on $\mathbf{R^n}$ as unconditional bases for Hardy spaces", *Proceeding of the conference in Analysis in Chicago in honor of A. Zygmund, 1981.* A. Wadsworth, 1983, 475 - 494.

[9] J.-O. Strömberg. "A modified Franklin system as the first orthonormal system of wavelets", *Proceedings of the conference on Wavelets and its application, Marseille, France, May 1989.* Researches en Mathématique Appliqueés / Research Notes in Applied Mathematics, 20 (1992). Masson / Springer Verlag

Representation Formulae for Infimal Convolution with Applications

Thomas Strömberg

Department of Mathematics
Luleå University of Technology
S-951 87 Luleå, Sweden

Abstract

This paper is concerned with the operation of *infimal convolution*. In particular, the following formulae for the infimal convolution $f \Box g$ of f and g are proved under the assumptions that f and g are strictly convex, coercive and Gâteaux-differentiable functions defined on a real reflexive Banach space:

$$\begin{aligned} f \Box g &= (f \circ f'^{-1} + g \circ g'^{-1}) \circ (f' // g'), \\ (f \Box g)' &= f' // g', \end{aligned}$$

where $f' // g' = (f'^{-1} + g'^{-1})^{-1}$ (the parallel sum of f' and g').

Some further results are presented including an application to a Hamilton–Jacobi equation and an example from mechanics.

1 Introduction

Let X denote a real vector space. To any function $f: X \to \bar{\mathbb{R}} = \mathbb{R} \cup \{-\infty, +\infty\}$ we associate the *epigraph* of f

$$\operatorname{epi} f = \{(x, \alpha) \in X \times \mathbb{R} | \ f(x) \leq \alpha\},$$

and the *strict epigraph*

$$\operatorname{epi}_s f = \{(x, \alpha) \in X \times \mathbb{R} | \ f(x) < \alpha\}.$$

Many properties of f can in a natural and appealing way be expressed in terms of epigraphs. For instance, f is convex if and only if $\operatorname{epi} f$ (or $\operatorname{epi}_s f$) is a convex set, whereas f lower semicontinuous is equivalent to $\operatorname{epi} f$ closed when X is a topological vector space.

Moreover, differentiability theory for convex and lower semicontinuous functions is most fruitfully developed in terms of supporting hyperplanes to the epigraphs, giving rise to the theory of subdifferentials.

A natural step is to apply set operations on epigraphs. The present paper deals with the Minkowski sum of strict epigraphs; Attouch and Wets [4] use the terminology "epigraphical sum": Given two extended real valued functions f and g defined on X, the addition

$$\operatorname{epi_s} f + \operatorname{epi_s} g =: \operatorname{epi_s} f \Box g$$

gives rise to the strict epigraph of a function that we denote by $f \Box g$ and call the *infimal convolution* or the *epigraphical sum* of f and g. The following functional identity justifies the name "infimal convolution": the function $f \Box g$ assigns to each $x \in X$ the extended real number

$$(f \Box g)(x) = \inf\{f(y) + g(z) \mid y, z \in X \text{ and } y + z = x\}. \tag{1}$$

(By convention, the sums $(+\infty) + (-\infty)$ and $(-\infty) + (+\infty)$ equal $+\infty$.) Infimal convolution $\Box$ is a commutative and associative binary operation on $\bar{\mathbb{R}}^X$ [23]. The function $\mathcal{I}$, defined as

$$\mathcal{I}(x) = \begin{cases} 0 & \text{when } x = 0, \\ +\infty & \text{when } x \neq 0, \end{cases}$$

acts as the identity element for $\Box$. Therefore, $(\bar{\mathbb{R}}^X, \Box)$ is a commutative monoid. Infimal convolution was introduced by Fenchel [11] and is discussed in the books [16], [18], [23] and [25] by Ioffe & Tihomirov, Laurent, Moreau and Rockafellar, respectively. An overview of the basic concepts of epigraphical calculus and analysis can be found in the paper [4] by Attouch and Wets.

Infimal convolution plays a basic role in optimization mainly because of the relation

$$(f \Box g)^* = f^* + g^* \tag{2}$$

which is valid for any proper functions f and g on X, see [16, page 178], X being a topological vector space. Here, $*$ denotes Legendre-Fenchel transformation, see equation (3) for its definition. Another reason for studying infimal convolution has traditionally been its—*sometimes*—regularizing effect. Namely, roughly speaking, if the function g possesses a regularity property, then, often under very mild assumptions on f, $f \Box g$ inherits this property, see [2], [3], [4], [6], [12], [13], [14] or [15] for results in this direction. Especially for convex functions defined on a Hilbert space, the most famous regularization and approximation technique by means of infimal convolution is called Moreau-Yosida approximation, consult for instance [2]. However, the smoothing effect is even in the convex case up to some limit; examples of situations when regularity is lost when performing infimal convolution can be found in [7], [8] and in [17].

Our aim is somewhat different; we focus on *formulae* for $f \Box g$, as well as for its subdifferential $\partial(f \Box g)$, under the assumptions that f and g are proper, convex, lower semicontinuous and coercive functions defined on a reflexive Banach space. One motivation for studying how infimal convolution acts on such "potential energy type" of functions is that the potential energy function of a mechanical system consisting of two springs connected in series is equal to the infimal convolution of the respective potential energy functions of the springs. The subdifferential of the potential energy function of a spring can be interpreted as a relation between force and elongation.

This paper is organized as follows. In section 2 we introduce two sets of functions especially considered: the cone $\mathcal{C}(X)$ [resp., $\mathcal{C}_w(X)$] of all proper lower semicontinuous convex and coercive [resp., weakly coercive] functions defined on a normed vector space X. In particular, assuming X to be a reflexive Banach space, it is proved that $\mathcal{C}(X)$ and $\mathcal{C}_w(X)$ are closed under infimal convolution and, moreover, that the subdifferential $\partial(f\Box g)$ of $f\Box g$ is equal to the parallel sum of ∂f and ∂g, when and f and g belong to $\mathcal{C}_w(X)$. In section 3 a representation formula for $f\Box g$ is presented for the case when f and g are Gâteaux-differentiable, strictly convex and coercive; X is again a reflexive Banach space. Section 4 contains a discussion of the mechanical example mentioned above. As another example, a special Hamilton–Jacobi equation is considered in section 5. In our final section we present a concluding remark concerning the relation between infimal convolution and an operation investigated by the author in the paper [26].

Throughout, an extended real valued function defined on X is called *proper* if it does not take the value $-\infty$ and if it is not identically equal to $+\infty$. By $\operatorname{dom} f$ we denote the *essential domain* of $f: X \to \bar{\mathbb{R}}$, that is,

$$\operatorname{dom} f = \{x \in X \mid f(x) < +\infty\}.$$

ACKNOWLEDGEMENTS. The author thanks Lars Erik Persson for his help and an anonymous referee for references and valuable valuable suggestions which have helped in improving the presentation.

The research reported here was supported in part by *Magnuson's fund* of *The Royal Swedish Academy of Sciences*.

2 General results for the infimal convolution of proper, lower semicontinuous, convex and coercive functions

Let X be a topological vector space with topological dual space X^*. The *subdifferential* $\partial f: X \to \mathcal{P}(X^*)$ of $f: X \to \bar{\mathbb{R}}$ is defined, for each $x \in X$, by setting $\partial f(x) = \emptyset$ if $f(x) = \pm\infty$ and

$$\partial f(x) = \{x^* \in X^* \mid \langle x^*, y - x\rangle \leq f(y) - f(x) \text{ for all } y \in X\}$$

if $f(x)$ is a finite number. The subdifferential generalizes the classical concept of a derivative in the case of proper, convex and lower semicontinuous functions. The subdifferential of a proper, convex and lower semicontinuous function defined on a reflexive Banach space is a maximal monotone mapping. In order to be able to formulate the following proposition we need a definition: for any mappings S and T from X into $\mathcal{P}(X^*)$ (that is, "multimappings" from X into X^*) we define the *parallel sum* $S//T: X \to \mathcal{P}(X^*)$ by

$$S//T = \{(x, x^*) \in X \times X^* |\ \text{there exist } y, z \in X \\ \text{such that } y + z = x \text{ and } x^* \in S(y) \cap T(z)\}.$$

We note that

$$(S//T)^{-1} = S^{-1} + T^{-1}.$$

Parallel addition of positive linear operators $H \to H$, with H a Hilbert space, has been studied in [1], [10], [22] and in [24].

We are now in position to state a general result concerning the subdifferential $\partial(f \Box g)$. Compare with [18].

Proposition 1. *Assume that f and g are proper functions defined on a topological vector space X. If the infimum in the formula (1) for $(f \Box g)(x)$ is attained for each $x \in X$, then*

$$\partial(f \Box g) = \partial f // \partial g.$$

We recall that if $f : X \to \bar{\mathbb{R}}$, then the *conjugate function* or the *Legendre-Fenchel transform* $f^*: X^* \to \bar{\mathbb{R}}$ of f is defined by

$$f^*(x^*) = \sup\{\langle x^*, x\rangle - f(x) |\ x \in X\} \tag{3}$$

for every $x^* \in X^*$. General properties of conjugate functions can be found in [9], [16], [25] and [28]. If f is a proper function, then

$$\partial f = \{(x, x^*) \in X \times X^* |\ \langle x^*, x\rangle = f(x) + f^*(x^*)\},$$

see [28, page 490].

Proof. Recall (2). The following equivalences hold for any $(x, x^*) \in X \times X^*$:

$$x^* \in (\partial f // \partial g)(x)$$
$$\Leftrightarrow$$
$$\exists y, z \in X \quad x = y + z \text{ and } x^* \in \partial f(y) \bigcap \partial g(z)$$
$$\Leftrightarrow$$
$$\exists y, z \in X \quad x = y + z \text{ and } \langle x^*, y\rangle = f(y) + f^*(x^*) \text{ and } \langle x^*, z\rangle = g(z) + g^*(x^*)$$
$$\Leftrightarrow$$
$$\exists y, z \in X \quad x = y + z \text{ and } \langle x^*, y + z\rangle = f(y) + g(z) + f^*(x^*) + g^*(x^*)$$
$$\Leftrightarrow$$
$$\langle x^*, x\rangle = (f \Box g)(x) + (f \Box g)^*(x^*)$$
$$\Leftrightarrow$$
$$x^* \in \partial(f \Box g)(x).$$

The proof is complete. □

A function $f: X \to \bar{\mathbb{R}}$, where $(X, \|\cdot\|)$ stands for normed vector space, will be called

- *coercive* if
$$\frac{f(x)}{\|x\|} \to +\infty \text{ as } \|x\| \to +\infty,$$
- *weakly coercive* if $f(x) \to +\infty$ as $\|x\| \to +\infty$.

Definition. *We shall denote by $\mathcal{C}(X)$ [resp., by $\mathcal{C}_{\mathrm{w}}(X)$] the cone of all proper lower semicontinuous convex and coercive [resp., weakly coercive] functions defined on X; X being a normed vector space.*

Note that $\mathcal{C}(X) \subset \mathcal{C}_{\mathrm{w}}(X)$. In the sequel we are concerned with the infimal convolution of functions belonging to $\mathcal{C}(X)$ or $\mathcal{C}_{\mathrm{w}}(X)$. In terms of epigraphs, f belongs to $\mathcal{C}_{\mathrm{w}}(X)$ if and only if its epigraph epi f is a non-empty closed convex subset of $X \times \mathbb{R}$ that contains no vertical lines, and such that the level set $\{x | \; (x, \alpha) \in \text{epi } f\}$ is bounded for each real number α.

When X is a reflexive Banach space, any function in $\mathcal{C}_{\mathrm{w}}(X)$ has a finite minimum (see [9, page 35]), and if $f \in \mathcal{C}(X)$, then $\bigcup_{x \in X} \partial f(x) = X^*$.

Now we are in position to formulate the main result of this section.

Theorem 1. *Let X be a reflexive Banach space.*

(a) *If f and g belong to $\mathcal{C}_{\mathrm{w}}(X)$, then the infimum in the formula* (1) *for $(f \Box g)(x)$ is achieved for each $x \in X$ and $\partial(f \Box g) = \partial f // \partial g$.*

(b) *The cones $\mathcal{C}(X)$ and $\mathcal{C}_{\mathrm{w}}(X)$ are closed under infimal convolution.*

(c) *$f \Box g = (f^* + g^*)^*$ for any elements f and g of $\mathcal{C}_{\mathrm{w}}(X)$.*

PROOF. Choose f and g in $\mathcal{C}_{\mathrm{w}}(X)$ arbitrarily. (a) We note that the problem of finding $(f \Box g)(x)$ for a fixed $x \in X$ can be reduced to finding the infimum of the function Ψ defined on X as follows:

$$\Psi(y) = f(y) + g(x - y) \text{ for all } y \in X.$$

We observe that Ψ is lower semicontinuous, convex and weakly coercive. From this follows that Φ has a minimum which is a number in $(-\infty, +\infty]$. According to Proposition 1, $\partial(f \Box g) = \partial f // \partial g$.

(b) It is not hard to show that $\min f \Box g = \min f + \min g$ and, consequently, $\min f \Box g \in \mathbb{R}$. Hence, $f \Box g$ is proper.

Lower semicontinuity. Let $(x_k)_{k \in \mathbb{N}}$ be a sequence in X that converges to x and let us set $\ell = \liminf_{k \to +\infty} (f \Box g)(x_k)$. Obviously, $-\infty < \ell \leq +\infty$. For each $k \in \mathbb{N}$ there exists

y_k in X such that

$$(f\Box g)(x_k) = f(y_k) + g(x_k - y_k).$$

We claim that $(f\Box g)(x) \leq \ell$ which is the case if $\ell = +\infty$. Assume $\ell < +\infty$. There exists a subsequence of (x_k), again denoted by (x_k), such that

$$f(y_k) + g(x_k - y_k) \to \ell$$

as $k \to +\infty$. Since f and g are weakly coercive and bounded from below, the sequence (y_k) must be bounded. Hence, since X is reflexive, we can extract a weakly convergent subsequence, for simplicity denoted by (y_k). Let y be its weak limit. Since f and g are convex and lower semicontinuous they are weakly sequentially lower semicontinuous, see [9, page 11]. Therefore,

$$\begin{aligned} \ell &= \liminf_{k\to+\infty}(f(y_k) + g(x_k - y_k)) \\ &\geq \liminf_{k\to+\infty} f(y_k) + \liminf_{k\to+\infty} g(x_k - y_k) \\ &\geq f(y) + g(x - y) \geq (f\Box g)(x). \end{aligned}$$

We conclude that $f\Box g$ is lower semicontinuous.

The *convexity* is clear. Indeed, $\mathrm{epi_s}\, f\Box g$ is a convex subset of $X \times \mathbb{R}$ since it is the Minkowski sum of the two convex sets $\mathrm{epi_s}\, f$ and $\mathrm{epi_s}\, g$.

Coerciveness. We show that $f\Box g$ is coercive when f and g are coercive. Take any sequence $(x_k)_{k\in\mathbb{N}}$ in X such that $\|x_k\| \to +\infty$ as k tends to infinity. We shall prove that

$$\lim_{k\to+\infty} \frac{(f\Box g)(x_k)}{\|x_k\|} = +\infty. \tag{4}$$

To each $k \in \mathbb{N}$ there correspond elements y_k and z_k of X such that

$$x_k = y_k + z_k \text{ and } (f\Box g)(x_k) = f(y_k) + g(z_k).$$

Choose, for each natural number k, u_k in $\{y_k, z_k\}$ in such a way that

$$\|u_k\| = \max\{\|y_k\|, \|z_k\|\}.$$

Then

$$\|x_k\| \leq \|y_k\| + \|z_k\| \leq 2\|u_k\|.$$

Moreover,

$$(f \wedge g)(x) := \min\{f(x), g(x)\} \geq \min\{\min f, \min g\} =: m$$

for all $x \in X$. We note that $\|u_k\| \to +\infty$ as k goes to infinity and that $f \wedge g$ is coercive. Therefore

$$\frac{(f\Box g)(x_k)}{\|x_k\|} \geq \frac{(f \wedge g)(u_k) + m}{2\|u_k\|} \to +\infty$$

as $k \to +\infty$ and (4) is proved.

(c) Since $f \Box g$ is convex and lower semicontinuous, $f \Box g = (f \Box g)^{**}$. Moreover, $(f \Box g)^{**} = (f^* + g^*)^*$, see (2). $\Box$

REMARK. The following subsets of $\bar{\mathbb{R}}^X$, where X stands for a reflexive Banach space, are also closed under $\Box$:

- The convex cone of all finitely valued functions in $\mathcal{C}(X)$.
- The convex cone of all finitely valued functions in $\mathcal{C}_w(X)$.

3 A representation formula

Let X be a normed vector space. We recall that an operator $A: X \to X^*$ is called

- *monotone* if
$$\langle A(x) - A(y), x - y \rangle \geq 0 \tag{5}$$
for all x and y in X, and *strictly monotone* if inequality (5) is strict for all x and y in X with $x \neq y$,
- *coercive* if
$$\frac{\langle A(x), x \rangle}{\|x\|} \to +\infty \text{ as } \|x\| \to +\infty,$$
- *demicontinuous* if $x_n \to x$ as $n \to +\infty$ implies $A(x_n) \rightharpoonup A(x)$ as $n \to +\infty$.

A formula for $f \Box g$ is presented in our next theorem. For its formulation we shall need some results from the theory of monotone operators. Assume that f is a Gâteaux-differentiable, convex and coercive function on X. Then $\langle f'(x), x \rangle \geq f(x) - f(0)$ for every x in X [28, page 247]. Hence,

$$\frac{\langle f'(x), x \rangle}{\|x\|} \geq \frac{f(x)}{\|x\|} - \frac{f(0)}{\|x\|} \to +\infty \text{ as } \|x\| \to +\infty,$$

which implies that f' is a coercive operator. If f is strictly convex and X is a reflexive Banach space, then $f': X \to X^*$ is strictly monotone, coercive and demicontinuous [28, page 247]. By the Browder-Minty theorem [27, page 557], the inverse operator $f'^{-1}: X^* \to X$ exists and is strictly monotone, coercive as well as demicontinuous.

Our *representation theorem* reads as follows.

Theorem 2. *Let X be a reflexive Banach space. Let f and g be Gâteaux-differentiable strictly convex coercive real valued functions on X. Put $P = (f'^{-1} + g'^{-1})^{-1}$. Then the following assertions hold.*

(a) *$P: X \to X^*$ is a strictly monotone, coercive and demicontinuous operator.*
(b) *$f \Box g$ is given by*

$$f \Box g = (f \circ f'^{-1} + g \circ g'^{-1}) \circ P. \tag{6}$$

(c) *$f \Box g$ is Gâteaux-differentiable with $(f \Box g)' = P$. Moreover,*

$$(f \Box g)(x) = (f \Box g)(0) + \int_0^1 \langle P(\lambda x), x \rangle \, d\lambda \tag{7}$$

for all $x \in X$.

PROOF. We know that the inverse operators f'^{-1} and g'^{-1} from X^* onto X exist and are strictly monotone and demicontinuous. The sum $f'^{-1} + g'^{-1}$ is also strictly monotone, coercive and demicontinuous and (a) holds.

(b) In order to prove (6), we note that $(f \Box g)(x)$, for a fixed $x \in X$, is equal to the minimum of the function Ψ defined on X as

$$\Psi(y) = f(y) + g(x - y) \text{ for all } y \in X.$$

Since f and g are strictly convex, so is Ψ. Therefore, the minimum of Ψ is uniquely attained at, say $\bar{y}$. Hence, $\Psi'(\bar{y}) = 0$. This is eqivalent to

$$f'(\bar{y}) = g'(\bar{z}) =: x^*,$$

where $\bar{z} = x - \bar{y}$. Consequently,

$$f'^{-1}(x^*) + g'^{-1}(x^*) = \bar{y} + \bar{z} = x$$

and thus $x^* = P(x)$. We arrive at

$$(f \Box g)(x) = f(\bar{y}) + g(\bar{z}),$$

where

$$\bar{y} = f'^{-1}(P(x)) \text{ and } \bar{z} = g'^{-1}(P(x)).$$

This completes the proof of (b).

(c) It easily verified that $(\partial f // \partial g)(x)$ equals the singleton $\{P(x)\}$ for each $x \in X$. By Theorem 1, $\partial(f \Box g)(x) = \{P(x)\}$ for each $x \in X$ which implies that $f \Box g$ is Gâteaux-differentiable and $(f \Box g)' = P$. Integration yields (7). □

4 A mechanical example: Two springs connected in series

Let us consider a mechanical system consisting of two springs connected in series. If the system is displaced x length units, then the two springs are elongated y and z length units, respectively, in such a way that $y + z = x$ and the total potential energy of the springs is minimized. Study Figure 1 below.

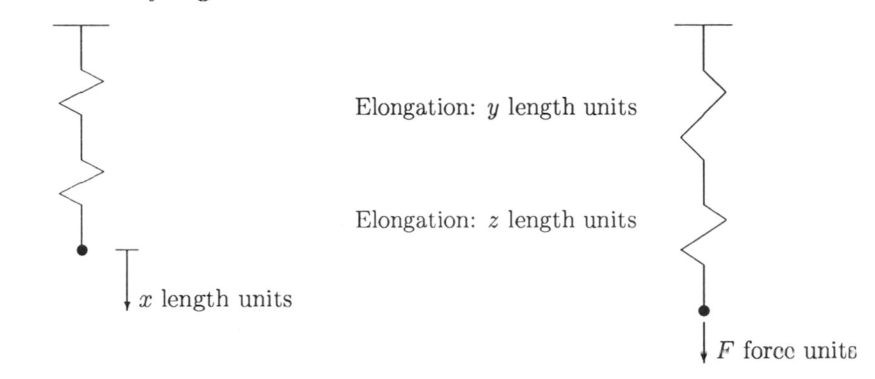

Unstretched position. *Stretched position.*

FIGURE 1. A mechanical spring system.

Let us denote the potential energy of the first spring due to the displacement y by $U(y)$, and by $V(z)$ the potential energy of the second spring due to the displacement z. The principle of minimal energy states that the potential energy $W(x)$ of the system is given as the solution to the following extremal problem.

$$\text{Minimize } U(y) + V(z) \text{ subject to } y + z = x.$$

In the terminology of infimal convolution, $W(x) = (U \Box V)(x)$, that is we have made the

OBSERVATION. The potential energy function of a system of two springs connected in series is equal to the infimal convolution of the respective potential energy functions of the springs.

Let us briefly discuss the infimal convolution of functions that are reasonable from the mechanical point of view, that is functions that are convex and have certain coerciveness and continuity properties. We introduce the convex cone $\mathcal{E}$ of "admissible potential energy functions" consisting of all convex coercive and lower semicontinuous functions U on $\mathbb{R}$ taking values only in $[0, +\infty]$ and such that $U(0) = 0$. We notice that $\mathcal{E} \subset \mathcal{C}(\mathbb{R})$. It is easily verified that $\mathcal{E}$ is closed under infimal convolution. For a potential energy function

$U \in \mathcal{E}$, the possible elongation of the corresponding spring is any number in the essential domain $\operatorname{dom} U$ and no other number. The set $\operatorname{dom} U$ is convex and contains 0. In the extreme case, $\operatorname{dom} U$ is equal to $\{0\}$ which means that $U = \mathcal{I}$. The spring corresponding to $\mathcal{I}$ admits *no* elongation—it is stiff. If U is a function in $\mathcal{E}$, then the subdifferential $\partial U : \mathbb{R} \to \mathcal{P}(\mathbb{R})$ is maximal monotone, $0 \in \partial U(0)$ and $\bigcup_{y \in \mathbb{R}} \partial U(y) = \mathbb{R}$. The relation between the spring force F and the displacement y reads $F \in \partial U(y)$. In particular, if $U(y) = \frac{1}{2}ky^2$, where $k \in (0, +\infty)$, then the mechanical law $F \in \partial U(y)$ takes the form $F = ky$—the case of linear spring theory. See Figure 2 below.

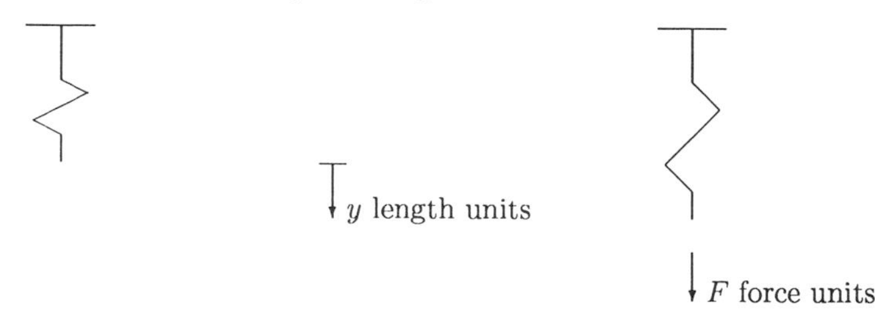

Unstretched position. *Stretched position.*

FIGURE 2. The mechanical law $F \in \partial U(y)$.

Our system in Figure 1 is displaced x length units due to a force of magnitude F force units; thus $F \in \partial(U \Box V)(x)$. By an elementary physical argument, we must have $F \in \partial U(y) \cap \partial V(z)$, where y and z $(y + z = x)$ denote the elongations of the first and second spring, respectively. Indeed, the two springs experience the same spring force (of magnitude F) and, therefore, $F \in \partial U(y)$ and $F \in \partial V(z)$. This is in agreement with assertion (a) of our Theorem 1: $\partial(U \Box V) = \partial U // \partial V$.

We close this section with an application of Theorem 2.

EXAMPLE. Let U and V be the following two functions in $\mathcal{E}$:

$$U(x) = \frac{|x|^p}{p} \text{ and } V(x) = \frac{|x|^q}{q},$$

where $1 < p < q < +\infty$. It is easily proved that $U \Box V$ is even. Denoting $P = (U'^{-1} + V'^{-1})^{-1}$, we have

$$P^{-1}(y) = \left(|y|^{1/(p-1)} + |y|^{1/(q-1)}\right) \operatorname{sgn} y$$

and, by Theorem 2,

$$\begin{aligned}(U \Box V)(x) &= U\left(U'^{-1}(P(x))\right) + V\left(V'^{-1}(P(x))\right) \\ &= \frac{1}{p}|P(x)|^{p/(p-1)} + \frac{1}{q}|P(x)|^{q/(q-1)}.\end{aligned}$$

Thus,

$$\lim_{x\to+\infty}\frac{(U\Box V)(x)}{x^p} = \lim_{y\to+\infty}\frac{(U\Box V)(P^{-1}(y))}{[P^{-1}(y)]^p}$$

$$= \lim_{y\to+\infty}\frac{\frac{1}{p}y^{p/(p-1)}+\frac{1}{q}y^{q/(q-1)}}{[y^{1/(p-1)}+y^{1/(q-1)}]^p} = \frac{1}{p}.$$

Therefore

$$\frac{(U\Box V)(x)}{|x|^p/p} \to 1 \text{ as } x\to\pm\infty.$$

In a similar way we get

$$\frac{(U\Box V)(x)}{|x|^q/q} \to 1 \text{ as } x\to 0.$$

5 An application to a Hamilton–Jacobi equation

Let $H : \mathbb{R} \to \mathbb{R}$ (the *Hamiltonian*) be convex and coercive. The conjugate function $L = H^*$ (the *Lagrangian*) is then finitely valued, convex and coercive. Let $f: \mathbb{R} \to \mathbb{R}$ be lower semicontinuous and bounded from below. We define a function u on $\mathbb{R}\times[0,+\infty)$ by

$$u(\cdot,t) = \begin{cases} f & \text{when } t=0,\\ f\Box L_t & \text{when } t>0.\end{cases}$$

Here, $L_t = tL(\cdot/t)$. Note that

$$\mathrm{epi_s}\, u(\cdot,t) = \mathrm{epi_s}\, f + t\,\mathrm{epi_s}\, L$$

for all $t \in [0,+\infty)$. The function u has the following properties (see Proposition 13.1 of [19] and [5, page 88]):

- $u(x,t) \to f(x)$ as $t\to 0+$ for each real number x.
- u is locally Lipschitzian on $\mathbb{R}\times(0,+\infty)$, hence differentiable almost everywhere.
- At each $(x,t) \in \mathbb{R}\times(0,+\infty)$ where u is differentiable, u satisfies the *Hamilton–Jacobi equation*

$$\dot{u}(x,t) + H(u'(x,t)) = 0. \tag{8}$$

  Here, the prime and the dot denote partial differentiation with respect to the first and second variable, respectively.

Actually, u is the *viscosity solution* of the Hamilton–Jacobi equation (8) with initial condition $u(\cdot,0) = f$, consult [19].

The following proposition gives properties of and an explicit formula for u in the purely convex case, that is, when also f is convex.

Proposition 2. *Let f be convex. Then*

(a) *u is a convex function and $\partial u(\cdot, t) = ((\partial f)^{-1} + t\partial H)^{-1}$ for each $t \in (0, +\infty)$.*
If in addition f is coercive, then (b) and (c) below hold.

(b) *$u(\cdot, t)$ is coercive for each $t \in (0, +\infty)$.*

(c) *If f and H are differentiable and strictly convex, then u is a classical solution of (8) supplied with the initial condition $u(\cdot, 0) = f$, and $u(x, t)$ is given by each of the following three formulae*

$$u(x,t) = f(f'^{-1}(\omega(x,t))) + tL(H'(\omega(x,t))), \tag{9}$$

$$u(x,t) = f(x) - \int_0^t H(\omega(x,\tau))\, d\tau, \tag{10}$$

$$u(x,t) = u(0,t) + \int_0^x \omega(\xi,t)\, d\xi, \tag{11}$$

for all $(x,t) \in \mathbb{R} \times (0, +\infty)$. Here, $\omega(x,t) = (f'^{-1} + tH')^{-1}(x)$.

PROOF. First we note that

$$u(x,t) = \min\{f(x-q) + tL(q/t) |\, q \in \mathbb{R}\}$$

for all $(x,t) \in \mathbb{R} \times (0, +\infty)$.

(a) We start by proving that the function $(q,t) \mapsto tL(q/t)$ on $\mathbb{R} \times (0, +\infty)$ is convex. From the definition of conjugate functions

$$tL(q/t) = (tH)^*(q) = \max\{pq - tH(p) |\, p \in \mathbb{R}\}.$$

The fact that $(q,t) \mapsto pq - tH(p)$ is linear for each real number p implies that the function $(q,t) \mapsto tL(q/t)$ is convex since it is a pointwise maximum of linear functions. Hence,

$$(x,t,q) \mapsto f(x-q) + tL(q/t)$$

is a convex function. Then u is convex since it is obtained by taking minimum over one variable in the above formula, see [9, page 50].

The subdifferential of $L_t = (tH)^*$ equals $(t \cdot \partial H)^{-1}$. By Proposition 1, $\partial u(\cdot, t) = ((\partial f)^{-1} + t\partial H)^{-1}$.

Assertion (b) follows directly from assertion (b) of Theorem 1.

(c) According to Proposition 15.1 in [19], u is a classical solution. Let us apply Theorem 2. Since the derivative of the function $L_t = (tH)^*$ equals $(tH')^{-1}$ we get $u' = \omega$

(see point (c) of Theorem 2) from which (11) follows. The formula (9) is a straightforward application of assertion (b) of Theorem 2. Finally,

$$
\begin{aligned}
u(x,t) &= u(x,0) + \int_0^t \dot{u}(x,\tau)\,\mathrm{d}\tau \\
&= f(x) + \int_0^t (-H(u'(x,\tau)))\,\mathrm{d}\tau \\
&= f(x) - \int_0^t H(\omega(x,\tau))\,\mathrm{d}\tau,
\end{aligned}
$$

and (10) is proved. □

6 A concluding remark

In [26], an operation $\oplus$ on the set Φ of all continuous and nondecreasing functions $\varphi\colon (0,+\infty) \to (0,+\infty)$ such that $\varphi(t) \to 0$ as $t \to 0+$ and $\varphi(t) \to +\infty$ as $t \to +\infty$, is considered. The operation $\oplus$ is defined by the formula

$$(\varphi \oplus \psi)(s) = \inf\{\varphi(t) + \psi(u) \mid t, u \in (0,+\infty) \text{ and } s = tu\}$$

for any $\varphi, \psi \in \Phi$ and every $s \in (0,+\infty)$. Clearly, given two functions φ and ψ in Φ, the *Young-type inequality*

$$(\varphi \oplus \psi)(tu) \leq \varphi(t) + \psi(u)$$

holds for all $t, u \in (0,+\infty)$. Consult [20] or [21] for possible applications of such inequalities in the theory of Orlicz spaces.

The following relation between $\oplus$ and $\square$ (with $X = \mathbb{R}$) holds: Given any elements φ and ψ of Φ we can define functions f and g on $\mathbb{R}$ by

$$f(y) = \varphi(e^y) \text{ and } g(z) = \psi(e^z)$$

for all real numbers y and z. Then

$$
\begin{aligned}
(f\square g)(x) &= \inf\{f(y) + g(z) \mid y, z \in \mathbb{R} \text{ and } y + z = x\} \\
&= \inf\{\varphi(e^y) + \psi(e^z) \mid y, z \in \mathbb{R} \text{ and } e^y e^z = e^x\} \\
&= (\varphi \oplus \psi)(e^x)
\end{aligned}
$$

for each real number x. In other words, the assignment $\varphi \mapsto \varphi \circ \exp$ establishes an isomorphism $(\Phi, \oplus) \simeq (\mathcal{M}, \square)$. Here, $\mathcal{M}$ denotes the set of all continuous nondecreasing functions $f : \mathbb{R} \to (0,+\infty)$ satisfying

$$\lim_{x\to-\infty} f(x) = 0 \text{ and } \lim_{x\to+\infty} f(x) = +\infty.$$

We conclude this paper by deducing from Theorem 1 in [26] the following result for the infimal convolution of monotone functions.

Proposition 3. *$\mathcal{M}$ is closed under infimal convolution and*

$$(f//g)(x) < (f\Box g)(x) \leq 2(f//g)(x) \tag{12}$$

for any f and g in $\mathcal{M}$ and all $x \in \mathbb{R}$.

For each $x \in \mathbb{R}$, let $y = \tilde{y}(x)$ be a solution to the equation $f(y) = g(x-y)$. By definition, the parallel sum $f//g$ assigns to x the positive real number $f(\tilde{y}(x)) = g(x-\tilde{y}(x))$. Note that $(f//g)(x)$ is well defined though the equation $f(y) = g(x-y)$ may have more than one root y. If f and g are strictly increasing, then $f//g = (f^{-1}+g^{-1})^{-1}$.

PROOF. We prove only the inequality (12). Choose $x \in \mathbb{R}$ arbitrarily. Clearly,

$$\max\{f(y), g(x-y)\} \geq (f//g)(x)$$

for every $y \in \mathbb{R}$. It is not difficult to verify that

$$(f\Box g)(x) = f(\bar{y}(x)) + g(x-\bar{y}(x))$$

for some $\bar{y}(x) \in \mathbb{R}$. Hence,

$$(f\Box g)(x) = f(\bar{y}(x)) + g(x-\bar{y}(x)) > \max\{f(\bar{y}(x)), g(x-\bar{y}(x))\} \geq (f//g)(x),$$

which is the left-hand side inequality in (12). In order to prove the right-hand side inequality in (12) take $\tilde{y}(x) \in \mathbb{R}$ such that $f(\tilde{y}(x)) = g(x-\tilde{y}(x))$. Then

$$(f\Box g)(x) \leq f(\tilde{y}(x)) + g(x-\tilde{y}(x)) = (f//g)(x) + (f//g)(x),$$

and the proof of inequality (12) is complete. □

References

[1] ANDERSON, W.N. JR. AND DUFFIN, R.J. *Series and Parallel Addition of Matrices*, J. Math. Anal. Appl. 26, 576-594 (1969).

[2] ATTOUCH, H. *Variational Convergence for Functions and Operators*, Applicable Math. Series, Pitman, London 1984.

[3] ATTOUCH, H. AND AZÉ, D. *Approximation and Regularization of Arbitrary Functions in Hilbert Spaces by the Lasry-Lions Method*, Annales de l'Institut H. Poincaré Analyse non linéaire (to appear).

[4] ATTOUCH, H. AND WETS, R. *Epigraphical Analysis*, in "Analyse non Linéaire", eds. H. Attouch, J.-P Aubin, F.H. Clarke, I. Ekeland, 73-99, Gauthiers-Villars Paris et C.R.M. Montreal 1989.

[5] BENTON, S.H. *The Hamilton–Jacobi Equation: A Global Approach*, Academic Press, New York-San Francisco-London 1977.

[6] BOUGEARD, M., PENOT, J.-P. AND POMMELLET, A. *Towards Minimal Assumptions for the Infimal Convolution Regularization*, J. Approx. Theory 64, 245-270 (1991).

[7] BOMAN, J. *The Sum of Two Plane Convex C^∞ Sets is not Always C^5*, Math. Scand. 66, 216-224 (1990).

[8] BOMAN, J. *Smoothness of Sums of Convex Sets with Real Analytic Boundaries*, Math. Scand. 66, 225-230 (1990).

[9] EKELAND, I. AND TEMAM, R. *Convex Analysis and Variational Problems*, North-Holland, Amsterdam-Oxford 1976.

[10] ERIKSSON, S.-L AND LEUTWILER, H. *A Potential-Theoretic Approach to Parallel Addition*, Math. Ann. 274, 301-317 (1986).

[11] FENCHEL, W. *Convex Cones, Sets, and Functions*, Lecture Notes, Princeton University, Princeton, New Jersey 1953.

[12] FOUGERES, A. AND TRUFFERT, A. *Régularisation S.C.I. et Γ-convergence: approximations inf-convolutives associées à un référentiel*, Annali di Matematica Puri ed Applicati, 21-51 (1986).

[13] HIRIART-URRUTY, J.-B. *Extension of Lipschitz Functions*, J. Math. Anal. Appl. 77, 539-544 (1980).

[14] HIRIART-URRUTY, J.-B. *Lipschitz r-Continuity of the Approximative Subdifferential of a Convex Function*, Math. Scand. 47, 123-134 (1980).

[15] HIRIART-URRUTY, J.-B. *How to Regularize a Difference of Convex Functions*, J. Math. Anal. Appl. 162, 196-209 (1991).

[16] IOFFE, A.D. AND TIHOMIROV, V.M. *Theory of Extremal Problems*, North-Holland, Amsterdam-New York-Oxford 1979.

[17] KISELMAN, C.O. *Smoothness of Vector Sums of Plane Convex Sets*, Math. Scand. 60, 239-252 (1987).

[18] LAURENT, P.-J. *Approximation et optimisation*, Hermann, Paris 1972.

[19] LIONS, P.L. *Generalized Solutions of Hamilton–Jacobi Equations*, Pitman, London 1982.

[20] MALIGRANDA, L. *Orlicz Spaces and Interpolation*, Seminars in Math. 5, Campinas 1989.

[21] MALIGRANDA, L. AND PERSSON, L.E. *Generalized Duality of some Banach Function Spaces*, Indagationes Math. 51, 323-338 (1989).

[22] MAZURE, M.-L. *Analyse variationelle des formes quadratiques convexes*, Thèse de doctorat, Université Paul Sabatier, Toulouse, 1986.

[23] MOREAU, J.J. *Fonctionelles Convexes*, Séminaire sur les équations aux dérivées partielles II, Collège de France 1966-1967.

[24] MORLEY, T.D. *Parallel Summation, Maxwell's Principle and the Infimum of Projections*, J. Math. Anal. Appl. 70, 33-41 (1979).

[25] ROCKAFELLAR, R.T. *Convex Analysis*, Princeton University Press, Princeton, New York 1970.

[26] STRÖMBERG, T. *An Operation Connected to a Young-Type Inequality*, Mathematische Nachrichten (to appear).

[27] ZEIDLER, E. *Nonlinear Functional Analysis and its Applications II*, Springer-Verlag, New York-Berlin-Heidelberg-London-Paris-Tokyo 1990.

[28] ZEIDLER, E. *Nonlinear Functional Analysis and its Applications III*, Springer-Verlag, New York-Berlin-Heidelberg-London-Paris-Tokyo 1985.

Free S-Algebras, Plethysms, and Noncommutative Invariant Theory

Torbjörn Tambour
Department of Mathematics
Stockholm University
S-106 91 Stockholm
Sweden

Abstract. - If V is a vector space and G a group acting linearly on V, then G also acts on the tensor algebra $T(V)$ and we get an algebra of G- invariants $T(V)^G$. This is an example of an S-algebra, i.e. a graded algebra where there are actions of the symmetric groups on the homogeneous parts. The characters of these parts as modules over the symmetric groups are in certain cases plethysms of representations. We will give an introduction to plethysms, and then discuss the structure of S-algebras and in particular of noncommutative rings of invariants, such as $T(V)^G$. Almost no proofs will be given.

1. **Plethysms.** All vector spaces in the sequel will be over the complex numbers **C**. Also all tensor products will be over **C** unless otherwise indicated. Let V be a finite-dimensional vector space and GL(V) the general linear group over V, i.e. the group of all invertible linear maps $V \to V$. All representations of GL(V) in the sequel will be polynomial, i.e. if $\Theta : \mathrm{GL}(V) \to \mathrm{GL}(W)$ is a representation, then if we choose bases for V and W, all matrix elements of $\Theta(g)$ will be polynomials in the matrix elements of $g \in \mathrm{GL}(V)$. All such representations are completely reducible. It is well known that the symmetric powers $S^m(V)$ and the exterior powers $\Lambda^m(V)$ are irreducible GL(V)-modules. But what can be said about things like $S^n(S^m(V))$ and $\Lambda^n(\Lambda^m(V))$ and so on?

Definition: Let $\Theta : \mathrm{GL}(V) \to \mathrm{GL}(W)$ and $\Psi : \mathrm{GL}(W) \to \mathrm{GL}(U)$ be two representations. The the composite $\Psi \circ \Theta : \mathrm{GL}(V) \to \mathrm{GL}(U)$ is a representation which we call the *plethysm* of Θ and Ψ.

So things like $S^n(S^m(V))$ are plethysms. An irreducible polynomial representation of GL(V) is necessarily homogeneous, i.e. the polynomials in the representing matrices are homogeneous and of the same degree. This degree is also called the degree of the representation. If $\Theta : \mathrm{GL}(V) \to \mathrm{GL}(W)$ is a (homogeneous) representation, then its character, i.e. the function $g \mapsto \mathrm{tr}(\Theta(g))$ is a symmetric function of the eigenvalues of g. We will use the following definitions and notation: The *complete symmetric functions* $h_m(x)$ (in a possibly countably infinite number of variables $x_1, x_2, \ldots$) are defined by

$$\sum_{m \geq 0} h_m(x) = \prod_{i \geq 1} (1 - x_i)^{-1}.$$

The *elementary symmetric functions* $e_m(x)$ are defined by

$$\sum_{m\geq 0} e_m(x) = \prod_{i\geq 1}(1 + x_i).$$

The power sums $p_m(x)$ are defined by $p_m(x) = x_1^m + x_2^m + \ldots$. If $\lambda = (\lambda_1, \lambda_2, \ldots)$ is a partition, we put $h_\lambda = h_{\lambda_1} h_{\lambda_2} \ldots$ and so on. Finally, when λ is a partition, we define the *Schur function* s_λ by $s_\lambda(x) = \det(h_{\lambda_i - i + j})$ (strictly speaking this is not the usual definition of the Schur function, but rather the Jacobi-Trudi identity; however for our purposes it will do). We have $s_{(m)} = h_m, s_{(1^m)} = e_m$. One can show that the h_λ, e_λ and s_λ form $\mathbf{Z}$-bases of the ring of symmetric functions (with integer coefficients) when λ runs over all partitions. We refer to [13] for the details of the theories of symmetric functions and representations of the general linear and symmetric groups. Now let $\Theta : \mathrm{GL}(V) \to \mathrm{GL}(W)$ be a representation and let $z_1, \ldots, z_n$ be the eigenvalues of $g \in \mathrm{GL}(V)$. Put $\chi(z) = \operatorname{tr}\Theta(g)$, so that χ is the character of Θ. Then we have

Theorem (Schur). *The representation Θ is irreducible if and only if $\chi(z) = s_\lambda(z)$ for some partition λ of length $\leq n$. Furthermore, for each such λ there is a representation with character s_λ and these are all non-isomorphic.*

Note: The length of a partition is the number of non-zero parts.

One should note that if W and U are $\mathrm{GL}(V)$-modules affording the characters f and g (so these are symmetric functions), respectively, then the tensor product $W \otimes U$ affords fg. The character of the symmetric power $S^m(V)$ is $h_m(z)$ and of the exterior power $\Lambda^m(V)$ it is $e_m(z)$. Hence the determinant $g \mapsto \det g$ has the character $e_n(z)$.

Let $\Theta : \mathrm{GL}(V) \to \mathrm{GL}(W)$, $\Psi : \mathrm{GL}(W) \to \mathrm{GL}(U)$ be two homogeneous representations with characters χ and ψ, respectively. One can show that the eigenvalues of $\Theta(g)$ are monomials in those of g of total degree equal to the degree of Θ. Let $z^\mu = z_1^{\mu_1} \ldots z_n^{\mu_n}$ be the eigenvalues of $\Theta(g)$, where μ runs over the multiindices such that $\sum \mu_i$ equals the degree of Θ, and let a_μ be the multiplicity of z^μ. Then the character of the plethysm $\Psi \circ \Theta$ is

$$\psi(z^\mu; z^\mu \text{ appears } a_\mu \text{ times}).$$

This defines plethysm on a subset of the ring of symmetric functions. The general definition is as follows:

Definition. Assume that f and g are homogeneous symmetric functions in variables $x_1, x_2, \ldots$ and write $g = \sum a_\mu x^\mu$ (where the coefficients a_μ are integers). Introduce new variables y_i by

$$\prod_i (1 + ty_i) = \prod_\mu (1 + tx^\mu)^{a_\mu}.$$

Then the *plethysm* of f and g is $f \circ g = f(y_1, y_2, \ldots)$.

The plethysm of f and g is a symmetric function whose degree is the product of the degrees of f and g. To see how this works in practice, we give an example:

Example: Let $f = e_2, g = h_2(x_1, x_2) = x_1^2 + x_1x_2 + x_2^2$. From the definition of the y_i we immediately get $(f \circ g)(x_1, x_2) = e_2(y) = x_1^3x_2 + x_1^2x_2^2 + x_1x_2^3$. As another example, take $f = p_2 = e_1^2 - 2e_2, g(x_1, x_2) = -x_1x_2$. Then

$$\prod(1 + y_it) = (1 + x_1x_2t)^{-1} = 1 - x_1x_2t + x_1^2x_2^2t^2 - \ldots,$$

so $e_1(y) = -x_1x_2, e_2(y) = x_1^2x_2^2$, whereof $(f \circ g)(x_1, x_2) = e_1(y)^2 - 2e_2(y) = -x_1^2x_2^2$.

One can prove the following: $(f + g) \circ h = f \circ h + g \circ h$ and $(fg) \circ h = (f \circ h)(g \circ h)$. In general plethysm is of course highly "non-commutative", but one has $f \circ p_n = p_n \circ f$ and also $p_n \circ p_m = p_{mn}$. The function $e_1 = h_1$ is a two-sided unit: $f \circ e_1 = e_1 \circ f = f$ for all f. Finally plethysm is associative: $(f \circ g) \circ h = f \circ (g \circ h)$.

The representation theories of the general linear groups and of the symmetric groups are closely related. We now describe this relation. Let f be a class function on the symmetric group S_m. For $\sigma \in S_m$, let $\rho(\sigma)$ denote the cycle type of σ (hence $\rho(\sigma)$ is a partition of m). We now define the *characteristic* of f to be the symmetric function

$$\mathrm{ch}(f) = \frac{1}{n!} \sum_{\sigma \in S_m} f(\sigma)p_{\rho(\sigma)}.$$

Note that $\mathrm{ch}(f)$ is homogeneous of degree m. Let R_m denote the set of virtual characters on S_m and put $R = \oplus_{m \geq 0} R_m$ ($R_0 = \mathbf{Z}$). Then R is a graded commutative and associative ring under the *induction product* defined by

$$f.g = \mathrm{ind}_{S_m \times S_n}^{S_{m+n}}(f \times g).$$

Here $f \in R_m, g \in R_n$ and we embed $S_m \times S_n$ into S_{m+n} by letting the first factor act on $\{1, 2, \ldots, m\}$ and the second on $\{m + 1, m + 2, \ldots, m + n\}$. Finally we note that the ring of symmetric functions carries a scalar product such that $\langle s_\lambda, s_\mu \rangle = \delta_{\lambda\mu}$ and that the ring R carries a scalar product inherited from the ordinary scalar products on the R_m.

Theorem (Frobenius). *The characteristic map is an isometric isomorphism from R to the ring of symmetric functions. An element $f \in R_m$ is an irreducible character on S_m if and only if $\mathrm{ch}(f) = s_\lambda$ for some partition λ of m.*

We denote the irreducible character corresponding to λ by χ^λ and we let M_λ be a module affording χ^λ. One can show that $\chi^{(m)}$ is the principal character and $\chi^{(1^m)}$ the sign character on S_m.

Consider the mth tensor power $T^m(V)$ of our vector space V. This is an S_m-module with action

$$\sigma(v_1 \otimes \cdots \otimes v_m) = v_{\sigma^{-1}(1)} \otimes \cdots \otimes v_{\sigma^{-1}(m)}$$

(place permutations). This action clearly commutes with the $\mathrm{GL}(V)$-action, so $T^m(V)$ is an $S_m \times \mathrm{GL}(V)$-module. Let V_λ be a $\mathrm{GL}(V)$-module affording the character s_λ. Then we have

Theorem (Schur-Weyl duality). *As an $S_m \times \mathrm{GL}(V)$-module, $T^m(V) \cong \oplus_\lambda V_\lambda \otimes M_\lambda$, where the sum is over all partitions of m of length at most $n = \dim V$.*

Another formulation of this duality is as follows: Let M be an S_m-module and let S_m act diagonally on $M \otimes T^m(V)$. The group $\mathrm{GL}(V)$ acts on the last factor and these two actions commute. Hence the space $E = (M \otimes T^m(V))^{S_m}$ of S_m-invariants is also a $\mathrm{GL}(V)$-module. If M is irreducible and corresponds to the partition λ of m of length at most $n = \dim V$, then E is also irreducible and affords the character s_λ. Note that m here is also the degree of the representation of $\mathrm{GL}(V)$ on E. One can show that if W and U are $\mathrm{GL}(V)$-modules corresponding (under this duality) to the modules M and N over the symmetric groups S_m and S_n then the tensor product $W \otimes U$ corresponds to the induction product

$$\mathrm{ind}_{S_m \times S_n}^{S_{m+n}}(M \otimes N)$$

of M and N. We now describe the operation corresponding to plethysm. The *wreath product* of S_m by S_n is a group with underlying set $S_m^n \times S_n$ and multiplication defined by

$$(\sigma_1, \dots, \sigma_n; \tau)(\sigma'_1, \dots, \sigma'_n; \tau') = (\sigma_1 \sigma'_{\tau^{-1}(1)}, \dots, \sigma_n \sigma'_{\tau^{-1}(n)}; \tau\tau').$$

We denote this group by $S_m \sim S_n$. If M and N are modules over S_m and S_n, respectively, then $N \otimes T^n(M)$ becomes an $S_m \sim S_n$-module by

$$(\sigma_1, \dots, \sigma_n; \tau)(x \otimes y_1 \otimes \cdots \otimes y_n) = \tau x \otimes \sigma_1 y_{\tau^{-1}(1)} \otimes \cdots \otimes \sigma_n y_{\tau^{-1}(n)}.$$

The wreath product of S_m by S_n is embedded into S_{mn} as the normalizer of S_m^n. With N and M as above, the plethysm of N and M (or of their characters) is

$$N \circ M = \mathrm{ind}_{S_m \sim S_n}^{S_{mn}}(N \otimes T^n(M)).$$

Example: When G is a subgroup of S_m the *cycle indicator polynomial* of G is the symmetric function

$$c(G) = \frac{1}{|G|} \sum_{\sigma \in G} p_{\rho(\sigma)} = \frac{1}{|g|} \sum_{\lambda} n_\lambda(G) p_\lambda,$$

where $n_\lambda(G)$ is the number of elements in G of cycle type λ and the sum is over all partitions of m. For instance, $c(S_m) = h_m$ and $c(A_m) = h_m + e_m$. Now let G be a subgroup of S_m and H a subgroup of S_n. Then we get in a natural way a subgroup $G \sim H$ of $S_m \sim S_n \subseteq S_{mn}$ and we have $c(G \sim H) = c(H) \circ c(G)$. One can also show that the cycle indicator polynomial of the direct product $G \times H \subseteq S_m \times S_n$ is $c(G)c(H)$.

For tensor products of (irreducible) $\mathrm{GL}(V)$-representations, or equivalently for induction products of (irreducible) representations of symmetric groups, or again equivalently for products of symmetric functions (Schur functions), there is a combinatorial rule, the Littlewood-Richardson rule, which gives the decomposition into irreducibles of such products: Let s_λ and s_μ be two Schur functions. There are non-negative integers $c^\rho_{\lambda\mu}$ such that

$$s_\lambda s_\mu = \sum_\rho c^\rho_{\lambda\mu} s_\rho,$$

where the sum is over the partitions of $|\lambda| + |\mu|$. Then $c^\rho_{\lambda\mu}$ is the number of tableaux T of shape $\rho \setminus \lambda$ and weight μ with the following property: If one reads the integers in tableau from the right to the left starting in the first row one gets a word $w_1w_2\ldots$. Then for every i the number of occurrences of 1 in the truncated word $w_1\ldots w_i$ is not less than the number of occurrences of 2, which is not less than the number of occurrences of 3, and so on. What one would like to have is of course some similar result for plethysms. However, such a rule seems to be very difficult to find. We will now describe a few of the known results about the decompositions of plethysms. These will all concern plethysms of complete symmetric functions.

Example: If we restrict to two variables x_1, x_2, then

$$(h_m \circ h_n)(x_1,x_2) = (h_n \circ h_m)(x_1,x_2) = \sum_{k=0}^{mn} A(k,m,n)x_1^k x_2^{mn-k},$$

where $A(k,m,n)$ is the number of ways of writing k as the sum of m non-negative integers all less that or equal to n. This result is easily proved by direct computation and is essentially the Cayley-Sylvester theorem from classical invariant theory. The fact that $(h_m \circ h_n)(x_1,x_2) = (h_n \circ h_m)(x_1,x_2)$ can also be formulated as follows: If V is two-dimensional, then

$$S^m(S^n(V)) \cong S^n(S^m(V))$$

as GL(V)-modules. In particular the dimensions of the spaces of SL(V)-invariants are the same. This result is usually called Hermite's reciprocity theorem. One can also prove that

$$(h_m \circ h_n)(1,q) = \begin{bmatrix} m+n \\ n \end{bmatrix}(1,q) = \frac{(1-q^{m+n})(1-q^{m+n-1})\ldots(1-q^{m+1})}{(1-q^n)(1-q^{n-1})\ldots(1-q)},$$

that is, a Gaussian polynomial.

Example: Let us say that a partition is *even* if all its parts are even. Then one has the following decompositions:

$$h_m \circ h_2 = \sum_{\substack{|\lambda|=2m \\ \lambda \text{ even}}} s_\lambda,$$

and

$$h_2 \circ h_m = \sum_{\substack{|\lambda|=2m, l(\lambda)\leq 2 \\ \lambda \text{ even}}} s_\lambda,$$

where $l(\lambda)$ is the length of λ. By summing over all m in the first formula one gets

$$\sum_{\lambda \text{ even}} s_\lambda = \sum_{m\geq 0} h_m \circ h_2 = \prod_{i\geq 1}(1-x_i^2)^{-1} \prod_{1\leq i<j}(1-x_ix_j)^{-1}.$$

This can be written

$$\sum_m e_m \sum_{\lambda \text{ even}} s_\lambda = \prod(1-x_i)^{-1}\prod_{i<j}(1-x_ix_j)^{-1}.$$

By Littlewood-Richardson's rule, the right-hand side is $\sum s_\lambda$, where now the sum is over all partitions, so finally we get

$$\sum_\lambda s_\lambda(x) = \prod_{i\geq 1}(1-x_i)^{-1} \prod_{1\leq i<j}(1-x_ix_j)^{-1}.$$

This in turn implies that the (infinite-dimensional) $\mathrm{GL}(V)$-module $S(V\oplus\Lambda^2V)$ contains all irreducible modules exactly once. R. Howe [6] has found formulae for the coefficients in the decompositions of $h_m \circ h_n$ also for $m=3$ and 4.

We close this section by mentioning a conjecture.

Foulkes's conjecture. *Assume that* $h_m\circ h_n = \sum a_\lambda s_\lambda$ *and that* $h_n\circ h_m = \sum b_\lambda s_\lambda$. *Then if* $m\leq n$, $a_\lambda \leq b_\lambda$ *for all* λ.

We saw above that this is true when $m=2$. An equivalent formulation is the following: There is a $\mathrm{GL}(V)$-equivariant embedding of $S^m(S^n(V))$ into $S^n(S^m(V))$ if $m\leq n$. It seems that M. Brion [2] has proved that for m fixed there is an n_0 such that if $n\geq n_0$, then Foulkes's conjecture holds. However, there seems to be no hope to find the explicit value of n_0.

There are many more results in the literature about the decomposition of plethysms. Most of these concern plethysms of complete symmetric functions and give the coefficient of s_λ for particular choices of the partition λ. It seems that plethysm was invented by D.E. Littlewood [12]. Algorithms for their computation have been given by Todd [21] and Duncan [5], by Robinson [16], and also by Chen, Garsia, and Remmel [3]. Some other references are [6], [7, Sect. 5.4], [14], [15], [20], and [22]. The word *plethysm* probably derives from a Greek word, *plethusmos*, meaning enlargement, cf. *plethora*. This word also appears in *pletysmograph*, which is an apparatus for measuring variations in the size of different parts of the human body, for instance when varying amounts of blood flow through them.

We will now discuss how plethysms turn up in invariant theory.

2. **Noncommutative invariant theory.** Let V be a finite-dimensional complex vector space and $T(V)=\oplus_{m\geq 0}T^m(V)$ and $S(V)=\oplus_{m\geq 0}S^m(V)$ the tensor and symmetric algebras over V, respectively. If $x_1,\ldots,x_n$ is a basis for V we make the identifications

$$T(V)=\mathbf{C}\langle x_1,\ldots,x_n\rangle \quad\text{and}\quad S(V)=\mathbf{C}[x_1,\ldots,x_n].$$

Let G be a subgroup of $\mathrm{GL}(V)$; then G acts on $T(V)$ and $S(V)$ as a group of graded algebra automorphisms, so we get algebras of G-invariants $T(V)^G$ and $S(V)^G$. We will first compare some properties of these.

Example: Assume that V is two-dimensional with basis x,y and let $G=\{1,g\}$ act by $gx=-x, gy=-y$. Then

$$T(V)^G=\mathbf{C}\langle x^2,xy,yx,y^2\rangle$$

and

$$S(V)^G=\mathbf{C}[x^2,xy,y^2]\cong \mathbf{C}[T_1,T_2,T_3]/(T_1T_3-T_2^2).$$

Note that both are finitely generated $\mathbf{C}$-algebras. Now let G act by $gx = y, gy = x$ instead. Then one can show that

$$S(V)^G = \mathbf{C}[x + y, xy]$$

is finitely generated, but

$$T(V)^G = \mathbf{C}\langle x + y, x^2 + y^2, x^3 + y^3, \dots\rangle$$

is not. Hence already a very small group can give a complicated ring of invariants $T(V)^G$.

Here are some results on the structure of the rings of invariants.

Theorem (Gordan, Hilbert, Noether, Nagata,...). *If G is reductive, then $S(V)^G$ is finitely generated.*

Theorem(Dicks-Formanek [4], Kharchenko [9]). *If G is finite, then $T(V)^G$ is finitely generated if and only if G acts as a group of scalar matrices. In particular G has to act as a cyclic group.*

The finite groups for which $S(V)^G$ is free have been classified (Shephard-Todd, Chevalley); they are the finite reflexion groups. Remarkably enough, Lane [11] and Kharchenko [8] have shown that $T(V)^G$ is always free, for any group and also in any characteristic!

Hence $T(V)^G$ is finitely generated only for rather uninteresting groups. To be able to say something more about their structure we now introduce the notion of an S-algebra.

Definition. Let $A = \oplus_{m\geq 0} A_m$ be a graded associative $\mathbf{C}$-algebra. Then we say that A is an *S-algebra* if there is an action of S_m on each A_m that is compatible with the ring structure in the following sense: Consider the usual embedding of $S_m \times S_n$ into S_{m+n} (see above). Denote it by $(\sigma, \tau) \mapsto \sigma \star \tau$. Then if $f \in A_m, g \in A_n$,

$$(\sigma \star \tau)(fg) = \sigma(f)\tau(g)$$

for all $\sigma \in S_m, \tau \in S_n$. We say that A is *S-commutative* if for all m, n and $f \in A_m, g \in A_n$ we have

$$\rho^n_{m+n}(fg) = gf,$$

where ρ_q is the q-cycle $(1\ 2\ \dots\ q) \in S_q$ (the choice of this particular element ρ_q has to do with our choice of embedding of $S_m \times S_n$ into S_{m+n}).

As we saw above there is an action of S_m on $T^m(V)$ and this action commutes with the action of GL(V). Hence the spaces of invariants $T^m(V)^G$ are also S_m-modules. The compatibility condition is easily checked to be fulfilled, so $T(V)^G$ is an S-algebra. Since the symmetric groups act by place permutations it is also clear that $T(V)$ is S-commutative. As an example of an S-algebra that is not S-commutative we can take $A = \oplus_{m\geq 0}\mathbf{C}[S_m]$ under the multiplication $\star$ and actions of the symmetric groups given by multiplication from the left. However, if we let the action be conjugation instead, then A is S-commutative.

If A is an S-algebra we say that it is S-generated by some set of elements f_i if it is generated by the f_i together with the action of the symmetric groups. We write $A = S\mathbf{C}\langle f_i\rangle$. In case A is S-commutative we write $A = S\mathbf{C}[f_i]$. We say that it is S-finitely generated if there is a finite set of S-generators.

Example: If V has basis x, y and $G = \{1, g\}$ acts by $gx = y, gy = x$, then

$$T(V)^G = S\mathbf{C}[x+y, x^2+y^2].$$

We now have a very nice result:

Theorem (Koryukin [10]). *If G is reductive, then $T(V)^G$ is S-finitely generated.*

This is a consequence of the following:

Theorem [18]. *The algebra $T(V)^G$ is S-finitely generated if and only if $S(V \otimes V)^G$ is finitely generated, where G acts on the first factor.*

Note: Teranishi has a similar theorem in [19]. Koryukin's proof does not depend on the latter result.

Given an S_{m_0}-module M (or a family of modules over different symmetric groups) one can define S-algebras generated by these, i.e. by their elements. We give some examples: Let M be an S_{m_0}-module. Then

$$S\mathbf{C}\langle M\rangle = \oplus_{m\geq 0}\mathbf{C}[S_{mm_0}] \otimes_{\mathbf{C}[S_{m_0}^m]} (M^{\otimes m})$$

is an S-algebra with multiplication

$$((\sigma \otimes \xi), (\tau \otimes \eta)) \mapsto ((\sigma \star \tau) \otimes (\xi \otimes \eta)).$$

See [17] for the details. This S-algebra is not S-commutative. An S-commutative S-algebra S-generated by M is

$$S\mathbf{C}[M] = \oplus_{m\geq 0}\mathbf{C}[S_{mm_0}] \otimes_{\mathbf{C}[S_{m_0}\sim S_m]} (M^{\otimes m})$$

and the multiplication (formally) looks the same as on $S\mathbf{C}\langle M\rangle$. Note that the homogeneous component of degree mm_0 is isomorphic to $1_{S_m} \circ M$ as an S_{mm_0}-module. The first one of these is free as an S-algebra and the second is free as an S-commutative S-algebra (for the definition see again [17]).

We will now discuss a generating function associated to S-algebras. In general, for a graded algebra $A = \oplus_{m\geq 0}A_m$ such that all A_m are finite-dimesional vector spaces, the Hilbert series is the formal power series $H(A,t) = \sum_{m\geq 0} \dim A_m t^m$. Many S-algebras and noncommutative rings of invariants have complicated or unpleasant Hilbert series [1]. For instance, the Hilbert series of $A = \oplus_{m\geq 0}\mathbf{C}[S_m]$ is $\sum_{m\geq 0} m!t^m$. Instead of the usual Hilbert series for S-algebras we want a series that reflects the "S-structure", so we define a formal power series $\hat{H}(A)$ by

$$\hat{H}(A) = \hat{H}(A, z, t) = \sum_{m\geq 0} \mathrm{ch}(\chi(A_m))(z)t^m.$$

Here $\chi(A_m)$ is the character of A_m as an S_m-module and ch is the characteristic map. Hence $\hat{H}(A)$ is an element of the ring of formal power series in t with coefficients in the ring of symmetric functions. We give some examples.

Example: Let V be two-dimensional with basis x, y and let $G = \{1, g\}$ act on V by $gx = y, gy = x$. Then

$$\hat{H}(T(V)^G, z_1, z_2, t) = \frac{1 - z_1^2 z_2^2 t^4}{(1 - z_1 t)(1 - z_2 t)(1 - z_1^2 t^2)(1 - z_1 z_2 t^2)(1 - z_2^2 t^2)}$$

(note that it is sufficient to restrict to two variables z_1, z_2 since V has dimension 2).

Example: If V has dimension n, then

$$\hat{H}(T(V)) = \prod_{i \geq 1} (1 - z_i t)^{-n}.$$

Example: If $A = \oplus_{m \geq 0} \mathbf{C}[S_m]$ with multiplication by S_m from the left, then

$$\hat{H}(A) = (1 - h_1(z))^{-1}.$$

We also have an analogue of Molien's theorem:

Theorem [18] *If G is a finite subgroup of* $\mathrm{GL}(V)$, *then*

$$\hat{H}(T(V)^G, z_1, \ldots, z_n, t) = \frac{1}{|G|} \sum_{g \in G} \frac{1}{\prod_{i=1}^n \det(1 - t z_i g)}.$$

It is almost obvious that if M is an S_{m_0}-module with characteristic

$$\mathrm{ch}(\chi(M)) = \sum_{|\alpha| = m_0} u_\alpha z^\alpha,$$

(where α is a multiindex), then

$$\hat{H}(S\mathbf{C}[M]) = \prod_{|\alpha| = m_0} (1 - z^\alpha t^{m_0})^{-u_\alpha}.$$

We also have

Theorem [18]. *Assume that A is S-commutative and S-generated by the homogeneous elements $f_1, \ldots, f_k$ of degrees $n_1, \ldots, n_k$, respectively, and that the characteristic of the S-module generated by f_i is $\sum_{|\alpha| = n_i} u_{i\alpha} z^\alpha$. Then*

$$\hat{H}(A) = \frac{p(z)}{\prod_{i=1}^k \prod_{|\alpha| = n_i} (1 - z^\alpha t^{n_i})^{u_{i\alpha}}},$$

where $p(z)$ is a (symmetric) polynomial.

Hence perhaps S-algebras can be useful in the study of plethysms. One can show that there are connections between S-algebras and various types of commutative (in the usual sense) algebras. We refer to [18] for a discussion of these connections.

We will conclude with a brief discussion of the non-commutative invariants of an interesting group. Let H_n be the subgroup of GL_n consisting of all diagonal matrices with determinant 1 and let G_n be the subgroup consisting of all monomial matrices such that the product of all non-zero elements is 0. Then one can show that (here V has dimension n)

$$\mathrm{ch}(\chi(T^m(V))) = h_m \circ (nh_1)$$
$$\mathrm{ch}(\chi(T^{mn}(V)^{H_n})) = h_m^n = h_{(m^n)}$$
$$\mathrm{ch}(\chi(T^{mn}(V)^{G_n})) = h_n \circ h_m.$$

Hence the $\hat{H}$-series of $T(V)^{G_n}$ is $\sum_m h_n \circ h_m$. Under the correspondence between non-commutative and commutative rings mentioned above $T(V)^{G_n}$ corresponds to $S(V \otimes V)^{G_n}$ with action of G_n on the first factor in $V \otimes V$. As $\mathrm{GL}(V)$-modules

$$S(V \otimes V)^{G_n} \cong \oplus_{m \geq 0} S^n(S^m(V))$$

(cf. [Howe]). Now let 1_{S_n} be the trivial S_n-module. One can show that there is a map of S-algebras $S\mathbf{C}[1_{S_n}] \to T(V)^{G_n}$ (see [17]). If this map were surjective from degree n^2 on we would have a proof of Foulkes's conjecture, but it seems to be difficult to prove this (if it is true). We have some facts on $T(V)^{G_n}$. By Koryukin's theorem it is S-finitely generated. If V is two-dimensional with basis x, y, then

$$T(V)^{G_2} = S\mathbf{C}[xy + yx].$$

One also has $T(V)^{H_n} = S\mathbf{C}[x_1 x_2 \ldots x_n]$ if V has basis $x_1, x_2, \ldots, x_n$. Another fact is that as an S_{mn}-module $T^{mn}(V)^{G_n}$ is generated by

$$\sum_{\sigma \in S_n} x^m_{\sigma(1)} x^m_{\sigma(2)} \cdots x^m_{\sigma(n)}.$$

Finally one can prove that

$$\sum_{m \geq 0} h_n \circ h_m = \frac{p(z)}{\prod_{|\alpha|=n}(1 - z^\alpha)},$$

where $p(z)$ is a symmetric polynomial. Whether these remarks are relevant for the study of Foulkes's conjecture or not is unknown.

3. **References.** [1] G. Almkvist, W. Dicks, E. Formanek, *Hilbert series of fixed free algebras and noncommutative classical invariant theory*, J. Alg. **93**(1985), 189-214.
[2] M. Brion, *On a conjecture of Foulkes*, preprint, Univ. of Grenoble, 1992.
[3] Y.M. Chen, A.M. Garsia, J. Remmel, *Algorithms for plethysms*, in Combinatorics and algebra (C. Greene ed.), pp. 109-154, Contemporary mathematics, Vol. 34, Amer. Math. Soc., Providence, RI, 1984.
[4] W. Dicks, E. Formanek, *Poincaré series and a problem of S. Montgomery*, Lin. multilin. alg. **12**(1982), 21-30.

[5] D.G. Duncan, *Note on a formula by Todd*, J. London Math. Soc. **27**(1952), 235-236.
[6] R. Howe, $(\mathrm{GL}_n, \mathrm{GL}_m)$*-duality and symmetric plethysm*, Proc. Indian Acad. Sci. (Math. Sci.), Vol. 97, Nos. 1-3, Dec. 1987, 85-109.
[7] G. James, A. Kerber, *The representation theory of the symmetric group*, Encyclopedia of mathematics and its applications, Vol. 16, Addison-Wesley, Reading, MA, 1981.
[8] V.K. Kharchenko, *Algebras of invariants of free algebras*, Algebra i logica **17**(1978), 478-487 (Russian). English translation: Algebra and logic **17**(1978), 316-321.
[9] V.K. Kharchenko, *Noncommutative invariants of finite groups and noetherian varieties*, J. Pure Appl. Alg. **31**(1981), 83-90.
[10] A.N. Koryukin, *Noncommutative invariants of reductive groups*, Algebra i logica **23**(1984), 419-429 (Russian). English translation: Algebra and logic **23**(1984), 290-296.
[11] D.R. Lane, *Free algebras of rank two and their automorphisms*, Ph.D. thesis, London, 1976.
[12] D.E. Littlewood, *The theory of group characters*, O xford University Press, 1950.
[13] I.G. Macdonald, *Symmetric functions and Hall polynomials*, Oxford University Press, 1979.
[14] F.D. Murnaghan, *The analyses of* $\{m\} \otimes \{1^k\}$ *and* $\{m\} \otimes \{k\}$, Proc. Nat. Acad. Sci. U.S.A., **40**(1954), 721-723.
[15] M.J. Newell, *A theorem on the plethysm of S-functions*, Quart. J. Math. **2**(1951), 161-166.
[16] G. de B. Robinson, *Induced representations and invariants*, Canad. J. Math. **2**(1950), 334-343.
[17] T. Tambour, *Free S-algebras*, Univ. of Lund preprint 1992:5, Lund, Sweden, 1992.
[18] T. Tambour, *S-algebras and commutative rings*, J. Pure Appl. Alg. **82**(1992) 289-313.
[19] Y. Teranishi, *Noncommutative invariant theory*, in Perspectives in ring theory (F. van Oystaeyen and L. LeBryun eds.), Kluwer Academic Publishers, 1988.
[20] R.M. Thrall, *On symmetrized Kronecker products and the structure of the free Lie ring*, Amer. J. Math. **64**(1942), 371-388.
[21] J.A. Todd, *A note on the algebra of S-functions*, Proc. Camb. Philos. Soc. **45**(1949), 328-334.
[22] S.H. Weintraub, *Some observations on plethysms*, J. Alg. **129**(1990), 103-114.

The Coupling Method and Regenerative Processes

Hermann Thorisson
Science Institute, University of Iceland
107 Reykjavik, Iceland
thoris@rhi.hi.is

0 Introduction

Coupling is a simple probabilistic method which has become increasingly popular in recent years. The general idea is the joint construction of two or more random elements (variables, processes) in order to deduce properties of the individual elements. Here we shall concentrate on the use of this method in investigating the asymptotic properties of stochastic processes. Particular attention will be given to a class of processes of basic importance in applied probability, so-called regenerative processes: processes starting from scratch at random times, like a Markov chain at the times of a visit to a fixed state.

In order to get a feeling for the method consider the following example, a version of the *classical coupling* dating back to a 1938 paper by Doeblin [1]. Suppose we wish to establish asymptotic stationarity of a Markov chain $X = (X_0, X_1, \ldots)$ (i.e. that $\mathbf{P}(X_n = i) \to \pi_i$ as $n \to \infty$ where π is the stationary distribution). For that purpose let a stationary version $X' = (X'_0, X'_1, \ldots)$ jump independently of X until the two chains meet, at a time T say. From T onward let X and X' jump together. Establish that T is finite with probability one to obtain that X and X' coincide in the end. Thus X behaves asymptotically as a stationary chain.

This idea of establishing limit results by pasting together the paths of two processes seems to have passed more or less unnoticed. During the last two decades, however, a revival of the method has lead to powerful new results and to simple proofs of known achievements in fields such as interacting particle systems, queueing, Markov theory, renewal theory and regeneration. For a survey, see Lindvall's new book [2]. The author of this paper is also working on a book with coupling as a main theme [6].

The plan of the paper is as follows. Section 1 sketches some general coupling theory for stochastic processes. The results in Section 1.1 are basically well-known, see [2], while those in Sections 1.2 and 1.3 are from [5]. Section 2 contains applications to processes regenerative in a weaker sense than tradionally: the "independence of the past at times of regeneration" assumption is relaxed, in fact in Section 2.1 we do not assume any independence at all. These results are from [4] and [5]. Finally, in Section 3 a coupling of regenerative processes is constructed.

1 Coupling and stochastic processes

Let $(\Omega, \mathcal{F}, \mathbf{P})$ be the common probability space supporting all random elements in this paper. Let $Z = (Z_s)_{s\in[0,\infty)}$ and $Z' = (Z'_s)_{s\in[0,\infty)}$ be stochastic processes on a Polish state space $(E, \mathcal{E})$ with right-continuous paths. Regard Z and Z' as random elements in $(H, \mathcal{H})$ where H is the set of all right-continuous functions $z = (z_s)_{s\in[0,\infty)}$ from $[0,\infty)$ to E and $\mathcal{H}$ is the smallest σ-algebra making all the projection mappings $\{z \mapsto z_t : t \in [0,\infty)\}$ $\mathcal{H}/\mathcal{E}$ measurable. Define the shift-maps θ_t, $t \in [0,\infty)$, by $\theta_t z = (z_{t+s})_{s\in[0,\infty)}$.

Let $||\cdot||$ be the total variation norm defined for bounded signed measures ν on $\mathcal{H}$ by

$$||\nu|| = \sup_{A\in\mathcal{H}} \nu(A) - \inf_{A\in\mathcal{H}} \nu(A) = \text{ mass of } \nu^+ + \text{ mass of } \nu^- .$$

Let $=_D$ denote identity in distribution.

Throughout the paper let U be uniformly distributed on $[0,1]$ and independent of Z, Z' and all other random elements introduced below.

1.1 Coupling

A pair of processes $\hat{Z}$ and $\hat{Z}'$ is a *coupling* of Z and Z' if

$$\hat{Z} =_D Z \quad \text{and} \quad \hat{Z}' =_D Z' .$$

An event C is a *coupling event* if

$$\hat{Z} = \hat{Z}' \quad \text{on } C .$$

Since $\mathbf{P}(Z \in \cdot) - \mathbf{P}(Z' \in \cdot) = \mathbf{P}(\hat{Z} \in \cdot\,; C^c) - P(\hat{Z}' \in \cdot\,; C^c)$ we have the follwing:

Coupling event inequality. *If C is a coupling event then*

$$||\mathbf{P}(Z \in \cdot) - \mathbf{P}(Z' \in \cdot)|| \leq 2\mathbf{P}(C^c) .$$

A random time T is a *coupling epoch* for Z and Z' if

$$\theta_T \hat{Z} = \theta_T \hat{Z}' \quad \text{on } \{T < \infty\}.$$

Clearly $\{T \leq t\}$ is a coupling event for $\theta_t Z$ and $\theta_t Z'$ and thus we obtain:

Coupling epoch inequality. *If T is a coupling epoch for Z and Z' then*

$$||\mathbf{P}(\theta_t Z \in \cdot) - \mathbf{P}(\theta_t Z' \in \cdot)|| \leq 2\mathbf{P}(T > t), \quad t \in [0, \infty).$$

This inequality is of basic importance for total variation asymptotics. For example, if T is a.s. finite then

$$||\mathbf{P}(\theta_t Z \in \cdot) - \mathbf{P}(\theta_t Z' \in \cdot)|| \to 0 \quad \text{as } t \to \infty. \tag{1.1}$$

If Z' is *time-stationary*

$$\theta_t Z' =_D Z', \quad t \in [0, \infty),$$

then (1.1) can be rewritten as

$$\theta_t Z \to_{t.v.} Z' \quad \text{as } t \to \infty.$$

If, for an $\alpha > 0, \mathbf{E}[T^\alpha] < \infty$ then (since $t^\alpha \mathbf{P}(T > t) \leq \mathbf{E}[T^\alpha; T > t]$)

$$t^\alpha ||\mathbf{P}(\theta_t Z \in \cdot) - \mathbf{P}(\theta_t Z' \in \cdot)|| \to 0 \quad \text{as } t \to \infty.$$

We may replace t^α by $\phi(t)$ provided the function ϕ is increasing and $\mathbf{E}[\phi(T)] < \infty$.

If there is a finite random variable T_1 such that for each pair Z, Z' in a class of pairs of processes there is a coupling with coupling epoch T which is stochastically dominated by T_1, i.e.

$$\mathbf{P}(T > t) \leq \mathbf{P}(T_1 > t), \quad t \in [0, \infty),$$

then, with $\sup_*$ denoting supremum over all pairs Z, Z' in the class,

$$\sup_* ||\mathbf{P}(\theta_t Z \in \cdot) - \mathbf{P}(\theta_t Z' \in \cdot)|| \to 0 \quad \text{as } t \to \infty.$$

In this case rates of convergence are obtained under $\mathbf{E}[T_1^\alpha] < \infty$ rather than $\mathbf{E}[T^\alpha] < \infty$.

In fact (1.1) holds *if and only if* there exists a coupling with a finite coupling epoch and *if and only if* $\mathbf{P}(Z \in \cdot) = \mathbf{P}(Z' \in \cdot)$ on the *tail* σ-algebra

$$\mathcal{T} = \bigcap_{t \in [0,\infty)} \theta_t^{-1} \mathcal{H}.$$

If Z is a Markov process with initial distribution λ and Z' is a version of Z with initial distribution λ' then $\mathbf{P}(Z \in \cdot) = \mathbf{P}(Z' \in \cdot)$ on $\mathcal{T}$ for all λ and λ' *if and only if* $\mathbf{P}(Z \in \cdot) = 0$ or 1 on $\mathcal{T}$ for all λ.

1.2 Shift-coupling

A shift-coupling of Z and Z' is a coupling $\hat{Z}$ and $\hat{Z}'$ and two random times T and T' such that $\{T < \infty\} = \{T' < \infty\}$ and

$$\theta_T \hat{Z} = \theta_{T'} \hat{Z}' \quad \text{on } \{T < \infty\}.$$

The times T and T' are the *shift-coupling epochs.* When $T < \infty$ then $T - T'$ is the *shift.* There is no shift if $T \equiv T'$ and then T is an ordinary coupling epoch. Finding an appropriate coupling event for $\theta_{Ut} Z$ and $\theta_{Ut} Z'$ yields:

Shift-coupling inequality. *If T and T' are shift-coupling epochs for Z and Z' then*

$$||\mathbf{P}(\theta_{Ut} Z \in \cdot) - \mathbf{P}(\theta_{Ut} Z' \in \cdot)|| \leq 2\mathbf{P}(T \vee T' > Ut), \quad t \in [0, \infty).$$

Note that this "picking a point Ut at random in $[0, t]$" formulation of the inequality can be rewritten on the following *time-average* form:

$$||\frac{1}{t}\int_0^t \mathbf{P}(\theta_s Z \in \cdot) ds - \frac{1}{t}\int_0^t \mathbf{P}(\theta_s Z' \in \cdot) ds|| \leq 2\mathbf{E}[\frac{T \vee T'}{t} \wedge 1].$$

If T is a.s. finite then the inequality yields

$$(1.2) \qquad ||\mathbf{P}(\theta_{Ut} Z \in \cdot) - \mathbf{P}(\theta_{Ut} Z' \in \cdot)|| \to 0 \quad \text{as} \quad t \to \infty.$$

If Z' is time-stationary then (1.2) can be rewritten as

$$\theta_{Ut} Z \to_{t.v.} Z' \quad \text{as } t \to \infty.$$

If, for an $\alpha \in [0, 1)$, both $\mathbf{E}[T^\alpha]$ and $\mathbf{E}[T'^\alpha]$ are finite then $\mathbf{E}[(\frac{T \vee T'}{U})^\alpha] < \infty$ and thus

$$t^\alpha ||\mathbf{P}(\theta_{Ut} Z \in \cdot) - \mathbf{P}(\theta_{Ut} Z' \in \cdot)|| \to 0 \quad \text{as } t \to \infty.$$

One can show that (1.2) holds *if and only if* there exists a shift-coupling with finite epochs and *if and only if* $\mathbf{P}(Z \in \cdot) = \mathbf{P}(Z' \in \cdot)$ on the *invariant* σ-algebra

$$\mathcal{I} = \{B \in \mathcal{H} : \theta_t^{-1} B = B \text{ for all } t \in [0, \infty) E >\}.$$

If Z is a Markov process with initial distribution λ and Z' is a version of Z with initial distribution λ' then $\mathbf{P}(Z \in \cdot) = \mathbf{P}(Z' \in \cdot)$ on $\mathcal{I}$ for all λ and λ' *if and only if* $\mathbf{P}(Z \in \cdot) = 0$ or 1 on $\mathcal{I}$ for all λ.

1.3 ϵ-coupling

Call a shift-coupling $\hat{Z}$ and $\hat{Z}'$ an *ϵ-coupling* if the epochs T and T' satisfy $|T - T'| \leq \epsilon$ on $\{T < \infty\}$. In this case one can find a coupling event for $\theta_{t+Uh} Z$ and $\theta_{t+Uh} Z'$ such that an application of the coupling event inequality yields:

ϵ-coupling inequality. *If T and T' are ϵ-coupling epochs of Z and Z' then for all $h > 0$*

$$||\mathbf{P}(\theta_{t+Uh}Z \in \cdot) - \mathbf{P}(\theta_{t+Uh}Z' \in \cdot)|| \leq 2\mathbf{P}(T \vee T' > t) + 2\epsilon/h, \quad t \in [0, \infty) .$$

This result can be rewritten on the following *moving* time-average form:

$$||\frac{1}{h}\int_t^{t+h} \mathbf{P}(\theta_s Z \in \cdot)ds - \frac{1}{h}\int_t^{t+h} \mathbf{P}(\theta_s Z' \in \cdot)ds|| \leq 2\mathbf{P}(T \vee T' > t) + 2\epsilon/h ,$$

The inequality yields

$$\limsup_{t\to\infty} ||\mathbf{P}(\theta_{t+Uh}Z \in \cdot) - \mathbf{P}(\theta_{t+Uh}Z' \in \cdot)|| \leq 2\epsilon/h .$$

If, for each $\epsilon > 0$, there is an ϵ-coupling of Z and Z' we can send $\epsilon \downarrow 0$ to obtain

$$\forall h > 0 : \ ||\mathbf{P}(\theta_{t+Uh}Z \in \cdot) - \mathbf{P}(\theta_{t+Uh}Z' \in \cdot)|| \to 0 \quad \text{as} \quad t \to \infty . \tag{1.3}$$

If Z' is time-stationary then this can be rewritten as

$$\forall h > 0 : \ \theta_{t+Uh}Z \to_{t.v.} Z' \quad \text{as } t \to \infty .$$

It turns out that (1.3) holds *if and only if* for each $\epsilon > 0$ there exists an ϵ-coupling with finite epochs and *if and only if* $\mathbf{P}(\theta_{Uh}Z \in \cdot) = \mathbf{P}(\theta_{Uh}Z' \in \cdot)$ on $\mathcal{T}$ for all $h > 0$.

Finally, let the state space $(E, \mathcal{E})$ be general but suppose the paths of Z and Z' are piecewise constant with only finitely many jumps in finite intervals. Suppose, that there is an ϵ-coupling of Z and Z' with epochs T and T'. This time finding an appropriate coupling event for Z_t and Z'_t yields

$$||\mathbf{P}(Z_t \in \cdot) - \mathbf{P}(Z'_t \in \cdot)|| \leq 2\mathbf{P}(T \vee T' > t) + 2\mathbf{P}(Z' \text{ jumps in } [t-\epsilon, t+\epsilon]) .$$

If Z' is stationary then $\mathbf{P}(Z' \text{ jumps in } [t-\epsilon, t+\epsilon]) = \mathbf{P}(Z' \text{ jumps in } [0, 2\epsilon]) \to 0$ as $\epsilon \downarrow 0$. Thus sending first $t \to \infty$ and then $\epsilon \downarrow 0$ in the inequality yields the following result: if Z' is stationary and for each $\epsilon > 0$ there exists an ϵ-coupling with a.s. finite epochs then

$$Z_t \to_{t.v.} Z'_0 \quad \text{as } t \to \infty .$$

2 Coupling and regenerative processes

Let $S = (S_n)_0^\infty$ be a sequence of random times such that

$$0 \leq S_0 < S_1 < \ldots \to \infty$$

and think of them as splitting Z into an initial *delay* $(Z_s)_{0 \leq s < S_0}$ and a sequence of *cycles* $(Z_{S_{n-1}+s})_{0 \leq s < X_n}$ where

$$X_n = S_n - S_{n-1} = \text{ the } \textit{cycle lengths} .$$

The pair (Z, S) is *cycle-stationary* if the sequence of cycles is stationary or, equivalently, if

$$(\theta_{S_n} Z, X_{n+1}, X_{n+2}, \ldots) =_D (\theta_{S_0} Z, X_1, X_2, \ldots), \quad n \geq 0 .$$

The pair (Z, S) is *regenerative in the traditional sense* if the cycles are i.i.d. and independent of the delay or, equivalently, if (Z, S) is cycle-stationary and

$$(\theta_{S_n} Z, X_{n+1}, X_{n+2}, \ldots) \text{ is independent of } ((Z_s)_{0 \leq s < S_n}, S_0, \ldots, S_n),\ n \geq 0 .$$

An example is a Markov process Z with a recurrent state together with the entrance times S to that state. A non-Markovian example is the stable GI/GI/1 queue length process together with times S of arrivals to an empty system.

The pair (Z, S) is *regenerative in the wide sense* if it is cycle-stationary and

$$(\theta_{S_n} Z, X_{n+1}, X_{n+2}, \ldots) \text{ is independent of } (S_0, \ldots, S_n),\ n \geq 0 .$$

In discrete time an example is a Harris recurrent Markov chain and in continuous time an example is a Markov process Z with a regeneration set: S is obtained by a certain extension of the underlying probability space. Another example is the stable GI/GI/k queue length process: even if the system never empties a sequence S can be found which makes Z regenerative in the wide sense.

If (Z, S) is regenerative in the wide sense then clearly S is a renewal process, i.e. $X_1, X_2, \ldots$ are i.i.d. and independent of S_0. If (Z, S) and (Z', S') are both regenrative in the wide sense (traditional sense) then they are *of the same type* if

$$(\theta_{S_0} Z, X_1, X_2, \ldots) =_D (\theta_{S_0} Z', X_1', X_2', \ldots) .$$

We also say that Z is regenerative in the wide (traditional) sense if there exists a sequence S such that (Z, S) has this property. The S_n are the *regeneration times.*

2.1 Cycle-stationarity

We shall take a brief look at cycle-stationarity without any independence assumption. Suppose (Z, S) is cycle-stationary with $S_0 = 0$ and that

$$\mathbf{E}[X_1 | \mathcal{I}_C] < \infty \text{ a.s.}$$

where $\mathcal{I}_C$ is the *cycle-shift-invariant* σ-algebra:

$$\mathcal{I}_C = \{A \in \mathcal{F} : \exists B \in \mathcal{H} \otimes \mathcal{B}_{[0,\infty)^\infty} \text{ s.th. } \{(\theta_{S_n} Z, X_{n+1}, X_{n+2}, \ldots) \in B\} = A \text{ for all } n\} .$$

Let (Z'', S'') have distribution defined by

$$\mathbf{E}[f(Z'', S'')] = \mathbf{E}[f(Z, S) X_1 / \mathbf{E}[X_1 | \mathcal{I}_C]] , \quad \text{for all measurable } f .$$

If V is uniformly distributed on $[0,1]$ and independent of (Z'', S'') then one can show that

$$\theta_{VX_1''}Z'' \text{ is time-stationary}$$

and that

$$\mathbf{P}(Z \in \cdot) = \mathbf{P}(\theta_{VX_1''}Z'' \in \cdot) \quad \text{on } \mathcal{I}\,.$$

Together with the equivalences at the end of Section 1.2 this yields the following result: there exists a shift-coupling of Z and $\theta_{VX_1''}Z''$ with finite epochs and

$$\theta_{Ut}Z \to_{t.v.} \theta_{VX_1''}Z'' \quad \text{as } t \to \infty\,.$$

2.2 Lattice and spread-out regeneration

From now on we shall assume that (Z,S) and (Z',S') are wide sense regenerative of the same type. In Section 3 we shall prove that if either

> both S and S' take values in a lattice $L_d = \{0, d, 2d, \ldots\}$ where $d > 0$ and X_1 is *aperiodic* on L_d, i.e. $\mathbf{P}(X_1 \in L_{d'}) < 1$ for all $d' > d$

or

> X_1 is *spread-out*, i.e. there is an integer r and a non-trivial sub-probability density f such that $\mathbf{P}(X_1 + \ldots + X_r \in dx) \geq f(x)dx$

then there exists a coupling of Z and Z' with a finite coupling epoch T and thus by (1.1)

$$||\mathbf{P}(\theta_t Z \in \cdot) - \mathbf{P}(\theta_t Z' \in \cdot)|| \to 0 \quad \text{as } t \to \infty\,.$$

If Z is regenerative in the wide sense then $\mathbf{E}[X_1|\mathcal{I}_C] = \mathbf{E}[X_1]$ a.s. and thus if $\mathbf{E}[X_1] < \infty$ then (Z'', S'') in Section 2.2 satisfies

$$\text{(2.1)} \qquad \mathbf{E}[f(Z'',S'')] = \mathbf{E}[f(\theta_{S_0}Z, (S_k - S_0)_0^\infty)X_1]/\mathbf{E}[X_1]\,, \quad \text{for all measurable } f.$$

Further, the time-stationary $\theta_{VX_1''}Z''$ is wide sense regenerative of the same type as Z. Hence if $\mathbf{E}[X_1] < \infty$ we have in the spread-out case

$$\theta_t Z \to_{t.v.} \theta_{VX_1''}Z'' \quad \text{as } t \to \infty$$

while in the lattice case $\theta_{[VX_1'']_d}Z''$ (where $[x]_d = \sup\{nd : nd \leq x\}$) is *periodically stationary*:

$$\theta_{nd}\theta_{[VX_1'']_d}Z'' =_D \theta_{[VX_1'']_d}Z'', \quad n \geq 0\,,$$

and has L_d valued regeneration times and thus

$$\theta_{nd}Z \to_{t.v.} \theta_{[VX_1'']_d}Z'' \quad \text{as } n \to \infty\,.$$

The above coupling result is established in Section 3.1 in the lattice case and in Section 3.2 in the spread-out case.

A coupling construction different from the one in Section 3 can be found in [4]. That construction relies on the condition that $\mathbf{E}[X_1] < \infty$ but yields a better coupling epoch T. For example, it is established in [4] that if

$$\mathbf{E}[\phi(S_0)], \mathbf{E}[\phi(S_0')], \mathbf{E}[\phi(X_1)] < \infty$$

where ϕ is an increasing function which is

(2.2) either concave (such as x^β where $\beta \leq 1$) or such that
$\phi(2x) \leq c\phi(x)$ for some $c < \infty$ and all $x \in [0, \infty)$
and ϕ has an increasing density which tends to ∞ as x tends to ∞
(such as x^α where $\alpha > 1$, but not e^{x^β} where $\beta > 0$)

or if

$$\mathbf{E}[\phi(S_0)], \mathbf{E}[\phi(S_0')], \mathbf{E}[\Phi(X_1)] < \infty$$

where Φ is a function with density ϕ and ϕ is an increasing function such that

(2.3) $\log \phi(x)/x \downarrow 0$ as $x \to 0$
(such as the ϕ in (2.2) or e^{x^β} where $\beta < 1$, but not $e^{\alpha x}$ for any $\alpha > 0$)

then $\mathbf{E}[\phi(T)] < \infty$ and thus

$$\phi(t)||\mathbf{P}(\theta_t Z \in \cdot) - \mathbf{P}(\theta_t Z' \in \cdot)|| \to 0 \quad \text{as } t \to \infty .$$

Also, if

$$\exists \alpha > 0 : \ \mathbf{E}[e^{\alpha S_0}], \mathbf{E}[e^{\alpha S_0'}], \mathbf{E}[e^{\alpha X_1}] < \infty$$

then there exists a $\beta > 0$ such that $\mathbf{E}[e^{\beta T}] < \infty$ and thus

$$\exists \beta > 0 : \ e^{\beta t}||\mathbf{P}(\theta_t Z \in \cdot) - \mathbf{P}(\theta_t Z' \in \cdot)|| \to 0 \quad \text{as } t \to \infty .$$

Moreover, in [4] it is shown that if S_0 and S_0' are stochastically dominated by a random variable Y_0 and X_1 by a random variable Y_1 with $\mathbf{E}[Y_1] < \infty$ then there is a random variable T_1 with distribution independent of the distributions of S_0, S_0' and X_1 and dominating the coupling epoch T stochastically. Thus

$$\sup_* ||\mathbf{P}(\theta_t Z \in \cdot) - \mathbf{P}(\theta_t Z' \in \cdot)|| \to 0 \quad \text{as } t \to \infty$$

where $\sup_*$ means supremum over all (Z, S) and (Z', S') which are wide sense regenerative of the same type with S_0 and S_0' stochastically dominated by Y_0 and X_1 by Y_1. If for ϕ increasing and satisfying (2.3) it holds that

$$\mathbf{E}[\phi(Y_0)] < \infty \ \text{ and } \ \mathbf{E}[\Phi(Y_1)] < \infty$$

then $\mathbf{E}[\phi(T_1)] < \infty$ and thus

$$\phi(t) \sup_{*} ||\mathbf{P}(\theta_t Z \in \cdot) - \mathbf{P}(\theta_t Z' \in \cdot)|| \to 0 \quad \text{as } t \to \infty$$

while if

$$\exists \alpha > 0 : \ \mathbf{E}[e^{\alpha Y_0}] < \infty \text{ and } \mathbf{E}[e^{\alpha Y_1}] < \infty$$

then there exists a $\beta > 0$ such that $\mathbf{E}[e^{\beta T_1}] < \infty$ and thus

$$\exists \beta > 0 : \ e^{\beta t} \sup_{*} ||\mathbf{P}(\theta_t Z \in \cdot) - \mathbf{P}(\theta_t Z' \in \cdot)|| \to 0 \quad \text{as } t \to \infty .$$

We finally remark that, with $(1-V)X_1''$ the length of the stationary delay, it is easily seen that $\mathbf{E}[\Phi(X_1)] < \infty$ implies $\mathbf{E}[\phi((1-V)X_1'')] < \infty$. Also if X_1 is stochastically dominated by Y_1 then $(1-V)X_1''$ is stochastically dominated by the length of the stationary delay we would get if Y_1 was a cycle length.

2.3 Non-lattice regeneration

Suppose (Z, S) and (Z', S') are wide sense regenerative of the same type and that X_1 is *non-lattice*, i.e. the distribution of X_1 is not supported by any lattice. In Section 3.3 we shall construct for each $\epsilon > 0$ an ϵ-coupling with a.s. finite epochs. Thus it holds that

$$\forall h > 0 : \ ||\mathbf{P}(\theta_{t+Uh} Z \in \cdot) - \mathbf{P}(\theta_{t+Uh} Z' \in \cdot)|| \to 0 \quad \text{as } t \to \infty .$$

If $\mathbf{E}[X_1] < \infty$ and $\theta_{VX_1''}Z''$ is the time-stationary version of Z then

$$\forall h > 0 : \ \theta_{t+Uh} Z \to_{t.v.} \theta_{VX_1''} Z'' \quad \text{as } t \to \infty .$$

Put

$$N_t = n \quad \text{if and only if} \quad S_{n-1} \le t < S_n .$$

Certainly, the paths of $(\theta_{S_{N_s-1}} Z)_{s \in [0,\infty)}$ are piecewise constant with only finitely many jumps in finite intervals. Since $(\theta_{S_{N_s-1}} Z)_{s \in [0,\infty)}$ is wide sense regenerative with regeneration times S and since $(\theta_{S''_{N''_s-1}} Z'')_{s \in [0,\infty)}$ is its time-stationary version we obtain the following result: if $\mathbf{E}[X_1] < \infty$ and X_1 is non-lattice then

$$\theta_{S_{N_t-1}} Z \to_{t.v.} Z'' \quad \text{as } t \to \infty$$

and in particular

$$X_{N_t} \to_{t.v.} X_1'' \quad \text{as } t \to \infty .$$

3 A coupling construction for regenerative processes

The coupling in [4] is in line with the classical coupling and relies on $\mathbf{E}[X_1]$ being finite. A main ingredient in the construction below is Ornstein's coupling idea in [3]: when coupling two random walks make the difference of their steplengths symmetric and bounded and then apply the recurrence (ϵ-recurrence) of such walks. This enables us to obtain a finite coupling (ϵ-coupling) epoch even when $\mathbf{E}[X_1] = \infty$.

Another basic ingredient in the coupling construction below is the following result which reduces shift-coupling of wide-sense regenerative processes to a study of the regeneration times. We need the following concept: a random time M in $\{0, \ldots, \infty\}$ is a randomized stopping time with respect to a sequence of random elements $(Y_k)_0^\infty$ if $\{M = n\}$ is conditionally independent of $(Y_k)_0^\infty$ given $(Y_k)_0^n$ for all $0 \leq n < \infty$.

Key coupling result for regenerative processes. *Suppose $\hat{X}$ and $\hat{X}'$ is a shift-coupling of the sequences $X = (S_0, X_1', X_2', \ldots)$ and $X' = (S_0', X_1, X_2, \ldots)$ with epochs $\hat{M}$ and $\hat{M}'$ which are randomized stopping times for $\hat{X}$ and $\hat{X}'$, respectively. Then the underlying probability space can be extended to support a shift-coupling $\hat{Z}$ and $\hat{Z}'$ of Z and Z' with epochs*

$$T = \hat{S}_{\hat{M}} \quad \text{and} \quad T' = \hat{S}'_{\hat{M}'} .$$

In order to establish this we need two results from [4]. The first is a regeneration analogue of the strong Markov property for randomized stopping times (see Proposition 1.1 in [4]).

Strong regeneration property. *Let (Z, S) be wide-sense regenerative. Suppose M is a finite randomized stopping time with respect to S and conditionally independent of Z given S. Then*

$$(\theta_{S_M} Z, X_{M+1}, X_{M+2}, \ldots) =_D (\theta_{S_0} Z, X_1, X_2, \ldots)$$

and

$$(\theta_{S_M} Z, X_{M+1}, X_{M+2}, \ldots) \textit{ is independent of } (S_0, \ldots, S_M) .$$

The second result we need is the following (see Construction 1.1 in [4]).

Conditional independence extension. *Let (Y_1, Y_2) and (Y_2', Y_3') be pairs of random elements such that $Y_2 \stackrel{D}{=} Y_2'$. If there exists a regular version of the conditional distribution of Y_3' given Y_2' then the underlying probability space can be extended to support a random element Y_3 such that*

$$(Y_2, Y_3) \stackrel{D}{=} (Y_2', Y_3')$$

and such that

$$Y_1 \textit{ and } Y_3 \textit{ are conditionally independent given } Y_2 .$$

Proof of the key result. Use the conditional independence extension thrice (Polish state space and right-continuous paths ensure existence of regular conditional distributions). First put

$$(Y_1, Y_2) = (Z', S') \text{ and } (Y_2', Y_3') = (\hat{S}', \hat{M}') \text{ to obtain } M' = Y_3 .$$

Applying the strong regeneration property yields that

$$S'_{M'} \text{ is independent of } \theta_{S'_{M'}} Z' \text{ and } \theta_{S'_{M'}} Z' \stackrel{D}{=} \theta_{S_0} Z . \tag{3.1}$$

Secondly, put

$$(Y_1, Y_2) = ((\hat{M}, \hat{S}', \hat{M}'), \hat{S}) \text{ and } (Y_2', Y_3') = (S, Z) \text{ to obtain } \hat{Z} = Y_3 .$$

Then $\hat{Z}$ depends on $(\hat{S}, \hat{M})$ only through $(\hat{S}, \hat{M})$ and thus $\theta_T \hat{Z}$ depends on T' only through $(\hat{S}, \hat{M})$. By the strong regeneration property $\theta_T \hat{Z}$ depends on $(\hat{S}, \hat{M})$ only through $(\hat{X}_{\hat{M}+1}, \hat{X}_{\hat{M}+2}, \ldots)$ and thus $\theta_T \hat{Z}$ depends on T' only through $(\hat{X}_{\hat{M}+1}, \hat{X}_{\hat{M}+2}, \ldots)$. By the strong regeneration property T' is independent of $(\hat{X}'_{\hat{M}'+1}, \hat{X}'_{\hat{M}'+2}, \ldots)$ and by the shift-coupling assumption we have $(\hat{X}'_{\hat{M}'+1}, \hat{X}'_{\hat{M}'+2}, \ldots) = (\hat{X}_{\hat{M}+1}, \hat{X}_{\hat{M}+2}, \ldots)$. Thus

$$T' \text{ is independent of } \theta_T \hat{Z} . \tag{3.2}$$

By the strong regeneration property $\theta_T \hat{Z} \stackrel{D}{=} \theta_{S_0} Z$ which together with $T' \stackrel{D}{=} S'_{M'}$ (3.2) and (3.1) yields $(T', \theta_T \hat{Z}) \stackrel{D}{=} (S'_{M'}, \theta_{S'_{M'}} Z')$. This allows us to carry out the third extension: put

$$(Y_1, Y_2) = ((\hat{Z}, T), (T', \theta_T \hat{Z})) \text{ and } (Y_2', Y_3') = ((S'_{M'}, \theta_{S'_{M'}} Z', Z') \text{ to obtain } \hat{Z}' = Y_3 .$$

Then $(T', \theta_T \hat{Z}, \hat{Z}') \stackrel{D}{=} (S'_{M'}, \theta_{S'_{M'}} Z', Z')$ which implies $\mathbf{P}(\theta_T \hat{Z} = \theta_{T'} \hat{Z}') = 1$. Removing a null set yields $\theta_T \hat{Z} = \theta_{T'} \hat{Z}'$ and completes the proof of the key coupling result.

In Section 3.1 below we construct a coupling with a finite coupling epoch in the lattice case and in Section 3.2 in the spread-out case. In Section 3.3 we construct an ϵ-coupling in the non-lattice case. An elementary proof of recurrence (ϵ-recurrence) of random walks with symmetric bounded steplengths is given in Section 3.4.

3.1 The lattice case

Proposition 1. *Let (Z, S) and (Z', S') be wide sense regenerative of the same type. If S and S' take values in a lattice $L_d = \{0, \pm d, \pm 2d, \ldots\}$ and X_1 is aperiodic with respect to L_d then there exists a coupling of Z and Z' with a finite coupling epoch T.*

Proof. First note that it is no restriction to assume that (Z, S) and (Z', S') are independent and that the distribution of $X_1 - X_1'$ is aperiodic on L_d: If this is not the case let $J_0, J_1, \ldots; J_0', J_1' \ldots$ be i.i.d., independent of (Z, S) and (Z', S') and taking only the values 1 and 2 with positive probabilities. Then $(Z, (S_{J_0+\ldots+J_k})_0^\infty)$ and $(Z', (S'_{J'_0+\ldots+J'_k})_0^\infty)$ are wide sense regenerative of the same type and the difference of two inter-regeneration times is aperiodic on L_d. Thus proceeding as below with (Z, S) and (Z', S') replaced by $(Z, (S_{J_0+\ldots+J_k})_0^\infty)$ and $(Z', (S'_{J'_0+\ldots+J'_k})_0^\infty)$ yields the desired result.

Let $V_1, V_2, \ldots; V'_1, V'_2, \ldots; W_1, W_2, \ldots; I_1, I_2, \ldots$ be independent random variables with distributions defined by

$$V_n =_D V'_n =_D [X_1 | X_1 \leq c]$$

$$W_n =_D [X_1 | X_1 > c]$$

$$\mathbf{P}(I_n = 0) = \mathbf{P}(X_1 > c) \quad \text{and} \quad \mathbf{P}(I_n = 1) = \mathbf{P}(X_1 \leq c)$$

where c is a finite constant large enough for $V_n - V'_n$ to be aperiodic. Put, for $n \geq 1$,

$$\hat{X}_n = \hat{X}'_n = W_n \quad \text{if} \quad I_n = 0$$

$$\hat{X}_n = V_n \text{ and } \hat{X}'_n = V'_n \quad \text{if} \quad I_n = 1\,.$$

Then both the $\hat{X}_n$ and the $\hat{X}'_n$ are i.i.d. with the same distribution as X_1. Thus with $\hat{X}_0$ and $\hat{X}'_0$ distributed as S_0 and S'_0, respectively, and independent of the $\hat{X}_n$ and $\hat{X}'_n$ we have that

$$\hat{S} = (\hat{X}_0 + \ldots + \hat{X}_n)_0^\infty \quad \text{and} \quad \hat{S}' = (\hat{X}'_0 + \ldots + \hat{X}'_n)_0^\infty$$

is a coupling of S and S'. Moreover,

$$R_n = (\hat{X}_0 - \hat{X}'_0) + \ldots + (\hat{X}_n - \hat{X}'_n), \quad 0 \leq n < \infty\,,$$

is a random walk on L_d with bounded aperiodic symmetric step-lengths and thus (see Section 3.4 below)

$$\hat{M} = \inf\{n \geq 0 : R_n = 0\}$$

is finite with probability one. Now $\hat{M}$ is a randomized stopping time for $\hat{S}$ and also for $\hat{S}'$. Certainly $\hat{S}_{\hat{M}} = \hat{S}'_{\hat{M}}$ and thus it only remains to replace $(\hat{X}'_{\hat{M}+1}, \hat{X}'_{\hat{M}+2}, \ldots)$ by $(\hat{X}_{\hat{M}+1}, \hat{X}_{\hat{M}+2}, \ldots)$ and apply the key coupling result.

3.2 The spread-out case

Proposition 2. *Let (Z, S) and (Z', S') be wide sense regenerative of the same type. Suppose X_1 is spread-out. Then there exists a coupling of Z and Z' with a finite coupling epoch T.*

Proof. We may assume that f is constant on an interval, $f = c1_{[a,b]}$, see Lemma 1 below. Moreover, we may assume that $r = 1$: if this is not the case note that $(Z, (S_{kr})_0^\infty)$ and $(Z', (S'_{kr})_0^\infty)$ are wide sense regenerative of the same type with inter-regeneration times having this property and thus proceeding as below with (Z, S) and (Z', S') replaced by $(Z, (S_{kr})_0^\infty)$ and $(Z', (S'_{kr})_0^\infty)$ yields the desired result.

Let $\hat{S}_0$ and $\hat{S}'_0$ be a coupling of S_0 and S'_0. For $s, s' \geq 0$ let $d = d(s, s')$ be the greatest number satisfying $0 < d \leq (a+b)/2$ and note that this implies

$$\mathbf{P}(X_1 \in dx) \geq c1_{[a,a+2d]}(x)dx\,, \quad s, s' \geq 0\,.$$

Let $V_1, V_2, \ldots; W_1, W_2, \ldots; I_1, I_2, \ldots$ be random variables that are conditionally independent given $(\hat{S}_0, \hat{S}'_0) = (s, s')$ with conditional distributions given by

V_n is uniformly distributed on $[a, a+d]$

W_n has the distribution $(\mathbf{P}(X_1 \in dx) - c1_{[a,a+2d]}(x)dx)/(1-2cd)$

$\mathbf{P}(I_n = 0) = (1-2cd)$ and $\mathbf{P}(I_n = -1) = \mathbf{P}(I_n = 1) = cd$.

Put, for $n \geq 1$,

$$\hat{X}_n = \hat{X}'_n = W_n \quad \text{if } I_n = 0$$

$$\hat{X}_n = V_n \text{ and } \hat{X}'_n = d + V_n \quad \text{if } I_n = -1$$

$$\hat{X}_n = d + V_n \text{ and } \hat{X}'_n = V_n \quad \text{if } I_n = 1\,.$$

Then $(\hat{X}_k)_1^\infty$ is independent of $(\hat{S}_0, \hat{S}'_0)$ and so is $(\hat{X}'_k)_1^\infty$. Further both $(\hat{X}_k)_1^\infty$ and $(\hat{X}'_k)_1^\infty$ are i.i.d. sequences of random variables distributed as X_1. Thus

$$\hat{S} = (\hat{S}_0 + \hat{X}_1 + \ldots + \hat{X}_n)_0^\infty \quad \text{and} \quad \hat{S}' = (\hat{S}'_0 + \hat{X}'_1 + \ldots + \hat{X}'_n)_0^\infty$$

is a coupling of S and S'. Moreover, conditionally on $(\hat{S}_0, \hat{S}'_0) = (s, s')$

$$R_n = (\hat{S}_0 - \hat{S}'_0) + (\hat{X}_1 - \hat{X}'_1) + \ldots + (\hat{X}_n - \hat{X}'_n),\ 0 \leq n < \infty\,,$$

is a random walk on $L_{d(s,s')}$ with steplengths $(\hat{X}_n - \hat{X}'_n) = d(s,s')I_n$. Thus (see Section 3.4)

$$\hat{M} = \inf\{n \geq 0 : R_n = 0\}$$

is finite with probability one. Now $(\hat{S}_k, \hat{S}'_k)_0^n$ is independent of $(\hat{X}_k)_{n+1}^\infty$ and the event $\{\hat{M} = n\}$ is determined by $(\hat{S}_k, \hat{S}'_k)_0^n$. Since $\hat{S}$ is determined by $(\hat{S}_k)_0^n$ and $(\hat{X}_k)_{n+1}^\infty$ this yields that $\hat{M}$ is a randomized stopping time with respect to $\hat{S}$. Similarily, $\hat{M}$ is a randomized stopping time with respect to $\hat{S}'$. Certainly $\hat{S}_{\hat{M}} = \hat{S}'_{\hat{M}}$ and thus it only remains to replace $(\hat{X}'_{\hat{M}+1}, \hat{X}'_{\hat{M}+2}, \ldots)$ by $(\hat{X}_{\hat{M}+1}, \hat{X}_{\hat{M}+2}, \ldots)$ and apply the key coupling result.

Lemma 1. *If there is an integer r and a non-trivial sub-probability density f such that*

$$\mathbf{P}(X_1 + \ldots + X_r \in dx) \geq f(x)dx$$

then there are constants a, b and c such that $0 \leq a < b < \infty, c > 0$ *and*

$$\mathbf{P}(X_1 + \ldots + X_{2r} \in dx) \geq c1_{[a,b]}(x)dx$$

Proof. Certainly

$$\mathbf{P}(X_1 + \ldots + X_{2r} \in dx) \geq f * f(x)dx$$

where

$$f * f(x) = \int f(s)f(x-s)ds\,.$$

It is no restriction to assume that f is ≤ 1 and has bounded support which can be seen to imply that $f * f(x)$ is continuous which we use as follows: take x such that $f * f(x) > 0$, put $c = f * f(x)/2$ and take a and b close enough to x for $f * f(y) \geq c$ when $a \leq y \leq b$.

3.3 The non-lattice case

Proposition 3. *Let (Z, S) and (Z', S') be wide sense regenerative of the same type. Suppose X_1 is non-lattice. Then, for all $\epsilon > 0$, there exists an ϵ-coupling of Z and Z' with finite epochs.*

Proof. Proceed as in the first part of the proof of Proposition 1 (introducing the J's if needed to make $X_1 - X_1'$ non-lattice) to obtain the random walk R on $(-\infty, \infty)$ with bounded non-lattice symmetric step-lengths. Thus for each $\epsilon > 0$ (see Section 3.4)

$$\hat{M} = \hat{M}(\epsilon) = \inf\{n \geq 0 : R_n \in [-\epsilon, \epsilon]\}$$

is finite with probability one. Now $\hat{M}$ is a randomized stopping time for $\hat{S}$ and also for $\hat{S}'$. Certainly, $|\hat{S}_{\hat{M}} - \hat{S}'_{\hat{M}}| \leq \epsilon$ and thus it only remains to replace $(\hat{X}'_{\hat{M}+1}, \hat{X}'_{\hat{M}+2}, \ldots)$ by $(\hat{X}_{\hat{M}+1}, \hat{X}_{\hat{M}+2}, \ldots)$ and apply the key coupling result.

3.4 Recurrence of R

Let $R = (R_k)_0^\infty$ be a random walk with bounded symmetric step-lengths. This means

$R_n - R_{n-1}$, $1 \leq n < \infty$, are i.i.d. random variables and independent of R_0;

$|R_n - R_{n-1}| \leq c$ for some finite constant c ;

$R_n - R_{n-1} =_D R_{n-1} - R_n$.

Certainly, R is a Markov chain. Put

$$R_n^\circ = R_n - R_0, \qquad 0 \leq n < \infty\,.$$

We shall give an elementary proof of the following well-known result which plays an essential role in the above coupling construction.

Proposition 4. (*a*) *Suppose R takes values in a lattice $L_d = \{0, \pm d, \pm 2d, \ldots\}$ and that R_1° is L_d-aperiodic, i.e. $\mathbf{P}(R_1^\circ \in L_{d'}) < 1$ for all $d' > d$. Then R is irreducible and recurrent and in particular*

$$K = \inf\{n \geq 0 : R_n = 0\} \text{ is finite with probability one.}$$

(*b*) *Suppose $\mathbf{P}(R_1^\circ \in L_d) < 1$ for all $d > 0$. Then, for each $\epsilon > 0, R$ is ϵ-irreducible and ϵ-recurrent and in particluar*

$$K(\epsilon) = \inf\{n \geq 0 : R_n \in [-\epsilon, \epsilon]\} \text{ is finite with probability one.}$$

Lemma 2. (*a*) *R is irreducible;* (*b*) *R is ϵ-irreducible.*

Proof. (*a*) We must prove that

$$D = \{x \in L_d : \mathbf{P}(R_n^\circ = x) > 0 \text{ for some } n > 0\}$$

equals L_d. Since R_1° is symmetric $x \in D$ implies $-x \in D$; since R_1° is L_d-aperiodic so is D; since

$$\mathbf{P}(R_{n+k}^\circ - x + y) \geq \mathbf{P}(R_n^\circ = x)\mathbf{P}(R_k^\circ = y)$$

we have that $x, y \in D$ implies $x + y \in D$. Thus D is an aperiodic additive subgroup of L_d which implies $D = L_d$.

(*b*) We must prove that

$$D = \{x \in (-\infty, \infty) : \forall \epsilon > 0 \ \exists n \geq 0 \text{ such that } \mathbf{P}(R_n^\circ \in [x - \epsilon, x + \epsilon]) > 0\}$$

equals $(-\infty, \infty)$. Since R_1° is symmetric $x \in D$ implies $-x \in D$; since R_1° is aperiodic so is D; since

$$\mathbf{P}(R_{n+k}^\circ \in [x + y - \epsilon, x + y + \epsilon]) \geq \mathbf{P}(R_n^\circ \in [x - \epsilon/2, x + \epsilon/2])\mathbf{P}(R_k^\circ \in [y - \epsilon/2, y + \epsilon/2])$$

we have that $x, y \in D$ implies $x + y \in D$. Thus D is an aperiodic additive subgroup of $(-\infty, \infty)$. Moreover, D is closed since if $x_k \in D$ and $x_k \to x$ then with k large enough for $|x - x_k| \leq \epsilon/2$ we have

$$\mathbf{P}(R_n^\circ = [x - \epsilon, x + \epsilon E >]) \geq \mathbf{P}(R_n^\circ = [x_k - \epsilon/2, x_k + \epsilon/2])$$

i.e. $x \in D$. Thus in particular $d = \inf\{x \in D : x > 0\} \in D$. If $d > 0$ we can for each $x \in D$ find an n such that $0 \leq x - nd < d$; but $x - nd \in D$ and thus by the definition of d we have $x - nd = 0$, i.e. $D = L_d$ which contradicts the aperiodicity of D. Thus $d = 0$ which yields the existence of strictly positive $x_k \in D$ such that $x_k \downarrow 0$. For an arbitrary $x \in (-\infty, \infty)$ let n_k be such that $0 \leq x - n_k x_k < x_k$. Then $n_k x_k \to x$ and since $n_k x_k \in D$ we have by closedness that $x \in D$, i.e. $D = (-\infty, \infty)$.

Lemma 3. *Either* $\mathbf{P}(\sup_{0 \leq n < \infty} R_n^\circ = \infty) = 0$ *or* 1.

Proof. Put $K_0 = 0$ and recursively for $n \geq 1$

$$K_n = \inf\{k \geq K_{n-1} : R_n^\circ \geq R_{K_{n-1}}^\circ + 1\} \text{ on } \{K_{n-1} < \infty\}.$$

Clearly $\mathbf{P}(K_n < \infty | K_{n-1} < \infty) = \mathbf{P}(K_1 < \infty)$ and $\{K_n < \infty \text{ for all } n < \infty\} = \{\sup_{0 \leq n < \infty} R_n^\circ = \infty\}$. Thus $\mathbf{P}(\sup_{0 \leq n < \infty} R_n^\circ = \infty) = 0$ if $\mathbf{P}(K_1 < \infty) < 1$ and $\mathbf{P}(\sup_{0 \leq n < \infty} R_n^\circ = \infty) = 1$ if $\mathbf{P}(K_1 < \infty) = 1$.

Lemma 4. $\mathbf{P}(\sup_{0 \leq n < \infty} |R_n^\circ| = \infty) = 1$.

Proof. For a given $x > 0$ let k be such that $\mathbf{P}(R_k^\circ > x) > 0$. Then

$$\mathbf{P}(\sup_{0 \leq n < \infty} |R_n^\circ| \leq x) \leq \mathbf{P}(\sup_{0 \leq i \leq n} |R_{ki}^\circ| \leq x) \leq \mathbf{P}(R_k^\circ \leq x)^n \to 0 \text{ as } n \to \infty.$$

Thus $\mathbf{P}(\sup_{0 \leq n < \infty} |R_n^\circ| \leq x) = 0$ and sending x to ∞ proves $\mathbf{P}(\sup_{0 \leq n < \infty} |R_n^\circ| < \infty) = 0$.

Lemma 5. $\mathbf{P}(\sup_{0 \leq n < \infty} R_n^\circ = \infty) = \mathbf{P}(\inf_{0 \leq n < \infty} R_n^\circ = -\infty) = 1$.

Proof. By symmetry $\mathbf{P}(\sup_{0 \leq n < \infty} R_n^\circ = \infty) = \mathbf{P}(\inf_{0 \leq n < \infty} R_n^\circ = -\infty)$, by Lemma 4 these probabilities must be strictly positive and Lemma 3 completes the proof.

Lemma 6. $\mathbf{P}(R_n \in [0, c]$ *for infinitely many* $n) = 1$.

Proof. With probability one R changes sign infinitely often due to Lemma 5. But R cannot change sign without visiting $[0, c]$, due to the boundedness.

Proof of Proposition 4. (a) By Lemma 2(a) there is for each $x \in L_d$ an n_x such that

$$\mathbf{P}(R_{n_x}^\circ = -x) > 0.$$

Put $m = \max_{x \in L_d \cap [0,c]} n_x$ and

$$p = \min_{x \in L_d \cap [0,c]} \mathbf{P}(R_k^\circ = -x \text{ for some } 0 \leq k \leq m)$$

and note that $p > 0$. Put $K_0 = 0$ and recursively for $n \geq 1$

$$K_n = \inf\{k \geq K_{n-1} + m : R_k \in [0, c]\}.$$

Due to Lemma 6 we have $K_n < \infty$ a.s. and thus

$$\begin{aligned} \mathbf{P}(R_k = 0 \text{ for some } k) &\geq \mathbf{P}(\mathbf{R_k = 0} \text{ for some } k \leq K_n + m) \\ &\geq 1 - (1 - p)^n \to 1 \text{ as } n \to \infty. \end{aligned}$$

Thus $\mathbf{P}(R_k = 0 \text{ for some } k) = 1$.

(b) Fix $\epsilon > 0$. By Lemma 2(b) there is for each x and n_x such that

$$\mathbf{P}(R_{n_x}^\circ \in [-x - \epsilon/2, -x + \epsilon/2]) > 0.$$

Put $m = \max_{x \in L_{\epsilon/2} \cap [0,c]} n_x$ and

$$p = \min_{x \in L_{\epsilon/2} \cap [0,c]} \mathbf{P}(R_k^\circ \in [-x - \epsilon/2, -x + \epsilon/2] \text{ for some } 0 \leq k \leq m)$$

and note that $p > 0$. Put $K_0 = 0$ and recursively for $n \geq 1$

$$K_n = \inf\{k \geq K_{n-1} + m : R_k \in [0, c]\}\,.$$

Due to Lemma 6 we have $K_n < \infty$ a.s. and thus

$$\begin{aligned} \mathbf{P}(R_k \in [-\epsilon, \epsilon] \text{ for some } k) &\geq \mathbf{P}(R_k \in [-\epsilon, \epsilon] \text{ for some } k \leq K_n + m) \\ &\geq 1 - (1-p)^n \to 1 \text{ as } n \to \infty\,. \end{aligned}$$

Thus $\mathbf{P}(R_k \in [-\epsilon, \epsilon] \text{ for some } k) = 1$.

References

[1] DOEBLIN, W.: Exposé de la theorie des chaînes simple constantes de Markov à un nombre fini d'états. *Rev. Mat. Union Interbalkan.* **2**, 77-105 (1938).

[2] LINDVALL, T.: *Lectures on the Coupling Method.* Wiley, New York, (1992).

[3] ORNSTEIN, D.: Random walks I, II. *Trans. Am. Math. Soc.* **138**, 1-42 and 45-60 (1969).

[4] THORISSON, H.: Coupling of regenerative processes. *Adv. Appl. Prob.* **15**, 531-561 (1983).

[5] THORISSON, H.: Shift-coupling in continuous time. Preprint, (1990).

[6] THORISSON, H.: *Coupling, Stationarity and Regeneration.* (In preparation).

Stochastic Partial Differential Equations–A Mathematical Connection Between Macrocosmos and Microcosmos

Bernt Øksendal
VISTA* /Dept. of Mathematics
University of Oslo
Box 1053 Blindern, N-0316 Oslo, NORWAY

§1. Introduction. What is a stochastic differential equation?

This paper gives a brief, non-technical introduction to some of the theory and methods of stochastic partial differential equations. No attempt has been made to be complete in any sense. The exposition is mainly based on recent joint works with a group of people, including H. Gjessing, H. Holden, T. Lindstrøm, N.H. Risebro, J. Ubøe and T.S. Zhang. These people are, however, not responsible for any errors produced in this article. Nor should they be blamed for the partly philosophical comments and interpretations I have added in this survey.

* VISTA is a research cooperation between the Norwegian Academy of Science and Letters and Den Norske Stats Oljeselskap A.S. (Statoil).

The purpose here is just to give a flavour of the theory and convince the reader of its usefulness and of the mathematical challenge that it represents. The reader is referred to the relevant publications listed in the back for proofs and further details.

The theory of *stochastic differential equations* (SDE) stems from, and is motivated by, the attempts to describe mathematically the stochastic dynamic phenomena that occur in physics, biology, economics etc. The term "stochastic phenomena" includes phenomena which may not really be stochastic in nature, but appear stochastic or random to us because of our lack of information. (What's the difference anyway? Is the outcome of a throw of a die really random or just appearing to be random because of our lack of information about the movement of the hand throwing it?)

When we describe a situation in physics or economics by means of a differential equation we are always making simplifying assumptions, for example about the coefficients of the equation. Coefficients which are assumed to be constant, may not really be so. They may be subject to fluctuations which to us appear random, or they may be constant but impossible to measure exactly. How will this randomness affect the solution of the differential equation? Questions like these lead to delicate questions about how to model various types of "noise" mathematically in such a way that it can be adopted as a proper (and rigorous) part of a differential equation.

Roughly speaking, a *stochastic differential equation* is a differential equation where some of the coefficients are subject to some properly defined versions of "noise".

In order to illustrate this, let us consider an example from population growth:

EXAMPLE 1.1

Let X_t denote the size of a population at time t. Suppose K is the carrying capacity of the environment for this population. Then a simple classical model for the growth of X_t is the differential equation

$$\frac{dX_t}{dt} = r(K - X_t), \quad \text{where } r \text{ is a constant.} \tag{1.1}$$

The solution of this equation is

(1.2) $$X_t = K + (X_0 - K)e^{-rt}.$$

Now suppose that r is not a constant but subject to random fluctuations due to unpredictable changes in the environment. Then we could try to put

(1.3) $$r = r_t = a + b \cdot W_t, \quad (a, b \text{ constants})$$

where $\{W_t(\omega);\ t \geq 0;\ \omega \in \Omega\}$, is some stochastic process modelling "noise". (The probability law P of this process is defined on a σ-algebra $\mathcal{F}$ of subsets of the given "set of outcomes" Ω). This gives - formally - the stochastic differential equation

(1.4) $$\frac{dX_t}{dt} = a(K - X_t) + b(K - X_t) \cdot W_t$$

In some cases it may be justified to assume that the noise W_t is *white*, in the sense that W_t has the following properties:

(i) If $t_1 \neq t_2$ then the random variables $W_{t_1}(\cdot)$ and $W_{t_2}(\cdot)$ are independent.

(ii) $\{W_t\}_{t \geq 0}$ is a *stationary* process, i.e. the law of

$$\{W_{t_1+h}, \ldots, W_{t_n+h}\}$$

is independent of $h > 0$ for all $t_1, \ldots, t_n$ and all n.

(iii) $E[W_t] = 0$ and $E[W_t^2] = 1$ for all t, where E denotes expectation (i.e. integration) with respect to the probability measure P.

It turns out that no (measurable) process W_t satisfying (i), (ii) and (iii) exists! To overcome this difficulty, the following approaches are natural:

Alternative 1. Weaken the requirement (i) to allow dependence between W_{t_1} and W_{t_2} if t_1 and t_2 are close.

Alternative 2. Interpret equation (1.4) in a weak sense, i.e. as an *integral* equation instead of a differential equation:

(1.5) $$X_t = X_0 + \int_0^t a(K - X_s)ds + \text{"} \int_0^t b(K - X_s)W_s ds\text{"}\ ; \quad t \geq 0$$

where the integral in quotation marks needs to be explained.

It turns out that there is not a big difference between these two approaches. We will return to Alternative 1 in §5-6. Let us first look at Alternative 2:

We now recall that *Brownian motion* $\{B_t(\omega)\}_{t\geq 0}$ is a stochastic process with independent, stationary *increments* of mean zero (in fact, it is the only t-continuous process with these properties). This indicates that the t-derivative of B_t, $\frac{dB_t}{dt}$, could have been a good model for W_t - if it existed! Unfortunately, the path $t \to B_t(\omega)$ of Brownian motion has infinite variation for almost all ω. Nevertheless, we could try to replace "$W_s ds$" by "dB_s" in (1.5) and then try to make sense out of integrals of the form

$$\int_0^t f(s,\omega)dB_s(\omega) \tag{1.6}$$

for a reasonably large class of integrands $f(s,\omega)$. For deterministic functions $f(s,\omega) = f(s)$ such integrals were defined by N. Wiener [Wie] in 1923 and the general construction for a class of random integrands $f(s,\omega)$ was carried out by K. Ito [I1] in 1944. Such integrals are now called *Ito integrals*. The most notable necessary requirement that such integrands must satisfy is that $f(t,\cdot)$ is measurable w.r.t. the σ-algebra $\mathcal{F}_t$ generated by $\{B_s(\cdot);\ s \leq t\}$. Such processes $f(t,\omega)$ are called $\mathcal{F}_t$*-adapted*. See e.g. [Ø] for more details.

In terms of Ito integrals (1.5) gets the form

$$X_t = X_0 + a\int_0^t (K - X_s)ds + b\int_0^t (K - X_s)dB_s\ ; \quad t \geq 0 \tag{1.7}$$

This gives a precise mathematical interpretation of the equation (1.4), but what is its solution? Using the *Ito formula* (see e.g. [Ø]) one can prove that if X_0 is independent of the σ-algebra $\mathcal{F}_\infty$ generated by $\{B_t(\cdot); t \geq 0\}$ then

$$X_t = X_t(\omega) = K - (K - X_0)\exp(-(a + \frac{1}{2}b^2)t - bB_t(\omega))\ ; \quad t \geq 0 \tag{1.8}$$

Note that this solution coincides with the solution (1.2) if $b = 0$ i.e. if the noise is zero.

Since we know a lot about the probabilistic behaviour of $B_t(\cdot)$, we can from (1.8) easily deduce interesting properties about X_t. For example, $X_t \to K$ as $t \to \infty$, a.s. (a.s.= "almost surely", i.e. with probability one) and if X_0 is a constant (deterministic) then

$$E[X_t] = K - (K - X_0)e^{-at}, \tag{1.9}$$

i.e. the expected value of X_t coincides with the solution (1.2) of the no-noise equation.

We will return to the solution of this equation in §6.

§2. Stochastic partial differential equations: The need for a more general framework.

Encouraged by the success in SDE let us now consider a stochastic *partial* differential equation (SPDE):

EXAMPLE 2.1 (Membrane in a sand storm)
If a membrane is exposed to a (vertical) force $F(t, x, y)$ at time t and at the point with horizontal coordinates (x, y), then the vertical coordinate $z = u(t, x, y)$ of the membrane will satisfy the wave equation

$$\frac{\partial^2 u}{\partial t^2} - \left(\frac{\partial^2 u}{\partial x^2} + \frac{\partial^2 u}{\partial y^2}\right) = F(t, x, y) \tag{2.1}$$

Now suppose the force is coming from the bombardement of infinitesimal sand particles. Then a natural mathematical model for $F(t, x, y)$ would be a *3-parameter white noise process* $W_{t,x,y}(\omega)$. Analogous to the 1-parameter case discussed in §1 we can relate this noise to *3-parameter Brownian motion* (or Brownian sheet) $B_{t,x,y}(\omega)$ as follows:

$$W_{t,x,y}(\omega) = \frac{\partial^3}{\partial t \partial x \partial y} B_{t,x,y}(\omega) \quad \text{(see §3)} \tag{2.2}$$

With this force (2.1) becomes

$$\frac{\partial^2 u}{\partial t^2} - \left(\frac{\partial^2 u}{\partial x^2} + \frac{\partial^2 u}{\partial y^2}\right) = \frac{\partial^3}{\partial t \partial x \partial y} B_{t,x,y} \tag{2.3}$$

Again this does not make sense as it stands, but it is natural to try the weak, integral interpretation:

$$\int_0^t \int_{\mathbf{R}^2} u \frac{\partial^2 \phi}{\partial s^2} ds dx dy - \int_0^t \int_{\mathbf{R}^2} u \Delta \phi ds dx dy = \int_0^t \int_{\mathbf{R}^2} \phi dB_{t,x,y} \tag{2.4}$$

for all test functions $\phi(t, x, y) \in C_0^\infty(\mathbf{R}^3)$ (the smooth functions with compact support in $\mathbf{R}^3$). Here the right hand side is a 3-parameter Ito integral, which can be defined in a similar way as in the 1-parameter case. We have put $\frac{\partial^2}{\partial x^2} + \frac{\partial^2}{\partial y^2} = \Delta$, the Laplace operator.

So far, so good. But now the surprise is that (2.4) has no 3-parameter stochastic process solution $u_{t,x,y}(\omega)$! This was proved by J. Walsh [Wa]. However, Walsh also showed that (2.4) has a solution u in a more general setting, namely as a *distribution valued* stochastic process $u_\phi(\omega)$.

The reader's first reaction to such a result might be disappointment or disbelief: The physical membrane must have a position at time t over (x, y), no matter what the mathematics says! But is that really the case? When we measure the position of the membrane we are really taking averages of microscopic quantities over small periods of time and space and the actual macroscopic measurement really depends on what "scale" or microscope we use. Mathematically, the process of taking averages corresponds to applying a test function $\phi(t,x,y)$ to the distribution $u(\omega)$. This is exactly how we interpret the Walsh solution.

The singular white noise force $W_{t,x,y}$ is really itself a distribution valued process (see §3), so from this point of view it is not so surprising that the solution u is also. We could also say that solving equation (2.3) really corresponds to trying to find what the *macroscopic* value u is if the membrane is exposed to the singular force W working on the *microscopic* level.

This mathematical connection between micro- and macro-cosmos will be illustrated again in connection with fluid flow in porous media (§4). But first we need to develop some mathematical machinery. Two basic questions need to be clarified:

Question 1. What is the *right* mathematical formulation of a "noisy" partial differential equation?

By "right" we mean both that it is mathematically rigorous and that it (or rather its solution) actually gives a realistic model of the situation we are studying.

Question 2. Having found the right mathematical formulation, how do we proceed to solve the corresponding stochastic partial differential equation? By "solve" we here mean proving the existence and uniqueness of the solution, finding some of its probabilistic properties and - if possible - obtaining an explicit solution formula like (1.8) in Example 1.1.

§3. White noise, chaos expansion, Wick products and Skorohod integrals

The previous examples illustrate the need for a rigorous mathematical framework for concepts like white noise and distribution valued processes. In this section we give a brief summary of such a framework. More details can be found, for example, in [HP] or [HKPS].

It was Hida's original idea [H] that the basic object to consider is not the classical Brownian motion, but rather the more troublesome concept of white noise. In §2 white noise was introduced as the (non-existing) derivative of Brownian motion. It is however, surprisingly simple to construct white noise directly - and rigorosly - as a distribution valued process:

The white noise probability space.

Fix a natural number d (the parameter dimension) and let $\mathcal{S} = \mathcal{S}(\mathbf{R}^d)$ denote the Schwartz space of rapidly decreasing smooth functions on $\mathbf{R}^d$. The dual of $\mathcal{S}$ is the space $\mathcal{S}' = \mathcal{S}'(\mathbf{R}^d)$ of *tempered distributions*. According to the Bochner-Minlos theorem [GV] there exists a probability measure μ on $\mathcal{S}'$ with the property that

$$(3.1) \qquad \int_{\mathcal{S}'} e^{i<\omega,\phi>} d\mu(\omega) = e^{-\frac{1}{2}\|\phi\|^2}\,; \quad \phi \in \mathcal{S}$$

where $< \omega, \phi >= \omega(\phi)$ is the result of applying $\phi \in \mathcal{S}$ to the distribution $\omega \in \mathcal{S}'$ and $\|\phi\| = (\int_{\mathbf{R}^d} |\phi(x)|^2 dx)^{1/2}$ is the classical $L^2(\mathbf{R}^d)$-norm of ϕ.
It is not hard to show that (3.1) implies that

$$(3.2) \qquad \int_{\mathcal{S}'} f(< \omega, \phi >) d\mu(\omega) = (2\pi\|\phi\|^2)^{-d/2} \int_{\mathbf{R}^d} f(x) e^{-\frac{x^2}{2\|\phi\|^2}} dx\,; \quad \phi \in \mathcal{S}$$

for all f such that the integral on the right hand side converges. (It suffices to show (3.2) for functions f which are the inverse Fourier transform of their Fourier transforms $\hat{f}$ and for such functions (3.2) follows from (3.1) and the Fubini theorem).

In particular, if we apply this to $f(x) = x^2$ we get

$$(3.3) \qquad \int_{\mathcal{S}'} |< \omega, \phi >|^2 d\mu(\omega) = \|\phi\|^2\,; \quad \phi \in \mathcal{S}$$

Using this we can extend the definition of $< \omega, \phi >$ from $\phi \in \mathcal{S}$ to $\psi \in L^2(\mathbf{R}^d)$ as follows:

$$(3.4) \qquad < \omega, \psi >= \lim_{n\to\infty} < \omega, \phi_n > \quad (\text{limit in } L^2(\mu))$$

where $\phi_n \in \mathcal{S}$ and $\phi_n \to \psi$ in $L^2(\mathbf{R}^d)$. (It follows from (3.3) that the limit exists in $L^2(\mu)$ and that it is independent of the actual choice of approximating sequence $\{\phi_n\} \subset \mathcal{S}$.)

In particular, for all $t_1, \ldots, t_d \geq 0$ we can choose

$$(3.5) \qquad \psi(x) = \chi_{[0,t_1]\times\cdots\times[0,t_d]}(x)$$

(i.e. $\psi(x) = 1$ if $x \in [0, t_1] \times \cdots \times [0, t_d]$ and 0 otherwise) and define

$$(3.6) \qquad \tilde{B}_{t_1,\ldots,t_d}(\omega) =< \omega, \chi_{[0,t_1]\times\cdots\times[0,t_d]} >$$

Then $\tilde{B}$ is a d-parameter Gaussian process with stationary, independent increments of mean zero. Moreover by (3.3) the covariance is (from now on $E = E_\mu$ means expectation with respect to the measure μ):

$$E[\tilde{B}_{t_1,\ldots,t_d} \cdot \tilde{B}_{s_1,\ldots,s_d}] = \int_{S'} < \omega, \chi_{[0,t_1]\times\cdots\times[0,t_d]} > \cdot < \omega, \chi_{[0,s_1]\times\cdots\times[0,s_d]} > d\mu(\omega) \tag{3.7.}$$

$$= \int_{\mathbf{R}^d} \chi_{[0,t_1]\times\cdots\times[0,t_d]} \cdot \chi_{[0,s_1]\times\cdots\times[0,s_d]} dx = \prod_{k=1}^{d} (s_k \wedge t_k).$$

One can prove that there exists a t-continuous stochastic process $B_{t_1,\ldots,t_d}$ which is a *version* of $\tilde{B}_{t_1,\ldots,t_d}$, in the sense that

$$\mu(\{\omega; B_t(\omega) = \tilde{B}_t(\omega)\}) = 1 \quad \text{for all } t = (t_1, \ldots, t_d). \tag{3.8}$$

In view of the properties of $\tilde{B}_t$ stated above, it is natural to call B_t the *d-parameter Brownian motion* (or Brownian sheet).

DEFINITION 3.1. The *white noise process* is the map $W : \mathcal{S} \times \mathcal{S}' \to \mathbf{R}$ defined by

$$W_\phi(\omega) =< \omega, \phi > \quad \text{for } \phi \in \mathcal{S},\ \omega \in \mathcal{S}' \tag{3.9}$$

By (3.4), (3.6) and the isometry (3.3) we see that if $\phi \in \mathcal{S}$ with $\operatorname{supp}\phi \subset (\mathbf{R}^+)^d$ and $\{\psi_n\}$ are step functions converging to ϕ in $L^2(\mathbf{R}^d)$ then

$$W_\phi(\omega) = \lim_n < \omega, \psi_n >= \lim \int_{\mathbf{R}^d} \psi_n(x) dB_x = \int_{\mathbf{R}^d} \phi(x) dB_x \tag{3.10}$$

where the limit is in $L^2(\mu)$ and the last term is the *d-parameter Ito integral* of ϕ. For general $\phi \in L^2(\mathbf{R}^d)$ formula (3.10) is used to *define* the Ito integral of ϕ over $\mathbf{R}^d$. The identity (3.10) may be regarded as a precise way of saying that $W_.(\omega)$ is the distributional derivative of $B_.(\omega)$:

$$W = \frac{\partial}{\partial x_1 \ldots \partial x_d} B_{x_1,\ldots,x_d} \tag{3.11}$$

as claimed earlier.

The Wiener-Ito chaos expansion.

Let h_n be the *n'th order Hermite polynomial* defined by

$$h_n(x) = (-1)^n e^{\frac{x^2}{2}} \frac{d^n}{dx^n}(e^{-\frac{x^2}{2}});\ n = 0, 1, 2 \ldots \tag{3.12}$$

Thus

$$h_0(x) = 1,\ h_1(x) = x,\ h_2(x) = x^2 - 1,\ h_3(x) = x^3 - 3x,$$
$$h_4(x) = x^4 - 6x^2 + 3,\ h_5(x) = x^5 - 10x^3 + 15x, \dots$$

For $n = 1, 2, \dots$ let ξ_n be *the Hermite function of order* n defined by

$$\xi_n(x) = \pi^{-\frac{1}{4}}((n-1)!)^{-\frac{1}{2}} e^{-\frac{x^2}{2}} h_{n-1}(\sqrt{2}x);\ x \in \mathbf{R} \tag{3.13}$$

Then $\{\xi_n\}_{n=1}^{\infty}$ forms an orthonormal basis for $L^2(\mathbf{R})$. Therefore the family of tensor products

$$e_\alpha := e_{\alpha_1,\dots,\alpha_d} := \xi_{\alpha_1} \otimes \cdots \otimes \xi_{\alpha_d}, \tag{3.14}$$

where α denotes the multi-index $(\alpha_1, \dots, \alpha_d)$, forms an orthonormal basis for $L^2(\mathbf{R}^d)$. With a slight abuse of notation let $e_1, e_2, \dots$ denote a fixed ordering of the family $\{e_\alpha\}_\alpha$ from now on. Put

$$\theta_j = \theta_j(\omega) = \int_{\mathbf{R}^d} e_j(x) dB_x(\omega) \tag{3.15}$$

and define, for each multi-index $\alpha = (\alpha_1, \dots, \alpha_m)$,

$$H_\alpha(\omega) = \prod_{j=1}^{m} h_{\alpha_j}(\theta_j) \tag{3.16}$$

The Wiener-Ito chaos theorem states that the family $\{H_\alpha\}_\alpha$ forms an orthogonal basis for $L^2(\mu)$. This gives that any $X \in L^2(\mu)$ has the (unique) expansion

$$X(\omega) = \sum_\alpha c_\alpha H_\alpha(\omega), \tag{3.17}$$

the sum being taken over all multi-indices of non-negative integers. Moreover, we have the isometry

$$\|X\|^2_{L^2(\mu)} = \sum_\alpha \alpha! c_\alpha^2, \tag{3.18}$$

where $\alpha! = \alpha_1!, \alpha_2! \dots \alpha_m!$ if $\alpha = (\alpha_1, \dots, \alpha_m)$.

EXAMPLE 3.2. For each $\phi \in L^2(\mathbf{R}^d)$ we have $X := W_\phi \in L^2(\mu)$. The expansion of W_ϕ is

$$W_\phi = \int \phi dB = \sum_j \int (\phi, e_j) e_j dB = \sum_j (\phi, e_j) h_1(\theta_j) = \sum_j (\phi, e_j) H_{(j)}, \tag{3.19}$$

where $(j) = (0, 0, \dots, 0, 1)$ with 1 in the j'th place and $(\cdot, \cdot)$ denotes inner product in $L^2(\mathbf{R}^d)$.

The Wick product

DEFINITION 3.3. If $X = \sum_\alpha a_\alpha H_\alpha$ and $Y = \sum_\beta b_\beta H_\beta$ and two functions in $L^2(\mu)$ we define their *Wick product* $X \diamond Y$ as follows

$$X \diamond Y = \sum_{\alpha,\beta} a_\alpha b_\beta H_{\alpha+\beta} = \sum_\gamma \Big(\sum_{\alpha+\beta=\gamma} a_\alpha b_\beta \Big) H_\gamma \quad \text{(when convergent)} \tag{3.20}$$

For general $X, Y \in L^2(\mu)$ this sum may or may not converge in L^p for some $p \geq 1$.

REMARKS

1) Note that if one of the two factors is constant (does not depend on ω) the the Wick product coincides with the ordinary product.
2) The Wick product is not local: It is not enough to know the value of $X(\omega_0)$ and $Y(\omega_0)$ in order to know the value of $(X \diamond Y)(\omega_0)$. In fact, not even the knowledge of $X(\omega)$, $Y(\omega)$ for ω in some neighborhood of ω_0 is sufficient in general. ([GHLØUZ]).

EXAMPLE 3.4.
If $X(\omega) = W_\phi(\omega) = \int_{\mathbf{R}^d} \phi(x) dB_x(\omega)$ and $Y(\omega) = W_\psi(\omega) = \int_{\mathbf{R}^d} \psi(x) dB_x(\omega)$ with $\phi, \psi \in L^2(\mathbf{R}^d)$ then it can be proved that

$$(X \diamond Y)(\omega) = \int_{\mathbf{R}^d \times \mathbf{R}^d} (\phi \hat{\otimes} \psi)(x, y) dB_{x,y}^{\otimes 2}, \tag{3.21}$$

where $\hat{\otimes}$ denotes *symmetrized tensor product* (i.e. $(\phi \hat{\otimes} \psi)(x, y) = \frac{1}{2}[\phi(x)\psi(y) + \phi(y)\psi(x)]$) and the right hand side of (3.20) is *the double Ito integral* (see e.g. [GHLØUZ] for details).

EXAMPLE 3.5. If $X(\omega) = B_t(\omega)$ for a fixed t, then (for example by (3.21))

$$B_t \diamond B_t := B_t^{\diamond 2} = B_t^2 - t$$

and

$$B_t^{\diamond 3} := B_t \diamond (B_t^{\diamond 2}) = B_t^3 - 3tB_t$$

(see e.g. [GHLØUZ] for extensions).

EXAMPLE 3.6. It is easy to see that $\diamond$ is an associative binary operation (when defined) so we can define, for $X \in L^2(\mu)$ and n a natural number, the *Wick power*

$$X^{\diamond n} = X \diamond X \diamond \cdots \diamond X \quad (n \text{ times}) \tag{3.22}$$

without specifying parenthesis on the right hand side (assuming the Wick products exist). In particular, for $X = W_\phi$ we can define the *Wick exponential* of W_ϕ, $\operatorname{Exp} W_\phi$, by

$$\operatorname{Exp} W_\phi = \sum_{n=0}^{\infty} \frac{1}{n!} W_\phi^{\diamond n}, \quad \phi \in L^2(\mathbf{R}^d). \tag{3.23}$$

One can in fact show that (see e.g. [GHLØUZ])

$$\operatorname{Exp} W_\phi = \exp(W_\phi - \frac{1}{2}\|\phi\|^2) \tag{3.24}$$

In particular, this shows that $\operatorname{Exp} W_\phi$ is *positive*, in the sense that

$$\operatorname{Exp} W_\phi(\omega) > 0 \quad \text{for all } \phi \in L^2(\mathbf{R}^d),\ \omega \in \mathcal{S}'.$$

This property makes it a natural model for certain "positive noises" occurring for example in fluid flow in porous media. See §6.

The Wick product (or slightly different version of it) was originally introduced by G.C. Wick [Wi] in 1950 in connection with quantum field theory, where it corresponds to a kind of renormalization. In a stochastic analysis context the Wick product was first used in 1965 by T. Hida and N. Ikeda [HI]. It is remarkable that the Wick product concept should prove natural in two so different contexts. The full reason for - and all implications of - this connection is not yet fully understood (at least not by me).

In order to explain why the Wick product is natural in stochastic analysis we need to define the following concept:

The Skorohod integral

Let us for a moment assume that $d = 1$ and consider a stochastic process $\{X_t\}_{t \geq 0} \subset L^2(\mu)$. Such a process can be represented by the expansion

$$X_t(\omega) = \sum_\alpha c_\alpha(t) H_\alpha(\omega). \tag{3.25}$$

Define *the Skorohod integral* of X_t, $\int_{\mathbf{R}} X_t \delta B_t$, by

$$\int_{\mathbf{R}} X_t(\omega)\delta B_t(\omega) := \sum_{\alpha,j}(\int_{\mathbf{R}} c_\alpha(t)e_j(t)dt)H_{\alpha+(j)}(\omega) = \sum_{\alpha,j}(c_\alpha, e_j)H_{\alpha+(j)}(\omega) \tag{3.26}$$

(when convergent), where as before $(j) = (0, 0, \ldots, 0, 1)$ with a 1 in the j'th place and $(\cdot,\cdot)$ denotes the usual $L^2(\mathbf{R})$ inner product.

Note that if X_t is deterministic, i.e. $X_t = c_0(t)$, then

$$\begin{aligned}(\int_{\mathbf{R}} X_t(\omega)\delta B_t(\omega) &= \sum_j (c_0, e_j)H_{(j)}(\omega) = \sum_j (c_0, e_j)\int_{\mathbf{R}} e_j dB \\ &= \int_{\mathbf{R}} c_0(t)dB_t = \int_{\mathbf{R}} X_t(\omega)dB_t.\end{aligned} \tag{3.27}$$

Thus the Skorohod integral coincides with the Ito integral in this case. In fact, the Skorohod integral can be shown to coincide with the Ito integral if the integrand is adapted [NZ].

For a general $\phi \in L^2(\mathbf{R})$, $t \in \mathbf{R}$ let $\phi_t(\cdot)$ denote the *t-shift of* ϕ, i.e.

$$\phi_t(s) = \phi(s - t) \tag{3.28}$$

The connection between Wick products, white noise and Skorohod integrals can now be formulated as follows:

$$\int_{\mathbf{R}} (\phi * X)_t \delta B_t = \int_{\mathbf{R}} X_t \diamond W_{\phi_t} dt; \; \phi \in \mathcal{S}(\mathbf{R}) \tag{3.29}$$

where $*$ denotes convolution with respect to Lebesgue measure on $\mathbf{R}$, i.e.

$$(\phi * X)_t(\omega) = \int_{\mathbf{R}} \phi(s + t)X_s(\omega)ds \tag{3.30}$$

(See [LØU 2], [AP], [ØZ]).

If we let $\phi \to \delta$ (the Dirac measure at 0) as measures on $\mathbf{R}$ one can prove that W_{ϕ_t} converges in a weak sense (in the space $(\mathcal{S})^*$ of Hida distributions) to an object which we - by an abuse of notation - denote by W_t (the "pointwise" version of white noise). The Wick product can be extended to $(\mathcal{S})^*$, so taking the limit in (3.39) we get

$$\int_{\mathbf{R}} X_t \delta B_t = \int_{\mathbf{R}} X_t \diamond W_t dt \tag{3.31}$$

In particular, if X_t is adapted this says that *Ito integration is equivalent to Wick multiplication by white noise followed by Lebesgue integration.* Here is the key to the fundamental importance of the Wick product in Ito stochastic calculus.

§4. The Hermite transform and its inverse

The Wiener - Ito expansion allows us to associate to any given $X \in L^2(\mu)$ a complex valued function $\mathcal{H}(X)$ of infinitely many complex variables $z_1, z_2, \ldots$:

DEFINITION 4.1. Let $X \in L^2(\mu)$ have the expansion

$$X(\omega) = \sum_\alpha c_\alpha H_\alpha(\omega) \tag{4.1}$$

Then the *Hermite transform* of X, denoted by $\mathcal{H}(X)$ or $\tilde{X}$, is the function defined on the space $\mathbf{C}_0^{\mathbf{N}}$ of all finite sequences of complex numbers $z_1, \ldots, z_n$ by

$$\mathcal{H}(X)(z_1, z_2, \ldots) = \tilde{X}(z_1, z_2, \ldots) = \sum_\alpha c_\alpha z^\alpha \tag{4.2}$$

where we again use the multi-index notation: If $\alpha = (\alpha_1, \ldots, \alpha_m)$, $z = (z_1, z_2, \ldots)$ then

$$z^\alpha = z_1^{\alpha_1} z_2^{\alpha_2} \ldots z_m^{\alpha_m}.$$

EXAMPLE 4.2. The Hermite transform of white noise is, by (3.19),

$$\mathcal{H}(W)(z) = \tilde{W}_\phi(z) = \sum_j (\phi, e_j) z_j\,; \qquad (z = z_1, z_2, \ldots) \tag{4.3}$$

A crucial feature of the Hermite transform is that it changes Wick products into ordinary complex products:

LEMMA 4.3 If X, Y, $X \diamond Y \in L^2(\mu)$ then

$$\mathcal{H}(X \diamond Y)(z) = \mathcal{H}(X)(z) \cdot \mathcal{H}(Y)(z)$$

Moreover, it is possible to recover X from $\tilde{X}$ by performing an integration with respect to an infinite product of Gaussian measures:

Let $d\lambda(y) = d\lambda(y_1, y_2 \ldots)$ be the probability measure on $\mathbf{R}^{\mathbf{N}} = \mathbf{R} \times \mathbf{R} \times \cdots$ defined by

$$\int_{\mathbf{R}^{\mathbf{N}}} f(y_1, \ldots y_n) d\lambda(y) = (2\pi)^{-n/2} \int_{\mathbf{R}^n} f(y) e^{-\frac{1}{2}|y|^2} dy \tag{4.4}$$

if f is a bounded function depending only on the first n coordinates $(y_1, \ldots, y_n)$ of y. Then we have

LEMMA 4.5. Let $X \in L^2(\mu)$. Then

$$X(\omega) = \int_{\mathbf{R}^{\mathbf{N}}} \tilde{X}(\theta_1 + iy_1, \theta_2 + iy_2, \ldots) d\lambda(y) \qquad (i = \sqrt{-1}) \tag{4.5}$$

where $\theta_j = \int e_j dB$ as before and the integral is interpreted as a limit of the integrals of the truncations of $\tilde{X}$. See [HLØUZ 1] for details.

EXAMPLE 4.6. Suppose $\tilde{X}(z) = z_1 = x_1 + iy_1$. Then

$$X(\omega) = \frac{1}{\sqrt{2\pi}} \int_{\mathbf{R}} (\theta_1 + iy_1) e^{-\frac{1}{2}y_1^2} dy_1 = \theta_1 = \int e_1 dB,$$

so $x = W_{e_1}$.

§5. A scheme for solving stochastic partial differential equations.

Example 2.1 illustrates that for a SPDE we cannot in general expect to find a solution which is an ordinary stochastic process. The reason is that white noise, albeit it has the ideal probabilistic properties, is too singular to produce solutions of this kind, even when the equation is interpreted in a weak, integrated sense. In view of this, and in view of equations (3.29) and (3.31), it is natural to adopt *Alternative 1* in §1 rather than Alternative 2 in the general setting.

Thus we replace the singular, pointwise white noise W_t by the "smeared out" version W_ϕ, where $\phi \in \mathcal{S}$ is fixed (at least for a while). We may regard ϕ as the "window" or the microcope we use to measure the noise. Then we consider translates of this window, $\phi_x(\cdot)$ and use $W_{\phi_x}(\cdot)$ as our approximate white noise in the equation. By integration by parts we see that distributional derivative with respect to ϕ is the same as ordinary derivative with respect to the shift x. Thus, with W_x replaced by W_{ϕ_x} we may regard the corresponding SPDE as a PDE in x for each ω, except that we use Wick products instead of ω-pointwise products (see below). In special cases the solution $X = X(x, \phi, \omega)$ of the SPDE may have limit as $\phi \to \delta$, but in general not. So in general we regard the equation as solved if we have found $X(\phi, x, \omega)$ for each $\phi \in \mathcal{S}$.

As we will illustrate in the next section this works well and gives us the same result as classical methods in the linear case. In the non-linear case, however, the question arises what kind of products one should use: Wick products or ordinary (pointwise) products. See [LØU2] for a more detailed discussion about this. As we have pointed out already, the Wick product is natural in the context of Ito integration. If we start with such a Wick SPDE, then in view of §4 the canonical solution procedure is the following:

1) Apply the Hermite transform to convert the original equation into a deterministic PDE with complex parameters $z_1, z_2, \ldots$

2) Solve this equation (if possible).
3) Apply the inverse Hermite transform to the solution in 2) to obtain a solution of the original equation.

§6. Applications

To illustrate the method outlined above let us use it to solve the SDE (1.7) of Example 1.1:

$$X_t = X_0 + a\int_0^t (K - X_s)ds + b\int_0^t (K - X_s)dB_s \tag{6.1}$$

Adopting the point of view of §5 we choose $\phi \in \mathcal{S}(\mathbf{R})$ and consider the approximate equation in $X_t = X(\phi, t, \omega)$:

$$X_t = X_0 + \int_0^t (K - X_s) \diamond (a + bW_{\phi_s})ds$$

Taking Hermite transforms we get

$$\tilde{X}_t = \tilde{X}_0 + \int_0^t (K - \tilde{X}_s)(a + b\tilde{W}_{\phi_s})ds$$

or

$$\frac{d\tilde{X}_t}{dt} = (K - \tilde{X}_t)(a + b\tilde{W}_{\phi_t}). \tag{6.2}$$

Recall that $\tilde{W}_{\phi_t} = \tilde{W}_{\phi_t}(z_1, z_2, \ldots) = \sum_k (\phi_t, \eta_k) z_k$ where $z_k \in \mathbf{C}$ so (6.2) is a differential equation with respect to t with complex parameters $z_1, z_2, \ldots$. It is easily verified that the solution of (6.2) is

$$\tilde{X}_t = K - (K - \tilde{X}_0) \cdot \exp(-\int_0^t (a + b\tilde{W}_{\phi_s})ds) \tag{6.3}$$

To find X_t we can apply the inverse Hermite transform. However, it is easier to note that by (4.4) we have

$$\mathcal{H}^{-1}(\exp(-b\int_0^t \tilde{W}_{\phi_s}ds)) = \operatorname{Exp}(-b\int_0^t W_{\phi_s}ds) \tag{6.3}$$

which gives, using (4.4) again,

$$X_t = K - (K - X_0) \diamond \mathrm{Exp}(-at - b \int_0^t (\int_{\mathbf{R}} \phi_s(u) dB_u) ds) \tag{6.4}$$

Now

$$\int_0^t (\int_{\mathbf{R}} \phi_s(u) dB_u) ds = \int_{\mathbf{R}} (\int_0^t \phi(u-s) ds) dB_u \to \int_{\mathbf{R}} \mathcal{X}_{[0,t]}(u) dB_u = B_t$$

as $\phi \to \delta$. Therefore (6.4) gives us

$$\lim_{\phi \to \delta} X(\phi, t, \omega) = K - (K - X_0) \diamond \mathrm{Exp}(-at - bB_t(\omega)) \tag{6.5}$$

This is the same as the solution (1.8) we gave earlier, because of the identity (3.24).

Note that not only did we get the solution quickly by this method, but we also got a more general result than in (1.8): Our method here did not assume that X_0 is independent of the σ-algebra $\mathcal{F}$ generated by $\{B_t(\cdot); t \geq 0\}$ and for general X_0 the solution is expressed by the Wick product in (6.4). This Wick product reduces to the ordinary product if X_0 is $\mathcal{F}$-independent.

The ability to handle such *non-adaptive* stochastic differential equations is an additional useful feature of this method. Non-adaptive SDE's occur for example in problems regarding economic investments under uncertainty. See [ØZ], where (adapted and non-adapted) stochastic Volterra equations are studied.

We proceed by giving some examples of SPDE's which can be handled by the method outlined above.

EXAMPLE 6.1. (Fluid flow in porous media).
In a porous rock the permeability k will often vary rapidly from point to point and it can be hard to measure. Accordingly, in the equation for the fluid pressure $p(x)$ at the point x for one phase flow,

$$\begin{cases} \mathrm{div}(k(x)\nabla p(x)) = -f(x)\,; & x \in D \subset \mathbf{R}^d \\ p(x) = 0\,; & x \in \partial D \end{cases} \tag{6.6}$$

(where f is the given source rate) it is natural to represent the quantity $k(x)$ by some *positive noise* $K(x,\omega); \omega \in \mathcal{S}'$. This makes the pressure stochastic, too: $p(x) = p(x,\omega)$. Representing $K(x,\omega)$ by $\mathrm{Exp}(W_{\phi_x}(\cdot))$ (see (3.23)) we arrive at the SPDE

$$\begin{cases} \mathrm{div}(\mathrm{Exp}(W_{\phi_x}(\cdot)) \diamond \nabla p(x,\cdot)) = -f(x)\,; & x \in D \\ p(x,\cdot) = 0\,; & x \in D \end{cases} \tag{6.7}$$

The question is now: How will the *microscopic* feature $k(x)$ of the medium affect the *macroscopic* properties of the flow? This equation is studied in [LØU 3].

EXAMPLE 6.2. (The stochastic Schrödinger equation)
A model for the Schrödinger equation with a random, positive potential is

$$
\begin{cases} \Delta u(x,\cdot) + V(x,\cdot) \diamond u(x,\cdot) = -f(x)\,; & x \in D \subset \mathbf{R}^d \\ u(x,\cdot) = 0\,; \quad x \in \delta D, \end{cases} \tag{6.8}
$$

where $\Delta = \sum_{k=1}^{d} \frac{\partial^2}{\partial x_k^2}$ is the Laplacian and $V(x,\cdot) = \epsilon \mathrm{Exp}(W_{\phi_x}(\cdot))$ is the positive potential, $\epsilon > 0$.

In [HLØUZ 1] it is proved that for ϵ small enough the unique solution u of (6.6) is given by

$$
u(x,\cdot) = \hat{E}^x[\int_0^{\tau_D} \mathrm{Exp}[\epsilon \int_0^t \mathrm{Exp}(W_{\phi_{b_s}})ds]f(b_t)dt], \tag{6.9}
$$

where $(b_t, \hat{P}^x)$ is a classical, 1-parameter Brownian motion in $\mathbf{R}^d$, $\hat{E}^x$ denotes expectation with respect to $\hat{P}^x$ and

$$
\tau_D = \inf\{t > 0; b_t \notin D\}
$$

is the first exit time from D of b_t.

EXAMPLE 6.3. (The transport equation in a turbulent medium)
If we model the turbulent motion of the medium by some d dimensional noise $\vec{v}(t,x,\omega)$ an equation modelling the transport of a substance in this medium is

$$
\frac{\partial}{\partial t}u(t,x,\cdot) + \vec{v}(t,x,\cdot) \diamond \nabla u(t,x,\cdot) = \frac{1}{2}\Delta u(t,x,\cdot) \tag{6.10}
$$

where the gradient ∇ and the Laplacian Δ work on the x-variable, $x = (x_1,\ldots,x_d)$.

Here $\vec{v}$ could be modelled as d-dimensional, $(d+1)$-parameter white noise

$$
\vec{v} = (W_1(t,x,\cdot),\ldots,W_d(t,x,\cdot)) \tag{6.11}
$$

where the W_i's are independent ([G]). This equation has been studied in [CP].

EXAMPLE 6.4. (SPDE's arising in non-linear filtering).
The Wong-Zakai equation for the unnormalized conditional density ρ_t of the filtered estimate has the form

(6.12) $$d\rho_t(x;\omega) = A^*\rho_t(x,\omega)dt + \rho_t(x,\omega)h^T(x)dB_t$$

where A^* is a semi-elliptic second order partial differential operator acting on the space variable $x \in \mathbf{R}^d$ (A^* is the adjoint of the generator of an Ito diffusion) and $h : \mathbf{R}^d \to \mathbf{R}^m$ is given.

A general existence and uniqueness theorem for SPDEs of this type has been given by Pardoux [Pa]. Using Wick calculus one can in fact obtain an explicit solution [B].

EXAMPLE 6.5. (The Burgers equation with a noisy force)
The Burgers equation (in dimension 1) is the non-linear partial differential equation

(6.13) $$\frac{\partial u}{\partial t} + \lambda u \cdot \frac{\partial u}{\partial x} = \nu \frac{\partial^2 u}{\partial x^2}$$

where λ, ν are constants. ($R = \frac{1}{\nu}$ is the Reynolds number.) It was orginally introduced as a model for turbulence but has later found many other applications as well.

With a noisy force the Burgers equation gets the form

(6.13) $$\frac{\partial u}{\partial t}(t,x,\cdot) + \lambda u(t,x,\cdot) \diamond \frac{\partial u}{\partial x}(t,x,\cdot) = \nu \frac{\partial^2 u}{\partial x^2}(t,x,\cdot) + M(t,x,\cdot),$$

where M is a (t,x)-parameter noise. By a certain Wick-substitution this equation can be transformed into a *linear* stochastic heat equation of the form

(6.14) $$\frac{\partial Y}{\partial t} = \nu \Delta Y + \frac{\lambda}{2\nu} Y \diamond N,$$

where N is another noise. This equation can be solved by Hermite transforms as outlined in §5. See [HLØUZ 2].

The Burgers equation is a special case of the conservation law

(6.15) $$\frac{\partial u}{\partial t} + \frac{\partial}{\partial x}(f(u)) = F$$

where f and F are given functions.

This equation appears for example in fluid flow in porous media, where $u(t,x)$ is the saturation of the fluid, f is the flux function and F is the source. Various stochastic versions of this equation appear naturally in applications. See [HR], [HLR] and the references there.

§7. Concluding remarks. Towards a random fractal calculus?

SPDE is a mathematical machinery developed to handle PDEs where some of the coefficients are subject to random fluctuations or noise (or modelled as noise because of lack of information). The point is that even though there is noise in the coefficients one can still say something about the probabilistic properties of the solution. Moreover, it is of interest to see explicitly how the noise in the coefficients affects the solution. Often the noise comes from (the basically unknown) *microscopic* properties of the medium (e.g. permeability) or the surroundings and one seeks the corresponding macroscopic properties of the solution (e.g. the fluid flow). The stochastic analysis that is used to handle these questions turns out to involve in a natural way the *Wick product*, a concept which has been developed earlier in connection with renormalization in quantum physics [S]. This confirms that there is a deep relation between stochastic analysis and quantum physics.

Another interesting aspect of this theory is that it seems to be able to handle analytically some classes of *random fractals*. This means that random fractals can be more than just a tool to describe certain phenomena; they can be adopted as rigorous parts of a stochastic differential equation: We can do calculus with it. More precisely, as soon as the random fractal can be represented as some kind of noise or white noise functional as explained in §3, the whole stochastic calculus machinery applies. For example, the Wick exponential of white noise appears to be a good model not just for permeability, but also for the multifractals appearing in connection with turbulence or in connection with oil distribution [M].

Acknowledgements

This work is supported by VISTA, a research cooperation between The Norwegian Academy of Science and Letters and Den Norske Stats Oljeselskap A.S. (Statoil)

REFERENCES

[AP] J. Ash and J. Potthoff: Ito's lemma without non-anticipatory conditions. Probab.Th.Rel.Fields 88 (1991), 17-46.

[B] F.E. Benth: An explicit solution of the non-linear filtering equation. (In preparation).

[CP] G. Cochran and J. Potthoff: Fixed point principles for stochastic partial differential equations. To appear in Proc. "Dynamics of Complex and Irregular Systems", Bielefeld 1991.

[GV] I.M. Gelfand and N.Y. Vilenkin: Generalized Functions, Vol.4: Applications of Harmonic Analysis. Academic Press 1964 (English translation).

[G] H. Gjessing: Some properties of the Wick product and the Hermite transform. Manuscript 1992.

[Gj] J. Gjerde: Multidimensional noise. (In preparation).

[GHLØUZ] H. Gjessing, H. Holden, T. Lindstrøm, J. Ubøe and T.-S. Zhang: The Wick product. To appear in A. Melnikov (editor): "New Trends in Probability and Statistics, Vol. IV". TVP Publishers, Moscow.

[H] T. Hida: Brownian Motion. Springer-Verlag 1980.

[HI] T. Hida and N. Ikeda: Analysis on Hilbert space with reproducing kernel arising from multiple Wiener integral. Proc. Fifth Berkeley Symp. Math.Stat.Probab. II, part 1 (1965), 117-143.

[HKPS] T. Hida, H.-H. Kuo, J. Potthoff and L. Streit: White Noise Analysis. Forthcoming book.

[HLØU] H. Holden, T. Lindstrøm, B. Øksendal and J. Ubøe: Discrete Wick calculus and stochastic functional equations. Potential Analysis 1 (1992), 291-306.

[HLR] H. Holden, T. Lindstrøm and N.H. Risebro: Conservation laws with a random source. Manuscript 1992.

[HR] H. Holden and N.H. Risebro: Stochastic properties of the Buckley-Leverett equation. SIAM J. Appl. Math. 51 (1991), 1472-1488.

[HLØUZ 1] H. Holden, T. Lindstrøm, B. Øksendal, J. Ubøe and T.-S. Zhang: Stochastic boundary value problems. A white noise functional approach. To appear in Probab. Th. Rel. Fields.

[HLØUZ 2] H. Holden, T. Lindstrøm, B. Øksendal, J. Ubøe and T.-S. Zhang: The Burgers equation with a noisy force. Manuscript 1992.

[HP] T. Hida and J. Potthoff: White noise analysis - an overview. In T. Hida, H.-H. Kuo, J. Potthoff and L. Streit (eds.): White Noise Analysis. World Scientific 1990.

[I1] K. Ito: Stochastic integral. Proc. Imp. Acad. Tokyo 20 (1944), 519.524.

[I2] K. Ito: Multiple Wiener integral. J.Math.Soc. Japan 3 (1951), 157-169.

[LØU 1] T. Lindstrøm, B. Øksendal and J. Ubøe: Stochastic differential equations involving positive noise. In M. Barlow and N. Bingham (editors): Stochastic Analysis. Cambridge Univ. Press 1991, 261-303.

[LØU 2] T. Lindstrøm, B. Øksendal and J. Ubøe: Wick multiplication and Ito-Skorohod stochastic differential equations. In S. Albeverio et al (editors): Ideas and Methods in Mathematical Analysis, Stochastics, and Applications. Cambridge Univ. Press 1992, pp. 183-206.

[LØU 3] T. Lindstrøm, B. Øksendal and J. Ubøe: Stochastic modelling of fluid flow in porous media. In S. Chen and J. Yong (editors): Control Theory, Stochastic Analysis and Applications. World Scientific 1991, pp. 156-172.

[M] B. Mandelbrot: Fractals - a geometry of nature. In N. Hall (editor): The New Scientist Guide to Chaos. Penguin Books 1991.

[NZ] D. Nualart and M. Zakai: Generalized stochastic integrals and the Malliavin calculus. Probab.Th.Rel.Fields 73 (1986), 255-280.

[NP] D. Nualart and E. Pardoux: Stochastic calculus with anticipating integrands. Probab.Th.Rel.Fields 78 (1988), 535-581.

[Ø] B. Øksendal: Stochastic Differential Equations. Springer-Verlag 1992 (Third edition).

[ØZ] B. Øksendal and T.-S. Zhang: The stochastic Volterra equation. Preprint, University of Oslo 1992.

[Pa] E. Pardoux: Stochastic partial differential equations and filtering of diffusion processes. Stochastics 3 (1979), 127-167.

[Po] J. Potthoff: White noise methods for stochastic partial differential equations. Manuscript 1991.

[S] B. Simon: The $P(\phi)_2$ Euclidean (Quantum) Field Theory. Princeton University Press 1974.

[Wa] J.B. Walsh: An introduction to stochastic partial differential equations. In R. Carmona, H. Kesten and J.B. Walsh (editors): École d'Été de Probabilités de Saint-Flour XIV-1984. Springer LNM 1180 (1986), 265-437.

[Wi] G.C. Wick: The evaluation of the collinear matrix. Phys.Rev. 80 (1950), 268-272.

[Wie] N. Wiener: Differential space. J. of Mathematics and Physics 3 (1924), 127-146.

Construction Methods for Mixed Covering Codes

Patric R. J. Östergård

Digital Systems Laboratory
Helsinki University of Technology
SF-02150 Espoo, Finland

Abstract

We consider construction methods for mixed covering codes. The constructions give upper bounds on $K_{q_1,\ldots,q_m}(n_1,\ldots,n_m;R)$, the minimum number of codewords in a code $C \subseteq F_{q_1}^{n_1} \cdots F_{q_m}^{n_m}$ with the property that all words in the space are within Hamming distance R from at least one codeword. The approach is general, no restrictions are set upon m and the arities q_i. The methods consist of generalizations of known construction methods for covering codes and some completely new constructions. They are divided into three classes: direct constructions, constructions of new codes from old, and the matrix method. Some results for nonmixed codes are also presented. These include a record-breaking binary code, proving $K_2(12,2) \leq 78$.

1 Introduction

Assume that somebody asks you to answer n questions, and q alternative answers are given for each question. Further assume that you do not possess any knowledge at all in the area of the questions, and that you are not limited to one n-tuple of answers.

If you want to give a set of answers that always contains one correct n-tuple whatever the correct answers are, all possible n-tuples have to be given. The number of these is q^n. However, assume that R incorrect answers are allowed, that is, there must in any case be at least one n-tuple with at least $n - R$ correct answers. This problem is a covering problem, and a set of n-tuples that fulfills the requirements is called a covering code with covering radius R.

Within the mathematical literature covering codes originated in 1948, when Taussky and Todd [45] discussed the problem group-theoretically. Even earlier, however, such codes had been constructed for football pools. Many results, obtained by "professional amateurs", have remained unknown to the scientific community for half a century [18, 48]. Over the years many contributions have appeared in different journals within the areas of coding theory and combinatorics. The main interest has concerned binary codes ($q = 2$) and ternary codes ($q = 3$) of covering radius 1 (the so-called *football pool problem* [27]). There has been considerable recent growth in interest in the area [10, 12, 16].

Traditionally only covering codes where all the coordinates are of the same arity have been considered. However, if the number of alternative answers for the questions mentioned in the beginning of this Introduction vary, or if we can exclude some of the possible answers, this case occurs. Recently some papers discussing these *mixed covering codes* generally have been published [18, 34, 36, 46]. Earlier research has mainly concentrated on *perfect* mixed covering codes, see [47] for a references to these contributions. Perfect codes are simultaneously both packing and covering; as for the aforementioned problem this means that there is then always exactly one n-tuple with at least $n - R$ correct answers.

In Section 2 simulated annealing (SA), an optimization method that has turned out to perform very well in the construction of covering codes, is discussed. SA is an indispensable aid in some of the construction methods, which are presented in Section 3. Finally, we conclude the paper in Section 4.

2 Simulated Annealing

Combinatorial optimization problems typically involve a cost function to be minimized (maximization is obtained by minimizing the negative function). A well known approach to such problems is named *iterative improvement*, working in the following way. Starting from an initial configuration one of the "neighbouring" configurations is explored. This new configuration is accepted if the value of the cost function is less for it than for the present configuration. The evaluation is repeated until no further improvements can be made.

The drawback of such a *greedy* algorithm is that it gets stuck in local minima, since only downhill moves are accepted. One way to cope with this problem is to repeat the algorithm a number of times from different initial configurations, and hope that at least one iteration will end up in a global minimum. Another way is to allow some uphill moves also; this is one of the basic features of the *simulated annealing* (SA) optimization method.

The SA algorithm was developed independently by Kirkpatrick, Gelatt Jr. and Vecchi [29], and Černy [8] in the early 1980's. The method is motivated by an analogy to physical processes. When a material is crystallized from the liquid phase, it tries to attain a state where the energy is at a minimum. This cooling (annealing) must be done very slowly to obtain a highly ordered, low-energy final state. At each temperature the likelihood of its being in a given state is given by the Boltzmann distribution for that temperature.

For an optimization problem we introduce a parameter T simulating the temperature of a physical process. The cost function $E(s)$ to be minimized corresponds to the energy of a process.

The SA algorithm has the following general form (in pseudo-code). For extensive consideration of the history and the mathematical theory of the algorithm the monographs [1, 31] are warmly recommended.

```
Start from T := T_0 and a random initial state s
while stop criterion not satisfied do
    begin
    while inner loop criterion not satisfied do
        begin
        Generate new state s'
        ΔE := E(s') − E(s)
        if ΔE ≤ 0 then s := s'
        if ΔE > 0 then s := s' with probability e^(−ΔE/T)
        end
    Update T
    end
```

The generation of new states and the energy function are both problem-specific. These parameters are discussed in Section 3 when the different design methods are described.

As for the general parameters, the following very simple *cooling schedule* has turned out to be surprisingly good.

- **Initial temperature**: The initial temperature is chosen such that the corresponding acceptance ratio (i.e. the percentage of the proposed new states that are accepted) is χ_0. We recommend that $\chi_0 \approx 0.3$.
- **Stop criterion**: Stop if the final energy is the same for a number (M) of consecutive temperatures.
- **Inner loop criterion**: The loop is executed a constant number of times (N).
- **Update of temperature:** $T := \lambda T$ (*geometric cooling*). In the literature values in the interval $0.5 \leq \lambda < 1$ have been used, we recommend $0.99 \leq \lambda < 1$.

Some experience is required to tune in appropriate values for χ_0, M, N and λ. All the time a trade-off has to be made between a too fast cooling with the risk that no solutions will be found, and a too slow cooling that is unnecessarily time-consuming. For a more extensive treatment of these parameters the reader is referred to [1, 31].

3 Construction Methods

3.1 Mixed Covering Codes

Let F_q denote the finite field $\mathcal{F}_q$ if q is a prime power, otherwise the integers modulo q. In this paper we consider arbitrary mixed spaces $H = F_{q_1}^{n_1} \cdots F_{q_m}^{n_m}, m \geq 1, n_i \geq 0, \sum_{i=1}^{m} n_i \geq 1, q_i \geq 2$. (In the rest of the paper, H always refers to this space.) If $m \geq 2$ and not all the q_i are the same, we have a *proper* mixed space, but unless specifically stated we do not require this further on.

The *Hamming distance* $d(x, y)$ between two words $x, y \in H$ is the number of coordinates in which they differ. A code $C \subseteq H$ is said to cover H with covering radius R if every word in H is within Hamming distance R from some codeword in C, and if this is not fulfilled for $R' = R - 1$. If such a mixed code has M codewords, we call it a $(q_1, \ldots, q_m; n_1, \ldots, n_m; M)R$ code. The set of codewords must be nonempty, so $M > 0$. We further denote

$$K_{q_1,\ldots,q_m}(n_1, \ldots, n_m; R) = \\ \min\{M \mid \text{there is a } (q_1, \ldots, q_m; n_1, \ldots, n_m; M)R \text{ code}\}.$$

By explicitly constructing covering codes upper bounds on this function are improved. Lower bounds are not further discussed in this paper; we confine ourselves to giving the *sphere covering bound*

$$K_{q_1,\ldots,q_n}(1, \ldots, 1; R) \geq (\prod_{i=1}^{n} q_i)/(1 + \sum_{j=1}^{R} \sum_{1 \leq i_1 \leq \cdots \leq i_j \leq n} (q_{i_1} - 1) \cdots (q_{i_j} - 1)).$$

The first paper on construction methods for mixed covering codes is a paper by Hämäläinen and Rankinen [18]. They, however, discuss only binary/ternary mixed codes, i.e., they give upper bounds on $K_{3,2}(n_1, n_2; R)$. These codes are interesting as they can be used in football pool systems. (In football pools there are three alternatives for each match: home team wins, home team loses, draw. Two alternatives remain if the gambler is convinced of the impossibility of one of these.)

Some of the constructions to be presented are generalizations of the results in [18]. In the consideration of the methods it is worth remembering that a permutation of the coordinates of a code gives an equivalent code.

3.2 Direct Constructions

For some parameters (especially for short codes) it is possible to give explicit general constructions for mixed covering codes of minimal cardinality. The proof of the following theorem is immediate.

Theorem 1 *If* $\sum_{i=1}^{m} n_i = R$, *then* $K_{q_1,\ldots,q_m}(n_1, \ldots, n_m; R) = 1$.

Minimal covering codes of length 1 and 2 are determined by Theorem 1, except for the case $n = 2$, $R = 1$. To deal with this case, we generalize [25, Lemma 1]. We first permute the coordinates so that $q_1 \leq \cdots \leq q_m$ ($\lceil x \rceil$ ($\lfloor x \rfloor$) stands for the smallest (largest) integer bigger (less) than or equal to x).

Theorem 2 *Let* $n = \sum_{i=1}^{m} n_i$. *If* $q_1 \leq \cdots \leq q_m$ *and* $R + 1 \leq n \leq R + \lceil n_1/q_1 \rceil$, *then* $K_{q_1,\ldots,q_m}(n_1, \ldots, n_m; R) = q_1$.

Proof: $K_{q_1,\ldots,q_m}(n_1, \ldots, n_m; R) \leq K_{q_1}(n_1, R - (n - n_1)) = q_1$ by [9, Theorem 6], so q_1 is an upper bound. On the other hand, for any code C with $q_1 - 1$ codewords there is always one word u such that $d(u, C) = n > R$. Thus $K_{q_1,\ldots,q_m}(n_1, \ldots, n_m; R) = q_1$. □

In [18], Theorem 2 can for example be used to prove that $K_{3,2}(1, 5; 3) = 2$. For $n_1 = 1$ we obtain the following result.

Corollary 1 *If* $q_1 \leq \cdots \leq q_m$ *and* $n = R + 1$, *then* $K_{q_1,\ldots,q_m}(1, n_2, \ldots, n_m; R) = q_1$.

3.2.1 Codes of Length 3

Now, moving on to length 3, minimal cardinalities are given by the theorems presented, except for the case $n = 3$, $R = 1$. Again, we generalize a method presented in [25] to obtain an upper bound for these codes. The idea behind the result in [25] (that treats nonmixed codes) is to partition the set of possible values of the coordinates into two sets, e.g., $F_3 = \{0, 1, 2\} = \{0\} \cup \{1, 2\}$. Then, for any word, at least two of the coordinates will have values that belong to one of these sets. Thus, for each of these two sets we have to construct codewords, such that for any two coordinates any pair of values (belonging to the set in question) will occur in some codeword. If a set has q values, then the number of codewords constructed with values from this set must apparently be at least q^2. As a matter of fact, there are always q^2 such codewords. Then, e.g., if $q = 3$, $K_3(3, 1) \leq 1^2 + 2^2 = 5$. The following general theorem can be proved ([25, Theorem 1]).

Theorem 3 $K_q(3, 1) = \lfloor (q^2 + 1)/2 \rfloor$.

Before presenting a result for the mixed case, we prove the following lemma.

Lemma 1 *If* $q_1' \leq \min\{q_2', q_3'\}$, *then there is a code* $C \subseteq F_{q_1'} F_{q_2'} F_{q_3'}$ *with* $q_2' q_3'$ *codewords, such that for any two coordinates, any pair of values occurs in some codeword.*

Proof: We show that the following code $C = \cup(a_i, b_i, c_i)$ fulfills the requirements. Take as codewords all $q_2' q_3'$ words such that all possible pairs of values in the last two coordinates occur in some word, and let $a_i \equiv b_i + c_i \pmod{q_1'}$. For a fixed c_i, as b_i goes through all its values cyclically, so does a_i, thus (since $q_1' \leq q_2'$) all possible pairs (a_i, c_i) occur. Similarly all pairs (a_i, b_i) occur. □

Actually, there is no code with fewer than $q_2'q_3'$ codewords that has the same property, since as any pair must occur, the number of codewords must be at least $\max\{q_1'q_2', q_1'q_3', q_2'q_3'\} = q_2'q_3'$.

Theorem 4 *If $q_1 \leq q_2 \leq q_3$ and $q_1 < q_3$, then* $\lceil q_1q_2/2 \rceil \leq K_{q_1,q_2,q_3}(1,1,1;1) \leq \min\{q_1q_2, \lfloor q_2q_3/2 \rfloor\}$.

Proof: Also in the case of mixed codes, we partition the sets of values of the coordinates into pairs of sets. That is, $F_{q_1} = F_1' \cup F_1'' = \{0, 1, \ldots, q_1' - 1\} \cup \{q_1', q_1' + 1, \ldots, q_1 - 1\}$, etc. The cardinalities of these sets satisfy the following equations: $q_1' + q_1'' = q_1, q_2' + q_2'' = q_2, q_3' + q_3'' = q_3$. Now, for any word, at least two of the coordinates, i and j, will have values that both belong to F_i' and F_j' or to F_i'' and F_j''. The construction in Lemma 1 can then be used to construct codes C_1' and C_2', such that both these values occur in some codeword.

Let $q_1' = \lfloor q_1/2 \rfloor$, $q_1'' = \lceil q_1/2 \rceil$, $q_2' = \lfloor q_2/2 \rfloor$, $q_2'' = \lceil q_2/2 \rceil$, $q_3' = \lceil q_3/2 \rceil$, and $q_3'' = \lfloor q_3/2 \rfloor$. Obviously $q_1' \leq \min\{q_2', q_3'\}$ and $q_1'' \leq \min\{q_2'', q_3''\}$ (since $q_1 < q_3$). Using Lemma 1 we can now construct a code with $M = \lfloor q_2/2 \rfloor \cdot \lceil q_3/2 \rceil + \lceil q_2/2 \rceil \cdot \lfloor q_3/2 \rfloor$ codewords and covering radius 1. Simplifying this expression we get that $M = \lfloor q_2q_3/2 \rfloor$.

By constructing codewords with all possible pairs of (all) values in the first two coordinates, we obtain the upper bound q_1q_2, which is better than $\lfloor q_2q_3/2 \rfloor$ for $q_3 \geq 2q_1 + 1$.

The lower bound can be proved by a slight modification of the proofs of [14, Lemmata 19c),d)]. This has also been pointed out in [15]. □

Better lower bounds can be obtained in special cases. Using this result we can prove the following bounds in [18]: $K_{3,2}(2, 1; 1) \leq 4$, $K_{3,2}(1, 2; 1) \leq 3$. It is essential that we do not allow all the cardinalities to be equal. If $q = q_1 = q_2 = q_3$ are odd, we get $(q^2 + 1)/2 \leq K_{q_1,q_2,q_3}(1,1,1;1) \leq (q^2 - 1)/2$, a contradiction.

Using the following recursive formula together with Theorem 4 enables us to calculate good upper bounds for most values of q_1, q_2 and q_3.

Theorem 5 $K_{q_1,\ldots,q_n}(1, \ldots, 1; n-2) \leq q_1 + K_{q_1,q_2-1,q_3-1,\ldots,q_n-1}(1, 1, 1, \ldots, 1; n-2)$.

Proof: Let C' be a $(q_1, q_2 - 1, q_3 - 1, \ldots, q_n - 1; 1, 1, 1, \ldots, 1; M')n - 2$ code. We now construct $C = C' \cup C''$, where $C'' = F_{q_1} \oplus (q_2 - 1, \ldots, q_n - 1)$. This code has $q_1 + M'$ codewords, so it remains to show that it has covering radius $n-2$. Take any $x = (x_1, \ldots, x_n) \in F_{q_1} \cdots F_{q_n}$. Now, if $x_i = q_i - 1$ for any $i \in \{2, \ldots, n\}$, then $d(x, C) \leq d(x, C'') \leq n - 2$. Otherwise, $x \in F_{q_1}F_{q_2-1}F_{q_3-1} \cdots F_{q_n-1}$, but then $d(x, C) \leq d(x, C') \leq n - 2$, so in any case $d(x, C) \leq n - 2$. □

Observe that we do *not* require the arities to be in order of magnitude, however, the best bounds are obtained when q_1 is the smallest of them. Now, e.g., $K_{3,4,5}(1, 1, 1; 1) \leq 3 + K_{3,3,4}(1, 1, 1; 1) \leq 3 + \lfloor 3 \cdot 4/2 \rfloor = 9$, which is better than the bound obtained using Theorem 4 with the original parameters ($\lfloor 4 \cdot 5/2 \rfloor = 10$).

Using simulated annealing (to be discussed further on) we have checked whether we could find better bounds than those given by Theorems 4 and 5. For $q_1, q_2, q_3 \leq 5$, one improvement could be made, viz. $K_{3,5}(1,2;1) \leq 10$. This bound is proved by the following code: $C = C_1 \cup C_2 \cup C_3$, where $C_1 = \{000, 100\}$, $C_2 = \{021, 012, 222, 211\}$, and $C_3 = \{243, 234, 144, 133\}$. The code is presented partitioned in order to reveal the structure of it. The construction is similar to that in the proof of Theorem 5. A slight improvement can be made since C_1 does not have to contain the word 200. This is due to the fact that for any $a \in \{1,2,3,4\}$ there is an $x \in F_5$ such that the codewords $(2, a, x)$ and $(2, x, a)$ are in $C_2 \cup C_3$.

3.2.2 Codes of Length at Least 4

In [44] the cases $n = 4, 5$, $R = 1$ are also considered. Unfortunately, we have not been able to give a nice generalization for these parameters in case of mixed covering codes. For $n \geq 4$, $R = n - 2$, we can use the same method as for $n = 3$, $R = 1$ to construct covering codes.

A $(q_1, \ldots, q_n; 1, \ldots, 1; M)R$ code C is called *s-surjective* if for any set of s coordinates $a_1, \ldots, a_s$ of C and any s-tuple $(b_1, \ldots, b_s) \in F_{q_{a_1}} \cdots F_{q_{a_s}}$ there is a codeword $(c_1, c_2, \ldots, c_n)$ such that for $1 \leq i \leq s$, $c_{a_i} = b_i$. In the nonmixed case, s-surjectivity has been treated, e.g., in [24, 41]; see also the references mentioned in those papers. The first discussion on s-surjectivity and covering codes appeared in [21]. An interesting function is $\mathrm{ms}_q(n, s)$ ($s_q(n, s)$ is used in [41]), denoting the smallest possible number of codewords in an s-surjective q-ary code of length n. In the mixed case we use $\mathrm{ms}_{q_1,\ldots,q_m}(n_1, \ldots, n_m; s)$. In Lemma 1 and the discussion subsequent to that lemma we proved the following.

Theorem 6 *If $q_1 \leq q_2 \leq q_3$, then $ms_{q_1,q_2,q_3}(1,1,1;2) = q_2 q_3$.*

Assume that $q_1 \leq \cdots \leq q_n$. It is easily seen that $\mathrm{ms}_{q_1,\ldots,q_n}(1, \ldots, 1; 1) = q_n$ and that $\mathrm{ms}_{q_1,q_2,\ldots,q_n}(1, 1, \ldots, 1; s) \geq \prod_{i=n-s+1}^{n} q_i$. In the nonmixed case this bound is exact for $s = 2$ if and only if there exist $n - 2$ mutually orthogonal Latin squares of order q, see [3, Theorem 5.2.1]. (For the first connection between Latin squares and covering codes, see [26]) The maximal number of mutual orthogonal Latin squares of order q is denoted by $N(q)$. If q is a prime power, then $N(q) = q - 1$ [3, Theorem 5.1.1]. It is further known that $N(6) = 1$ and that $N(q) \geq 2$, for $q > 6$. In the binary case we know the exact values for $\mathrm{ms}_2(n, 2)$ (see [2, Corollary 5.2.4] and [30]).

Theorem 7 $ms_2(n,2) = \min_M n \leq \binom{M-1}{\lfloor (M-2)/2 \rfloor}$.

We now prove the following general theorem.

Theorem 8 *$K_{q_1,\ldots,q_n}(1, \ldots, 1; n-2) \leq \sum_{i=1}^{n-1} ms_{q_{i,1},\ldots,q_{i,n}}(1, \ldots, 1; 2)$, where for all j, $\sum_{i=1}^{n-1} q_{i,j} = q_j$.*

Proof: For any $n > 3$, the sets of possible values of the coordinates are partitioned into $n-1$ subsets. Now for any $x \in H$, at least two of the values of the coordinates must belong to the same subset. The code can now be constructed so that for any two coordinates any pair of values from the same subset will occur in some codeword, that is, the code is a union of $n-1$ 2-surjective subcodes, and the theorem follows. □

We first consider some results in the nonmixed case. For $n = 4$ we get the following sequence of bounds ($\mathrm{ms}_2(4,2) = 5$, $\mathrm{ms}_q(4,2) = q^2$ for $q \neq 2,6$): $K_q(4,2) \leq 3,7,11,15,19,23,27$ for $q = 3,4,5,6,7,8$, and 9, respectively. (This is not an arithmetic series!) Thus, we have an alternative proof for $K_5(4,2) \leq 11$, which was proved in [37] by explicitly listing the codewords of a corresponding covering code. The bound $K_4(4,2) \leq 7$ has been proved to be exact by Honkala [9]. An incorrect value (8) for these parameters has been reported in [6]. Another incorrect value in the same paper is $K_3(5,2) = 9$, it is currently known that $6 \leq K_3(5,2) \leq 8$ [37].

For $n = 5$, some values of $\mathrm{ms}_q(5,2)$ are unknown, the most interesting $\mathrm{ms}_3(5,2)$ (> 9). The following 2-surjective code proves that $\mathrm{ms}_3(5,2) \leq 11$: {00000, 10111, 21012, 11220, 01121, 20222, 11102, 22110, 02212, 12021, 22201}. The following bounds are then obtained ($\mathrm{ms}_2(5,2) = 6$): $K_q(5,3) \leq 4,9,14,19,24,29$ for $q = 4,5,6,7,8$, and 9, respectively. $K_5(5,3) \leq 9$ is as a matter of fact a new record! The previous best known upper bound (10) was also obtained by a combinatorial construction [37].

For lower bounds the reader is referred to [40], where it is shown that $K_q(n,n-2) \geq q^2/(n-1)$, and that this bound is exact if and only if $(n-1)$ divides q, and $N(q/(n-1)) \geq n-2$. Thus, e.g. $K_9(4,2) = 27$.

For results concerning proper mixed codes we need the following theorem.

Theorem 9 *If* $q_1 \leq q_2 \leq q_3 \leq q_4$, $q_1 \neq q_4$, *then* $ms_{q_1,q_2,q_3,q_4}(1,1,1,1;2) = q_3q_4$.

Proof: We have earlier pointed out that q_3q_4 codewords are necessary, so we have to show that the number is sufficient. We first assume that $q_3 \notin \{2,6\}$. If $q_3 = q_4 = q$, we take a set of words that prove $\mathrm{ms}_q(4,2) = q^2$ and replace values in the first two coordinates that do not belong to F_{q_1} and F_{q_2} with 0. If $q_3 < q_4$, we start with a set of words that prove $\mathrm{ms}_{q_3}(4,2) = q_3^2$, again doing the same replacement as in the first case if necessary. To this set we add the words $(i \pmod{q_1}, i \pmod{q_2}, i \pmod{q_3}, j)$, where $i \in F_{q_3}$ and $j \in \{q_3,\ldots,q_4-1\}$. The codes have in both cases q_3q_4 codewords and they are 2-surjective.

For $q_3 = 2$ we start from the following code proving $\mathrm{ms}_{2,2,2,3}(1,1,1,1;2) = 6$: {0000, 1111, 1001, 0111, 0102, 1012}, adding more codewords in the same way as in the first part of the proof if necessary ($q_4 > 3$). For $q_3 = 6$, there is a code proving $\mathrm{ms}_{6,6,6,7}(1,1,1,1;2) = 42$ (the construction is not difficult and is left to the reader). 2-surjective codes with $q_3 = 6$ can again be constructed from this code by replacement of values in the first two coordinates and by adding codewords if $q_4 > 7$. □

In Table 1 we present bounds for codes of length at most 4 with $2 \leq q_i \leq 5$. Some of the bounds for codes of length 4 with covering radii $R = 1$ and $R = 2$ have been found using simulated annealing (see next subsection). By examining the codes found in this way it turned out that many of them can be obtained by nice combinatorial constructions. Unfortunately, most of these constructions cannot be generalized but are applicable only to some special cases. The entries are marked as follows (some of the theorems from the next section are used): a) Corollary 1 (exact), b) Theorem 3, c) Theorem 4, d) Theorem 5, e) Theorem 11, f) Theorem 12(1), g) Theorem 12(2), h) Theorem 8, i) Theorem 10, j) [44], and s) found using simulated annealing.

3.2.3 Using SA in the Construction of Mixed Covering Codes

Constructing covering codes by hand is generally possible only as for codes with up to 5–6 codewords; exhaustive computer searches can manage somewhat larger codes. In any case, the number of inequivalent codes grows extremely fast as the length and the number of codewords increase. One possibility to deal with this problem is to use some kind of probabilistic methods.

Simulated annealing is such a method. The first attempts to use SA in the search for covering codes were made by Wille [49]. The results turned out to be very promising, and SA has been used to construct many record-breaking covering codes during the last few years [32, 36, 37].

The implementation of the direct approach is quite straightforward. The energy function is simply the number of words not covered by the codewords. When we are searching for a code C that covers H with radius R, the energy function is

$$E = |\{x \in H \mid d(x, C) > R\}|.$$

For covering codes, the value of this function is apparently 0. In the annealing process we consider the codewords one at a time. A neighbourhood configuration is obtained by replacing a codeword c with a randomly chosen word c', such that $1 \leq d(c, c') \leq R$.

The bigger the radius and the number of codewords to find are, the more time-consuming the SA algorithm becomes. Finally, a point where the algorithm is not feasible any more is reached, and other methods have to be considered. The possibility of using old codes to construct new ones is often utilized in coding theory.

3.3 Constructing New Codes from Old

3.3.1 Simple Constructions

A classic method to combine codes is the *direct sum construction* [35, p. 76]. The direct sum of two codes $C_1 = (q_1, \ldots, q_m; n_1, \ldots, n_m; M_1)R_1$ and $C_2 = (q_1, \ldots, q_m; n'_1, \ldots, n'_m; M_2)R_2$ is $C_1 \oplus C_2 = (q_1, \ldots, q_m; n_1 + n'_1, \ldots, n_m + n'_m; M_1 M_2)R_1 + R_2$. This construction can be used to prove the following.

				$K_{q_1,q_2,q_3,q_4}(1,1,1,1;R)$			$K_{q_2,q_3,q_4}(1,1,1;R)$		$K_{q_3,q_4}(1,1;R)$
q_1	q_2	q_3	q_4	$R=1$	$R=2$	$R=3$	$R=1$	$R=2$	$R=1$
2	2	2	2	4^i	2^e	2^a	2^b	2^a	2^a
2	2	2	3	6^f	2^e	2^a	3^c	2^a	2^a
2	2	2	4	7^s	2^e	2^a	4^c	2^a	2^a
2	2	2	5	8^f	2^e	2^a	4^c	2^a	2^a
2	2	3	3	6^s	3^e	2^a	4^c	2^a	3^a
2	2	3	4	8^f	3^e	2^a	5^d	2^a	3^a
2	2	3	5	10^f	3^e	2^a	6^c	2^a	3^a
2	2	4	4	8^s	4^e	2^a	6^d	2^a	4^a
2	2	4	5	10^f	4^e	2^a	7^d	2^a	4^a
2	2	5	5	12^f	4^e	2^a	8^d	2^a	5^a
2	3	3	3	9^f	3^g	2^a	5^b	3^a	
2	3	3	4	12^f	4^e	2^a	6^c	3^a	
2	3	3	5	13^s	4^e	2^a	7^c	3^a	
2	3	4	4	12^s	4^g	2^a	8^c	3^a	
2	3	4	5	15^f	5^e	2^a	9^d	3^a	
2	3	5	5	17^s	5^s	2^a	10^s	3^a	
2	4	4	4	16^f	4^s	2^a	8^b	4^a	
2	4	4	5	20^f	5^s	2^a	10^c	4^a	
2	4	5	5	22^s	6^g	2^a	12^c	4^a	
2	5	5	5	26^s	6^d	2^a	13^b	5^a	
3	3	3	3	9^j	3^h	3^a			
3	3	3	4	12^f	4^f	3^a			
3	3	3	5	15^f	5^f	3^a			
3	3	4	4	16^f	5^h	3^a			
3	3	4	5	20^f	6^f	3^a			
3	3	5	5	24^s	6^s	3^a			
3	4	4	4	21^f	6^d	3^a			
3	4	4	5	25^s	7^d	3^a			
3	4	5	5	31^f	8^d	3^a			
3	5	5	5	36^s	8^s	3^a			
4	4	4	4	24^j	7^h	4^a			
4	4	4	5	30^f	8^g	4^a			
4	4	5	5	37^f	8^s	4^a			
4	5	5	5	42^s	10^h	4^a			
5	5	5	5	51^j	11^h	5^a			

Table 1: Upper bounds for mixed codes of length at most 4.

Theorem 10 $K_{q_1,q_2,\ldots,q_m}(n_1+1,n_2,\ldots,n_m;R) \le q_1 \cdot K_{q_1,\ldots,q_m}(n_1,\ldots,n_m;R)$.

Proof: Follows from the direct sum construction if C_1 is a code with the previously mentioned parameters and C_2 is the $(q_1,1,q_1)0$ code consisting of all words in the space F_{q_1}. □

Theorem 11 $K_{q_1,q_2,\ldots,q_m}(n_1+1,n_2,\ldots,n_m;R+1) \le K_{q_1,\ldots,q_m}(n_1,\ldots,n_m;R)$.

Proof: Follows from the direct sum construction when C_1 is the same code as in the previous theorem and C_2 is a $(q_1,1,1)1$ code. □

Constructions (a) and (b) in [18] are special cases of Theorem 10. From Theorem 10 it is clear why we accept $n_i = 0$ in the definition of mixed covering codes. The generalization of constructions (c) and (d) in [18] is as follows.

Theorem 12 $K_{q_1,q_2,q_3,\ldots,q_m}(n_1,n_2+1,n_3,\ldots,n_m;R) \le$

1. $K_{q_1,q_2,\ldots,q_m}(n_1+1,n_2,\ldots,n_m;R) + (q_2-q_1)\cdot\lfloor K_{q_1,q_2,\ldots,q_m}(n_1+1,n_2,\ldots,n_m;R)/q_1\rfloor$, *if* $q_2 > q_1$,
2. $K_{q_1,q_2,\ldots,q_m}(n_1+1,n_2,\ldots,n_m;R)$, *if* $q_2 \le q_1$.

Proof: (1) Assume that C is a $(q_1,q_2,\ldots,q_m;n_1+1,n_2,\ldots,n_m;M)R$ code of length n. As in [18] we define $C'_a = C \cup \{(b,x_2,\ldots,x_n) \in F_{q_1}^{n_1}\cdots F_{q_m}^{n_m} \mid (a,x_2,\ldots,x_n) \in C, b \in \{q_1,\ldots,q_2-1\}\}$ for $a \in F_{q_1}$. For all these a the $(q_1,q_2,q_3,\ldots,q_m;n_1,n_2+1,n_3,\ldots,n_m;M_a)R_a$ codes C'_a satisfy $R'_a = R$. There is further an $i \in F_{q_1}$ such that the number of codewords in C with value i in the first coordinate is less than or equal to $\lfloor |C|/q_1 \rfloor$. Then $|C'_i| \le |C| + (q_2-q_1)\cdot\lfloor |C|/q_1\rfloor$.

(2) The second part of the theorem is proved by changing the values of the first coordinate of C, so that values in the interval $q_2,\ldots,q_1$ are changed to 0's (or any other value in the interval $0,\ldots,q_2-1$). The covering radius is then $\le R$, and the code has become q_2-ary in the first coordinate. This completes the proof. □

We could also have used $q_2/q_1 \cdot K_{q_1,q_2,\ldots,q_m}(n_1+1,n_2,\ldots,n_m;R)$ as an upper bound in the first part of Theorem 12, but the expression used gives tighter bounds in some cases. Certainly, an investigation of the values of the coordinates of a code C could give even better codes, in cases the distribution of the values is not even.

3.3.2 Partitioning Codes

Many good construction methods require that the codes have certain *partitioning properties.* The *amalgamated direct sum* (ADS) construction can e.g. efficiently be applied to two *normal* codes, see [12, 16].

We here give the definitions of (k,t)-normal and (k,t)-subnormal (mixed) covering codes, following the approaches in [22, 23, 37, 38], which can also be consulted for further details.

Definition 1 SUBNORMALITY.
Suppose that C is a $(q_1,\ldots,q_m;n_1,\ldots,n_m;M)R$ code. Then C has (k,t)-subnorm (S,R') if there is a partition of C into k nonempty subsets $C_0,\ldots,C_{k-1}$ such that $\min_a d(x,C_a)+\max_a d(x,C_a)\leq S$, whenever $R-t\leq d(x,C)\leq R$ and for all x, $\max_a d(x,C_a)\leq R'$.

Such a partition is called acceptable. *If the subnorm is given as a single value and not as a tuple, it refers to the S parameter. If C has (k,t)-subnorm $2R+1$, then it is called* (k,t)-subnormal. *If C is not (k,t)-subnormal, then it is called* (k,t)-absubnormal. *If C is $(k,0)$-subnormal, then it is called* k-seminormal. *If C is $(2R+1,R+1)$-subnormal, then it is called* strongly k-seminormal.

The following two theorems, which show how subnormal codes can be used, are given without proofs. The interested reader is referred to [39] for details.

Theorem 13 *Let A be a $(q_1,\ldots,q_m;n_1,\ldots,n_m;M_A)R_A$ code, and let B be a $(q'_1,\ldots,q'_{m'};n'_1,\ldots,n'_{m'};M_B)R_B$ code. Assume further that A has (k,R_A)-subnorm (S_A,R'_A) with the partition $A=A_1\cup\cdots\cup A_k$ acceptable, and that B has (k,t_B)-subnorm S_B with the partition $B=B_1\cup\cdots\cup B_k$ acceptable. Then the* blockwise direct sum *of A and B*

$$BDS(A,B)=\bigcup_{i=1}^{k}(A_i\oplus B_i)$$

is a $(q_1,\ldots,q_m,q'_1,\ldots q'_{m'};n_1,\ldots,n_m,n'_1,\ldots,n'_{m'};M)R$ code with $M=\sum_{i=1}^{k}|A_i|\cdot|B_i|$ and $R\leq\max\{R_1,R_2\}$, where

$$\begin{aligned} R_1 &= \lfloor (S_A+S_B)/2\rfloor \\ R_2 &= \begin{cases} 0 & \text{if } t_B\geq R_B \\ R_B-t_B-1+R'_A & \text{otherwise.}\end{cases}\end{aligned}$$

By using a punctured Hamming code as code A in Theorem 13, the following result can be obtained.

Theorem 14 *If there exists a q-seminormal $(q_1,\ldots,q_m;n_1,\ldots,n_m;M)R$ code, where q is a prime power, then $K_{q,q_1,\ldots,q_m}(q,n_1,\ldots,n_m;R+1)\leq q^{q-2}M$.*

Again, a special case of this result is proved in the following result, first presented by Honkala in [20, Theorem 8].

Theorem 15 *If $n_1\geq 1$ and there is an $i\geq 2$ such that $n_i\geq 1$, then $K_{2,q_2,\ldots,q_m}(n_1+3,n_2,\ldots,n_m;2)\leq K_{2,q_2,\ldots,q_m}(n_1+1,n_2,\ldots,n_m;1)$.*

The theorem is closely related to a conjecture first presented by Cohen et. al. [11]. They conjectured that $K_2(n+2, R+1) \leq K_2(n, R)$ for $n \neq R$. It has been proved that this inequality holds in many special cases, but a complete proof is still missing. We extend the conjecture to concern mixed covering codes with at least one binary coordinate.

Conjecture 1 *If* $n_1 \geq 1$ *and* $\sum_{i=1}^{m} n_i \neq R$*, then* $K_{2,q_2,\ldots,q_m}(n_1+2, n_2, \ldots, n_m; R+1) \leq K_{2,q_2,\ldots,q_m}(n_1, n_2, \ldots, n_m; R)$.

Let C be a $(q_1, \ldots, q_m; n_1, \ldots, n_m; M)R$ code with $C_a^{(i)\prime} = \{(c_1, \ldots, c_{i-1}, c_{i+1}, \ldots, c_n) \mid (c_1, \ldots, c_{i-1}, a, c_{i+1}, \ldots, c_n) \in C\}, a \in F_q$, where F_q is the arity of the ith coordinate of C. These sets are assumed to be nonempty for all a.

Definition 2 NORMALITY.
The code C *has* (q,t)*-norm* $(N^{(i)}, R_i')$ *with respect to the* i^{th} *coordinate if* $\min_a d(x, C_a^{(i)\prime}) + \max_a d(x, C_a^{(i)\prime}) \leq N^{(i)}$ *whenever* $R-t \leq d(x, \bigcup_a C_a^{(i)\prime}) \leq R$*, and for all* x $\max_a d(x, C_a^{(i)\prime}) \leq R_i'$. *If* $N^{(i)} \leq N, R_i' \leq R'$ *for at least one coordinate* i*, we say that* C *has* (q,t)-norm (N, R') *(or* N *alone, cf. the definition of* (q,t)*-subnorm), and the coordinates* i *for which this equation holds are called* acceptable. *If* C *has* (q,t)*-norm* $2R$ *it is called* (q,t)-normal. *If* C *is not* (q,t)*-normal it is called* (q,t)-abnormal.

The following theorem has turned out to be useful in many constructions (cf. [38, Theorem 3], and the discussion following that theorem).

Let A be a $(q_1, \ldots, q_m; n_1, \ldots, n_m; M_A)R_A$ code and let B be a strongly q_m-seminormal $(q_1', \ldots, q_{m'}'; n_1', \ldots, n_{m'}'; M_B)R_B$ code ($n_m > 0$, $B = B_0 \cup \cdots \cup B_{q_m-1}$ is an acceptable partition). We then define the ADS of A and B as

$$C = A\dot{\oplus}B = \bigcup_{i=0}^{i=q_m-1} \{(a,b) \mid (a,i) \in A, b \in B_i\}.$$

Theorem 16 $C = A\dot{\oplus}B$ *is a* $(q_1, \ldots, q_{m-1}, q_m, q_1', \ldots, q_{m'}'; n_1, \ldots, n_{m-1}, n_m - 1, n_1', \ldots, n_{m'}'; M)R_C$ *code with* $M = \sum_{i=0}^{q_m-1} |A_i^{(n)}| \cdot |B_i|$ *codewords and* $R_C \leq R_A + R_B$.

See [39, Theorem 20] for a proof of this theorem. Here $A_i^{(n)}$ denotes the set of codewords that have value i in the nth coordinate.

As mentioned in [38], there is a strongly 4-seminormal $(2,3,8)0$ code, whose acceptable partition consists of the 4 cosets of the perfect $[3,1]1$ Hamming code over F_2. Generally, the following result can be obtained ([14, Theorem 23]).

Lemma 2 *If* q *is a prime power and* $k \geq 2$*, then there exists a strongly* q^k*-seminormal* $(q, (q^k-1)/(q-1), q^{(q^k-1)/(q-1)})0$ *code.*

The following theorem generalizing [36, Theorem 1] can now be obtained.

Theorem 17 *If q is a prime power and $k \geq 2$, then* $K_{q_1=q,q_2=q^k,\ldots,q_m}(n_1 + (q^k - 1)/(q - 1), n_2, \ldots, n_m; R) \leq q^{(q^k-1)/(q-1)-k} \cdot K_{q_1=q,q_2=q^k,\ldots,q_m}(n_1, n_2 + 1, \ldots, n_m; R)$.

Proof: Follows from Theorem 16 and Lemma 2. □

Actually, Heden presented a similar construction in [19], but he restricted his discussion to the construction of perfect codes of covering radius one starting from other (mixed and nonmixed) perfect codes of covering radius one.

[36, Theorem 2] is simply obtained by applying Theorem 17 several times. Theorems 12 and 16 can be used to prove construction (e) in [18]. One ternary coordinate of the original code is first made quaternary (with unchanged R). The ADS of this code and the aforementioned strongly 4-seminormal $(2,3,8)0$ code concludes the construction. Construction (f) and (i) in the same paper can be seen as "going the other way" in Theorem 17. The original code in this construction consists of codewords that are direct sums of shorter words and cosets of the $[3,1]1$ Hamming code over F_2. These cosets are now replaced by a coordinate with values in F_4. Finally, in (f) this coordinate is made ternary (Theorem 12), and in (i) the ADS of this code and a strongly 4-seminormal $(3,4,12)1$ is taken.

Theorem 17 can be used to explain the results by Fernandes and Rechtschaffen in [13]. They proved that there are 25 and 63 codewords covering $F_9F_3^3$ and $F_9F_3^4$, respectively, with covering radius 1. This gives $K_3(7,1) \leq 225$ and $K_3(8,1) \leq 567$. Significantly better upper bounds for $K_3(7,1)$ and $K_3(8,1)$ (186 and 486) have, however, recently been found using other methods [32].

Theorem 18 $K_q(3(q+1),1) \leq q^{3(q-1)}\lfloor (q^4+1)/2 \rfloor$, *for q a power of a prime.*

Proof: From Theorem 3 we know that $K_q(3,1) = \lfloor (q^2+1)/2 \rfloor$, so we get that $K_{q^2}(3,1) = \lfloor (q^4+1)/2 \rfloor$. The desired result now follows by applying Theorem 17 three times. □

Corollary 2 $K_3(12,1) \leq 29889$.

This corollary gives an upper bound for the football pool problem for 12 matches. The bound is first mentioned by Hämäläinen and Rankinen in [18], the idea of the construction was, however, presented in a Finnish football pool magazine as early as in 1949.

3.3.3 Obtaining Coverings From Good Near-Coverings

The method to be discussed here is not widely applicable, but gives interesting and record-breaking codes in some special cases. Consider an arbitrary code C with covering radius R. Assume that the first coordinate is q-ary. We then obtain $C' = \bigcup_{a \in F_q} C'_a = \bigcup_{a \in F_q} \{c \mid (a,c) \in C\}$ by puncturing it once. Let us now have a look at C'_0. It is easily seen that those words that are at a distance bigger than R from C'_0 must be within distance $R-1$ from one of $C'_1, \ldots, C'_{q-1}$.

The simplest case is $R = 1$, $q = 2$. Then C'_1 must contain all words w, such that $d(w, C'_0) > 1$. We can now simply try to find the best possible covering C'_0, let $C'_1 = \{w | d(w, C'_0) > 1\}$, and check whether the extended code $C = \{(a, c) | c \in C'_a\}$ has covering radius 1.

There is a code consisting of 36 codewords that covers $F_3^5 F_2 \backslash Q$ ($|Q| = 60$) with radius 1 ($K_{3,2}(5, 1; 1) \leq 54$, so the near-covering is very good). By proceeding according to the previous discussion, the code that is obtained turns out to cover $F_3^5 F_2^2$ with radius 1. (We here talk about *the* code, which may not be correct. However, all attempts (using SA) to find this near-covering has ended up with the same code (up to equivalence), so we conjecture that there is only one such code.) As it has 36+60 codewords, $K_{3,2}(5, 2; 1) \leq 96$, explaining the bound in [18]. From this code it is further possible (using Theorem 12 and the discussion after that) to construct codes proving that $K_{3,2}(6, 1; 1) \leq 132$ (best known) and $K_3(7, 1) \leq 198$ (best known for short time in the late 1960's [17]).

3.3.4 A New Method

In the following theorem we are going to make use of the fact that appending a parity check bit to a $(2, n, M)R$ binary code C, and thereafter deleting *any* coordinate of the new code, gives a code, which has covering radius R (see, e.g., [4, Fact 2]).

Theorem 19 *If there is a* $(2, n_1, M_1)R_1$ *code* C', *a* $(2, n_2, M_2)R_2$ *code* C'', *and* $R_1 \leq R_2 - (n_2 - n_1 - 2)$, *then* $K_{3,2}(n_2 - n_1 + 1, n_1; R_2 + 1) \leq (M_1 + M_2)$.

Proof: We show that the code $C = (C' \oplus (2, 2, \ldots, 2)) \cup C'''$, where the all-2 vector is of length $n_2 - n_1 + 1$ and C''' is obtained from C'' by adding an overall parity check bit, proves the upper bound.

Consider any word $u \in F_2^{n_1} F_3^{n_2 - n_1 + 1}$. If the values of at least 2 of the last $n_2 - n_1 + 1$ coordinates of u are 2, then $d(u, C) \leq R_1 + (n_2 - n_1 + 1 - 2) = R_1 + (n_2 - n_1 - 2) + 1 \leq R_2 + 1$. Otherwise the value of at most one of the last $n_2 - n_1 + 1$ coordinates is 2, so the values of these coordinates (except for at most one) belong to F_2. We then get that $d(u, C) \leq R_2 + 1$, since (according to the discussion preceding the theorem) the deletion of one of the last $n_2 - n_1 + 1$ coordinates of C''' gives a $(2, n_2, M_2)R_2$ code. □

This principle has been used by many constructors of football pool systems. For example, let C' be a $(2, 5, 7)1$ code and C'' the $(2, 7, 16)1$ Hamming code. Theorem 19 now gives $K_{3,2}(3, 5; 2) \leq 23$, explaining the bound in [18].

3.4 The Matrix Method

The method to be presented in this section was first considered by Kamps and van Lint [28] and later generalized by Blokhuis and Lam [5]. Carnielli [7] and van Lint Jr. [33]

have independently generalized it to arbitrary covering radii. We here show how it can be applied to the mixed case.

Let $A = (I; M) = (a_1, \dots, a_n)$ be a $r \times n$ matrix where I is the $r \times r$ identity matrix and M is a $r \times (n-r)$ matrix. The following restriction has to be made to the A matrix. Two functions f_1 and f_2 associate the rows and the columns of A to arities, i.e. these are functions $f_1 : \{1, 2, \dots, r\} \to \{q_1, q_2, \dots, q_m\}$ and $f_2 : \{1, 2, \dots, n\} \to \{q_1, q_2, \dots, q_m\}$. Due to the I matrix part we have that for all $1 \le i \le r$,

$$a_{i,j} = \begin{cases} 1, & \text{if } i = j, \\ 0, & \text{otherwise,} \end{cases}$$

and we require that $f_1(i) = f_2(i)$. For all $1 \le i \le r$ and $r+1 \le j \le n$ we require that $a_{i,j} \in F_q$, if $f_1(i) = f_2(j)$ and that $a_{i,j} = 0$, if $f_1(i) \ne f_2(j)$.

For $s \in H = F_{q_1}^{n_1} F_{q_2}^{n_2} \cdots F_{q_m}^{n_m}$ (so $f_1(1) = \cdots = f_1(n_1) = q_1$, etc.) we define

$$S_{A,R}(s) = \{s + \sum_{j=1}^{n} \alpha_j a_j \mid \alpha_j \in f_2(j), |\{j \mid \alpha_j \ne 0\}| \le R, 1 \le j \le n\},$$

and we say that s R-covers $S_{A,R}(s)$ using A. Consequently, $A = I$ corresponds to covering in the traditional sense. A subset S of H *R-covers H using A* if

$$H = \bigcup_{s \in S} S_{A,R}(s).$$

For the following theorem we define a space $G = F_{f_2(1)} F_{f_2(2)} \cdots F_{f_2(n)}$, the length of which is n. In practice, the space can be expressed in a short form by grouping coordinates of equal arity together.

Theorem 20 *If S R-covers H using a $r \times n$ matrix $A = (I; M)$, then $W = \{w \in G \mid Aw \in S\}$ R-covers G. $|W| = |S| \cdot \prod_{i=r+1}^{n} f_2(i)$.*

Proof: (Follows the proof of Theorem 2.1 in [5].) Take any $x \in G$. Then $Ax \in H$, so there are s, $\alpha_1, \dots, \alpha_R$ and $i_1, \dots, i_R$ such that $Ax = s + \sum_{j=1}^{R} \alpha_j a_{i_j}$. Then $A(x - \sum_{j=1}^{R} \alpha_j e_{i_j}) = s \in S$, where e_j is the jth unit vector in G. Hence $x - \sum_{j=1}^{R} \alpha_j e_{i_j} \in W$ and $d(x, W) \le R$.

To find W, we start by denoting an n-tuple in G by (u, v), dividing the word into the first r and the last $n - r$ components. Then $A(u; v)^T = (I; M)(u; v)^T = u + Mv \in H$. For each $s \in S$, v can first be arbitrarily chosen and then we can choose $u = s - Mv$. We can choose v in $\prod_{i=r+1}^{n} f_2(i)$ ways, so $|W| = |S| \cdot \prod_{i=r+1}^{n} f_2(i)$. □

In the definitions we have had a very general approach, in the sense that we do not require coordinates of equal arity to be grouped, that is we could well have $H = F_2^2 F_3^3 F_2$. However, by requiring that, we get a matrix M that has the following equivalent form:

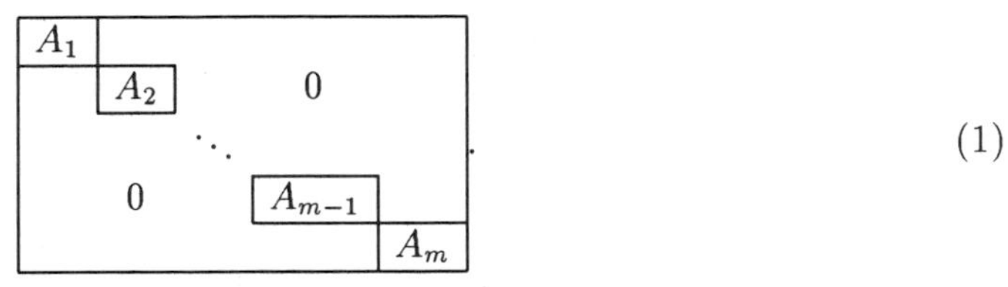

$$(1)$$

The values within an area A_i belong to F_{q_i}. We have for example found 80 codewords (these can be obtained from the author on request) covering $F_2^5 F_3^3$ with radius 1 using

$$A = \begin{bmatrix} 1 & 0 & 0 & 0 & 0 & 0 & 0 & 0 & 1 & 1 \\ 0 & 1 & 0 & 0 & 0 & 0 & 0 & 0 & 1 & 1 \\ 0 & 0 & 1 & 0 & 0 & 0 & 0 & 0 & 0 & 1 \\ 0 & 0 & 0 & 1 & 0 & 0 & 0 & 0 & 0 & 0 \\ 0 & 0 & 0 & 0 & 1 & 0 & 0 & 0 & 0 & 0 \\ 0 & 0 & 0 & 0 & 0 & 1 & 0 & 0 & 0 & 0 \\ 0 & 0 & 0 & 0 & 0 & 0 & 1 & 0 & 0 & 0 \\ 0 & 0 & 0 & 0 & 0 & 0 & 0 & 1 & 0 & 0 \end{bmatrix}.$$

Here $f_1(i) = 2$ for $1 \le i \le 5$, $f_1(i) = 3$ for $6 \le i \le 8$, $f_2(i) = 2$ for $1 \le i \le 5$, $i = 9, 10$, and $f_2(i) = 3$ for $6 \le i \le 8$, Theorem 20 then gives that there are $80 \cdot 2^2 = 320$ codewords covering $F_2^7 F_3^3$ with covering radius 1. Thus $K_{3,2}(3, 7; 1) \le 320$. This improves on the previous best known upper bound (332) [18].

Now SA can be used to find the codewords (S) and the matrix **A** (i.e. the **M** part), after first fixing the parameters of the spaces and $|S|$, r and n. The energy function is $E = |H \setminus \bigcup_{s \in S} S_{A,R}(s)|$ (cf. [32]). If the value of this function reaches 0, a solution is found. Since we now have to find both codewords in the set S, and the matrix **M**, it is not immediately clear how to use SA in the search for a covering. In [32] the approach is to change both the codewords in S and the matrix **M** during the annealing process. This method has turned out to perform well only when S is a small set.

One of the most promising approaches to deal with the problem seems to be the following. If there is a small number of nonequivalent matrices **A**, all these matrices can be considered in finding a covering set S. The number of nonequivalent matrices of size $r \times n$ over F_q having no repeated columns and no columns of zero is denoted by $\Phi_q(n, r)$ (cf. [43]). Tables on $\Phi_2(n, r)$ were first published by Slepian in [42], some values on the same function can also be found in [43]. Slepian uses $\bar{S}_{nr} = \Phi_2(n, r)$ in his paper. In Table 2 we present values of $\Phi_2(l + r, r)$ for small values of l and r.

The task of producing $\Phi_q(n, r)$, $n > r$, inequivalent matrices having no repeated columns and no columns of zero is much more difficult than just enumerating them. Only for the simplest case (i.e. **M** has only one column) an easy explicit description of the classes can be given. $\Phi_q(n, n-1) = n - 2$; **M** is then a single column vector with 2 to $n - 1$ non-zero positions. This fact was used in the search of a covering code improving on the bound $K_2(12, 2) \le 80$ [18]. That is, this is not a *proper* mixed code.

$l\backslash r$	2	3	4	5	6	7	8	9
1	1	2	3	4	5	6	7	8
2	-	1	4	8	14	22	32	44
3	-	1	5	15	38	80	151	266
4	-	1	6	29	105	312	821	1948
5	-	-	5	46	273	1285	5098	17934

Table 2: $\Phi_2(l+r,r), 1 \leq l \leq 5, 1 \leq r \leq 9$.

Theorem 21 $K_2(12,2) \leq 78$.

Proof: Let $r = 11$ and $n = 12$. For $\mathbf{M} = (11111111000)^T$, the following 39 codewords cover F_2^{11} with radius 2 using $(\mathbf{I};\mathbf{M})$:

00001011110	01000000011	10001011011	11000010000
00001100000	01000010101	10001110100	11011101010
00010001000	01011101111	10010011111	11011111100
00010110110	01011111001	10010100101	11100010001
00101001111	01100000010	10010110011	11100101001
00101100011	01100101100	10101011000	11100111111
00101110101	01100111010	10101100110	11110010100
00110001011	01111000100	10110001110	11111000001
00110011101	01111010010	10110110000	11111010111
00110100111	10001001101	11000000110.	

Theorem 20 gives that there are $39 \cdot 2 = 78$ codewords covering F_2^{12} with covering radius 2. Thus $K_2(12,2) \leq 78$. □

If the mixed spaces $H = F_{q_1}^{n_1} \cdots F_{q_m}^{n_m}$, and $G = F_{q_1}^{n'_1} \cdots F_{q_m}^{n'_m}$ have the property that $q_i \neq q_j$, when $i \neq j$, then $\mathbf{A}$ has the form displayed in (1), and the number of nonequivalent matrices is

$$N = \prod_{i=1}^{i=m} \Phi_{q_i}(n'_i, n_i).$$

In the previous example, where we showed that $K_{3,2}(3,7;1) \leq 320$, $q_1 = 2$, $q_2 = 3$, $n_1 = 5$, $n'_1 = 7$, and $n_2 = n'_2 = 3$. Then $N = \Phi_2(7,5) = 8$, so 8 different matrices have to be used in the attempts to find a covering with this method.

4 Conclusions

In this paper an overview of methods for constructing mixed covering codes has been given. The approach has been very general considering any possible arities and combinations of these. Earlier work in this area has mainly concentrated on binary/ternary mixed codes, which are interesting due to possible applications to football pool systems, and perfect mixed codes.

It has been our intention to cover the area as well as possible, including all main methods that have been used in the construction of football pool systems. A few very special methods have been omitted due to their lack of generality. These are of two types: complicated combinatorial constructions, and constructions starting from near-coverings trying to cover the uncovered words in some way, using simulated annealing or some combinatorial method (cf. [38]). The near-coverings can be obtained for example by the general methods presented in this report.

In the paper a table of upper bounds for short codes was given. The most interesting of all the bounds presented in the paper are $K_2(12,2) \leq 78$ and $K_5(5,3) \leq 9$. Many old bounds could also be explained using the constructions.

Acknowledgment

The author would like to thank Heikki Hämäläinen and Iiro Honkala for many rewarding discussions. The work has been financially supported by The Engineering Society in Finland TFiF (Tekniska Föreningen i Finland) and the Oscar Öflund Foundation.

References

[1] E. Aarts and J. Korst, *Simulated Annealing and Boltzmann Machines: A Stochastic Approach to Combinatorial Optimization and Neural Computing*, Wiley, 1989.

[2] I. Anderson, *Combinatorics of Finite Sets*, Clarendon Press, 1987.

[3] I. Anderson, *Combinatorial Designs: Construction Methods*, Ellis Horwood, 1990.

[4] E. F. Assmus Jr. and V. Pless, *On the covering radius of extremal self-dual codes*, IEEE Trans. Inform. Theory 29 (1983), 359–363.

[5] A. Blokhuis and C. W. H. Lam, *More coverings by rook domains*, J. Combin. Theory Ser. A 36 (1984), 240–244.

[6] W. A. Carnielli, *On covering and coloring problems for rook domains*, Discrete Math. 57 (1985), 9–16.

[7] W. A. Carnielli, *Hyper-rook domain inequalities*, Stud. Appl. Math. 82 (1990), 59–69.

[8] V. Černy, *Thermodynamical approach to the traveling salesman problem: An efficient simulation algorithm*, J. Opt. Theory Appl. 45 (1985), 41–51.

[9] W. Chen and I. S. Honkala, *Lower bounds for q-ary covering codes*, IEEE Trans. Inform. Theory 36 (1990), 664–671.

[10] G. D. Cohen, M. G. Karpovsky, H. F. Mattson, Jr., and J. R. Schatz, *Covering radius—Survey and recent results*, IEEE Trans. Inform. Theory 31 (1985), 328–343.

[11] G. D. Cohen, A. C. Lobstein, and N. J. A. Sloane, *On a conjecture concerning coverings of Hamming space*, in A. Poli (ed.), *Applied Algebra, Algorithmics and Error-Correcting Codes*, LNCS 228, Springer-Verlag, 1986, 79–89.

[12] G. D. Cohen, A. C. Lobstein, and N. J. A. Sloane, *Further results on the covering radius of codes*, IEEE Trans. Inform. Theory 32 (1986), 680–694.

[13] H. Fernandes and E. Rechtschaffen, *The football pool problem for 7 and 8 matches*, J. Combin. Theory Ser. A 35 (1983), 109–114.

[14] S. W. Golomb and E. C. Posner, *Rook domains, latin squares, affine planes, and error-distributing codes*, IEEE Trans. Inform. Theory 10 (1964), 196–208.

[15] S. W. Golomb and E. C. Posner, *Hypercubes of non-negative integers (Research problem 10)*, Bull. Amer. Math. Soc. 71 (1965), 587.

[16] R. L. Graham and N. J. A. Sloane, *On the covering radius of codes*, IEEE Trans. Inform. Theory 31 (1985), 385–401.

[17] H. Hämäläinen, personal communication.

[18] H. Hämäläinen and S. Rankinen, *Upper bounds for football pool problems and mixed covering codes*, J. Combin. Theory Ser. A 56 (1991), 84–95.

[19] O. Heden, *A new construction of group and nongroup perfect codes*, Inform. and Control 34 (1977), 314–323.

[20] I. S. Honkala, *On the normality of codes with covering radius one*, in *Proc. Fourth Joint Swedish-Soviet Int. Workshop Inform. Theory*, Studentlitteratur, 1989, 223–226.

[21] I. S. Honkala, *Modified bounds for covering codes*, IEEE Trans. Inform. Theory 37 (1991), 351–365.

[22] I. S. Honkala, *On (k, t)-subnormal covering codes*, IEEE Trans. Inform. Theory 37 (1991), 1203–1206.

[23] I. Honkala, *On $(q, 1)$-subnormal q-ary covering codes*, submitted for publication.

[24] I. Honkala, *A Graham-Sloane type construction for s-surjective matrices*, submitted for publication.

[25] J. G. Kalbfleisch and R. G. Stanton, *A combinatorial theorem in matching*, J. London Math. Soc. 44 (1969), 60–64; and 1 (1969), 398.

[26] J. G. Kalbfleisch and P. H. Weiland, *Some new results for the covering problem*, in W. T. Tutte (ed.), *Recent Progress in Combinatorics*, Academic Press, 1969, 37–45.

[27] H. J. L. Kamps and J. H. van Lint, *The football pool problem for 5 matches*, J. Combin. Theory 3 (1967), 315–325.

[28] H. J. L. Kamps and J. H. van Lint, *A covering problem*, in *Colloq. Math. Soc. János Bolyai; Hung. Combin. Theory and Appl.*, Balantonfüred, 1969, 679–685.

[29] S. Kirkpatrick, C. D. Gelatt Jr., and M. P. Vecchi, *Optimization by simulated annealing*, Science 220 (1983), 671–680.

[30] D. J. Kleitman and J. Spencer, *Families of independent sets*, Discrete Math. 6 (1973), 255–262.

[31] P. J. M. van Laarhoven and E. H. L. Aarts, *Simulated Annealing: Theory and Applications*, Reidel, 1987.

[32] P. J. M. van Laarhoven, E. H. L. Aarts, J. H. van Lint, and L. T. Wille, *New upper bounds for the football pool problem for 6, 7 and 8 matches*, J. Combin. Theory Ser. A 52 (1989), 304–312.

[33] J. H. van Lint Jr., *Covering Radius Problems*, M.Sc. thesis, Eindhoven University of Technology, The Netherlands, 1988.

[34] J. H. van Lint Jr. and G. J. M. van Wee, *Generalized bounds on binary/ternary mixed packing- and covering codes*, J. Combin. Theory Ser. A 57 (1991), 130–143.

[35] F. J. MacWilliams and N. J. A. Sloane, *The Theory of Error-Correcting Codes*, North-Holland, 1977.

[36] P. R. J. Östergård, *A new binary code of length 10 and covering radius 1*, IEEE Trans. Inform. Theory 37 (1991), 179–180.

[37] P. R. J. Östergård, *Upper bounds for q-ary covering codes*, IEEE Trans. Inform. Theory 37 (1991), 660-664; and 37 (1991), 1738.

[38] P. R. J. Östergård, *Further results on (k, t)-subnormal covering codes*, IEEE Trans. Inform. Theory 38 (1992), 206–210.

[39] P. R. J. Östergård, *Constructions of mixed covering codes*, Research Report A 18, Digital Systems Laboratory, Helsinki University of Technology, 1991.

[40] E. R. Rodemich, *Coverings by rook domains*, J. Combin. Theory 9 (1970), 117–128.

[41] G. Roux, *k-propiétés des tableaux de n colonnes; cas particulier de la k-surjectivité et de la k-permutivité*, Ph.D. thesis, Univ. Paris 6, France, 1987.

[42] D. Slepian, *Some further theory of group codes*, Bell System Tech. J. 39 (1960), 1219–1252.

[43] N. J. A. Sloane, *A new approach to the covering radius of codes*, J. Combin. Theory Ser. A 42 (1986), 61–86.

[44] R. G. Stanton, J. D. Horton, and J. G. Kalbfleisch, *Covering theorems for vectors with special reference to the case of four and five components*, J. London Math. Soc. 1 (1969), 493–499.

[45] O. Taussky and J. Todd, *Covering theorems for groups*, Ann. Soc. Polon. Math. 21 (1948), 303–305.

[46] G. J. M. van Wee, *Bounds on packings and covering by spheres in q-ary and mixed Hamming spaces*, J. Combin. Theory Ser. A 57 (1991), 117–129.

[47] G. J. M. van Wee, *On the non-existence of certain perfect mixed codes*, Discrete Math. 87 (1991), 323–326.

[48] G. J. M. van Wee, *Covering Codes, Perfect Codes, and Codes from Algebraic Curves*, Ph.D. thesis, Eindhoven University of Technology, The Netherlands, 1991.

[49] L. T. Wille, *The football pool problem for 6 matches: A new upper bound obtained by simulated annealing*, J. Combin. Theory Ser. A 45 (1987), 171–177.